From the Ground Up

From the Ground Up

The History of Mining in Utah

Edited by
Colleen Whitley

Foreword by
Philip F. Notarianni

Utah State University Press
Logan, UT

Utah State University Press
Logan, Utah 84322–7800
www.usu.edu/usupress/

Maps of Utah counties printed herein are reproduced from the Utah Centennial County History Series, courtesy of the series editor, Allan Kent Powell, and copublisher, the Utah State Historical Society.

All illustrations unless otherwise credited were provided by the author of the chapter they illustrate.

Publication of this book was supported by subventions from the following organizations:

The Charles Redd Center for Western Studies
Utah Mining Association
Andalex Resources, Inc.
Brush Resources, Inc.
Weyher Construction Company
Wheeler Machinery Company

Manufactured in the United States of America
Printed on acid-free paper

Library of Congress Cataloging-in-Publication Data

From the ground up : the history of mining in Utah / edited by Colleen Whitley.
p. cm.
Includes bibliographical references and index.
ISBN-13: 978-0-87421-639-4 (pbk. : alk. paper)
ISBN-10: 0-87421-639-7 (pbk. : alk. paper)
1. Mines and mineral resources--Utah--History. 2. Mining engineering--Utah--History.
I. Whitley, Colleen, 1940-
TN24.U8F76 2006
338.209792--dc22
2006020257

Contents

Foreword xi
Philip F. Notarianni

Preface xiii

Part I The Ground of Utah Mining

1 Geology and Utah's Mineral Treasures 3
William T. Parry

2 Generating Wealth from the Earth, 1847–2000 37
Thomas G. Alexander

3 General Patrick Edward Connor, Father of Utah Mining 58
Brigham D. Madsen

4 The Stories They Tell 81
Carma Wadley

Part II Some Mineral Industries

5 Saline Minerals 101
J. Wallace Gwynn

6 Coal Industry 126
Allan Kent Powell

7 Uranium Boom 142
Raye C. Ringholz

8 Beryllium Mining 166
Debra Wagner

Part III Major Mining Regions

9 Iron County 197
Janet Burton Seegmiller

10 Bingham Canyon 220
Bruce D. Whitehead and Robert E. Rampton

11 Silver Reef and Southwestern Utah's Shifting Frontier 250
W. Paul Reeve

12 Alta, the Cottonwoods, and American Fork — 272
Laurence P. James and James E. Fell, Jr.

13 Park City — 318
Hal Compton and David Hampshire

14 Tintic Mining District — 342
Philip F. Notarianni

15 San Francisco Mining District — 359
Martha Sonntag Bradley-Evans

16 Uinta Basin — 378
John Barton

Glossary of Geologic and Mining Terms — 392

Notes — 418

Resources and Bibliography — 456

About the Authors — 483

Index — 490

Illustrations

Geologic Figures

Plate tectonic system 4
Normal fault 7
Thrust fault 7
Kinds of folds 9
Early Paleozoic western North America 12
Pennsylvanian time, western North America 13
Utah in early Jurassic time 16
Utah in mid Jurassic time 18
Alluvial plain and coal swamps 19
Three dominant trends 21
Hydrograph of Lake Bonneville 23
Schematic of volcano 26
Beryllium mining 169

Maps

Faults and physiography of Utah 8
Dry Canyon, Ophir Mining District 97
Great Salt Lake 103
Components of the West Desert Pumping Project 114
Saline resource areas of Utah 116
Carbon County 127
Southeastern Utah, setting for uranium boom 143
Iron County 198
Bingham Canyon 221
Washington County 251
Big Cottonwood Mining District 273
Cottonwoods and American Fork Mining Districts 274
Topographic map of Park City District 319
Juab County 343
San Francisco Mining District 360–61
Duchesne County 380
Uintah County 381

Photos

View of Zion Canyon	17
Patrick Edward Connor	94
Johanna Connor	95
Miners in Stockton	96
Steamboat on Great Salt Lake	96
Stockton	98
Joe Hill	98
Mining camp home, Sunnyside	171
Miners loading coal	171
Coal miners	172
Huntington Canyon mine	172
Winter Quarters disaster at Scofield	173
Scofield disaster burial service	173
Castle Valley Coal Company store and office	174
Interior, Castle Valley Coal Company store	174
Castle Valley Coal Company tram	175
Independent Coal and Coke Company officials	175
Independent Coal and Coke Company housing	176
Independent Coal and Coke Company hotel	176
Independent Coal and Coke Company store	177
Lion Coal Company workers, Wattis	177
Independent Coal and Coke Company tram	178
Uranium prospector	179
Charlie Steen and companions underground	179
Vernon Pick	180
Duncan Holaday and Dr. W. F. Bale in uranium mine	180
Shot Harry nuclear test	181
Lindsey Hill Pit near Iron Springs	182
Blowout Pit on Iron Mountain	182
Frame scaffolding at Jennie Mine	183
Corry Mine on Lone Tree Mountain	183
Enos Wall	184
Daniel Cowen Jackling	184
Bingham, 1903	185
Smelter near Great Salt Lake	185
White Elephant Saloon, Bingham	186
Shay steam engines moving copper	186
Steam shovels in Bingham Canyon	187
Main Street, Magna, 1915	187
Dinkey engine, Bingham Mine	188

Bingham, 1919 188
Carr Fork, 1925 189
Mallet "110" engine 189
Dry Fork railroad viaduct 190
Stack at Kennecott's smelter 190
Kennecott's power plant, 1944 191
Diesel-electric ore truck 191
Daniel C. Jackling statue 192
Kennecott Utah Copper mine 193
Women working at Kennecott during WWII 194
Wells Fargo building, Silver Reef 299
Silver Reef ruin, 1976 300
Main Street, Silver Reef 301
Savage shaft, Stormont Mine 301
Margaret Grambo, Cosmopolitan Restaurant 302
Mining equipment remains, Cardiff Fork 303
Ruins of Wasatch Power Company plant 303
Remnants of Davis horse whim, Cardiff Mine 304
Charles H. Malmborg 305
Terminal of Michigan–Utah aerial tramway 306
Cornish pump, Ontario Mine 307
Mill, Ontario Canyon 307
Machine shop, Park City 308
R. C. Chambers 308
Ontario Mine workers 309
Track bikes to travel inside mines 309
Tipple, Park City 310
Miners in Mayflower Mine 310
Silver King Mine, Park City 311
Hard-rock miners, Park City 311
Emil J. Raddatz 312
Eagle and Blue Bell surface plant near Eureka 312
Head frame of Bullion Beck Mine 313
San Francisco Mining District beehivee kilns 313
Cowboy vein of gilsonite, Dragon 314
Hauling gilsonite from Bonanza 314
Gilsonite mining crew, Bonanza, 1906 315
Gilsonite explosion 315
Black Diamond Mine, Fort Duchesne 316
Dyer Mine above Vernal 317

Tables

Geological ages and events in Utah 10
Statistics of Utah's bituminous coal production, 1880–1945 40
Percent of national production and total value of gold, silver, copper, lead, and zinc produced in Utah, 1865–1945 40
Coal mine fatalities in Utah and relevant statistics relating to them, 1913–19 45
Fatalities per 1,000 workers in metal mines of Utah, other western states, and the United States, 1915–19 45
Utah and United States workforce, percent engaged in mining and number of Utahns engaged in mining, 1850–2000 47
Number of establishments and average nonagricultural monthly wages in Utah and mining industries, 1960–2000 48
Percent of gross state product accounted for by major industries, 1963–2000 49
Utah gross state product and its mining components, selected years, 1963–2000 50
Average monthly wages for selected nonagricultural industries, 1960–2000 51

Foreword

Philip F. Notarianni
Director, Utah State Historical Society

Mining played a vital role in diversifying both the economy and population of Utah. These factors in turn exerted an impact upon geography, architecture, business activity, and social movements. The tandem industries of mining and railroading combined to change the face of Utah—changes that remain evident in all aspects of the state's history. The ethnic and geographical landscapes of Utah continue to be profoundly influenced by these forces.

Mormons relied on an agrarian economy for survival. Ironically, prospectors en route to the gold-rush fields of California and Nevada provided needed monies to support the fledgling Mormon community. Yet Mormon efforts at mining concentrated on coal and iron, industries that proved ephemeral in nature. Brigham Young discouraged mining precious metals as a forerunner of moral and spiritual decay.

The onslaught of the United States Civil War ushered in the era of the "soldier prospector" in Utah Territory. Colonel Patrick Connor and the Third California Volunteers entered Salt Lake Valley in 1862, established Fort Douglas high on the east bench, and led the vanguard of miners who prospected and discovered outcroppings of precious ores mainly in the Wasatch and Oquirrh Mountains. The West Mountain Quartz Mining District, established in 1863, became the area's first one, but transporting ore proved expensive to these soldiers, still serving in the military. Commercial mining of these resources, therefore, awaited the coming of the railroad in 1869. With the joining of the rails on 10 May 1869, mining in Utah began to emerge as a vital industry. Changes in the economy, the physical environment, business, and the peopling of the area intensified.

Geographical changes appeared quickly. Railroad spur lines raced to the burgeoning mining towns cropping up in many areas. These towns—Alta, Park City, Ophir, Mercur, Rush Valley, Tintic, Bingham Canyon, and others—altered a landscape dominated by orderly Mormon villages into one where towns curved, rose, and dipped due to the geography of mineralization. Ore bodies mined close to the ground surface by individuals gave way to massive plants operated by large mining companies chasing veins of ore down thousands of feet.

As these mining towns sprang up, they were populated by a diverse ethnic and cultural labor force. In the push-pull of immigration evident in Europe and the United States during the late nineteenth and early twentieth centuries, Utah's mining and railroading endeavors proved a solid magnet. New arrivals from northern, southern, and eastern Europe, Asia, the Middle East, and Mexico soon populated these towns. The rhythms and tones of Utah life changed.

This altering of Utah life was echoed in the late 1880s and early 1890s by the developing coal industry. The discovery of coal in the rich area of Carbon County by the Denver and Rio Grande Western Railroad forever changed that landscape. Diverse towns, peoples, and economies followed coal. Mining as an industry also changed as coal substituted for charcoal in firing the smelting furnaces of Salt Lake and other areas.

The business of mining gave rise to a diversity of other endeavors. Utah mining stocks early were traded on the London Stock Exchange. During the early twentieth century, mining entrepreneurs, such as Samuel Newhouse, sought to turn Salt Lake City into the "Wall Street of the West." Others such as David Keith, Thomas Kearns, and Enos Wall transformed Salt Lake's Brigham Street, now South Temple, into a promenade of mining mansions. Suddenly, architecture really mattered. Both residential homes and business blocks in many mining areas reflected the architectural styles in vogue throughout the country. At the same time, miners' cottages and bungalow-style homes made a unique Utah contribution. Mining money altered the architectural landscape.

Economic upturns and downswings affected mining. With prosperity came continued growth, but misery often followed depression. Such was, and is, the nature of mining. The Great Depression of the 1930s, coupled with the restrictive immigration legislation of the 1920s, left an indelible mark on the industry. Utah suffered along with most of the United States. World War II reinvigorated the economic devastation of the 1930s. Bingham Canyon, which had continued to produce minerals from its beginnings in the 1860s, rose to new heights as one of the world's largest copper producers. Kennecott Utah Copper, formerly Utah Copper, supplied much metal to aid the Allied efforts. With most males in the U.S. military, women and workers from Puerto Rico and other South American areas were highly sought. Again, the face of Utah's economy and her population altered.

Hard-rock and coal mining have exerted an incredible impact on the economic landscape, the natural landscape, and the population landscape of the state, and that legacy continues. New technologies have led to new practices, often requiring less labor and diminishing demand. In the United States, the costs of mining minerals, once manageable, have increased. Mining is no longer the economic giant of the past. Renewed interest in mining occurred from the 1960s to the new millennium, but world markets and the use of fossil fuels have curbed Utah's impact. Nonetheless, mining has changed Utah in many ways, and the history of that industry is a significant part of the present and future.

Preface

Colleen Whitley

Mining is Utah's oldest nonagricultural industry. It is also the largest. The mining industry has directly employed thousands in mining, milling, refining, and transporting ores. That employment has in turn created thousands more jobs in the support sector: people who provide groceries, clothing, homes, and the dozens of other goods and services needed to maintain a population. Mining has contributed so much to the state that when the Salt Lake Olympic Organizing Committee chose three mascots for the 2002 Winter Games, two of them honored mining.[1] Despite mining's importance to Utah's people and their economy, a single-volume history of the industry in the state does not exist. This book fills that void, providing an overview of major mining ventures and guideposts to further research.

The book is organized into three sections. The first covers global issues that impact the entire state, beginning with the geology of the region that produced such a remarkable range of mineral wealth, from plate tectonics through the volcano at the bottom of the Bingham Copper Mine. The mineral wealth produced by that geology in turn has impacted the economic well-being (and occasionally ill-being) of Utahns ever since Patrick E. Connor, commercial mining-exploration pioneer, sent his soldiers out from Fort Douglas looking for mineral wealth. Though the folklore of mining and miners may not impinge directly on the state's history or economy, nonetheless the tales miners told each other enliven all of us, whether we have ever gone searching for the lost Rhoades Mine or listened for Tommy Knockers haunting the depths of the caverns.

The second section of the book is devoted to particular mineral industries, some centered in one area, some found across the state, some extending over centuries, and others operating only for a short time, with each chapter focused on a specific mineral or ore: salt, coal, uranium, and beryllium.

The third part of the book explores Utah's major mining regions, organized chronologically by the first major or commercial ventures in each one. Metal mining began in Utah in the late 1840s, and the first mining district was created in 1863, but when Congress passed a general mining law in 1872, it opened vast stretches of federal land to exploration and development, and within a few years Utah Territory had more than 90 mining districts.[2] Obviously this book does not contain histories of all of those, but most, whether or not they are specifically named, are encompassed within these chapters.[3] Each chapter gives an overview of the exploration and development of mining in a specific geographic area. Along the way, we are introduced to some of Utah's most fascinating people, from the miners who descended into tunnels carrying pickaxes to those who stayed in offices wielding pens—individuals like Tom Kearns, who parlayed his mining fortune into a political and publishing career; Joe Hill, who organized unions until he was executed for a murder he may or may not have committed; and Jesse Knight, who ran the only saloon-free, brothel-free mining town in the West. Some people we meet as groups, immigrants from many nations who brought their own customs, religions, foods, and traditions that add to the state's diversity and widen our viewpoints.

Finally, this book provides readers with some tools for further research into those areas they may want to examine in more detail. The glossary of geologic and mining terms includes both historic and current usages and guides to Web sites that are constantly being updated. The research resources extend far beyond the works cited in this volume, listing research centers and organizations, reference works, and a broad sampling of histories, biographies, technical analyses, and government publications.

Mining in the state of Utah impacts our economy and ecology; it is part of our heritage and will be part of our future. We need to understand mining operations to make knowledgeable decisions in public policy and balance our demand for minerals with our desire for wilderness. This book provides a starting point for that understanding.

A book of this kind is possible only because a great many busy and talented people were willing to share their time and expertise to prepare it. Each chapter has been researched and written by an authority with access to information on a specific topic. I greatly appreciate the extensive research and concise presentations from each of these generous authors: geologists J. Wallace Gwynn and William T. Parry; historians, both by profession and passion, Thomas G. Alexander, John Barton, Martha Bradley, Hal Compton, James Fell, David Hampshire, Larry James, Brigham D. Madsen, Philip Notarianni, Kent Powell, W. Paul Reeve, and Janet Seegmiller; company insiders Robert Rampton, Debra Wagner, and Bruce Whitehead; and writers Raye Ringholz and Carma Wadley.

Thanks also to several others whose contributions have greatly enriched this volume. The idea for the book came from the fertile brain of Judy Dykman, who recognized the need for a single volume to provide a comprehensive view of this important factor in the state's history. Several people provided essential background information or otherwise aided writers and researchers: Douglas Alder, Louis Cononelos, J. Kenneth Davis, Richard Sadler, and Kathryn Shirts. The Utah Mining Association and the Charles Redd Center at Brigham Young University generously helped underwrite printing costs. Kennecott Utah Copper, the Utah State Historical Society, and the *Deseret Morning News* graciously opened their photo archives, providing a range of pictures, some never before published. My husband, Tom, has helped with technical problems, and John Alley of Utah State University Press has offered his customary excellent support and advice. I am grateful to all of them.

Part I

The Ground of Utah Mining

1

Geology and Utah's Mineral Treasures

William T. Parry

Introduction

The occurrence of valuable mineral resources in Utah and the West is not accidental, but rather the result of understandable geological processes. However, the original discoverers did not understand these processes. Copper is at Bingham, coal at Price, and uranium at Moab because of the geological histories of these areas. A search for coal at Bingham or copper in Carbon County would be, and probably was, futile. Not only did the discoverers not understand the diverse processes, but neither did the scientists of the day. Much of what we know has been learned in the past few decades. The geological events that shaped present-day Utah formed the host rocks that contain the minerals and initiated the processes that formed the diverse assemblage of deposits that are Utah's mineral heritage. The geologic history of mineral resources is long and complex.[1]

This essay examines the geology of mining in Utah. The first section describes the general theoretical background. The second outlines the geological development of the West. And the third part explores the geology of each of the regions covered in the other chapters in this book.

Geologic Time

The range of geologic time involved exceeds by many orders of magnitude any time span with which we are familiar. We need some way to deal with the enormity of time involved in the events that shaped Utah's geology. One approach is to divide geologic time into successively smaller units, each with its own name, so that we can refer to them without the number of years getting in the way. Next, because of the technological discovery of radioactive age dating, we can specify the age and duration of each of the named units in millions of years. The named age blocks, together

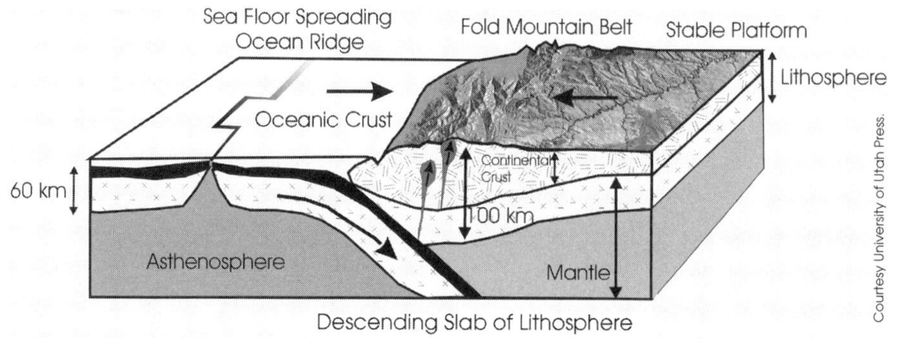

Figure 1. Schematic diagram of a plate tectonic system. Here a continental plate moving to the left converges with an oceanic plate moving to the right in response to formation of new oceanic lithosphere at a mid ocean ridge. The oceanic lithosphere is more dense than continental lithosphere and sinks downward. As it does so, the increased temperature, pressure, and changing chemical environment cause melting to occur. The melted rocks are buoyant and rise toward the surface. The convergent continental margin forms a fold mountain belt. All photographs and drawings in this chapter are by the chapter author.

with their absolute ages and durations in millions of years, appear in table 1 (see page 10). The table also includes the specific geologic event important to Utah's mineral resources that is associated with each block of time.

Plate Tectonics

The geological story of Utah is a journey through time and around the globe, for the geological architecture of Utah is not only ancient, but also much of it developed in distant locations. How is this all possible? Explaining the sequence of events that have shaped the geological architecture of Utah relies on the theory of plate tectonics. This theory holds that the outer, rigid layer of the Earth is divided into a series of separate plates and that these plates are free to move about the surface of the Earth in response to driving forces from within. The energy that drives the motion is heat; the Earth is a giant heat engine. Movement of the plates results in separation or divergence of some plates such as the divergence that produced the opening of the Atlantic Ocean. Plates may also converge and collide such as the impact that produced the spectacular Himalayas and closed the ocean that had separated India from Asia. This type of convergence is shown schematically in figure 1. Some plates may also slip past adjacent plates such as in the case of the San Andreas Fault in California.

As a consequence of convergence, oceans disappear and continents collide, producing spectacular mountain ranges. Interaction of crustal plates produced Utah's mountains, volcanoes, and mineral deposits. Geologists have known for nearly a century that vertical movements of the earth's crust were common. Thus, shells of marine snails, clams, and other sea creatures are found in rocks well above sea level.

Only during the last few decades, however, have scientists discovered that horizontal motions were just as common as vertical ones. Horizontal movements and the resulting plate interactions produced many of the features of present-day Utah.

To understand the lateral movement of continents and the consequences for Utah, we must review some geological theory. First, the earth is layered concentrically. The surface layer is the crust, then comes the mantle, and then the core. The crust and uppermost mantle form a rigid layer with a soft layer immediately underneath. It is this rigid layer of crust and mantle, called the *lithosphere*, that moves. The rigid layer consists of separate plates, and accommodating motion of one plate with respect to another on the spherical earth requires some plate boundaries to converge and others to diverge. Modern tools such as global positioning satellites confirm the direction and magnitude of plate motions.

Geological and Mountain-Building History

Precambrian (4,600 to 570 million years ago)

Now we are ready for the journey that accounts for the geological features of the region encompassed by Utah. We begin with the basement, the foundation on which the younger rocks are placed. Horizontal motion of plates has fragmented continents and reassembled others. North America is composed of a number of fragments of ancient continents that have been welded or sutured together by plates converging at various periods in geologic history. The ancient basement rock of Utah is the assemblage of at least two continental fragments called *cratons* (Greek for "shield"). In this region, two separate continental masses of different age were sutured together near central Utah. Following accretion or addition of largely volcanic rocks to the southern margin of what was to become North America 1.8 to 1.4 billion years ago, the Utah region was mountainous and had been eroding for many millions of years, producing raw material for a new generation of rocks. At that time the ocean shoreline (now west, but the orientation of the continent was different than today), lay at about the present-day location of Elko, Nevada. On the north was the ancient Wyoming craton with rock ages in excess of 2.5 billion years, possibly as old as 3.5 billion years, and on the south was the Arizona craton, largely volcanic in nature, which collided with and was sutured to Wyoming about 1.6 to 1.8 billion years ago in a mountain-building event.

The ocean basin that separated these two lithospheric plates gradually disappeared as they approached one another. The continents finally collided, forming a suture zone that trends from Cheyenne, Wyoming, through the Uinta Mountains and west across the Oquirrh Mountains to the Deep Creek Mountains on Utah's western border. The collision resulted in the formation of a mountain range that has long since eroded away; only roots and volcanic rocks remain. This ancient suture zone marks the locus of present mineral deposits in overlying younger rocks at Park

Kinds of Faults

There are basically three kinds of faults: a normal fault shown in figure 2A; a thrust fault, like the one shown in figure 2B; and a strike-slip fault. Mineral deposits in Utah are often associated with normal or thrust faults. The present landscape is reflective of normal faults. Utah has two systems of great faults. The first system is the oldest and is the result of compressive forces directed from the west during an episode of lithospheric plate convergence. The North American lithospheric plate moving west converged with an oceanic plate.

Utah's great thrust faults—the Charleston-Nebo, the Willard-Paris, and other related thrust faults shown in figure 3—have their beginning in Cretaceous time. These faults are exposed to view due to the large displacements on normal faults that bound the mountain blocks such as the Wasatch. In the Pahvant Mountains and nearby Canyon Range and in the Wasatch, we see rocks as old as Precambrian thrust over much younger rocks such as Paleozoic and Mesozoic sedimentary rocks. Crustal shortening can amount to as much as 100 miles or more on these faults. Compressive forces from converging lithospheric plates produce thrust faults and folds. Anticline and syncline folds are shown in figure 4.

The second system of faults are normal. These normal faults resulted from extension of the region to the west. The major normal faults that affect the present landscape are the Wasatch Fault in northern Utah and the Sevier and Hurricane Faults in southern Utah and northern Arizona shown in figure 3. Each of these faults is generally down to the west. Normal faults also bound each of the mountain ranges in the Great Basin west of the Wasatch.

The normal faults are younger than the thrust faults. They have their beginning in Tertiary time and are no older than 35 million years or so. Historical earthquake activity in Utah indicates that displacement is still taking place on some of these normal faults and that the Great Basin is still growing wider. The Wasatch Fault in northern Utah is the longest. Total displacements on the faults can be established by observing offset formations on their hanging and footwall sides. The Hurricane Fault has a displacement of 8,000 feet. When hanging-wall rocks are covered by valley fill, such as with the Wasatch Fault, displacements can be estimated by reconstructing material removed from the footwall by erosion. The Wasatch Fault probably has a displacement of at least 30,000 feet.

Figure 2a. Normal Fault

Figure 2. Kinds of faults. First a normal fault example from the Moab fault at Bartlett Wash in southeastern Utah. Here, the Cretaceous age Cedar Mountain formation on the right has been faulted down with respect to the older Jurassic age Entrada sandstone on the left. Next, a thrust fault example from the Alta thrust fault in the Wasatch Mountains. Here the Early Paleozoic Tintic Quartzite above has been thrust over the younger Mississippian limestone beneath. Note, the view has been rotated to remove the uplift on the Wasatch fault.

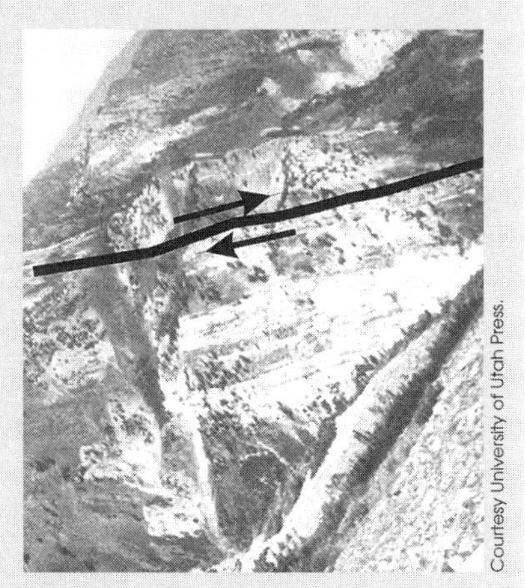

Figure 2b. Thrust Fault

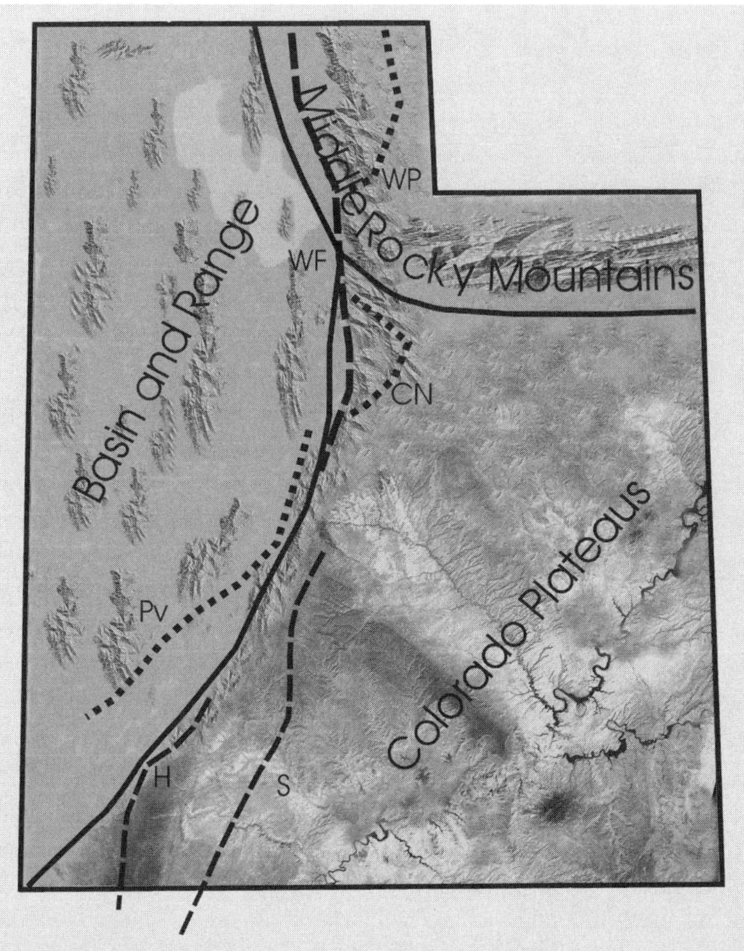

Figure 3. Faults and physiography of Utah showing each of the physiographic provinces and examples of the major normal and thrust faults. Dotted lines represent thrust faults and dashed lines are normal faults. WP is the Willard-Paris thrust fault, CN is the Charleston-Nebo thrust fault, PV is the Pahvant thrust fault, WF is the Wasatch normal fault, S is the Sevier normal fault, and H is the Hurricane normal fault.

Utah's Triple Personality

Examination of the geologic and physiographic map of Utah shown in Figure 3 shows three major characters to the state. First, the western half appears somewhat like the wrinkled skin on drying fruit with its north-south trending mountain ranges and intervening basins. This part of Utah is known as the Basin and Range Province. Second, the smoother topography in the southeastern portion of the state is known as the Plateaus province. Third, the rugged mountains that make the central spine of Utah together with the craggy peaks of the Uinta Mountains make up the Rocky Mountains province. Each of these regions has characteristic geology.

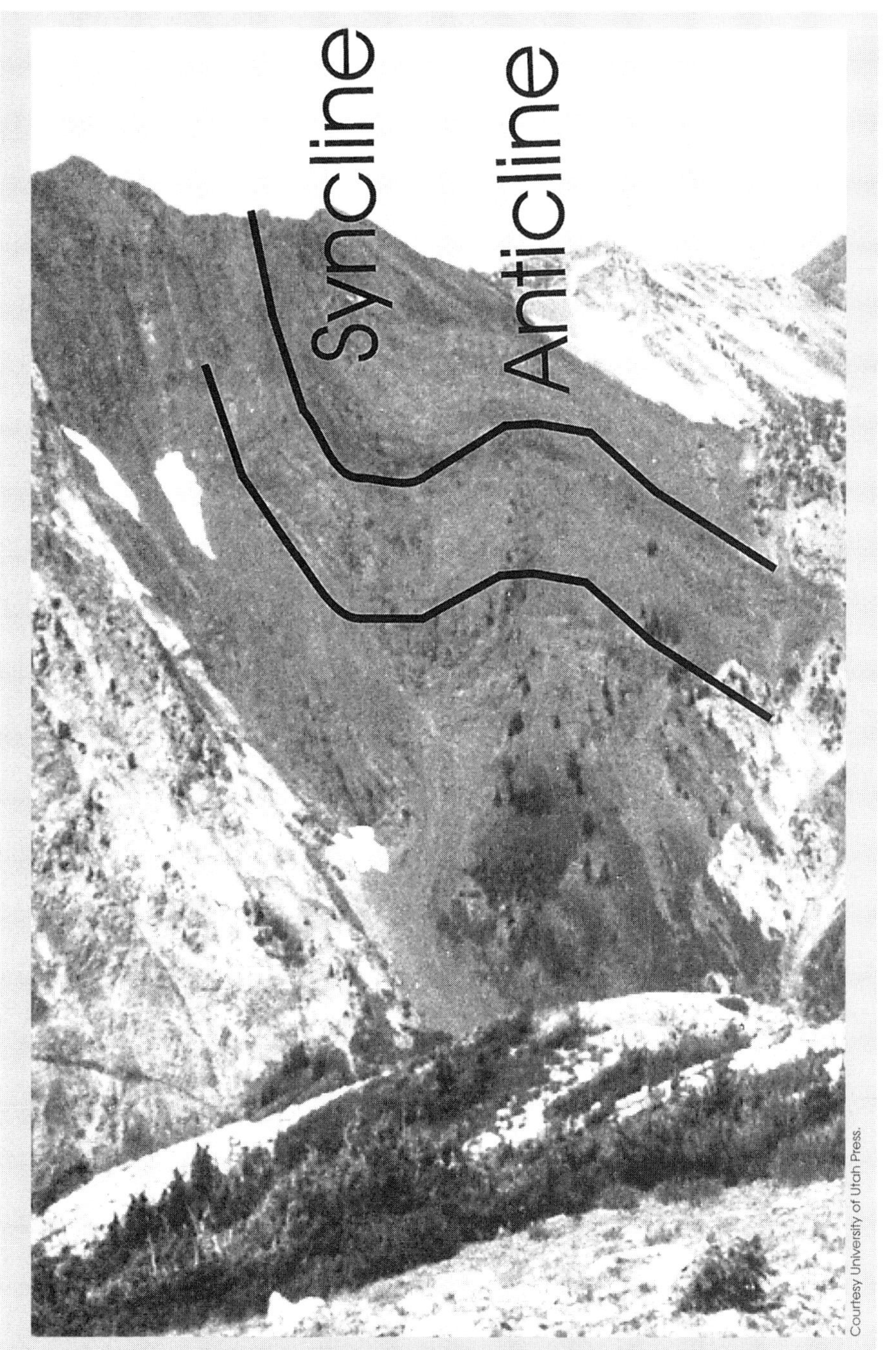

Figure 4. Kinds of folds. The compressive forces from converging lithospheric plates produces thrust faults and folds. An example of an anticline and a syncline shown here from the Precambrian Big Cottonwood formation in Little Cottonwood Canyon.

Table 1
Geological Ages and Events in Utah

Age Millions of Years	Era	Period	Epoch	Event	Resource
		Quaternary	Holocene		Great Salt Lake
			Pleistocene		
	Cenozoic				
66.4–1.6		Tertiary		Laramide	Bingham copper, lead, and zinc ore
144–66.4		Cretaceous		Sevier	Coal
208–144	Mesozoic	Jurassic		Nevadan	Morrison formation uranium host
245–208		Triassic			Chinle formation uranium host
					Homestake formation host for iron ore at Cedar City
					Host for silver at Silver Reef
					Thaynes host for silver at Park City
286–245		Permian		Sonoma	
320–286		Pennsylvanian			Limestone host for lead, zinc, and silver at Bingham
360–320		Mississippian			Great Blue limestone gold host at Mercur
					Humbug, Deseret, and Great Blue host for ore at Park City
408–360	Paleozoic	Devonian		Antler	
438–408		Silurian			Bluebell dolomite host for ore at Tintic
505–438		Ordovician			
570–505		Cambrian			Middle Ophir limestone, copper, lead, zinc, silver host at Tintic and Wasatch Mountains
2500–570	Precambrian	Proterozoic		Accretion of Arizona to Wyoming	Basement control on Wasatch Igneous trend
4600–2500		Archean			

City, Bingham, and Gold Hill. Evidently the suture had an effect on later geological processes.

Assembly of continental plates resulted in the formation of large supercontinents. One of these, formed long ago and far away, had formed by 750 million years ago in the Southern Hemisphere. This distant land is called Rodinia. Rodinia included North America nestled against eastern Antarctica.

The Grouse Creek, Raft River, and Albion ranges in northwestern Utah expose Archean rocks 2,500 million years old. The Farmington Canyon complex in the Wasatch Mountains east of Farmington is Archean in age, exposed there due to displacement on the Wasatch Fault. The Red Creek quartzite in the eastern Uinta Mountains is also probably of Archean age. These old rocks represent the eroded roots of an ancient mountain range.

Beginning about 1.1 billion years ago, the edge of the continent was split in a great east-west rift where astonishing thicknesses of shale and sand accumulated, rocks that are now exposed in the Uinta Mountains and Big Cottonwood Canyon in the Wasatch Mountains. These sediments were exposed to the action of tidewater glaciers and are in places covered with deposits left by melting ice. A veneer of additional sandstone and shale covers these glacial deposits. Proterozoic sedimentary rocks 2,500 to 570 million years old, mostly only weakly metamorphosed, include the Uinta Mountain group and the Big Cottonwood series that overlie this ancient basement rock.

Paleozoic (570 to 245 million years ago)

Rodinia broke apart, and fragments drifted in various directions across the globe. North America set off on a journey north, and by 530 million years ago, the western coast was near the equator. Salt Lake City is presently 700 miles from the Pacific Ocean, but earlier history places Utah much nearer the ocean and at times under water. Land areas of western Nevada and California were added to the North American continent much later. The northern margin of North America, which was to become its western part, was a broad, shallow, warm sea similar to the Coral Sea and the coast of northeastern Australia. The floor of the sea had numerous coral-building organisms living on it.

From 570 to 360 million years ago, an uneventful accumulation of sedimentary rocks formed in Utah and across Nevada, shown in figure 5. Sediments are thickest and most complete in western Utah. The sedimentary layers thin eastward and are sometimes even absent approaching the central portion of the continent. A great sandstone beach marched across Utah and the West, forming the first sedimentary deposits of the Paleozoic era. Sands and carbonate rocks, now exposed in the Grand Canyon, accumulated on a broad continental shelf in a tropical ocean near the equator that extended into Nevada and south into Arizona. These early Paleozoic rocks in Utah are mostly carbonate (limestone and dolomite) with lesser amounts of sandstone, quartzite, shale, and conglomerate. Sandstone and quartzite formed near shore on a passive continental margin. The belt of sandstone shifted east

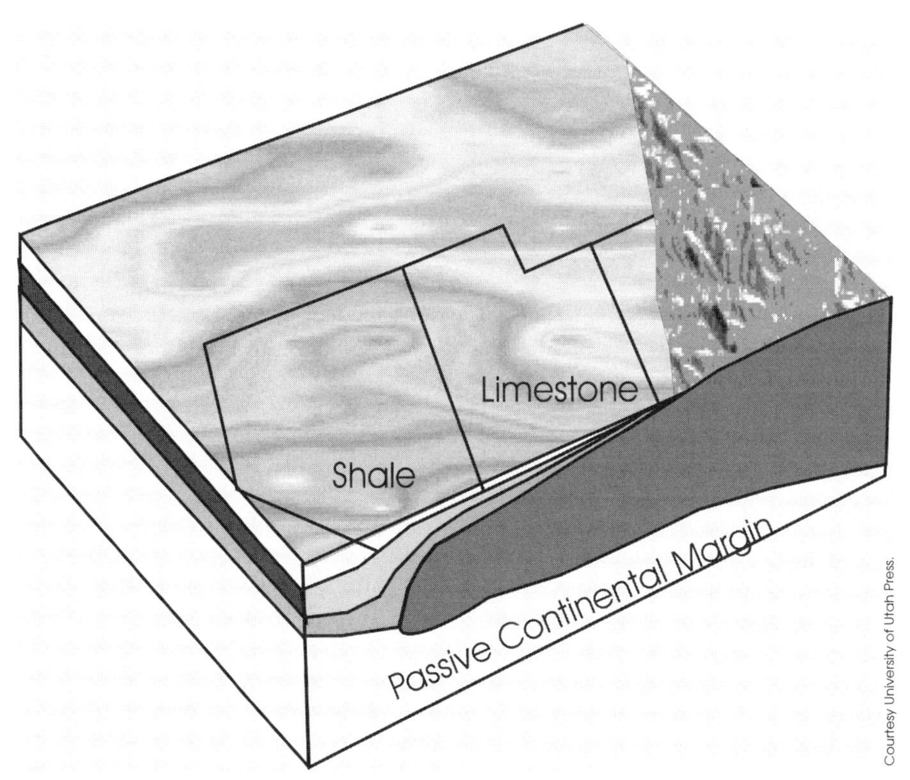

Figure 5. Early Paleozoic western North America. The western continental margin was passive, not converging with the adjacent oceanic lithosphere. Quiet accumulation beach sand, limestone near the shoreline and shale in deeper water characterizes this period of time.

with the shoreline. Shallow-water offshore deposits consist of shales that contain one important limestone: the Middle Ophir. Farther from shore in deeper waters of the continental shelf, limestone deposits precipitated from seawater with the aid of biologic activity. Deepwater shales accumulated in central and western Nevada. Though quiet accumulation of sedimentary rocks on the continental shelf may seem uninteresting, these rocks had important consequences for Utah's mineral deposits, for they are the host rocks for the rich copper, lead, zinc, silver, and gold deposits in the Park City, Tintic, Big Cottonwood-American Fork, and other mining districts in Utah.

This period of quiet accumulation of sedimentary rocks came to a close with the beginning of an eventful mountain-building history that shaped Utah geology for the next 300 million years up until about 50 million years ago. The western edge of the continent became the focus of mountain-building events as the Antler Mountains formed a north-south chain near the continental margin near Elko, marking the growth of the continent into western Nevada. Erosion from these mountains provided debris for the ocean basin to the east in western Utah. Deposition of marine

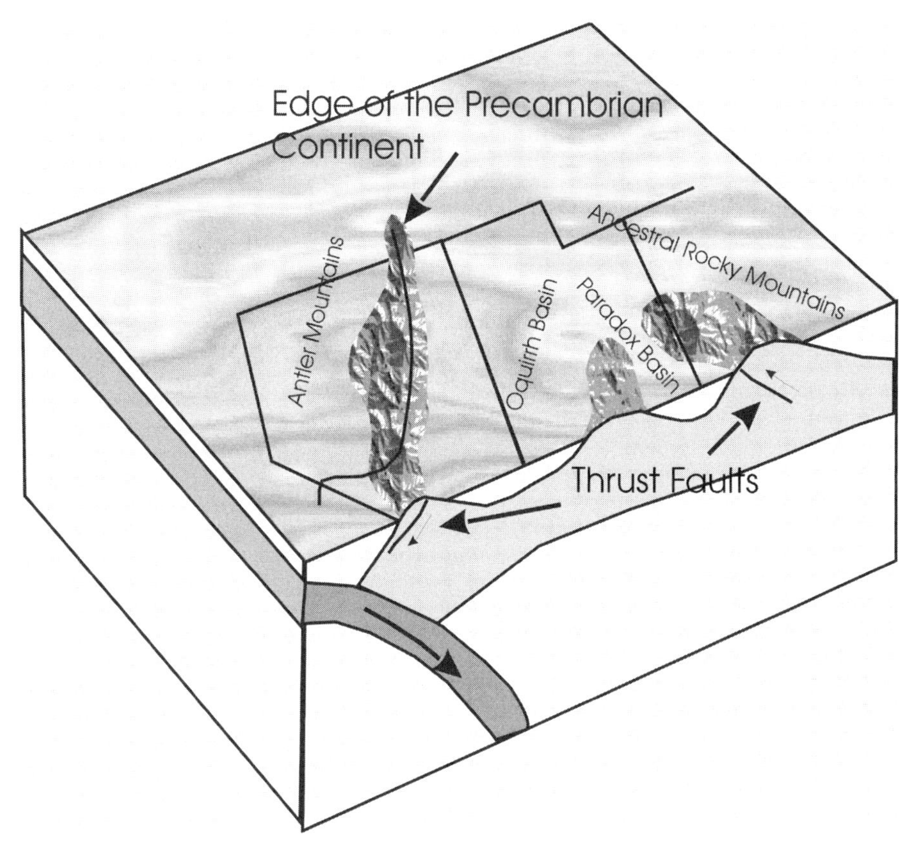

Figure 6. By Pennsylvanian time shown in this figure, the Antler mountains had formed in Nevada, the Ancestral Rocky mountains had formed in Colorado, and the Paradox and Oquirrh basins had formed in Utah.

carbonate, sandstone, and shale continued in a more restricted marine basin between the Antler highland and the continent.

Fragments of Rodinia continued to drift about on the globe until about 320 million years ago when the supercontinent of Pangea assembled. Assembly of Pangea was accompanied by collisions of continents and more mountain building. Next in the sequence of mountain-building events came a series of mountains in Colorado and adjacent southeastern Utah that occupied the present location of the Rocky Mountains and are thus called the Ancestral Rockies. These mountains formed when South America collided with North America in the Gulf of Mexico region. Accompanying the formation of the Utah portion of the Ancestral Rockies, two deep, fault-bounded marine basins formed, shown in figure 6. The first—in southeastern Utah near the present Green, Colorado, and San Juan Rivers—is called the Paradox Basin. Erosion from these mountains helped fill this basin. Because access to the open ocean was restricted and the climate may have been arid, much seawater evaporated, leaving behind thick accumulations of salt and related minerals. The physical properties

Definitions of Rock Types

Igneous rocks: rocks that have been melted within the earth and have been emplaced in the subsurface or on the surface and solidified as they cooled.

Sedimentary rocks: rocks that are deposited by water, either running water in rivers or quiet water in oceans and lakes. Limestone is chemically precipitated from the water, sandstone, shale, and conglomerate are transported as particles.

Metamorphic rocks: rocks that have been transformed by heat and pressure.

Crystalline Basement: ancient igneous and metamorphic rocks upon which younger sediments are deposited.

of the salt (low density and plastic deformation) led to later salt-intrusion anticlines and faulting in southeastern Utah.

The name "Paradox" comes from much later geological processes that resulted from movement of salt that had accumulated in the Basin. The movement produced great northwest-to-southeast trending anticlines that collapsed as the salt dissolved and was removed forming valleys such as the valley where Moab is located. The Colorado River paradoxically flows directly across the north end of Moab Valley, and the Dolores River cuts Paradox Valley in Colorado at nearly right angles. The mountains formed near the Utah/Colorado border are now called the Uncompahgre Plateau, a northwest-to-southeast-trending mountain range with crystalline roots.

The second basin, formed in northwestern Utah and known as the Oquirrh Basin after the mountain range west of Salt Lake City, is composed largely of sediment that accumulated. The Oquirrh Basin accumulated an enormous thickness of sediment with a cyclic repetition of limestone, sandstone, and shale possibly because of sea-level fluctuation related to glacier advances and retreats. More than three miles of these sediments accumulated in water depths that sometimes exceeded 1,000 feet. The lower 4,000 feet of these sediments are now exposed on Mount Timpanogos.

Sediments in these two basins had important consequences for Utah's mineral resources. In the Paradox Basin, both salt and potash accumulated along with much organic matter that later became petroleum. In the Oquirrh Basin, the limestones became a favorable host for the rich lead and zinc deposits in the Bingham Mining District.

With the Antler and nearby mountains to the west and the Ancestral Rockies to the east, most of Utah was still under water from 286 to 245 million years ago except for its southeastern corner. More mountains and terrain were added to western Nevada with the Sonoma mountain-building event. Sedimentary rocks that accumulated in the shallow ocean included sandstones, shales, and limestones. The Paradox Basin was completely filled with sediment, and wind-blown sand dunes covered the area. Deposition of carbonate and sandstone continued in the Oquirrh Basin of northwestern Utah. The limestones became the economically important host for rich silver-lead ores at Park City, Milford, and the Deer Trail Mine near Marysvale and an important source of phosphate mined on the flanks of the Uinta Mountains and tar sands located in Wayne and Garfield Counties. The limestone also formed a prominent rimrock of the Grand Canyon. The top of these sediments is an erosion surface that represents an interval of more than 5 million years before deposition of any younger sediment.

The Golconda thrust fault and accretion of associated Sonomia in a mountain-building event closed the Permian and ushered in the next major episode in Utah geology. For 325 million years, Utah had escaped the direct effects of mountain building. Episodes had produced mountains in Nevada and Colorado and sedimentary basins in Utah, but only the tip of the Ancestral Rockies had penetrated Utah. The Mesozoic era saw Utah pummeled by major mountain-building events.

Mesozoic (245 to 66.4 million years ago)

The Mesozoic history of Utah includes several mountain-building events that changed the course of rivers and the location of oceans. The shoreline of the Pacific Ocean curved into Utah during the early Triassic, and much of the state was under water, once again separating Utah from the Sonoma and Antler Mountains in Nevada. A shelving lowland rose gently toward the Ancestral Rockies in eastern Utah. Sediments thickened from eastern Utah near the Ancestral Rockies west toward Nevada. The Woodside shale and Thaynes formations of the Wasatch Mountains were deposited at this time along with the Moenkopi formation of southern Utah.

Exposed rocks were then eroded for more than 10 million years. A large river flowed from the Ancestral Rockies west to the sea somewhat like the Nile River now flows into the Mediterranean, and, like the Nile, this ancient river crossed fields of sand dunes. That river transported floating logs that are now preserved in the Petrified Forest in Arizona. The river began in Colorado's Ancestral Rockies. River channels were filled with sands and gravels that became important sites for uranium mineralization in southeastern Utah and adjacent states.

Late Triassic sediments accumulated between Sonomia and the remnant Ancestral Rockies. Many of these Triassic rocks are red. Late Triassic and early Jurassic rocks included wind-deposited sand, forming the impressive sandstones of the Wingate, Navajo, and Entrada formations. Middle Jurassic rocks were deposited in a narrow marine belt in central Utah in front of the Nevadan mountain belt forming in eastern Nevada and western Utah. Late Jurassic sediments included the conglomerates and stream deposits of the Morrison formation deposited by east-flowing streams arising in the Jurassic mountains in eastern Nevada and western Utah. Igneous intrusions in western Utah are also Jurassic.

Pangea followed the fate of supercontinents such as Rodinia. Pangea began to break up with the opening of the South Atlantic Ocean. South America separated from Africa, and the western margin of South America pushed against the Pacific Ocean plate, beginning the formation of the Andes Mountains. North America

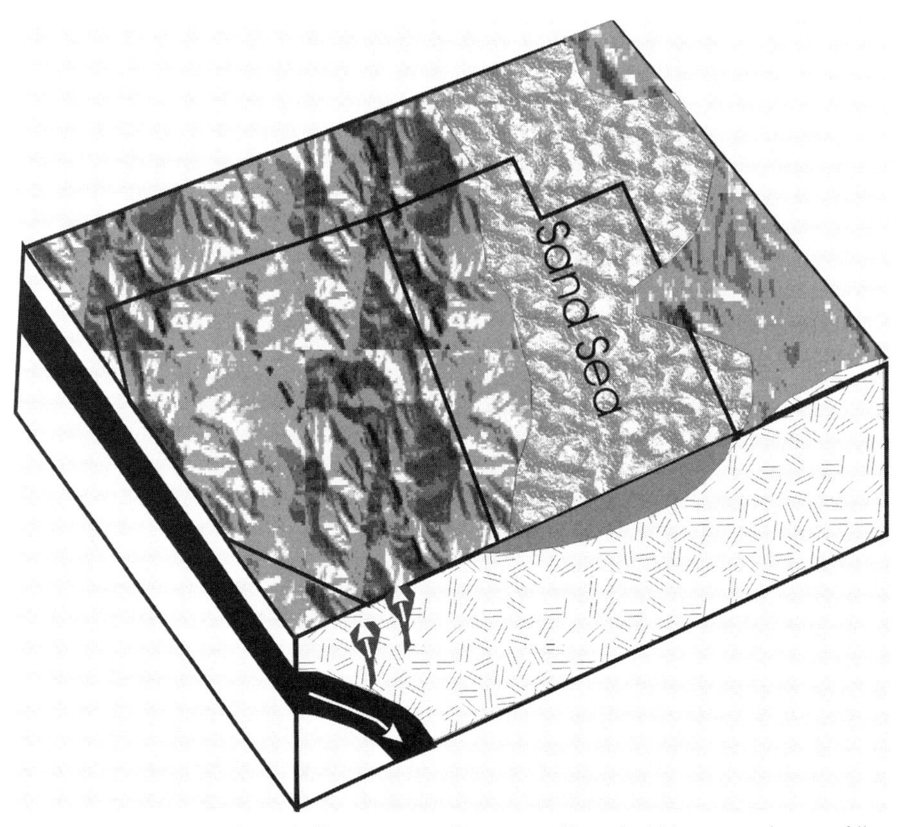

Figure 7. Utah in early Jurassic time was mostly covered by wind blown sand, one of the largest sand seas in the world. To the west were mountains formed by convergence of North America and the oceanic plate to the west. This convergence produced igneous rocks that are located in western Utah and Nevada.

separated from Europe as the North Atlantic Ocean opened, and North America collided with the Pacific plate, initiating the start of an Andean-type mountain range in Utah and Nevada. The volcanic activity associated with these events was first offshore, and the rivers followed westerly courses into the ocean.

North America continued to evolve with the opening of the North Atlantic Ocean and convergence of North America with the Pacific plate. The Andean-type mountain range continued to rise, causing a reversal in the drainages so that rivers flowed to the east from the mountain highland into a large sea that had developed connecting the Gulf of Mexico with the Arctic. The region was still a desert located about 15 degrees north of the equator with sand dunes similar to the Sahara today. The windblown sand accumulated in sedimentary formations that are widely distributed in Utah and adjacent states and spectacularly exposed in the cliffs of Zion Canyon in southern Utah today (figures 7, 8).

Continued convergence of North America with the Pacific plate and the attendant horizontal compressive forces and volcanic activity formed a very impressive

Figure 8. View of Zion Canyon from the East Rim Overlook looking southwards down the Virgin River. The cliffs are composed of Navajo sandstone, a rock that formed from accumulated wind-blown sand.

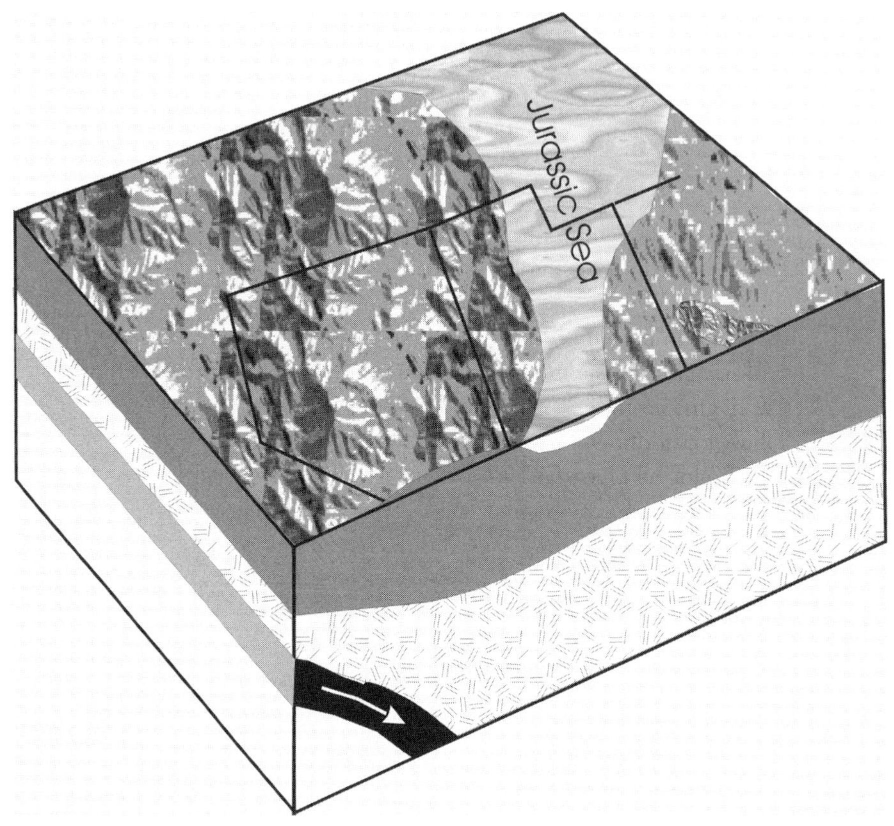

Figure 9. Utah in Mid Jurassic time with an arm of the ocean in central Utah where evaporation produced the salt deposits at Redmond and the gypsum that is mined at Nephi and Sigurd.

mountain range to the west, and streams continued to flow into the interior seaway. The horizontal compressive forces formed thrust faults, shortened the region, and thickened the crust. Rocks that had been deposited in the Oquirrh Basin, for example, were pushed tens of miles east. The rocks on Mount Timpanogos were among those thrust east, and the thrust fault lies in the subsurface beneath the mountain.

Utah, or at least western Utah, was very much like the Altiplano in southern Bolivia and northern Argentina. Dinosaurs roamed the stream floodplains and shorelines.

Continued mountain building in California and Nevada began to affect Utah more strongly so that west-flowing streams were interrupted by highlands that were forming and rivers were obliged to reverse their flow to the east. These mountain-building events continued to add terrain to North America. In fact, much of western Nevada and California became part of North America at this time. These mountain-building events produced some topographic changes in Utah. At times portions of the state were covered by a narrow arm of the ocean that invaded through Canada (figure

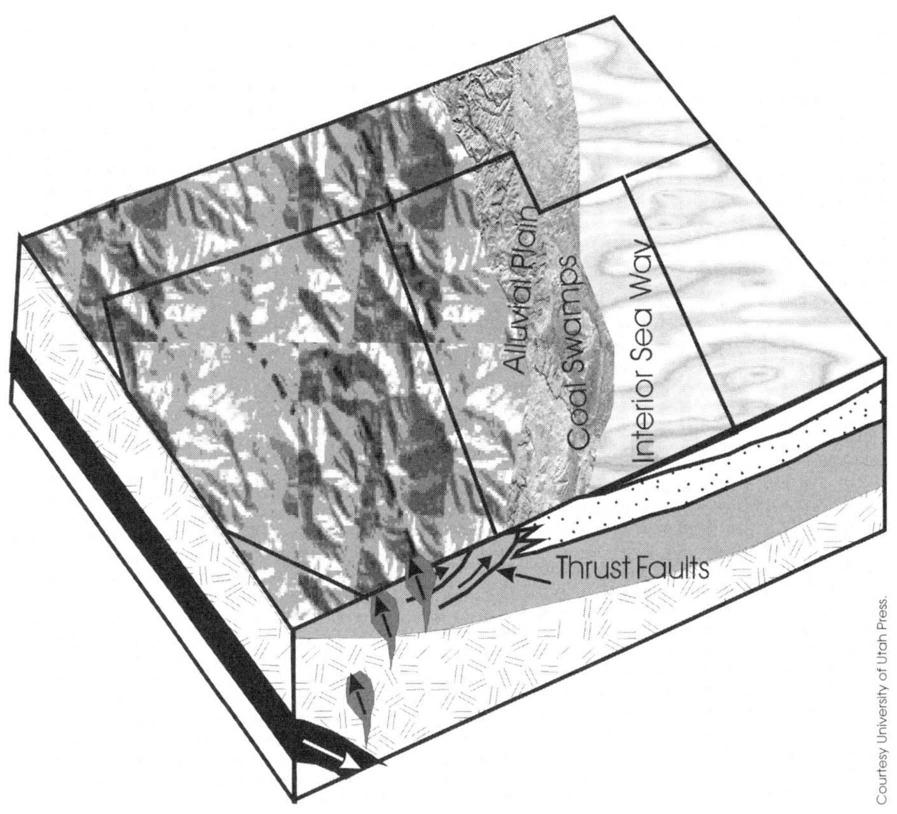

Figure 10. Continued convergence of western North America with the oceanic plate to the west resulted in subsidence from eastern Utah into Colorado and the formation of an interior sea that extended from about the Gulf of Mexico to the Gulf of Alaska. The alluvial plain and coastal areas were swampy and accumulated plant debris that was to become the coal deposits of Utah.

9). Evaporation of seawater left behind the salt and gypsum deposits in central Utah at Sigurd and Redmond. Marine limestone (Homestake member of the Carmel) is host to Iron County iron ores. Sediments from east-flowing rivers and large lakes covered these marine deposits. Abundant dinosaur remains are preserved in the rivers' channels.

The first igneous intrusions in Utah in more than 1 billion years formed in western Utah, and volcanic activity to the west deposited ash layers in Utah sediments. East-flowing streams continued to build deposits while mountain building and associated volcanic activity supplied the region with ash.

Cretaceous marine and nonmarine sediments were deposited to the east of the Sevier Mountain belt in western Utah as the area to the west rose and that to the east sank. Cretaceous coal deposits formed near the shoreline of the seaway shown in figure 10. At the end of the Cretaceous, eastern Utah was occupied by shifting river systems, swamps, and alluvial plains. Thick conglomerates were deposited at the foot of the Sevier Mountains in the west.

Cenozoic (66.4 to 0 million years ago)

The Cenozoic era from 65 million years ago to the present consists of two geologic periods: the Tertiary and the Quaternary. During the Cenozoic, the present landscape developed, and many of the metallic ore deposits of Utah were formed. Events in the Tertiary period led to the growth of the Uinta Mountains, uplift of the San Rafael Swell, and formation of the rich petroleum deposits in the Uinta Basin and elsewhere. Continued uplift finally obliterated the interior seaway, causing the last vestiges of the ocean to disappear. With no ocean to drain the east-flowing rivers, lakes began to form. These lakes held predominantly fresh water where abundant tropical plants and animals were preserved as fossils. The Tertiary, beginning about 65 million years ago, was perhaps the most important period of time for forming economically valuable deposits of copper, silver, gold, lead, and zinc in Utah. A belt of volcanic activity swept south and west through Utah in response to heating and melting far below the surface due to converging plates.

The high-elevation plain of western Utah and eastern Nevada that had formed during earlier mountain building collapsed under its own weight, and the whole region began to extend to the west, widening into what is now the Great Basin. The extension took place in response to horizontal extensional forces that resulted in normal faults. These normal faults formed the typical north-south-trending mountain ranges and intervening basins that we see today from the Wasatch Mountains west to the Sierra Nevada, for example, the Snake Range in Nevada.

Tertiary (66.4 to 1.6 million years ago)

The last compressional mountain-building event to affect Utah, called the Laramide, followed close on the heels of the Sevier Mountain belt and lasted from about 70 to 50 million years ago. The Laramide mountain-building event is responsible for producing the Uinta Arch and changing the drainage system. In Cretaceous time, an integrated drainage system eroded the fold mountain belt, and sediments from it were deposited in the Cretaceous interior seaway. This Laramide thick-skinned, mountain-building event involving the Precambrian basement moved east and elevated the Cretaceous seaway.

The combined effect of igneous activity, accumulation of thick sequences of sedimentary rock, and compression, folding, and thrust faulting thickened the crust of western North America enormously. The thickened crust had a deep, hot, and weak root. When the compressive forces were relieved as North America overrode the Mendocino triple junction, the area began to extend in response to body forces from the thickened crust, convection in the underlying asthenosphere, and movement on the San Andreas Fault. As extension proceeded, the former area of thickened crust was segmented into a series of north-south-trending mountain ranges and intervening basins bounded by normal faults. The Oquirrh Mountains display their geology as a consequence of displacement on these range-bounding normal faults.

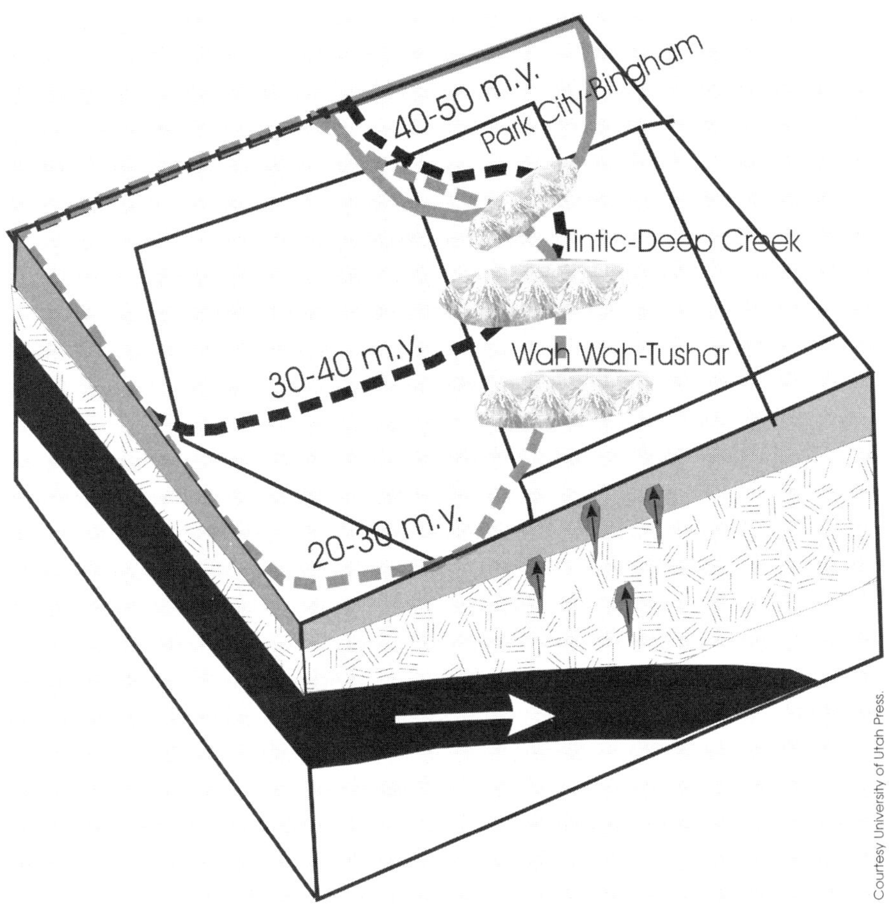

Figure 11. Igneous activity and volcanoes swept southward and westward through Utah from about 50 million years ago to 20 million years ago. Three dominant trends are recognized: The Park City-Bingham trend, the Tintic-Deep Creek trend, and the Wah Wah-Tushar trend.

Heat from within the Earth warmed the mantle, causing an elevation of the Cretaceous seaway. As a result, the integrated drainage system that had been eroding the Sevier Mountains was completely disrupted, and there was nowhere for the water to go. The rivers flowed into lakes.

After the Tertiary lakes disappeared, a period of intensive igneous activity swept through Utah. The rate of convergence of North America with the Pacific plate decreased again, and the inclination angle of the descending lithospheric slab increased. Extrusive and intrusive igneous rocks formed extensively for about 25 million years. A volcanic arc swept south and west across Utah as the convergent edge of western North America changed to a transform fault margin.

The major trends of igneous activity shown in figure 11 are the trend from Park City through the Wasatch to Bingham, a trend from Eureka (East Tintic Mountains)

Volcanoes

Volcanoes form from igneous rock, rock that has been melted. Melted rocks are less dense than solid, cooler rocks so the melted rock is buoyant and rises from the source in the mantle through the crust and erupts on the surface. Crustal rocks through which the melted rock rises are saturated with water. The water near the melted rocks is heated, rises, and is replaced by cooler water, thus forming a convection cell. The heated water can then extract metals and sulfur from the igneous rocks and nearby crustal rocks.

west to the Deep Creek Mountains, and a third trend from the Tushar Mountains west to the Wah Wah Mountains. Each trend is associated with significant economic deposits of precious and base metals. The igneous rocks of the Wasatch belt are aligned along the east-west Uinta-Little Cottonwood lineament, a relic of the ancient suture of the Arizona and Wyoming cratons. Several base and precious metal deposits are associated with the Wasatch igneous belt, including those in the Park City, Little Cottonwood, and Big Cottonwood Mining Districts. These deposits include silver, lead, zinc, copper, molybdenum, and gold. These are the ages of the igneous rocks from east to west: Ontario, 30 million years old; Clayton Peak, 35.5 million years old; Alta, 33.4 million years old; and Little Cottonwood, 30.5 million years old. Bingham lies farther west and is 38.55 million years old.

Quaternary (1.6 to 0 million years ago)

Quaternary events include the volcanic activity spectacularly displayed in Yellowstone Park and continued faulting and climate changes that advanced the glaciers and associated lakes in the Great Basin. Lakes accumulated because of the formation of north-south-trending mountain ranges and intervening valleys with no outlet for water. Lake Bonneville has a complex history that began about 30,000 years ago when the lake started to rise (figure 12). The lake rise was interrupted about 23,000 years ago at the Stansbury level at an elevation of 4,500 feet above sea level. Three additional interruptions in the rise of the lake have been recognized. The lake reached the Bonneville level, its highest point of 5,100 feet, about 15,000 years ago, when it overflowed at Red Rock Pass north of Preston, Idaho, and flowed down Marsh Creek into the Portneuf River, the Snake River, and the Columbia River. At its maximum elevation, Lake Bonneville was comparable in depth and area to Lake Michigan or Lake Huron.

The Bonneville flood at Red Rock Pass was catastrophic and involved nearly five trillion cubic meters of water with a peak flow of one million cubic meters per second. The flood lasted about 300 days. Erosion of the threshold to 4,740 feet produced the Provo level 14,000 to 13,000 years ago. Threshold control ceased, and the lake continued to fall due to evaporation to near its present level from 13,000 years ago to the present. As a consequence of unloading, the lithosphere has risen by as much as 70 meters in the central Bonneville basin with lesser amounts of uplift toward the edges of the ancient lake.

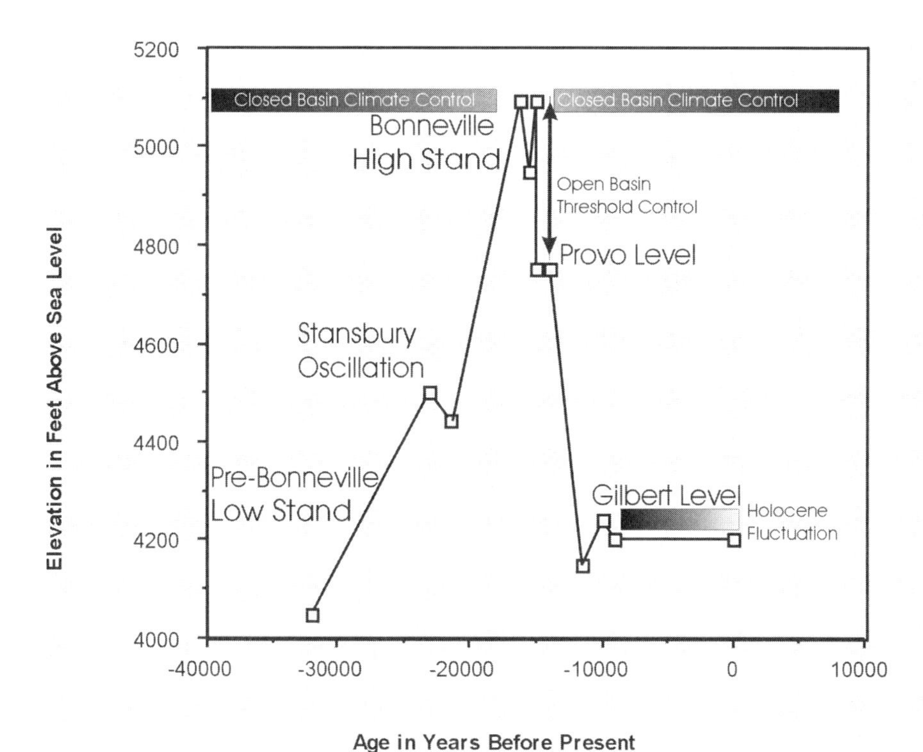

Figure 12. Hydrograph of Lake Bonneville showing lake levels in feet above sea level and age in years before the present.

The animals that lived on the shoreline included the extinct bear, horse, peccary, camel, bison, deer, musk ox, mastodon, mammoth, wolf, fox, and mountain sheep. The remains of these large mammals seem to indicate open parklands or grasslands situated in coniferous forests, very different from the present shoreline vegetation of cheatgrass, sagebrush, and scrub oak.

The present Great Salt Lake is a salty remnant of Lake Bonneville. The water inflow to the present lake comes from the Bear River (39 percent), precipitation (31 percent), the Weber River (13 percent), the Jordan River from Utah Lake and Provo River drainage (9 percent), and springs (3 percent). These inflows also carry dissolved salts that accumulate in the lake during evaporation.

The dissolved salts include sodium, calcium, potassium, magnesium, chloride, sulfate, and bicarbonate. Springs contain the highest concentrations and precipitation the lowest. The 30,000-year history of the lake has seen the addition of 0.69×10^{11} tons of chloride from rivers, 0.15×10^{11} tons of chloride from precipitation, and 6.16×10^{11} tons of chloride from springs. Some of the chloride now resides in the lake brine, some in water in the pore spaces in lake sediment, and some as salt beds that formed when the lake was so saline that the salt precipitated.

The climate changes that produced the lakes were linked to climate changes that caused ice to accumulate in the high peaks so that glaciers formed, similar to

Building Mountains

The mountainous areas of Utah bear the record of repeated episodes of mountain building. The oldest rocks in Utah—exposed in the Wasatch above Farmington and in the Raft River Mountains—have the longest record of repeated mountain building.

All of the mountains, except the ones that decorate our present topography, resulted from collisions of lithospheric plates. The mountains we see now, though they expose folding and faulting from earlier collisions, resulted from extension and collapse of thickened mountainous crust.

Mountains formed at eight intervals in the development of the western United States. While all of the events did not take place in Utah, they influenced the type of sedimentary rocks that accumulated here. Each of the mountain-building events possesses the following characteristics to some degree: folding, thrust faulting, intrusive igneous rocks, volcanic rocks, and a wedge of clastic sediments from erosion of the mountains.

the glaciers that are present in the Chugach Mountains in Alaska today. The climate changes that produced the lake and then allowed it to dry up are indicated by the relationship of the lake levels to glaciation. During the lake rise, a large continental glacier covered the northeastern part of Canada and the northern part of the United States. This ice sheet had a profound effect on the climate of Lake Bonneville. Valley glaciers also occupied the canyons of the Wasatch Mountains and formed the present topography of Mount Timpanogos with its horns and cirques. Of these glaciers, two reached the level of Lake Bonneville: the Little Cottonwood and Bells Canyon glaciers. The maximum glacier advance occurred about 22,000 years ago, long before the maximum rise of the lake.

Computer modeling of climate suggests that if the mean annual temperature decreased by seven degrees centigrade, evaporation would decrease and permit the lake to rise. Similar modeling studies of glacier dynamics suggest that the glacial climate was as much as 15 degrees cooler than today and that the glacier could be maintained at the mouth of Little Cottonwood Canyon if precipitation were 75 percent of the present amount. Maximum glacier advance occurred during a time when the climate was cooler and drier. The lake rose to the Bonneville level during a receding glacier period when the climate may have been warmer and wetter. The lake then fell below the Provo level as the climate changes permitted evaporation to exceed precipitation and inflow. When the lake was at its maximum capacity, the lake effect added substantially to precipitation and helped maintain the level.

Economic Deposits

A metallic mineral deposit is a concentration of chemical compounds containing gold, silver, copper, lead, and zinc. To be economical, a deposit requires concentration by many orders of magnitude over average abundance. The average abundance

Mountain-Building Events

Precambrian mountain-building events affecting Utah include these:

First, mountain building produced the metamorphic rocks of the Wyoming craton 2.5 to 3.5 billion years ago. Second, mountain building produced the metamorphic rocks of the Matzatzal craton in Arizona. Third, mountain building resulted from accretion of the Arizona craton to the Wyoming craton 1.6 to 1.8 billion years ago, called the Cheyenne event.

Paleozoic mountain building events affecting Utah include the following:

1. The Antler Mountains in Nevada are Devonian. The thrust faulting associated with this event is known as the Roberts Mountain thrust and is exposed in eastern Nevada. The volcanic and intrusive igneous rocks are part of the Sierra Nevada now. The sediments derived from this mountain range were deposited in the Devonian sea in western Utah and eastern Nevada, for example, the Stansbury formation on Stansbury Island. This mountain-building event was associated with the accretion of a volcanic island arc to the edge of the North American craton.

2. The Ancestral Rockies in eastern Utah, Colorado, and New Mexico are Pennsylvanian. The thrust faulting here was probably deep seated and involved much of the Precambrian crystalline basement. The sediments derived from these mountains filled the Pennsylvanian Paradox Basin in southeastern Utah.

3. The Sonoma mountain-building event involved accretion of a second island-arc complex to the western edge of North America in Permian to Triassic time. The thrust-fault system associated with this event is the Golconda in central Nevada, and the igneous rocks are also part of the Sierra Nevada. Sediments from Sonomia probably helped fill the Oquirrh Basin of northwestern Utah.

Mesozoic and Cenozoic mountain building events include these:

1. The Nevadan mountain-building event created the Sierra Nevada in Jurassic time. Igneous rocks of Jurassic age core the Sierra Nevada and are distributed across western Utah in the House Range and Newfoundland Mountains, for example. The thrust fault associated with this event appears in Utah as the Manning Canyon Detachment. Debris eroded from the mountains appears in eastern Utah as the Jurassic Morrison formation.

2. The Sevier mountain-building event strongly affected Utah in Cretaceous time, producing the Willard-Paris and the Charleston-Nebo thrust faults. The igneous rocks are part of the Sierra Nevada. The debris eroded from this mountain range occurs in the Indianola conglomerate in central Utah and the Echo Canyon conglomerate in northern Utah.

3. The Laramide mountain-building event took place at the end of the Cretaceous and the beginning of the Tertiary (80 to 40 million years ago). The Laramide's most prominent mountain range is the Uinta Mountains. Less prominent features include the San Rafael Swell, the Waterpocket Fold, and the Monument uplift.

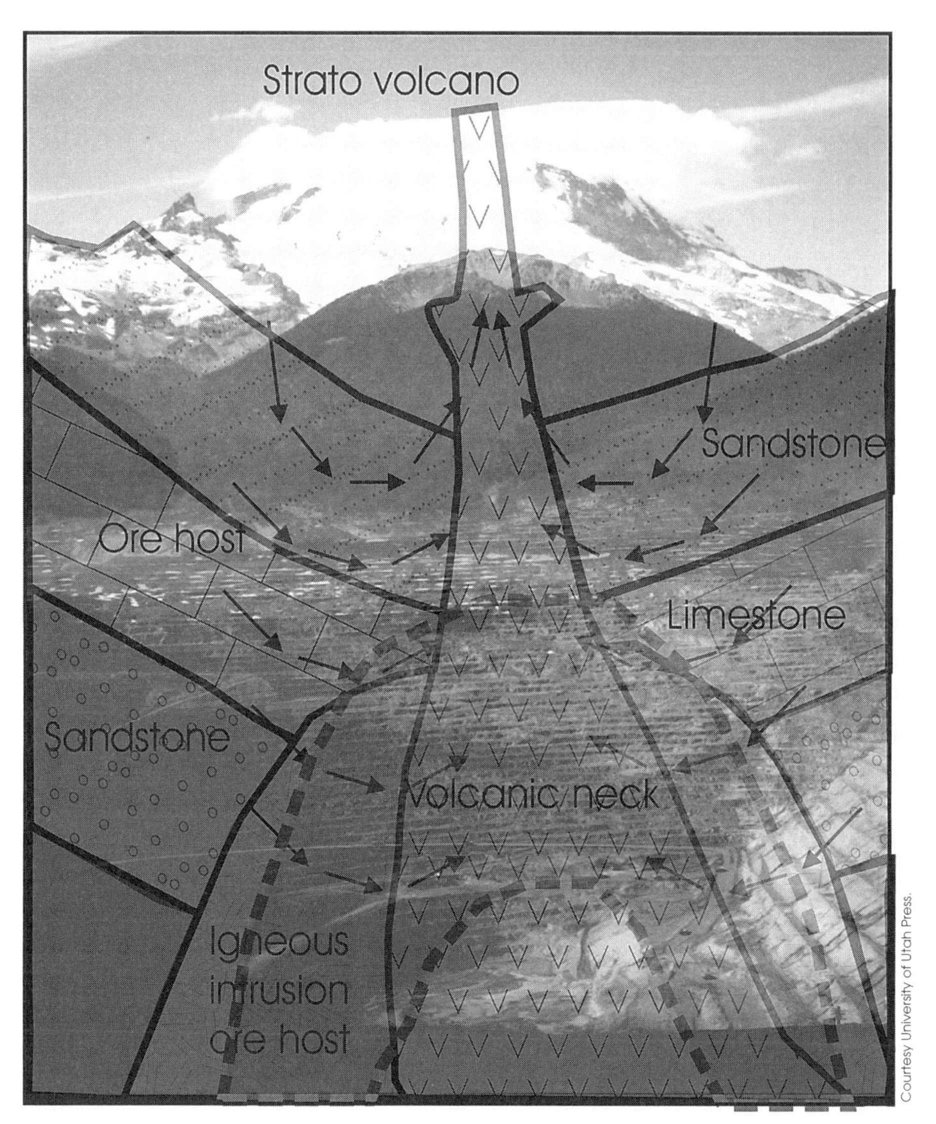

Courtesy University of Utah Press.

Figure 13. A schematic model shows the formation of deposits of valuable metals such as Bingham, Utah. Here, in a reconstruction of what Bingham was like, a large strato volcano similar to Mount Ranier was formed and in its roots circulating water indicated by the arrows extracted metals from the igneous rocks and possibly surrounding sediments and deposited them in an inverted, saucer shaped deposit within the intrusive igneous rocks at depth. Mining of the deposits at Bingham has produced the open-pit mine benches. The volcano no longer exists at Bingham; it was eroded away long ago.

of gold in granite is about two parts per billion parts of rock. The concentration required for an economical ore deposit is approximately two parts per million parts of rock, a thousand times more. So geological processes must concentrate the gold by a thousand times and must then transport the gold or copper or other valuable metal to someplace where it can be deposited in concentrated form, for example, Bingham, or Eureka, or Kimberly in the Tushar Mountains in southern Utah.

The transporting medium is a fluid. It can be water or magma. Water can come from magma or the ocean or from precipitation. Water is invariably involved in Utah ore deposits. The mechanism for depositing the ores in concentrated form can be cooling of the fluid, mixing fluids of diverse compositions, or chemical reactions as the water interacts with the rocks. These mechanisms change the ability of the water to maintain materials in solution.

Figure 13 shows a subsurface igneous intrusive that has reached the surface and formed a volcano, a big edifice such as the steep-sided composite volcano at Mount Rainier. The subsurface rocks include a sequence of sedimentary rocks deposited on the Precambrian basement. The pore spaces within the sedimentary rocks are filled with water, and the igneous rock heats up that water. Hot water is less dense than cold, and so it rises. Arrows indicate the hot water rising to the surface; it may produce hot springs similar to hot springs in Yellowstone Park. Lines of equal temperature are elevated toward the surface. At the surface, some boiling springs may occur. Cold meteoric water that falls on the ground circulates down because it is cold and dense. It gets heated; the density decreases so it rises to the surface. Water circulating through the rocks concentrates and precipitates valuable heavy metals. These kinds of deposits are associated with volcanic activity at lithospheric convergent margins.

Such processes have left Utah an important legacy of valuable deposits of copper, silver, gold, and other metals. Examples include the large copper deposit at Bingham and the productive silver, lead, and gold deposits in the East Tintic Mountains near Eureka. Ore deposits in the Tintic, Bingham, Park City, Big Cottonwood, Little Cottonwood, and American Fork Districts are variations on a three-part theme: carbonate sediments, faulting and folding, and igneous intrusion. The deposits differ only in detail.

Geology and Ore Deposits of the Wasatch Mountains

The Wasatch Mountains near Salt Lake City display much of the geological history of Utah. The oldest rocks are Precambrian crystalline basement exposed east of Farmington (the Farmington Complex), and a small exposure occurs near the mouth of Little Cottonwood Canyon (the Little Willow Series). Younger Precambrian rocks of the Big Cottonwood Canyon series and still-younger Precambrian glacial deposits of Mineral Fork Tillite overlie these rocks. Shoreline and shallow-water continental-shelf deposits of Paleozoic time begin with Tintic Quartzite (Cambrian), Ophir Shale, and then a thick limestone, the Maxfield Limestone, indicating rising sea and migration of the shoreline eastward. Immediately east of Salt Lake City, Ordovician,

Silurian, and Devonian rocks are missing, and the Maxfield Limestone has been variably eroded away so that a thick sequence of Mississippian limestones directly overlies the Cambrian rocks. Pennsylvanian and Permian rocks in the Wasatch are much thinner than to the west in the Oquirrh Basin. Mesozoic rocks include marine limestones and the windblown sand deposit (Nugget) so prominent in southern Utah. Cretaceous rocks are clastics eroded from the thrust-fault sheets of the Sevier mountain-building event. All of these rocks are folded and thrust faulted. Later Laramide events are superimposed on the Sevier events. A chain of igneous intrusive rocks from Park City to Little Cottonwood Canyon postdates the Laramide. The Wasatch Mountains were later uplifted by the prominent Wasatch Fault.

One of the largest and most recent ore producers in this area was the Cardiff Mine. The greatest production came from the Deseret Limestone. Ore mineralization occurred where northeast fissures intersected the limestones in the footwall of the Alta thrust fault. The Cardiff Mine lies well up a glaciated fork of Big Cottonwood Canyon, which was itself glaciated above the confluence with Cardiff Fork. Hike up the fork on the old mining road littered with baseball-sized cobbles of limestone and quartzite. You reach the Price tunnel, where no ore was found but much water flows into the creek. Above here, at the Cardiff Mine site, spectacular cliffs of blue-gray limestone with bleached zones tower over the varied wood and metal debris from mining. Just above the mine, the Alta thrust fault can be seen placing older Cambrian Tintic Quartzite on top of younger Mississippian limestone and putting Precambrian tillite on top of the quartzite so the rocks appear in reverse order of age. The thrust fault played a leading role in forming the ores. The Alta stock was the heat engine that forced water to circulate through channels in the rock; the thrust fault and nearby northeast-trending fractures provided the fluid pathways, and the limestone neutralized the acid waters and precipitated the ores.

The famous Emma Mine in Little Cottonwood Canyon produced lead and silver from replacement of the Fitchville and Gardison limestones. Rich ore shoots occurred along northeast-trending fissures and within rocks that were especially permeable due to earlier fracturing along a thrust fault. Ore in the Emma and nearby Flagstaff Mine contained 60 ounces of silver per ton and 40 percent lead.

Park City

The three of us—a mining engineer, a lawyer, and I—straddled the bench behind the gasoline-powered locomotive for the three-mile ride into the Ontario Number-Two Drain Tunnel. We headed straight into the core of the Park City anticline. We first passed through Keetley volcanic rocks, then through the Thaynes limestone that sometimes hosts ore, and then the Mayflower and Ontario igneous intrusives. The tunnel paralleled the Hawkeye-McHenry fault. At times tunnels bearing water entered from the south where lead-silver ore had been extracted along the fault. Tunnels used to explore side veins entered from the north. Where fluids and faults had destroyed the rock strength, heavy wood, steel, and concrete supported the overlying layers of rock. Three miles into the mountain, we came upon the vertical shafts through which ore, men, and supplies were hoisted or lowered. The light at the tunnel mouth had long since shrunk to a pinpoint as we crouched down to avoid banging hard-hatted heads on the timbers, and cold water dripped down our necks. Our only illumination came from the battery-powered lamps attached to our helmets. As we quietly spoke, the water-saturated air caused our breath to condense in streams of steaming vapor.

The ore deposits of Park City occur in limestones and on faults that fed the metal-rich fluid into them. Little value is present in the igneous rocks whose heat circulated the water and whose heritage produced much of the ore. One particular limestone has been exceptional for its production of lead and zinc: the Jenny bed within the Park City formation. This limestone is about 20 feet thick and is a principal ore host in the Silver King, Ontario, Daly-West, and Judge Mines. The Weber quartzite is chemically less reactive than limestone, but it is brittle and fractures easily, and the fractures provide good conduits for ore-forming solutions that deposit ore along the way.

Bingham and the Geology of the Oquirrh Mountains

The geologic history of the Oquirrh Mountains resulted in the formation of the Bingham copper deposit, the gold ores at Mercur, and the base and precious-metal deposits of Ophir and Stockton. Most of the mountain range is composed of sedimentary rocks: shales, limestones, and sandstones. These sediments were deposited in an ocean, but their position varied from near-shore sandstones to shales and limestones deposited far out to sea. A thick sequence of limestones and sandstones was deposited in an inland sea perhaps similar to the Black Sea today. Collision of South America with North America sent repercussions through western North America that included formation of the Ancestral Rocky Mountains and the creation of two important inland seas: the Paradox Basin and the Oquirrh Basin. The Oquirrh Basin in northwestern Utah was a deeply subsiding basin bounded by faults where a great amount of sediment accumulated, more than 20,000 feet. Thinner, younger deposits that also appear in the nearby Wasatch Mountains capped the thick basin fill.

The sequence of sedimentary rocks was folded and thrust faulted as a consequence of the convergence of North America and the Pacific plate during the breakup of Pangea. A series of anticlines and synclines is a prominent feature of the Oquirrh Mountains, and the accompanying Midas and North Oquirrh thrust faults are displayed in the northern part of the range.

Very early on Saturday morning, I mounted my horse Shorty for the ride from the end of the road in Ophir Canyon. I had driven up Ophir Canyon past the Ophir Hill Mine through the Ophir anticline. We were on the east limb of the Ophir anticline, and we rode across easterly inclined rocks, turned up Picnic Canyon, and continued riding to the crest of the mountain range. Lowe and Lewiston peaks were shrouded in thunderclouds, gale-force wind swept west across the ridges, and West

Canyon was obscured by more clouds far below us. Shorty walked carefully down the steep switchbacks and dugway trail covered with a few inches of snow to bring me to our lunch stop in the brilliant autumn-colored aspen grove. We saw a few elk and dozens of deer.

The Oquirrh Mountains are the richest hills in Utah and possibly in the world. Their richness stems from ores at Bingham, Stockton, Ophir, and Mercur. This unusual concentration is due to the sedimentary-rock host for the ore, faults that conduct the water that deposited the ores, and volcanic activity that supplies ore constituents and the heat engine that circulates the water.

The presence and action of a descending lithospheric slab underneath Utah melted rock that buoyantly rose in the vicinity of Bingham, creating a large stratovolcano and the associated volcanic rocks that erupted from the volcano. Water circulating from the igneous rock, water present in the rocks at the time of volcanic action, and water that precipited on the mountains extracted metals from the igneous rock and surrounding sedimentary rocks to form the ores at Bingham and elsewhere.

The copper, molybdenum, and gold deposits at Bingham are largely contained within and related to an igneous-intrusive rock formation known as the Bingham stock. The Bingham stock has in turn been intruded by younger igneous rock in its northwestern part. Both of these igneous intrusions are cut by still younger and smaller igneous intrusions. The copper, molybdenum, and gold mineralization is disseminated throughout the igneous intrusions. The fluids that deposited the ore also converted the igneous rocks to a variety of minerals unlike the originals. Before mining the Bingham deposit contained 3.1 billion tons of ore averaging 0.73 percent copper, 0.043 percent molybdenum disulfide, and 0.013 ounces of gold per ton.

Thin limestone beds within the Pennsylvanian Oquirrh formation contain rich lead and zinc ore. The entire 8- to 15-foot thickness of the Lark limestone was replaced by lead and zinc minerals along a bedding plane fault. In the U.S. Mine, one of the major lead-zinc mines in the Bingham district, most of the mineralization existed within northeast-striking fissures in quartzite and igneous-intrusive rocks. Where these fissures crossed limestones, the ore mineralization extended laterally for variable distances. A third type of ore mineralization is also present at Bingham. Limestones in contact with the igneous-intrusive rock have been replaced by large quantities of constituents from the igneous-hydrothermal system, including copper.

Ophir

Most of the ore bodies in the Ophir area are lead and zinc replacements of limestones. The ore bodies in the Ophir Hill Mine near the entrance to the town of Ophir occur in the limestone portions of the Ophir formation along fissures and faults.

Stockton

The ores at Stockton occur at the intersection of the Great Blue limestone with north-south-trending fissures. The ore shoots are inclined to the northwest and may contain 40 percent lead, about 20 ounces of silver per ton, and a small amount of gold. An igneous intrusive is present with an age of 39.4 million years.

Mercur

The Mercur District extends from Ophir Canyon south to Five Mile Pass. The district has produced more than 3.5 million ounces of gold. The gold deposits occur in limestones of upper Paleozoic age that have been deformed into a series of northeast-trending folds. The gold is concentrated in a mineralized sequence of the lower member of the Mississippian Great Blue limestone. Although 32- and 37-million-year-old igneous rocks are present in the district, minerals reveal surprisingly old ages for the mineralization, more than 100 million years.

Tintic and East Tintic

The town of Eureka lies at the foot of Eureka Ridge. A hike along the ridge from west to east traverses the sedimentary rocks that host the ores of the main Tintic District. To the north lies the small mining community of Eureka and to the south the smaller Mammoth town site. Mining structures of wood and metal and rock debris decorate the north-south-trending ore runs. The two principal mining areas are the Tintic and the East Tintic Districts. These two, though they are adjacent, differ in some important ways.

We were lowered a couple of thousand feet down the Chief Consolidated shaft; then, as the hoist cable stretched and the man cage bobbed up and down, we got off and walked to the nearest mine area, where rich lead-silver ore of the main Tintic District was being mined. The native silver along with the lead minerals glinted in the light of our headlamps. The ventilation air was cold, and we were told that so much water must be pumped out by the electric pumps that in case of a power failure, indicated by the occasional lightbulb going out, we had limited time to escape before the mine workings were completely submerged. In this main Tintic District, the major portion of the ore occurs as several northeast-trending ore runs that descend to the north. The most favorable host is the Bluebell dolomite of Silurian age.

Volcanic rocks, some extensively altered by circulating hot water from below much like Yellowstone Park, mostly cover the East Tintic District. Many of the mines in this area are hot from circulating water. Discovering the rich lead-zinc-silver deposits in the favorable sedimentary rocks beneath the volcanic cover provided a formidable challenge. Exploration was guided by the characteristics of the alteration in volcanic rocks and the complex assemblage of thrust, tear, and normal faults that acted as fluid-flow pathways.

To examine some of these deposits, we recruited a hoist operator and a couple of guides and descended the Eureka Lilly shaft. The air became hot and moist as we went down, and upon leaving the man cage, one of our guides burned his hand on the ringing handle to the telephone. (These telephones had a handle that must be cranked to ring the phone at the surface, hence a ringing handle.) The air temperature was at least 120 degrees Fahrenheit, the top reading of the thermometer that I carried. We walked through the old mine tunnels to the Tintic Standard Number-One shaft, where no ore was discovered. We then walked to the Number-Two shaft, which was in the middle of the Tintic Standard ore body. Here lead-silver ore has completely replaced the Middle Ophir limestone along a complex system of faults related to thrust faulting known as the *pothole structure*. Fluids set into convective circulation by intrusions of hot igneous rocks reacted with the first limestone bed in the Paleozoic sequence. Then the limestone reacted to cause precipitation of the ore minerals.

Pine Valley and Silver Reef

Interstate 15 descends toward St. George south of Cedar City. The high cliffs on the east mark the Hurricane Fault, and the dark mountains to the west are the Pine Valley range. Soon the core of the Virgin anticline appears with Quail Creek Reservoir at its southern tip. Rocks are tilted in opposite directions away from the crest of the anticline. It is this anticline and the thrust faults that accompanied its formation that affected the sandstone host for the silver deposits at Silver Reef. The ore horizon has been repeated three times by faulting as one layer of sandstone pushed over another. These are the same sandstone beds that host uranium ore to the east, but here only small amounts of uranium are present. Silver is the main attraction.

At Silver Reef and Leeds, the Pine Valley Mountains dourly preside over the lavishly colored red rocks to the south and east. These mountains are cored by an igneous intrusive formed 20.9 million years ago whose role in forming the silver deposits remains uncertain. The sandstone host for the silver ores is the Triassic-age Chinle. The silver was carried in solution in groundwater that moved through the porous and permeable river-channel sands. Chemical interaction with plant debris and associated bacteria resulted in precipitation of silver compounds that at times replaced tree limbs and trunks.

The Interior Seaway and Coal Deposits

As you drive on U.S. Highway 6 from Spanish Fork Canyon to Price and beyond, imagine the scenery when the coal deposits were formed. The coarse red conglomerate in the Red Narrows of the canyon near the ghost town of Thistle and the junction with U.S. Highway 89, though younger than the coal, mark vigorous mountain streams eroding the folded and faulted older rocks of the mountains to the west. Such streams flowed east across a broad floodplain toward the arm of the ocean known as the Cretaceous interior seaway. Dinosaurs traversed this floodplain of swamps and forests covered by plants such as the ancient giant sequoia. Flood stages in the rivers covered the downed trees and prevented their decomposition. The result was layers of coal enclosed by beds of sandstone.

Major deposits of coal in Utah occur in Cretaceous rocks. The two most important areas where coal is mined are the Book Cliffs, where coal is found in several zones in the Blackhawk formation, and the Wasatch Plateau, a continuation of the

Book Cliffs to the southwest. The principal coal bed lies in the Blackhawk formation in the Mesa Verde group of the Upper Cretaceous. The coal beds formed about the same time as the Echo Canyon conglomerate along Interstate 80 east of Coalville and the Indianola conglomerate of central Utah. Dinosaurs, magnolias, fig cypress, palms, water lilies, and sequoias are major contributors to coal deposits. A third major accumulation of Cretaceous-age coal occurs in the Kaiparowits Plateau of Garfield and Kane Counties. The favorable circumstances that led to coal formation from middle to late Cretaceous time were luxuriant plant growth, favorable climate, low-lying coastal areas on the shoreline of the Cretaceous interior seaway, and relative subsidence of the area that led to the changing position of the shoreline. Coal deposits form on extensive delta plains adjacent to large rivers and along alluvial coastal plains where numerous small rivers shift their courses.

The prominent Andean-type mountain range in western Utah was associated with a broad downwarp in eastern Utah and western Colorado. This depression formed an arm of the ocean running from about the Gulf of Mexico to the Gulf of Alaska. Much fine-grained sediment accumulated in this ocean, but the shorelines had a different history. Vigorous mountain streams flowing east from the mountains produced thick accumulations of gravel and then meandered across a broad floodplain to reach the ocean. Coal beds formed from the dense accumulation of tropical plants that proliferated on this coastal plain.

Vigorous mountain streams like those in the Wasatch deposited conglomerates at the foot of the mountains during spring floods. After leaving their load of gravel at the foot of the mountains, the streams wandered across the floodplain watering a tropical jungle and depositing sand until they reached the sea. Some spring floods covered the plants, and succeeding layers of sediment preserved them as coal.

Oil and Natural Gas

Oil formation requires a source bed, a reservoir rock, a trap, and a seal. The source bed must contain from 0.3 to 0.5 weight percentage of organic carbon with a high hydrogen to carbon ratio (0.7 to 1.5) and a low oxygen to carbon ratio (less than 0.1). The source bed must have been heated to a temperature sufficient to form petroleum, usually in the range of 130 to 140 degrees Centigrade, but time is important, too. Longer times require lower temperatures. If the temperature is too high, then only gas is produced. The reservoir rock must be both porous and permeable; the trap, for instance an anticline, permits the petroleum to accumulate because it is less dense than water, and an overlying impermeable bed such as shale prevents farther upward migration of the petroleum.

The oldest-known oil- and gas-bearing rocks in Utah are Devonian, although potential oil- and gas-bearing rocks may occur in the Chuar group of Precambrian age. Three geological areas have been major producers of oil and gas. They are the Paradox Basin in southeastern Utah, the Uinta Basin in northeastern Utah, and the overthrust belt of Utah and Wyoming. Oil is found in the Paradox and Honaker Trail formations of Pennsylvanian age in the Greater Aneth field in the Paradox Basin, from the Green River formation of Tertiary age in the Altamont and Bluebell fields in the Uinta Basin, and from the Nugget and Twin Creek formations of Jurassic age in the overthrust belt in Summit County. Potential source beds include Twin Creek and Thaynes limestones, Mancos and Arapien shales, the Mesa Verde group, the Phosphoria (Park City formation), and lake sediments of the Green River formation. Other suggested source rocks are the Oquirrh formation and Kirkman and Flagstaff limestones.

Uranium Deposits of Southeastern Utah

Uranium deposits in southeastern Utah are so distinctive that they are named for the region: Colorado Plateau uranium deposits. The uranium minerals have concentrated in sandstone bodies that filled river channels in the west-flowing Chinle rivers or the east-flowing Morrison rivers. The uranium accumulated in an idealized crescent-shaped deposit. Chemical reactions between the surrounding rocks and flowing groundwater mobilized the trace amounts of uranium that they contained. The uranium-charged water then moved easily through the sands and gravels of the ancient rivers until chemical reactions with plant debris or hydrocarbon precipitated the uranium.

Mesozoic rocks are the host for uranium-ore deposits that were intensively mined following World War II. The most famous mine in Triassic rocks is probably the Mi-Vida in Lisbon Valley southeast of Moab. This deposit, discovered in 1952, produced $40 million worth of uranium. The major host rocks for uranium include the Triassic Chinle formation and the Jurassic Morrison formation. Creating uranium deposits in these rocks requires extraction from some source rock, transportation in porous and permeable rocks, and final deposition. Uranium occurs in two oxidation states: U^{+4} (reduced) and U^{+6} (oxidized). Uranium is soluble in the oxidized form, where it occurs in solution as uranyl ion UO_2^{+2}, and insoluble in the reduced form, where it occurs as the mineral uraninite UO_2. Uranium is extracted from granitic source rocks that contain an average of four parts uranium per million parts of rock by oxidized rainwater. The uranium is then transported in groundwater through porous and permeable rocks. Stream-channel sands in the Chinle and Morrison formations are ideal transportation pathways. The uranium is then precipitated by chemical reduction when the groundwater encounters organic matter such as logs or dinosaur bones in the buried paleo stream channel. Other important reducing agents include natural gas or petroleum-type hydrocarbons.

The early morning sun rises near the LaSal Mountains to the east and picks out the tips of red sandstone cliffs at Courthouse Rock. Green and purple sage still in shadow provides an aromatic accompaniment to the growing illumination of Arches National Park. To the west the huge, red, Wingate cliffs are like the sides of an enormous ship sailing in the red sea of the underlying Chinle formation. Jeep roads mark the uranium-bearing beds in the Chinle. The Chinle was also a profilic uranium producer in the San Rafael Swell to the northwest.

Matt and Bill camped on the bank of the Muddy River near the heart of the San Rafael Swell in the shadow of Tomsich Butte. Small uranium mines and prospects decorated the Chinle exposures beneath the Wingate and Navajo sandstone cliffs. Early the next morning, they put on wading shoes and, with day packs filled with spare clothing, food, and a first-aid kit, set out down the Muddy toward the steep-walled and narrow canyon known as the Chutes of Muddy Creek. The trail guide had said that water was seldom more than ankle deep through the chutes. They alternately waded the river and walked on the sand banks until the canyon became so narrow that they were in the water all of the time. The water gradually deepened until it was chest high, and small irregularities in the footing caused them to sink deeper. Without flotation gear, they decided to turn back before reaching the end of the chutes at the Delta Mine.

The Muddy River originates in the high Wasatch Plateau, where it has cut its canyon into Tertiary-aged lake sediments. It flows south toward the San Rafael Swell, that large, bean-shaped, anticlinal fold that formed during the last episode of plate-convergence-driven mountain building. The river system has exhumed the swell by removing the overlying cover of lake sediments and superimposing its meanders. Continued exhumation has cut through the Mesozoic and into Paleozoic sediments. Before the river emerges from the swell, it flows through a high-walled, narrow slot canyon called the Chutes, cut through Navajo and Wingate sandstone. The river exits from the chutes into a softer, more mellow terrain and there, 400 feet above, on the cliffs on the left side of the canyon, is the Delta or Hidden Splendor Mine. The Muddy then joins the Dirty Devil River and finally the Colorado.

In its journey, the Muddy cuts through the Chinle and Moenkopi formations and deep into the late Paleozoic rocks. As it flows from the canyon gash cut into the Paleozoic into the more subdued terrain of the Moenkopi, high Wingate and Navajo cliffs above shelter the Delta Mine in the underlying Chinle sandstone. The uranium, vanadium, and copper ores are contained in ancient river-channel sand fillings, elongated, as are the river channels themselves, in a northwestern direction. Groundwater here extracted uranium and other metals from volcanic ash, flowed in the subsurface along easy pathways in the channel sands, and deposited the uranium in the trunks and limbs of fossil trees and other plant debris.

Summary

The rich and diverse geological history of Utah has provided source rocks for mineral wealth, host rocks for placing minerals, pathways to conduct the fluids that place the minerals, and driving mechanisms that produce the mineral deposits. The early marine history created the limestone host rocks that contain the rich metal deposits at Bingham, Tintic, Park City, Big Cottonwood-Little Cottonwood-American Fork, Mercur, and other places around the state.

Later, streams flowing from high mountains deposited gravels and sands that are the hosts for uranium in southeastern Utah and silver at Silver Reef. Rivers running west from the Ancestral Rockies to the sea deposited the Chinle formation while rivers running east from the Nevadan Mountains into the Sundance interior sea laid down the younger Jurassic Morrison formation. River channels are filled with coarse, permeable sand. These sands provided pathways to move groundwater and form uranium deposits in Triassic Chinle and Jurassic Morrison formations.

The younger Sevier Mountains supplied the water for rivers flowing east into the Cretaceous interior sea. These rivers watered the lush vegetation that was buried in flood stages to form coal deposits.

Igneous activity accompanying mountain building created the metal deposits at Bingham, Park City, Eureka, and elsewhere. Finally, the area was fragmented and extended by great north-south-trending normal faults, and erosion of the mountain ranges of the Great Basin and the Wasatch exposed the ore deposits for study and exploitation.

2

Generating Wealth from the Earth 1847–2000

Thomas G. Alexander

The increase in wealth of all advanced nations has depended to an appreciable extent on extracting and utilizing minerals from the Earth's treasure trove. People have used precious minerals like gold and silver to manufacture such diverse items as coins, jewelry, and electronic parts. Prosperity has depended upon fashioning useful and decorative objects from copper, iron, zinc, molybdenum, chrome, beryllium, aluminum, potash, and magnesium. Building materials fashioned from clay, gravel, stone, gypsum, dolomite, cement, and limestone have made life more comfortable. Societies have obtained energy for domestic, commercial, and manufacturing needs from coal, oil, and natural gas. Builders have paved roads and walkways with concrete and hydrocarbons like gilsonite.

Each of these minerals has contributed to the growth of national and individual wealth in a similar way. Each mineral has an intrinsic value which, however, those who uncover it can exploit only by changing it into a form for which they can find either a market or a personal use. After locating and extracting the minerals, workers add to its value by separating it from the unwanted minerals mixed with it. Next, they fashion the valuable minerals into a form they or someone else can use.[1]

On the whole, the story of the wealth generated from mineral resources in Utah parallels the experience of other areas, except that nature has blessed Utah with a wider variety and a richer abundance of minerals than most other states. We see the replication of the general pattern as we examine the years between 1847, when pioneers belonging to the Church of Jesus Christ of Latter-day Saints first entered the Salt Lake Valley, and the end of World War II in 1945, the time period during which Utah's mines were maturing, and mining's period of mature growth between 1946 and 2000.

Progression to Maturity, 1847–1945

Almost immediately after the new immigrants arrived in the Salt Lake Valley, they recognized the economic value of mining for their comfort and prosperity. Their earliest ventures included recovering salt from the Great Salt Lake, undertaking missions to California to mine for gold, iron mining and manufacturing near Cedar City, lead mining near Las Vegas, and coal mining near Coalville in Summit County, Wales in Sanpete County, and Cedar City in Iron County.[2] Of these ventures, salt and coal mining grew into operations of persisting importance; the gold mined in California helped provide a local circulating medium; and, following early failures, iron mining eventually furnished raw materials for a ferrous-metals industry.[3]

Shortly after arriving in the Salt Lake Valley, a committee left for the Great Salt Lake to mine and manufacture salt. In a short time, the workers produced 125 bushels of coarse white salt and a barrel of table salt.[4] Production of salt continued generally by evaporating the lake's supersaturated water, and although we do not know the volume between 1847 and 1879, in 1880 salt miners on the Great Salt Lake and at other locations recovered 12,000 short tons with a market value of $60,000.[5]

Recognizing the value of this commodity, companies sold salt in Utah and nearby territories. Smelters in Utah, Nevada, Idaho, Colorado, and Montana used the chlorine from Utah salt in milling operations to recover silver.[6] In 1887 James Jack and a group of Utah business people organized the Inland Salt Company, one of three companies that extracted salt from the Great Salt Lake. Production increased considerably after a syndicate of midwestern capitalists purchased the Inland Salt Company and changed its name to Inland Crystal Salt Company in 1891.[7] In 1892 Utah's salt production reached nearly 181,000 short tons, with a value of more than $340,000, or an average of $1.88 per ton.[8] During the depression of the 1890s, output declined, and the volume of salt produced did not exceed 100,000 tons again until 1941. Significantly, however, in spite of the decline in production, by 1917 the gross income earned by the salt companies had exceeded the 1892 level as the value of the salt increased to $4.35 per ton in 1900 and $6.16 per ton in 1917. Salt remained a commodity more valuable than it had been during the 1890s, even throughout the Great Depression of the 1930s.[9]

Even more than salt mining, coal mining required an infusion of outside capital. Although Mormon pioneers discovered coal in Summit, Sanpete, and Iron Counties, production remained relatively insignificant until 1882, when the Denver and Rio Grande Western Railroad laid its tracks into Carbon County. The railroad's subsidiary, the Utah Fuel Company, successfully challenged the Union Pacific Coal Company, whose mines in Wyoming had dominated the territory's markets. In 1881 Utah produced a scant 52,000 tons of coal. In 1883, the first full year of production from Carbon County, Utah miners recovered 200,000 tons of coal. Coal production reached 2.3 million tons in 1909 and 3 million tons in 1912. After peaking at 6 million tons in 1920, annual coal production did not decline below 4 million tons until the depression year of 1931, when Utah mined only 3.4 million tons. Most importantly, the value of that production increased from $1.4 million in 1900 (the first year the value of annual production grossed more than $1 million) to $19 million in 1920. Thereafter, the annual value of coal production never declined below $10.3 million until 1931. Even during the Depression, however, the annual value of coal production remained above $5.1 million, and the gross value exceeded that of every year prior to America's entry into World War I (1917) except 1913, the year before the war began, and 1916, the year before America's entry into the war.[10] (See table 1.)

An understanding of the methods used to increase profit from the production of salt and coal offers a window into the reasons for the wealth generated from other types of mining ventures. In both cases the presence of minerals available for exploitation alone did not guarantee their usefulness. The salt had remained dissolved in the waters of Great Salt Lake for thousands of years, and the coal had rested under eastern Utah soil for millions of years before miners extracted and marketed it. In the late nineteenth century, however, the introduction of appropriate technology and the investment of capital allowed entrepreneurs and workers to profit from mining and distributing the two minerals.

Although operators in a number of eastern and midwestern mining districts began introducting machines to cut coal from the face of underground seams and load it in cars as early as 1899, Utah companies resisted adopting this technology for a time. In 1913, however, companies installed 60 electric cutting machines in Utah's coal mines.[11]

Utah's hard-rock mining companies tended to introduce technology more rapidly, and these mines benefited even more quickly from improvements in transportation than the coal mines did.[12] Since Utah is a landlocked state, distant from significant national markets and blessed with neither ocean nor commercially viable river transportation, access to markets depended on the development of the transcontinental railroad. In 1866, the first full year of peace after the Civil War, Utah grossed a mere $85,826 from the sale of the principal nonferrous metals: gold, silver, copper, lead, and zinc. Just four years later in 1870, the first full year after the wedding of the rails at Promontory Summit, Utah grossed $1.4 million from the same metals. Moreover, in 1870 the territory produced 23.4 percent of the nation's lead and 3.9 percent of its silver.[13] (See table 2.)

Thereafter, the state continued to prosper from the extraction, manufacture, and sale of nonferrous minerals. Production fluctuated with markets and business cycles, but over the years from 1870 to 1945, Utah produced a high of 45.5 percent of the nation's lead in 1872 to a low of 10.5 percent in 1945. Utah produced the nation's silver at a rate ranging from 3.9 percent in 1870 to 31.9 percent in 1925. Utah's percentage of the nation's gold production varied from 0.1 percent in 1873 to 34.5 percent in the war year of 1944 (see table 2). Over the years between 1864 and 1945, Utah produced more than $2.7 billion worth of the principal nonferrous metals: gold, silver, copper, lead, and zinc.[14]

Table 1
Statistics of Utah's Bituminous Coal Production, 1880–1945

Year	Total Production (short tons)	Value (dollars)	Average Value per Ton	Employees	Average Tons per Employee per Day
1880	14,748	33,645	2.28	NA	NA
1885	213,120	426,000	2.00	NA	NA
1890	318,159	522,390	1.74	429	2.57
1895	471,836	617,349	1.31	670	3.47
1900	1,147,027	1,447,750	1.26	1,308	3.54
1905	1,332,372	1,793,190	1.35	1,361	3.24
1910	2,517,809	4,422,556	1.68	3,053	3.46
1915	3,108,715	4,916,916	1.58	3,564	3.91
1920	6,005,199	19,350,000	3.22	4,504	5.29
1925	4,690,342	11,991,000	2.56	4,441	5.90
1930	4,257,541	10,515,000	2.47	3,504	7.23
1935	2,946,918	6,091,000	2.07	2,752	5.70
1940	3,575,586	7,831,979	2.20	2,590	7.58
1945	6,644,000	22,922,000	3.45	NA	NA

University of Utah Bureau of Economic and Business Research, "Measures of Economic Changes in Utah, 1847–1947," *Utah Economic and Business Review:* 77; U.S. Bureau of Mines, *Minerals Yearbook*, 1945, 79.

Table 2
Percent of National Production and Total Value of Gold, Silver, Copper, Lead, and Zinc Produced in Utah, 1865–1945

Year	Gold	Silver	Copper	Lead	Zinc	Total Value $
1865	0.1	NA	NA	NA	NA	56,416
1870	0.6	3.9	0.4	23.4	NA	1,449,566
1875	0.5	9.3	1.8	31.9	NA	5,504,401
1880	0.5	12.1	0.1	15.6	NA	5,918,252
1885	0.6	13.1	0.1	19.8	NA	7,793,867
1890	2.1	14.7	0.4	23.8	NA	12,308,915
1895	3.0	13.4	0.6	18.4	NA	8,464,549
1900	5.0	16.1	3.0	17.7	NA	16,992,869
1905	5.8	19.7	6.4	17.2	0.8	26,046,948
1910	4.3	18.2	11.7	15.6	2.5	32,199,185
1915	3.6	17.0	12.6	17.8	2.1	55,105,070
1920	4.0	23.2	9.6	13.7	0.7	49,744,334
1925	7.7	31.9	14.1	22.4	3.7	82,701,394
1930	9.8	27.5	12.8	20.7	7.5	48,653,464
1935	5.0	18.9	17.0	19.2	6.0	31,651,571
1940	5.9	17.0	26.4	16.6	6.6	86,585,499
1945	29.3	21.0	29.3	10.4	5.4	90,018,641

University of Utah Bureau of Economic and Business Research, "Measures of Economic Changes in Utah, 1847–1947," *Utah Economic and Business Review:* 68–69.

In the early 1870s, many of those who built smelters in Utah failed because they lacked technical skill. In his 1873 report to the federal government, Rossiter W. Raymond pointed out that most of Utah's smelters had proved unprofitable, in part, because of depressed economic conditions but also because of management "without technical ability," so that with the notable exception of the Germania smelter in Murray, "fortunes were there lost in slags, dust, and matte which might easily be saved."[15] Subsequent technological improvements, especially the introduction of reverberatory furnaces and oil flotation systems of concentration, increased efficiency, and with it the wealth generated by the operations.

Most importantly, the introduction of rail transportation into Utah that nearly coincided with significant technological developments allowed more efficient extraction of mineral wealth. In 1866 Alfred Nobel invented dynamite by stabilizing nitroglycerine in sawdust and earth called *kieselguhr.* Several times more powerful than black powder, dynamite had the advantage of greater stability than nitroglycerine, which tended to explode without warning.[16]

At the time of Nobel's invention, miners still had to hammer a double or single jack to drill holes into which they placed the black powder, nitro, or dynamite. In 1870, the year after the wedding of the rails, however, Charles Burleigh changed that by inventing the pneumatic drill. Since drilling accounted for 75 percent of the cost, when paired with dynamite, a drill that could strike 300 blows per minute made mining much more productive.

After 1870, as the exposure and recovery of minerals became increasingly efficient, mining companies began to invest in lifts to take miners deeper into the earth and hoist minerals from greater depths. Mine cages traveling at 800 feet per minute lowered the miners into deep shafts, where they used pneumatic drills and dynamite to recover mineral-bearing ores.

To dynamite, mechanical drills, and elevator cages, mining companies added numerous inventions designed to remove and process the ore. These included railways, trams, cable systems, concentrators, electric power plants, and smelters.[17] In some cases the investments offered such short-term advantage that the investors may not have recovered either capital or much interest. The development of such facilities in Little Cottonwood Canyon offers an example. In the spring of 1903, Anton "Tony" Jacobson, working with his brother Alfred, filed notice of appropriating water in Little Cottonwood Creek to build a concentrating mill and electric power plant.[18] After constructing the power plant in the canyon about four and a half miles west of Alta, the brothers began concentrating ores in November 1904. The electrically powered concentrator transformed ores valued at $15 per ton into concentrate sold to smelters at $45 to $60 per ton. The mine that fed the concentrator suffered from a buildup of excessive water which hampered operations, but the concentrator continued to process ores from other mines in the canyon for some time. After passing through receivership and witnessing the closing of most of the mines that the operation served, the company closed its mill and sold the electric plant to the Utah Power and Light Company.

To move the minerals to smelters, operators experimented with various methods of transportation, a number of which proved expensive but only marginally successful. Henry M. Crowther of the Continental Mines and Smelters Company constructed a tramway down Little Cottonwood Canyon. In the face of environmental hazards, the tramway—with a number of realignments—operated from 1905 until March 1910, when an avalanche destroyed it.[19] Between September 1915 and June 1918, Walter K. Yorston operated the Alta-Cottonwood Railway Company, which ran trains over tracks that eventually extended from Wasatch at the mouth of the canyon to Alta. The operators incurred such great cost in constructing and operating the railroad, however, that they failed to cut into wagon traffic and eventually trucks replaced both means of transportation.[20]

In spite of these examples of marginal technology, some developments proved eminently successful. These included transmission of 40,000-volt electricity by Lucien L. Nunn's Telluride Power Company from a plant in Provo Canyon to gold mines at Mercur in Tooele County.[21] V. C. Heikes of the U.S. Geological Survey considered the operation at Mercur "a pioneer" in the development of the cyanide-electric process, "which transformed the operations of the Mercur camp from a forlorn hope to an important commercial enterprise."[22] At Park City, California capitalists George Hearst and James Ben Ali Haggin purchased the Ontario Mine from Rector Steen. They invested in large-scale technological improvement and mined more than $13.9 million dollars worth of silver-lead ore between 1873 and 1904.[23]

In 1914 the federal report on mineral industries recorded a wide variety of technologies to concentrate and recover precious and useful metals. Mills recovering gold and silver in Beaver, Box Elder, Juab, and Washington Counties used mercury amalgamation. Mills in Beaver, Piute, Salt Lake, Juab, Tooele, and Utah Counties used cyanidation. The Knight-Christiansen mill at Silver City in the Tintic District and the Ontario mill at Park City treated silver-lead ores through chloridizing/roasting. Other Park City mills also used leaching to convert metals into chlorides, then treated the chlorides with scrap iron to precipitate the metals out.[24]

Perhaps the most extraordinary technological successes took place at Bingham on the east slope of the Oquirrh Mountains under the aegis of the Boston Consolidated Mining Company and the Utah Copper Company. As Table 2 indicates, Utah produced a relatively insignificant percentage of the nation's copper until about 1900.

The development at Bingham owes its success to several people who recognized the possibilities hidden in a relatively large mass of low-grade copper ore.[25] The Oquirrh Mountains in southern Salt Lake County harbored a vast store of igneous rock containing from as much as 2 percent to as little as 0.4 percent copper. Instead of depositing the copper in compact bodies, volcanic eruptions had scattered it relatively uniformly in small flecks throughout the rock mass as porphyry ore. After constructing a concentrator and other facilities, Samuel Newhouse, born in New York City to a Russian-Jewish family; Thomas Weir, a Leadville, Colorado, associate of Newhouse; and William Rockefeller and Henry H. Rogers, who had earned a fortune in Standard

Oil, began using steam shovels to scrape off the overburden and recover the porphyry ore en masse at the Boston Consolidated Mine at Bingham on 24 June 1906.

Two months later the Utah Copper Company, under the leadership of Daniel C. Jackling, Charles M. MacNeill, and Spencer Penrose, who had worked together in Colorado, began a similar operation at Bingham. With financial backing from Swiss-born Philadelphia smelting entrepreneur Meyer Guggenheim, and the continued financial interest of Meyer's family—especially his son Daniel—Utah Copper applied industrial technology such as steam shovels, railroad cars, and dump cars to scrape off the overburden and remove and transport the ore to a concentrator.[26] Relying at first on steam shovels, the company began to use electric shovels with caterpillar tracks by December 1923. The operators eventually replaced all of the steam shovels with electric equipment, and electric trains replaced steam locomotives.[27]

Engineers attributed much of the success to the operation of the mills. With the technical expertise of George O. Bradley and Frank G. Janney, the company developed an experimental concentrator at Copperton and a larger mill at Magna. After concentrating the ore, they shipped it to a smelter at Midvale, and after 1906 to the ASARCO smelter at Garfield, which the company later purchased.

Technological improvement helped make the company even more profitable. In 1918 Utah Copper introduced the flotation process at the Arthur mill. Between 1905 and 1917, the company recovered an average of 61 percent of copper from the porphyry mass. Initial runs increased the recovery through flotation to 73 percent, and by 1923 efficiency had improved to 81 percent. Improvements in efficiency at the concentrators reached 85 percent by the 1920s and 90 percent by the 1960s.[28]

Other companies adopted the flotation process as well, often using it to treat tailings still rich in recoverable metals from previous less-efficient mill operations. The Utah Apex mill at Bingham, the Daly-Judge and two other mills at Park City, the Horn Silver mill at Frisco, and a dump at Newhouse all treated ores or tailings with flotation. At Eureka in the Tintic District, the Utah Ore Concentrator Company used flotation on dumps from the Chief Consolidated, Eagle and Blue Bell, and Victoria Mines.[29] By 1925, though most companies still used crushing and grinding machines based on nineteenth-century designs, many had adopted flotation to concentrate copper, lead, and zinc ores with "only slight changes . . . in chemicals and oils used" in the process.[30] Moreover, by the mid-1920s virtually all gold and most of the silver produced in Utah resulted "as a by-product in the smelting of copper, lead, and lead-zinc ores and concentrates."[31]

Utah Copper also improved the technology of its smelter operations. Between 1948 and 1950, the company spent $17 million to install electrolytic copper-refining units at a new facility near the ASARCO smelter at Garfield. Electrolysis produced copper at 99.96 percent purity.[32] The company also prospered from the collateral recovery of such minerals as gold, silver, and molybdenite.

Historians can make a similar point with regard to operations at other mining sites. Some of these include Mercur in Tooele County, the Tintic District in Utah and Juab Counties, the San Francisco District near Frisco in Beaver County, Silver Reef in Washington County, Big and Little Cottonwood and American Fork Canyons on the western slope of the Wasatch Mountains in Salt Lake and Utah Counties, and Park City on the eastern slope in Summit County.[33]

In calculating the value of mining to Utah's economy and people, we ought to subtract what we may justifiably call the bads from the goods produced. By the early twentieth century, the crops grown in central Salt Lake County had begun to suffer from the sulfur fumes exuded by the lead and copper smelters in nearby towns.[34] Sulfur dioxide (SO_2), which resulted from the combination of sulfur with oxygen, reacts with airborne water vapor to create sulfuric acid (H_2SO_4). Both the sulfur dioxide and sulfuric acid attacked and corroded crops, and farmers living near the smelters appealed to the United States District Court for relief.

After hearing their appeal, Judge John A. Marshall granted an injunction which forced the smelters to close until they could control the release of sulfur fumes. The United States Smelter at Murray reopened only after its engineers developed a system of 2,080 canvas bags, each 18 inches in diameter and 33 feet long. When the operators coated the inside of the bags with zinc oxide, they captured a sizable proportion of the sulfur dioxide without burning through the canvas. With this process, the smelters not only reduced "the damaging effects of the gasses," but they also recovered "products" that went "a long way toward defraying the cost of operating the bag house."[35]

Another major "bad" in Utah's mining industry in the late nineteenth and early twentieth centuries was the large number of accidents, many leading to disability, dismemberment, or death. The introduction, for instance, of pneumatic drills increased the level of particulates in the underground air, and, as their lungs coated with the tiny particles, miners developed silicosis, often called "miner's consumption." Rapidly moving lift cages often posed hazards to their occupants since miners stood only a few inches from stationary walls as the elevators whizzed by at 800 feet per second.

Studies of death often focus on explosions, which take numerous lives at one time. These include the Scofield Mine disaster in 1900, when 200 miners died, and the Castle Gate explosion of 1924, when 172 died. Both blasts resulted from the failure to water the coal adequately to control the buildup of explosive dust. Additional deaths occurred from carbon monoxide poisoning.[36] Between 1892, when the federal government began mine inspections for Utah, and 1913, 355 coal miners died. The majority (200) perished in the Scofield explosion, but the next largest number (95) died from falling rocks or coal.[37]

The rules under which miners worked contributed to these dangerous conditions. Until the passage of Utah's Workers Compensation Act in 1917, mine employees, like those in other industries, labored under what were called the assumed risk (which meant that workers understood and accepted the dangers of their jobs when they hired on), contributory negligence (they may have done something to

TABLE 3
COAL MINE FATALITIES IN UTAH AND RELEVANT STATISTICS RELATING TO THEM, 1913–19

Year	Deaths	Fatalities per 1,000 Worker Hours	Tons Produced per Fatality	Tons Produced per Worker per Day	Average No. Days Mines Open
1913	20		165,236		
1914	22	6.37	141,047	3.60	210
1915	11	3.71	282,618	4.19	208
1916	23	8.03	155,106	5.00	228
1917	22	7.19	187,510	5.41	219
1918	19	4.40	270,359	4.77	259
1919	27		171,471		

1913: Utah Bureau of Immigration, Labor, and Statistics, *Second Report of the State Bureau of Immigration, Labor and Statistics for the Years 1913–1914,* 220; 1914–19: Utah Industrial Commision, *Report of the Industrial Commission of Utah, Period July 1, 1918 to June 30, 1920,* 290.

TABLE 4
FATALITIES PER 1,000 WORKERS IN METAL MINES OF UTAH, OTHER WESTERN STATES, AND THE UNITED STATES, 1915–19

Year	Underground (U) and Total (T)	Utah	Calif.	Colo.	Idaho	Montana	Nevada	U.S.
1915	U	4.46	5.38	9.69	6.69	3.84	4.14	
	T	3.78	4.65	7.61	5.41	4.53	3.13	3.89
1916	U	2.82	4.84	5.52	3.84	4.62	5.25	
	T	3.07	3.34	5.84	2.84	3.88	4.13	3.62
1917	U	3.80	5.38	6.47	4.86	14.61	5.07	
	T	3.02	3.86	6.23	4.33	13.20	4.52	4.44
1918	U	3.59	8.46	4.89	4.77	3.74	7.39	
	T	3.58	5.54	4.63	4.11	3.50	4.83	3.57
1919	U	2.76						
	T	3.00						

Utah Industrial Commission, *Report of the Industrial Commission of Utah, Period July 1, 1918 to June 30, 1920,* 269.

help cause the accident), and fellow servant (one of the other employees may have contributed to the accident) rules. These rules meant that employees or their families could almost never recover compensation from companies for injury or death.[38]

More significantly, these rules provided disincentives to companies to make certain that they offered safe workplaces to miners. Under these circumstances, as the Industrial Commission put it, "it not infrequently happens that a foreman or fire boss simply suggests to the miner that he put up a prop, instead of insisting that it be done and done at once."[39] Such laxity contributed to falls of coal or rock. Moreover, many of the companies left holes unprotected, inviting accidents and even death.[40]

Since workers assumed the risks of a dangerous occupation, many people considered any remuneration given to injured employees or their survivors as charity rather than compensation for an accident or death while working. In the Scofield disaster, for instance, the company provided the dead men with burial clothes and a coffin, gave each dead miner's family $500, and forgave $8,000 in accumulated debt at the company store. In the case of the Castle Gate disaster, even though Utah had adopted workers' compensation by that time, at the instigation of Utah Governor Charles R. Mabey, a committee collected $132,445.13 for the 417 dependents left behind without any support.

Before the creation of the Utah Industrial Commission in 1917, the state coal mine inspector reported statistics of deaths and injury almost as though they constituted a normal part of the work year. In 1913, for instance, accidents occurred at the rate of 20 fatal, 34 serious, and 191 minor. One miner died for every 165,236 tons of coal mined.[41] (See tables 3 and 4.)

Even after the establishment of workers' compensation in 1917, the state of Utah hardly made munificent payments for accidents or death. In the case of death, the law allowed a maximum payment to the deceased's dependents of medical and burial expenses plus 55 percent of the weekly wage or $15 per week—whichever was less—to a maximum of $4,500. Payments from the fund or the company's insurance carrier came monthly to the survivors at a rate of $15 per week. This means that the family of a worker who died in an industrial accident while earning $6 per day for a six-day week received 42 percent of his or her income for 25 years.[42] To support their families under these circumstances, many widows had to find day care for their minor children while they entered the workforce.

Even after the establishment of the Industrial Commission, death continued to be a frequent visitor in Utah's mines. During the 1918–1920 biennium, 59 employees died in accidents in mines, quarries, and smelters. Of these accidents, 25 took place in underground coal mines and 26 in metal mines. In spite of their large output, the open-pit mines of Utah Copper Company were statistically less dangerous than underground mines.[43] (For the size of Utah's mining workforce during these years, see table 5.)

We should note, however, that in the years immediately before and after the establishment of the Industrial Commission and the Workers' Compensation Fund,

TABLE 5
UTAH AND UNITED STATES WORKFORCE, PERCENT ENGAGED IN MINING AND NUMBER OF UTAHNS ENGAGED IN MINING, 1850–2000

Year	Utah			United States	
	Persons in UT Workforce	Persons Engaged in Mining in UT	Percent of UT Workforce in Mining	Persons in U.S. Workforce	Percent of U.S. Workforce in Mining
1850	3,135	11	0.4	5,371,876	1.5
1860	8,431	4	0.2	8,287,043	0.6
1870	21,517	575	2.7	12,505,923	1.3
1880	40,055	2,648	6.6	17,392,099	1.5
1890	66,901	3,819	5.7	22,735,661	1.8
1900	84,604	7,028	8.3	29,073,670	2.1
1910	131,540	10,019	7.6	38,167,336	2.5
1920	149,201	10,117	6.8	41,614,248	2.6
1930	170,000	10,514	6.2	48,829,920	2.0
1940	148,886	10,102	6.8	45,166,083	2.0
1950	228,822	12,081	5.3	56,225,340	1.6
1960	302,147	13,202	4.4	64,639,247	1.0
1970	384,760	11,215	2.9	77,308,792	0.8
1980	585,921	18,128	3.1	97,639,355	1.1
1990	736,059	9,473	1.3	115,681,202	0.6
2000	1,068,000	8,001	0.7	140,863,000	0.4

1850–1950: Leonard J. Arrington, *The Changing Economic Structure of the Mountain West, 1850–1950;* 1960–70: U.S. Census Bureau, *Census of Population, Characteristics of the Population, United States Summary,* and *Part 46, Utah* (1970 data for those 14 and older); 1980–90: U.S. Census Bureau, *Census of Population, General Social and Economic Characteristics, United States Summary,* and *Part 46, Utah* (data for both years are for those 16 and older); 2000: U.S. Bureau of Census *Statistical Abstract of the United States, 2001,* 367, 370; Utah Department of Workforce Services, "Industry Profile, Mining."

the rate of accidents in Utah tended to be lower than most other western states, and either lower or about equal to the remainder of the United States. We do not know all the things that contributed to Utah's better record, but the Industrial Commission noted that a number of Utah's coal companies had adopted more stringent safety regulations than the law required.[44]

Improvement in safety paralleled the growth of union organization in the mines during the 1930s. Until the passage of the labor provisions of the National Industrial Recovery Act of 1933 and the Wagner Labor Relations Act of 1935, employers had no legal obligation to bargain with their employees. The legal requirement of collective bargaining led to demands for a safe work place as part of union contracts. By 1936 approximately 94 percent of all Utah coal miners worked under union contracts. Unions demanded safer working conditions, and the companies complied.[45]

Table 6
Number of Establishments and Average Nonagricultural Monthly Wages in Utah and Mining Industries, 1960–2000
(Wages in Current Dollars)

Date	Ave. Utah Wage	Ave. Mine Wage	Mine Estab.	Ave. Metal Mine Wage	Metal Mine Estab.	Ave. Bit. Coal Mine Wage	Bit. Coal Mine Estab.	Ave. Other Mine and Quarry Wage	Other Mine and Quarry Estab.
1960	370	505	629	520	252	504	47	466	330
1965	431	598	460	619	170	522	36	578	254
1970	529	774	433	781	152	792	27	704	254
1975	731	1,169	416	1,209	110	1,219	21	1,056	285
1980	1,111	1,911	590	2,020	157	2,008	40	1,666	393
1985	1,440	2,614	629	3,078	118	3,008	32	2,161	479
1990	1,644	2,976	375	3,161	67	3,545	16	2,325	292
1995	1,936	3,384	354	4,041	48	3,892	18	2,712	288
2000	2,401	4,043	346	4,646	41	4,822	17	3,316	288

Utah Department of Workforce Services, Economic Data and Analysis Unit, *Annual Report of Labor Market Information, 2000,* table 19.

Growth and Concentration, 1946–2000

In spite of the problems caused by disease and accidents, mining companies continued to build on the foundation established in the early years and to contribute to the state's prosperity after World War II. Between 1946 and 2000, significant changes in markets, mineral troves, and technology, coupled with a tendency for small firms to leave the field to larger national corporations, altered the structure of Utah's mining industry as well as its relative importance in Utah's economy. (On the number of companies involved in various mining industries, see table 6.) As in the period before World War II, during the years up to the new millennium, firms that adopted improved technology and took advantage of changing markets succeeded. Those that did not moldered, sold out, or closed. Most importantly, the number of firms declined as many smaller companies either closed or sold out to larger operations, and, as a result, ownership in the minerals industry became increasingly more concentrated.

Between the end of World War II and 2000, the comparative importance of the mining industry in relation to other sectors of Utah's economy changed considerably as well. In 1963 income from mining constituted 8.7 percent of Utah's gross state product (GSP). In that year, as a major contributor to the value of goods and services produced in Utah, mining easily exceeded agriculture, forestry, and fisheries; and construction. Moreover, it held a place only slightly lower than services; and transportation, communications, and utilities (see table 7).

As the structure of Utah's economy changed, however, mining's contribution to the GSP declined. By the late 1960s, mining's portion of the GSP exceeded only

Table 7
Percent of Gross State Product Accounted for by Major Industries, 1963–2000

Year	Agriculture Forestry Fisheries	Mining	Construction	Manufacturing	Transportation Communication Utilities	Wholesale Retail Trade	Finance Insurance Real Estate	Services	Government
1963	2.8	8.7	5.3	20.7	9.9	16.3	12.6	9.3	14.4
1967	3.3	3.9	4.7	11.8	10.6	17.4	13.7	10.1	19.2
1972	2.9	3.6	6.4	11.2	10.4	17.9	13.9	10.9	18.9
1977	2.1	5.1	7.6	15.3	10.4	17.7	13.3	12.1	16.2
1982	2.1	5.9	5.2	15.8	12.5	16.0	14.6	13.0	14.9
1986	1.5	4.1	5.2	14.2	11.2	16.9	13.9	15.8	17.3
1990	1.6	4.9	4.0	14.8	10.0	15.3	13.1	18.5	17.8
1994	1.3	3.0	5.5	14.0	9.5	16.7	14.0	19.9	16.2
1996	1.1	2.5	6.0	15.8	8.9	16.4	15.4	19.1	14.8
1998	1.1	1.8	6.4	13.5	8.9	17.2	16.6	20.0	14.4
2000	1.0	1.8	6.4	12.5	8.6	16.2	18.5	20.8	14.1

1963–72: Boyd L. Fjeldsted, "Utah's Gross State Product for the Years 1963–1986," *Utah Economic and Business Review:* 8–9; 1977–82; Boyd L. Fjeldsted, "Utah's Gross State Product for the Years 1977–1989," *Utah Economic and Business Review:* 10–11; 1986–2000: calculated from State of Utah, *2003 Economic Report to the Governor* (www.governor.Utah.gov/dea), 73.

agriculture, forestry, and fisheries, and though it was greater or nearly greater than construction's contribution at times during the 1980s and early 1990s, during the mid-to-late 1990s, it began to fall behind. Replacing mining, the major growth sectors of Utah's economy were purveyors of services, especially financiers, insurers, realtors, engineers, health providers, and business consultants. By 1998 mining was contributing a mere 1.8 percent of the state's GSP, larger only than agriculture, forestry, and fisheries (1.1 percent). (See table 7 and its sources.)

Nevertheless, for the miners and those counties that depended upon mineral extraction, these industries made an exceedingly important contribution; and mining's benefits for Utah's economy exceeded its benefits nationally. In 1963 mining contributed $259 million to Utah's $2.98 billion GSP. In 1990 it contributed $1.5 billion to Utah's $31.4 billion GSP, and in 2000 it contributed $1.2 billion to the state's $68.5 billion GSP (see table 8). In 2000 mining employed 0.7 percent of Utah's workforce, compared with only 0.4 of workers nationally (see table 5).

Moreover, secondary impacts of basic industries like mining and manufacturing tend to exceed those of industries like the service area, which includes jobs with low wages and thus less extensive purchases of goods and services. The Utah Mining Association estimates that mining may generate as many as three jobs in subsidiary industries for every actual mining job. Mining companies purchase goods and services

Table 8
Utah Gross State Product and Its Mining Components, Selected Years, 1963–2000 (Millions of Current Dollars)

Year	Gross State Product	Mining Product	Metal Mining	Coal Mining	Oil and Gas	Non-metallic Minerals	Percent of Utah GSP from Mining
1963	2,979	259	141	23	86	9	8.7
1967	3,479	137	52	17	49	19	3.9
1972	5,465	197	104	24	56	13	3.6
1977	10,116	520	141	148	199	32	5.1
1982	18,018	1,058	170	350	491	47	5.9
1986	24,473	1,001	142	255	583	22	4.1
1990	31,359	1,534	382	210	858	84	4.9
1994	42,236	1,256	448	286	484	37	3.0
1996	51,523	1,296	411	409	423	53	2.5
1998	59,084	1,074	237	335	416	86	1.8
2000	68,549	1,208	265	335	517	91	1.8

1963–72: Boyd L. Fjelsted, "Utah's Gross State Product for the Years 1963–1986," *Utah Economic and Business Review:* 4, 8; 1977–82: Boyd L. Fjelsted, "Utah's Gross State Product, 1977–1989," *Utah Economic and Business Review:* 6, 10; 1986–2000: State of Utah, *2003 Economic Report to the Governor*, 73.

from other businesses, and they borrow money from banks. Miners patronize stores, gas stations, and public utilities as well.[46]

In addition, without mining Utah's workers would have received lower wages, and a larger number in counties away from the Wasatch Front would undoubtedly have suffered from the poor pay offered by agriculture, retail sales, and tourist services. On the average, Utah's miners earned more annually than workers in any other industrial sector. At an average salary of $46,140 per miner in 2000, mining wages in Utah stood 69 percent above the average pay of all other industries (see tables 6 and 9). Most particularly, Emery, Carbon, and Uintah Counties, and to a lesser extent Duchesne, and Sevier Counties, relied on income from mining to support a sizeable portion of their population. Moreover, these counties, together with Daggett, San Juan, Morgan, and Juab, depended upon mineral industries for a significant portion of their tax revenue.[47]

Beyond the changes in the relationship between mining and other components of Utah's GSP, the earning profiles among the various sectors of the mining industry changed considerably between 1946 and 2000. In 1963 metal mining outperformed the other mineral industries as a component of GSP (table 8). Beginning in the 1970s, the relationship between coal and metal mining seemed like alternating loops on a roller coaster. During the late 1970s and early 1980s, coal mining surpassed metal mining, only to fall behind from the early to mid-1990s. During the late 1990s and into the early 2000s, however, coal mining again outdid metal mining.

Table 9
Average Monthly Wages for Selected Nonagricultural Industries, 1960–2000 (Wages in Current Dollars)

Year	State	Mining	Construction	Manufacturing	Transportation Communications Utilities	Trade	Finance Insurance Real Estate	Services	Government
1960	370	505	442	453	448	303	340	218	380
1965	431	598	539	533	544	342	397	270	453
1970	529	774	690	614	687	410	509	350	589
1975	731	1,169	941	848	1,028	561	677	534	793
1980	1,111	1,911	1,386	1,305	1,628	838	1,073	882	1,118
1985	1,140	2,614	1,668	1,811	2,107	1,038	1,502	1,196	1,517
1990	1,644	2,976	1,843	2,066	2,424	1,173	1,818	1,458	1,735
1995	1,936	3,484	2,042	2,348	2,703	1,141	2,303	1,789	2,054
2000	2,401	4,043	2,477	2,944	3,223	1,779	3,069	2,301	2,455

Utah Department of Workforce Services, Economic Data and Analysis Unit, *Annual Report of Labor Market Information,* 2000, table 19.

By contrast, oil and gas extraction experienced a steady rise. By the late 1970s, the value of oil and gas extraction had surged ahead of both metal and coal mining, and it continued to hold the lead early in the twenty-first century (see table 8).

Alterations in the coal industry resulted from a shift to larger companies, the introduction of improved mining technology, and a change in markets. Between 1945 and the late 1960s, Utah's and America's coal industry completely lost the residential-heating and steam-railroad markets upon which they had traditionally relied. Railroads converted to diesel locomotives, and people changed to natural gas or, to a lesser extent, oil or electricity.[48]

Moreover, over time Utah mines lost markets for coke. Utah provided coke for steel manufacturing at Kaiser Steel's Fontana, California, plant until the company declared bankruptcy and closed in 1983. Utah coal mines continued to supply coking coal to the Geneva Steel plant near Orem until it failed to reorganize under bankruptcy and shut its doors in November 2002.

Undeterred by these losses, coal companies sought new markets. They found these by providing coal to generate electric power, and by 2000 Utah companies were shipping more than 50 percent of their coal to electric power plants.[49] As this market was expanding, however, Utah's underground coal was losing its market share to more cheaply strip-mined coal from Wyoming and Montana.[50]

Fighting back, Utah's underground mines actually ended up producing more coal than previously by adopting improved technology. During the early 1950s and 1960s, continuous mining machines and long-wall technology began to replace less-efficient means of coal removal. In 1965 66 percent of the coal mined in Utah was loaded by "continuous miners."[51] Replacing workers with technology, the continuous miner uses a "rotating drum-shaped cutting head studded with carbide-tipped teeth" to chop coal from the mineral face. Gathering arms on the machines scoop coal onto conveyor belts and eventually into shuttle cars. During the 1950s companies also began to introduce long-wall technology, which allowed continuous cutting and caving of a seam of coal several hundred feet wide.[52]

Mechanization, however, came at a cost. Although it lowered the cost of underground mining, it also reduced the number of miners need to produce larger quantities of coal. For instance, the number of coal miners declined from 4,296 in 1982 to 1,535 in 2002. Over the same period, tons per miner hour increased from 2.05 to 6.00 as prices declined from $29.42 to $17.33 per ton.[53]

Such changes in technology and competition altered the economics of Utah's coal industry enormously. Continuous miners cost more than a million dollars each, but because coal is relatively heavy and bulky, transportation to markets constitutes the major cost of mining and marketing. By lowering these transportation costs, Utah mines succeeded in capturing some of the nearby electric-utility market from strip-mined coal.

Clearly improved technology and new markets moved coal companies out of the doldrums to increased production. During the late 1940s and early 1950s, Utah mined more than 6.1 million tons of coal annually.[54] By the late 1950s, production had begun to decline, and in 1962 Utah produced only 4.3 million tons of coal, largely because of the loss of home-heating and railroad markets. By 1982, largely because of increased demand by electric utilities, Utah production had increased to 16.91 million tons. Then, as the electric-utility business ramped up and Utah mines found markets abroad, the state experienced record years between 1996 and 2002, never producing fewer than 26.4 million tons per year.

At the same time, the number of mining companies declined as larger, more cost-effective operations captured the field. In 1959 46 mines in Utah produced 1,000 tons of coal or more. By contrast in 2002, only 13 mines were operating in the state.[55] (See table 6.) In 2001 Utah's largest operation, Canyon Fuel Company's SUFCO mine in Sevier County, produced more than a quarter of all the coal mined in the state.[56]

As Utah's coal industry advanced technologically, companies extracting coal, metal, and hydrocarbons opened new markets. By 1999, for instance, Japan had become the largest foreign market for Utah's coal. At the same time, primary metals had become Utah's leading export commodity, accounting for 24 percent of the value of all of the Beehive State's exports. Switzerland, followed by the United Kingdom and Canada, offered the largest international markets for primary metals.[57]

As Utah's mining companies found new markets, the mineral industry changed considerably. Immediately after World War II, as America ratcheted up for the cold war and the Korean War, it appeared that the need for atomic weapons would create a permanent market for uranium. In 1952 three plants—the Vanadium Corporation at Hite, the Atomic Energy Commission plant in Monticello, and the Vitro Chemical Plant in Salt Lake City—were processing uranium ore, much of it mined in southeastern Utah.[58] In fiscal 1956 Utah produced 772,000 tons of uranium ore, or 35 percent of the domestic total. Most of it came from San Juan, Emery, Piute, and Grand Counties.[59]

After the heady days of the 1950s and early 1960s, markets failed. Uranium operations declined markedly after the federal government ended its purchase program in 1971, and companies operated only intermittently with the decline in uranium and vanadium prices in the 1980s. [60]

At the same time, the need for petroleum and natural gas to feed motor vehicles, warm homes, and supply industry accelerated. Plants in northern Salt Lake and southern Davis Counties refined most of the petroleum. By 1952 Uinta Basin fields were supplying much of the petroleum.[61] Then, the discovery of oil in the Aneth field in southern San Juan County in 1953 expanded oil's prospects, and since that time the Uinta Basin and San Juan County have been the major sources of oil. Modern geologic, drilling, and extraction technology helped find and pump out the oil, and pipelines flowed it to the refineries. Construction during the 1950s of pipelines from the Aneth field to California and the Gulf Coast and from Rangley, Colorado, to Salt Lake City helped boost Utah's oil industry.[62]

During the last half of the twentieth century to the early twenty-first, Utah continued to mine and market significant quantities of other minerals. In 1953, for instance, Utah produced the most gilsonite in the United States; ranked second in copper, gold, molybdenum, and silver; placed third in lead and potash; stood fourth in iron ore; and held ninth place in coal and zinc.[63] In 1992 Utah was the only source of mined beryllium in the U.S., and it ranked second in copper and magnesium production; third in gold, iron ore, and molybdenum; and sixth in silver.[64]

After 1959 beryllium from west central Utah became extremely important.[65] Following exploration by a number of companies, the Brush Wellman Company in October 1962 and afterward purchased substantial ore deposits in the Spor Mountains in Juab County, northwest of Delta.[66] In 1969 the company completed construction of a mill 10 miles north of Delta to process ore into beryllium hydroxide, which it shipped to a plant at Elmore, Ohio, for additional processing.[67] By 2002, however, Brush Wellman had reduced its operations considerably.[68]

Kennecott Utah Copper, by every measure Utah's principal metal producer, increased its contribution through a number of technological innovations. In 1967, for instance, it reduced the cost of handling smelter slag by installing new granulation facilities and increased its copper output from leaching by installing new precipitation cones.[69] In 1971 adding new grinding and flotation units to the Magna and Arthur plants increased copper recovery to more than 90 percent.[70]

In the depths of a depressed copper market in 1985, Kennecott announced a four-year plan to modernize by mechanizing operations that were previously more labor intensive. These included constructing an in-pit crushing system, a modern ore-conveying system, and a new grinding plant at Copperton, and laying pipelines to carry slurry from Copperton to the Magna concentrators.[71]

In addition to the innovations to improve productivity, companies invested in technology to clean up previous ground and water pollution and to mitigate pollution from current operations. Among the most serious problems were the tailings left from uranium mills. In 1962 uranium mills operated in Moab, Mexican Hat, and Salt Lake City.[72] In 1964 the Salt Lake and Mexican Hat plants closed, and only the Atlas plant in Moab remained open.[73] Atlas closed its mill in 1987, and legislation in 1992 required the federal government to reimburse companies for cleanup when government contracts had helped generate the waste.[74] By the 1990s the Atlas site had clearly become a major hazard as groundwater was leaching radioactive waste into the nearby Colorado River. In May 2005, the U.S. Department of Energy announced a decision to move the tailings to a site near Crescent Junction off I-70. (*High Country News*, 2 May 2005.)

Tailings from mining and milling operations in Salt Lake Valley also contributed to pollution. Between 1984 and 1986, the U. S. Department of Energy and the Utah Department of Health paid Argee Corporation of Denver $50 million to remove 2.9 million tons of uranium-vanadium tailings from the Vitro Chemical site in south Salt Lake City to a dump at Clive in Tooele County.[75] In 1984 the EPA placed the Sharon Steel Plant in Midvale, abandoned in 1971, on its superfund list. Among other companies, Sharon's successors, ARCO and UV Industries, had to fund part of cleanup in addition to testing the area for soil and groundwater contamination and monitoring people who lived nearby for high levels of lead in their blood.[76]

Polluted groundwater resulted from other operations as well. In 1983 Kennecott funded a five-year study to determine groundwater contamination from the Bingham Canyon operation. Studies showed that runoff had contaminated the Bingham Canyon Reservoir, and Kennecott spent $10 million to build structures to capture the runoff from contamination in the Dry Fork area five miles north of the open pit.[77] ARCO spent $5 million to "cover and revegetate the tailings pond on its . . . Carr Fork property" in Tooele County.[78] In the mid-1990s Kennecott spent $510 million rehabilitating its tailings pond and, as part of the mitigation, constructed a 2,500-acre wildlife preserve.[79] Still, complaints arose over blowing dust from drying tailings ponds, and in 1985 Kennecott spent $3.5 million to control the dust. In 1992 the company agreed to spend $4.75 million to remove contaminated soil along Bingham Creek to a permanent disposal site in the Oquirrh foothills.[80] By March 2003, for cleaning up contaminated sites and removing more than 25 million tons of mining waste in southwestern Salt Lake County, Kennecott had spent $290 million and ARCO $37 million.[81]

Companies invested to mitigate air pollution as well. Following the adoption of standards in 1977 under the Clean Air Act, Kennecott installed a 1,200-foot smokestack and spent $280 million to convert its reverberatory furnaces to the continuous-copper-smelting Noranda process. No more cost effective than reverberatory furnaces, the Noranda system nevertheless controlled sulfur-dioxide emissions by capturing 86 percent of the sulfur as compared with 55 percent in the reverberatory furnaces.[82] In the mid-1980s Kennecott's improved technology removed 92 percent of the sulfur dioxide from the emissions.[83]

Unsatisfied with the environmental pollution still remaining with Noranda system, Kennecott again modernized its smelter between 1992 and 1995 by installing an Outokumpu flash furnace and converter. Unlike the Noranda system, the Outokumpu furnace and converter increased capacity. At the same time, they recovered 99.9 percent of sulfur-dioxide emissions.[84]

Air pollution also constituted a major problem at other operations. In its 1989 report, the Environmental Protection Agency cited the AMAX Magnesium Corp. (later MagCorp) plant at Rowley, Tooele County, as the "single largest air polluter in the Nation," even though in 1988 the company had installed equipment to reduce chlorine emissions by 50 percent.[85] At the present time, MagCorp remains the largest air polluter.

Abandoned mines constituted an additional hazard. The federal government provided funding to reclaim abandoned mines through a coal-production tax. A survey indicated that the government might need to reclaim nearly 17,000 mine openings in Utah at a potential cost of $174 million. In 1991 workers reclaimed coal mines in Sevier, Sanpete, Carbon, Garfield, Duchesne, Emery, Kane, and Iron Counties.[86]

As companies and governments continue to recover from past inattention to environmental damage, newer regulations require mitigation of such harm. The days of companies running roughshod over the land have ended. Before any mining operation can begin, companies must file a reclamation plan with the Utah Division of Oil, Gas, and Mining.[87]

Beyond the environmental hazards, injury and death in mining accidents remained a significant problem during the 1960s. In the mid-1960s, in accidents per million miner hours, coal mines remained the least safe, though because of the larger number of workers, more people actually died in metal mines. In 1965 13 deaths occurred in mining operations. In 1966 there were 21 deaths.[88] In an attempt to prevent accidents from coal falls and gas-ignited explosions the companies installed safety systems. In 1966, for instance, Independent Coal and Coke constructed a new system of hydraulic jacks to provide "adequate roof support." At the same time, to comply with new mining regulations, the company installed monitors to measure gas concentrations more reliably.[89] In spite of the care that some companies took, some major disasters occurred. A fire in 1984 at the Wilberg Coal Mine in Emery County took 25 lives.[90]

Although many companies tried to reduce the incidence of fatalities, some of the uranium companies seemed absolutely criminal in their mistreatment of employees. In February 1985, for instance, families of 24 miners received $1.19 million in partial compensation for the cancer they had contracted from breathing radon 222 gas between 1949 and 1968 in the Vanadium Corporation of America's poorly vented mines near Marysvale. The miners had tried to sue the federal government because it had not warned them about high levels of the gas, but the U.S. District Court held the government immune from negligence claims.[91]

Despite the history of accidents and fatalities, mining had become a much safer occupation by the 1990s. In 1993, among 66 fatal accidents in various Utah occupations, 5 occurred in mining. In 1995, of 51 fatal accidents in Utah occupations, 8 took place in the mines. In 2000 mining caused 6 of 61 fatal accidents. Deaths still occurred in Utah's mines, but mining was clearly no longer the least-safe occupation.[92]

Assessing Mining's Contribution

With all of the evidence we have reviewed, how can we assess mining's economic contribution to Utah? Clearly mining has contributed substantially to Utah's economic development. Between the arrival of the railroad in 1869 and the end of World War II, miners extracted more than $2.7 billion worth of minerals from Utah's soil. In the period from World War II to 2000, mining constituted an appreciable component of Utah's GSP, ranging from 8.7 percent in 1963 to 1.8 percent in 2000. Far from resulting simply from the discovery of mineral deposits, these economic contributions ensued from the introduction of capital, ever-improving technology, marketing skill, and the labor and sacrifices of Utah's workers. Development, jobs, and wealth resulted in part from the investment of nationally active capitalists like George Hearst, James Ben Ali Haggin, Simon Guggenheim, and William Rockefeller. In addition to outsiders, Utahns who invested, learned mineralogy and geology, and introduced technology, like Daniel Jackling, Thomas Kearns, David Keith, and Jesse Knight contributed to Utah's economic growth. Most significantly, Utah's economic development also owed a great deal to thousands of dedicated, but largely anonymous, miners.

From 1870 to the Great Depression, mining was arguably the most significant growth industry in Utah's private sector. Since World War II, mining's relationship to other industrial sectors has changed considerably. During the 1930s, World War II, and the cold war, public-sector employment in defense-related industries challenged mining's dominance. Increasingly since the mid-1970s, while surpassing agriculture, mining has shrunk in importance compared with other economic sectors such as services, manufacturing, and governmental agencies.

This is not necessarily a negative situation. During the years to the Great Depression, when extractive industries like mining and agriculture reigned as Utah's most important occupations, the state suffered more severely than the remainder of the nation from the boom and bust of economic cycles. During the depression of the 1930s, while the national economy experienced 25 percent unemployment, Utah bore a greater burden as more than 35 percent of its workers had no jobs.

By the 1970s, by contrast, Utah had developed a more balanced economy which did not rely as much on extractive industries. As a result, during the national recession of 1987–92, while the nation experienced unemployment that reached more than 9 percent, Utah's rate remained below 5 percent. Utah suffered more during the recession which began in 2001 since its unemployment rate, though slightly lower than the national level, exceeded 5.5 percent.

Although Utah's economy experienced more distress during the 2001 recession, its mining component may well explain part of its slightly better performance than the national economy. Mining jobs pay more than other types of hourly work, and mining contributes more than its share to income from foreign sources (see tables 6 and 9). As an article in *USA Today* suggested, Utah might be even better off if it had large fields of strip-mineable coal like Wyoming and Montana, but it does not—or at least such resources are not currently available.[93]

At the same time, we should recognize that these developments have come with a price. One is the enormous environmental damage caused by soil, water, and air pollution. The entire community paid for these as they occurred, while responsible companies like Kennecott and ARCO have undertaken extensive restoration and remediation projects to clean up damage and prevent its reoccurrence.

Another negative development is the number of accidents and death from mining operations. The miners and their families paid for these with disability and grief. Fortunately, in recent years mining companies have installed equipment and initiated procedures which have reduced mining accidents, and mining has become less dangerous than a number of other industries.

Beyond these problems, we should understand that in part the inability to discover new ore bodies or exploit those that miners have newly discovered is due to environmental regulations and opposition to mining in environmentally sensitive places. Clearly environmental regulations have been salutary. There is really no reason why the entire community should pay the price of environmental damage caused by mining, smelting, or any economic activity. Part of the cost of engaging in any business is investing in air, water, and ground pollution-control equipment and providing a safe working environment for employees. Moreover, those who have neglected such investments in the past should rightly pay to set things right in the present.

Beyond the need to maintain a healthy ecosystem and population, however, competing interests influence and dictate decisions about the relative importance of environmental change versus the recovery of mineral wealth. Personal choices and ideology impact controversies over oil and gas exploration, battles over mining on the Kaiparowits Plateau, and decisions about mining precious metals. Ordinarily such disputes end up pitting the relative value of scenery, wildlife protection, and the potential for service-industry jobs against high-paying mining jobs and supplying industrial goods.

Such controversies will undoubtedly continue. As they persist, however, Utahns should recognize the extraordinary economic contributions of mining. Moreover, they should value the reasons for those contributions: technological innovation, capital investment, marketing, and well-paid labor.

3

General Patrick Edward Connor Father of Utah Mining

Brigham D. Madsen

General Patrick Edward Connor was the prime mover in the start of mining operations in Utah Territory. His soldiers, sent under his command from Camp Douglas (later Fort Douglas), established the first large-scale mining districts in Utah in the early 1860s. It is important to emphasize that most of his California Volunteers, sent to Utah in 1862 to keep the mail lines open during the Civil War, had been recruited by Connor from the gold camps in California and often considered themselves miners first and soldiers second. Wherever they went on Indian-hunting expeditions in Utah Territory, they spent almost as much time looking for silver and other metal outcrops as they did on their primary mission, hoping to return to these promising mining prospects after their military service was over. This chapter describes the activities of Connor's soldier/miners in opening mining areas to production and then examines Connor's own mining enterprises as he competed with other mine operators to develop the rich lodes in Utah.[1]

The peace that ensued following the Bear River Massacre of 29 January 1863 and completion of the 1863 treaties with the Shoshoni left time for the Volunteers to explore the mineral riches of the area. Connor's encouragement of this pursuit only added to tensions with Brigham Young, who had publicly opposed searching for mineral wealth. He told the Saints that farming was the most advantageous pursuit: "Go to California if you will; we will not curse you, we will not injure or destroy you, but we will pity you. People who stay will in ten years be able to buy out four who go."[2] Young also feared that mining camps would introduce unsavory elements and entice church members into abominations that would destroy their souls. However, Young also secretly sent two companies of young men on a "Gold Mission" in the fall of 1849 to gather gold dust to benefit church coffers.[3]

Mining in Bingham Canyon

Stories of rich veins of ore in Bingham Canyon in the Oquirrh Mountains across the valley from Salt Lake City had circulated. On 27 September 1863, George B. Ogilvie and several other Mormon cattle herders came upon an outcrop that contained silver. Ogilvie took a sample to Connor to have it assayed, and the general asked Ogilvie to locate the West Jordan Mine in the name of the Jordan Silver Mining Company at that site. Ogilvie received two shares, and the other 24 people listed each got one share, including Connor.[4] Ogilvie was at once denounced by church officials as an apostate.[5]

Other claims previously had been filed on 7 September and 17 September 1863. On the latter date, Mrs. Robert K. Reid, wife of the physician at Camp Douglas, located a vein near the Ogilvie find while on a picnic. A notice was made out designating Mrs. Reid as "original discoverer" and listing 19 other shareholders, including Johanna Connor, the general's wife. Another claim, with 26 shareholders, was called the Vedette.[6] As soon as the claims were staked, the excited participants adjourned to the Jordan Ward House on the Jordan River near Bishop Archibald Gardner's mill to organize the West Mountain Quartz Mining District. Gardner was elected recorder, and seven bylaws were passed.

Mormon reaction to this ore discovery on the very doorstep of the Mormon capital came most forcefully from Brigham Young, who, on 6 October 1863, asked who fed, clothed, and supplied the prospectors. "Were they really sent here to protect the mail and telegraph lines, or to discover, if possible, rich diggins in our immediate vicinity. . . ?"[7] William Clayton, in a letter to Jesse A. Smith, added, "A tremendous effort is now being made to bring to light the rich minerals and the enemy has already partially succeeded. The greatest trial to the integrity of the saints is now before them, viz. to prove whether their religion or wealth is of most value to them."[8]

The coming of spring in 1864 did not produce the gold rush to Utah that Connor expected. A group of California miners returning from Bannack in Montana did discover a little placer gold in Bingham Canyon west of the city, which sparked temporary interest, but the excitement was short lived.[9] In Utah ore bodies were hidden deep in the earth, requiring money and technical expertise to extract the precious metals. Initially Connor believed that the slow progress in mining was due to Mormon hostility. On 1 January 1863, the *Union Vedette*, the newspaper for the Union soldiers billeted at Camp Douglas, stated that Brigham Young's efforts to discourage mining were as futile as trying to "dam up the waters of the Nile with bulrushes." However, from 1863 to 1869, the high cost of transportation, the price of labor, the scarcity of charcoal, and the inexperience of these would-be miners led to only limited development.[10] Eventually, the dearth of free gold and expense of investing in men and machines convinced Connor that mining in Utah was not easy.

General Connor began to receive information that nefarious tactics were being used to jeopardize his efforts to open up the closed society of Utah. A group of 26 miners wrote to him from Franklin, Utah Territory, on 5 February 1864, that town residents had made certain threats "that we shan't prospect for gold in the country."[11] In response to this message and other indications of Mormon opposition, Connor issued a circular on 1 March stating that any offenders "will be tried as public enemies, and punished to the utmost extent of the law."[12] The *St. Louis Democrat* applauded Connor's actions because he did not indulge "in covert insinuation or secret threats. He approaches his subject in plain old Saxon words, and says just what he means and means what he says." On the other hand, Mormon William Clayton was particularly disdainful of the soldiers' attempts to raise excitement about gold because so far "all they have done has been a failure." He added, in another letter, "I do not enquire after gold, neither shall I trouble myself about it. . . . To me one thing is *certain*. If rich mines are opened in Utah, the Priesthood and honest saints will soon have to leave for some other region."[13]

By March 1864 the Jordan Silver Mining Company reported construction of a tunnel 60 feet into the mountain, and P. Edward Connor was listed as one of the trustees of the company. Another company was organized to build the first silver quartz mill in Utah to crush and work the ore of the West (Oquirrh) Mountains. New discoveries of minerals were announced in Carr's Fork of Bingham Canyon.[14] On 14 March 1864, the Kate Connor Gold and Silver Mining Company was organized, named for Patrick and Johanna Connor's daughter. Connor was granted one of the 13 shares in the mine.

A unique mining claim called the "Woman Lode" was recorded on 7 May 1864. Nine women, six of them the wives of Fort Douglas officers, announced in the filing statement,

> We the undersigned 'Strong Minded Woman [Women?],' do hereby determine and make manifest our intention and right to take up 'Felt' ore [*sic*] anything else in our names, and to work the same independent of any other man . . . 1000 One Thousand feet with all its dips, Spurs, and angles, and Variations and Whatever other Rights and priviledges [*sic*] the laws or guns of this district give to Lodes so taken up.[15]

The first name on the list was Mrs. Genl. P. Edw. Connor, and it seems obvious that she took the lead in producing this remarkable document. It was an indication of Johanna's independence and her determination to exercise her own judgment.

Mining in Meadow Valley

General Connor was not only directly involved in opening mines; he also directed his officers, as surrogates, to prospect for minerals. He sent Captain Samuel P. Smith and Company K of the Second Cavalry to Raft River in Idaho on 9 May 1864 to protect the immigrants and to "thoroughly prospect the country for precious metals, particularly placer gold, and report from time to time the result to this office." To Captain N. Baldwin, Company A, Second Cavalry, Connor's orders on 11 May were to proceed to Uinta Valley to "afford ample protection to prospectors and miners . . . [and] cause the valley and vicinity to be thoroughly prospected by your men, and will report from time to time the result to this office." On 13 May he dispatched Captain David J. Berry to the Meadow Valley Mining District in southwestern Utah to "afford protection to miners from Mormons and Indians. . . . You will thoroughly explore and prospect the country over which you travel, and if successful in finding placer diggings, you will at once report the fact to these headquarters."[16] Connor kept his superiors informed about directing "soldiers to prospect the country and open its mines . . . [to] peacefully revolutionize the obvious system of church domination which has so long bound down a deluded and ignorant community."[17]

The Meadow Valley area, where Captain Berry was headed, was of interest to Connor at this time and after his army tour of duty ended. Located 100 miles west of Cedar City and about 10 miles southeast of present Pioche, Nevada, the Panaca silver mines were first discovered in the winter of 1863–64 by Mormon settlers from Santa Clara. The settlements near St. George then organized a party to go and examine the prospects to determine whether or not LDS church members should locate claims.[18] When Brigham Young was informed of the mining possibilities, he wrote Bishop Edward Bunker on 6 February 1864, advising the Saints to occupy Meadow Valley as grazing area for their stock "and also to claim, survey and stake off as soon as possible, those veins of ore that br. [William] Hamblin is aware of, . . . all that are sticking out, or likely to be easily found and profitably worked."[19] Again, Young clearly had a private view on the development of the area's mineral wealth that contrasted with his public stand about mining. He was eager to oppose Connor's desire to open mines but willing to seize an opportunity for the church—especially when it was away from Salt Lake City and could be kept under wraps.

But the Mormon Church was not destined to have exclusive exploitation of these diggings. A group of Gentile (the Mormons' name for those not of their faith) miners left Salt Lake City in February 1864, led by Stephen Sherwood, to share in the wealth Hamblin and the Mormons had discovered. On 16 March Hamblin agreed to guide the Sherwood group to the new mines. Connor had supplied Sherwood with a copy of the bylaws of the West Mountain Quartz Mining District before the group had left the Mormon capital, and Sherwood persuaded Hamblin and the Mormon prospectors to adopt these regulations.[20]

Local Mormon Church officials were disgruntled when Hamblin accepted Sherwood's regulations because they had devised their own laws to exclude Gentile proprietors. But Hamblin had been given the impression "that the intentions of these men were honestly for the upbuilding of the church as they claimed to be Saints etc." LDS leaders were now caught in the embarrassing position of opposing mining development in northern Utah while actively, though secretly, engaging in it in the southern part of the territory. The Sherwood party returned to Salt Lake City with samples of ore that

Connor had assayed to reveal a value of $300 per ton. Sherwood returned to Meadow Valley by 18 May to develop the ore bodies with a larger group of eager miners.[21]

The race was on between Gentile prospectors and the local Mormons to grab the best claims. As indicated, Captain Berry's Company A had already been dispatched to the region. On 22 May Captain Charles Hempstead, editor of the *Vedette*, left Camp Douglas with a detachment of soldiers bound for the new mines. A third force under Captain George F. Price had left Salt Lake City to try to open a new road to Fort Mojave, Arizona Territory, near present Needles, California. While Price's orders had nothing to do with the Panaca mines at Meadow Valley, his troops patrolled through the area and represented still another military force to worry the Mormon miners.[22]

To counter the threat of all these troops, a party of Mormons under Bishop Erastus Snow left St. George on 20 May for the mining district.[23] On their return, the Snow group met Captain Hempstead at Mountain Meadows. Hempstead later recorded that "the party were all in high glee and wonderfully elated at the success of their mining enterprises. It was more than intimated to us that we were 'a day after the fair,' for the Saints had been before us at the new Dorado; gobbled up the prize, and left little for ungodly sinners, like unto us."[24]

Soon Mormon leadership at St. George, faced with aggressive prospecting by Connor's soldiers, began to have second thoughts about competition. At a High Council meeting on 11 June 1864, Bishop Snow stated that "he was satisfied it was the intention of Gen. Connor and Gentiles to settle in there and not only claim the mines of silver in that vicinity, but also the farming lands, water privileges, etc."[25] Agriculture was still of paramount importance to the Saints, and Mormon officials agreed to give up their claims. By midsummer, 417 claims had been filed on 33 veins, but little development was accomplished because local Indians became so hostile that most of the miners were forced out of the district. No significant mining activity took place again until 1869. William Hamblin did take some of the ore to a smelter in Rush Valley to have it processed into 26 "bars of metal—silver and lead," which were exhibited in Salt Lake City.[26]

Throughout the period of the Meadow Valley mining excitement, there was almost a "news blackout" in northern Utah on the claims being filed on the Panaca ledge. The Mormons in the St. George area were certainly involved, but church leaders or the *Deseret News* gave the matter little publicity. One southern correspondent to that paper wrote, "There is some talk about silver, etc., in this part of the country; but the people have but little faith in it; samples of the ore exhibited here are pronounced worthless." The LDS leaders also did not wish to lend any support to Connor's efforts at developing mines.[27]

Mining in Rush Valley and Tooele

In addition to Meadow Valley, the Rush Valley region nearer Salt Lake City became important in Connor's postwar activities. Lieutenant Colonel E. J. Steptoe had established a grazing camp near Rush Lake in September 1854, which later became a military reservation by executive order on 4 February 1855. When General Albert Sidney Johnston's army arrived, the reserve was being used as a forage area for animals from Camp Floyd. The government sold the site in 1861 to a Mr. Standish, who sold the land and buildings to General Connor's command for $1,100 in April 1864. Connor established Camp Relief east of the lake, then only a pond due to dry conditions. Many acres of natural hay and good grazing for the cavalry horses of the Volunteers were available in the dry lake bed. Lieutenant Colonel William Jones and Captain Samuel P. Smith had already directed companies A, H, K, and L of the Second Cavalry to the camp in March, and all at once there were hundreds of former California miners-turned-soldiers within sight of the beckoning west side of the mineral-rich Oquirrh Mountains.[28]

On 11 March James W. Gibson of Company L of the Second Cavalry established the first claim, and the rush was on. The Rush Valley lode was located about two miles east of the north end of the valley. Trooper Gibson granted Connor one share in the mine.[29] The silver ore from this claim and the other early discoveries assayed from $81.50 to $97.50 per ton.[30] On 11 June the new miners organized the Rush Lake Valley Mining District with Andrew Campbell as recorder. As early as 9 June there were more than 30 ledges being worked that showed "from $80 to $350 per ton of silver" and enough lead to pay the expenses of extracting the ore, according to the exuberant editor of the *Vedette*. In another article, the editor did point out what soon became common knowledge: the ore bodies were deep in the earth and required capital and technological know-how to get the riches out—"a stern reality, and where . . . for mining purposes poor men are not wanted." Connor was to have a close association with the Great Basin, the Silver Queen, and the Quandary Mines in the area.[31]

The ambitious and audacious Connor now initiated a project to build a new town, the first Gentile settlement in Utah. He chose a site 7 miles south of the Mormon town of Tooele and about 16 miles from the south end of the Great Salt Lake. It was a strategic location, lying at the midpoint of the west side of the mineral-rich Oquirrh Mountains and close enough to the lake to allow wagon shipments of ores to a possible landing for steamboats on the inland sea. Connor had visions of a great mineral empire being developed in the area. The new settlement was named after Stockton, his home city in California, and was surveyed by Joseph Clark on 19 May 1864. The town had 811 lots and was large enough to accommodate 10,000 people. Connor revealed his penchant for the military and his high regard for Republican government officials by naming the streets Grant, Sherman, Sheridan, and Lincoln, Johnson, Wright, Doty, Seward, and Silver.[32]

The *Vedette* editor visited Stockton and reported on 13 July that a hotel/restaurant/saloon building was under construction and timber was being hauled from nearby canyons for other structures and houses.[33] It was obvious that experienced contractor P. Edward Connor was very much involved in establishing the new town. Connor made frequent visits to Camp Relief, where he could observe the development of

Stockton. In one instance, Connor also went to inspect his new smelters, accompanied by investors from New York and California.[34] Eventually Connor invested $80,000 of his own money in the town, in the Rush Valley mines, and in the new smelters, a pioneering contribution from which he received inadequate returns.[35]

Connor led the way by constructing the Pioneer Smelting Works and a reverberatory furnace in 1864.[36] Other furnaces followed. The Knickerbocker and Argenta Mining and Smelting Company of New York City invested $100,000 in machinery to separate the lead from the valuable silver and gold. It seemed that a smelting process would work for the Rush Valley ores, and an exuberant Connor wired Lieutenant Colonel Drum on 1 September, "The furnaces in Rush Valley are a decided success. Much rejoicing among miners."[37] But developments were defeated by the scarcity of charcoal and the high cost of wagon transport.[38]

Discharged Volunteers and hopeful miners flocked into Stockton as a base for their prospecting and mining. The grateful citizens sponsored a "grand opening ball at the [two-story, adobe] Emporium of regenerated Utah" for Connor and his guest, Warren Leland, in late October; later, the *Vedette* reported that there were 38 buildings, a completed sawmill, and two furnaces in actual operation.[39] By the end of 1864, hundreds of mining claims had been recorded in the Rush Valley District, and Connor owned shares in 29.[40]

Connor reported to his commanding officer that he was spending "every energy and means" he possessed, "both personal and official, towards the discovery and development of the mining resources of the Territory, using without stint the soldiers of my command, whenever and wherever it could be done without detriment to the public service."[41]

Connor continued to find time to maintain his interest in mining and with others became involved in locating some oil springs on the overland route about 30 miles from Fort Bridger. The *Union Vedette* kept up its constant hammering to entice miners to Utah, announcing such exciting prospects as a New York company being organized to invest $100,000 in Utah mines, and reporting that Connor, a few of his officers, and some private citizens had gone on a visit to the Rush Valley Military Reservation and the town of Stockton.[42] To make the *Vedette* more a part of the business community of Salt Lake City, and no doubt to further show his independence from the Mormons, Connor moved the office of the paper in November 1865 from Camp Douglas downtown to an adobe building on the northwest corner of Second South and Main Street with Adam Aulbach as printer.

On 30 April 1866, Connor was discharged from the military service, breveted as a major general, and free to devote his entire time to his mining operations.[43] Now back in Salt Lake City, he and Brigham Young continued their "cold war" feud. Young wrote two of his missionary sons in England on 11 August 1866 that "Connor is out of the service, and is here now as plain 'Pat,' engaged in mining business, which, as Government pap has been withdrawn, will very likely, if he pursue it diligently, break him up financially."[44]

Brigham Young also deprecated Connor and other would-be mining entrepreneurs for wasting their time trying to develop the new silver mines in Rush Valley. Samuel Bowles noted that the Gentiles were wasting money and labor in the new mining district and quoted Brigham Young as saying "that for every dollar gained by it, four dollars have been expended."[45] Connor was still enthusiastic about the prospects in the Stockton area and spent a lot of time traveling back and forth the 40 miles from his home in Salt Lake City to his mining claims in Rush Valley and the new reduction works he was constructing to process the ores.[46]

By the end of 1866, Connor's investment in the Rush Valley Mining District probably made him the leading mining promoter there. During this year he purchased 27 transfers of 51,465 2/3 feet for $5,326 and invested in 26 new mines, owning 6,700 feet of the claims. Some of the new purchases revealed his connections with friends—the General Dodge and the Hempstead lode.[47] His relationship with Grenville Dodge was extended further, as revealed in a 4 July 1866 letter. Dodge was by then working as chief engineer for the Union Pacific Railroad and wrote confirming a verbal agreement which the two men had entered into earlier. Connor was to engage two prospecting parties to locate any valuable silver, gold, coal, or iron mines and then was to bill Dodge, by voucher, for the expenses. In addition, Connor agreed to locate suitable sites in Utah, close to coal and mineral lands, on which to erect foundries in the names of companies submitted by Dodge. Finally, Connor could establish as one of the foundry companies a firm under his own name and those of Dodge and Judge W. H. Carter. With such outside contracts and his personal investments in Rush Valley, Connor was well launched on a career as a mining entrepreneur.[48]

By the first of 1867, the Connors had become convinced that their safety and certainly any kind of comfortable living circumstances were threatened in Salt Lake City. As Episcopal Bishop Daniel S. Tuttle remembered on his arrival in the City of the Saints in 1867, "The little company of the Gentiles was practically ostracized as if they had been in the heart of Africa."[49] This was particularly true of the bitterly hated founder of Camp Douglas. So on 8 January the Connors left their comfortable home on Fifth South between First East and East Temple (Main Street) and moved to a house Connor had constructed in Stockton, Utah. The change must have been traumatic for Mrs. Connor—leaving the tree-lined streets of Salt Lake City for the stark desert atmosphere of a new, rough mining town where, by the end of the year, the camp had "dwindled down to ten or twelve men."[50]

Connor attempted to live in Stockton because he had invested so much in developing his silver mines there that he could not leave "without much financial sacrifice."[51] He was also committed to the new town he had helped found and, with John Paxton, owned most of the lots in the settlement.[52] His reduction works were operating part of the time, and a government survey of mineral resources in the territories indicated some optimism for Utah, and especially for Rush Valley, where Connor's smelters were revealing a wealth of silver and lead ores.[53]

Uncomfortable with his situation in Stockton, Connor sought advice from his friend Grenville Dodge. In a long letter of 16 January, he assured Dodge that he could make a success of his mining ventures if he had, in addition to his present reduction works, $15,000 in working capital. But the Mormon Church authorities were throwing every obstacle in his way. "Brigham Young and his satelites [*sic*] in the pulpit and through the press have been grossly abusing me since my return from the East, indeed, so much so that my friends feared that some of his fanatical followers would assassinate me." He added that Brigham Young's hatred of him was "intense, caused by my making him behave himself while I commanded here." Unless the government provided help and security for him and other Gentiles, all of them would be forced to leave Utah in the spring. In fact, Young was making every effort to drive them from the territory and had, through his agents in Washington, succeeded in "prejudicing the President against me." Connor concluded his long letter by stating that his chief hope was that Dodge would be able to complete the Union Pacific Railroad to Salt Lake City by summer or begin to manufacture iron in Utah. If not, Connor would be forced to leave his property, "$351,000, worth," and move to California.[54]

No doubt to the relief of Johanna Connor, the general moved his family from the frontier town of Stockton, Utah, in May 1867 first to a hotel in San Francisco but eventually to a permanent home in Redwood City, where Mrs. Connor and her children lived for the rest of their lives while the general pursued his dreams of a fortune in the Utah mines and visited his California family as often as he could. He could not leave without a final blast at his nemesis, Brigham Young, in a letter to U.S. Secretary of War Edwin M. Stanton, resigning his post as sutler at Camp Douglas. Connor wrote Stanton that, "having in the performance of my duty as district commander in this Territory incurred the deadly hostility of Brigham Young and his fanatical followers, I do not deem it prudent to remain with my family in Utah in the present unfortunate aspect of affairs." Connor explained that Young had "absolute and unchecked" power over the lives and property of all citizens, Mormon as well as Gentile, that the Mormon leader was "arrogant and vindictive," that Gentile lives were held "at the pleasure or caprice of an autocratic church leader," and "that they remain and pursue their several avocations only by his permission."[55]

As for Connor's mining holdings in Rush Valley, the 1868 official report of the federal government on the mineral resources of the territories did not sound encouraging. The smelters, including Connor's, "failed to extract the metal in a satisfactory manner." The writer of the report made a very important point which Connor and other mining operators of the Stockton area did not heed, and they suffered financially over the years as a result: "Silver occurs in galena in the same irregular manner as in quartz. Many suppose that if a vein of galena assays well in one part it will do the same in all; an erroneous idea, as miners frequently find to their cost." The article concluded that the mines in Rush Valley would become valuable when transportation and labor were cheaper and fuel more abundant.[56]

Trains and Ships

The approach of the Union Pacific and Central Pacific Railroads to Utah by late 1868 promised cheaper and speedier transportation necessary for successful mining operations. In anticipation of this fact, Connor, in November 1868, launched a steamboat on the Jordan River named the *Kate Connor* after his daughter. He had built the boat during the summer, obtaining the machinery from California but using local materials for everything else. He planned to use the 90-ton craft to tow ties and telegraph poles across Great Salt Lake to sell to the Union Pacific during the coming winter, probably through a contract arranged with his friend Grenville Dodge.

Connor also had a grand design for recouping and enhancing his modest finances while striking a blow at the monolithic Mormon establishment. He would construct and would encourage other entrepreneurs to provide steamboats to ply the Great Salt Lake from Lake Point on the south shore to a connection with the Central Pacific near a spot on the Bear River. Ore from his many claims in Rush Valley would have to be moved only 20 miles by wagon to the lake, where it could be picked up by boat and delivered, at very low water rates, to the railroad.[57]

During 1869, and with rail transportation providing easier, faster, and more comfortable access from Salt Lake City, Connor established a pattern of constant travel that was to dominate the rest of his life. He practically lived in railroad cars and hotels with intervening visits to his family in California as often as he could conveniently leave his business interests. At this point mention needs to be made that from 1877 to 1885, Connor spent almost as much time in Eureka, Nevada, in a mining venture as he did with his property in Utah, but that is a part of his life that has little relevance here.

Despite his interests in Nevada, Connor concentrated most of his mining efforts in the Rush Valley District. Whereas a national survey the year before had reported mining developments in all of Utah to be "very slight and unimportant," by 1870 conditions had changed.[58] Miners who had left the depressed White Pine area of Nevada discovered rich deposits of ore at the southern end of the Oquirrh Mountains in Rush Valley. On 1 October 1870, a new Ophir Mining District was organized. Connor participated and was partly responsible for the renewed vigor.[59] He purchased footage in two new claims in the Rush Valley District in June and in nine mines at Ophir during the months of October and November.[60]

Connor's most valuable mines were the Silver King, Great Basin, Quandary, and Silveropolis. By May he was taking out more than $3,000 worth of ore each week from the Silver King.[61] The new *Mormon Tribune* of 13 August 1870 described the mine as a "true fissure vein of argentiferous galena, five feet wide between walls, depth of shaft 100 feet, length of incline tunnel 150 feet at present shipping ore to California for reduction."[62] The *Utah Reporter* was even more enthusiastic: "The Silver King, the great work of the Stockton District . . . is General Connor's great lode, and occupies a central position on what is thought to be 'the mother ledge' of this district."[63] The Great Basin, destined to be one of the richest strikes in the area, at this time was a "gash vein" only two feet wide with a 50-foot tunnel and "some ore on the dump."[64]

The Silveropolis, owned by Connor and the Walker brothers, seemed to be the most promising of all. They had paid $10,000 for the property but by October had shipped 500 sacks of ore, worth from $50 to $100 per sack. About two weeks later, the *Utah Reporter* crowed, "General Connor has just been offered $12,000 for two hundred feet of the Silveropolis, and has refused it! Bully for P. Edward!" On the other hand, William Clayton sounded a little discouraged: "Gold and silver mines are being opened upon every hand, but it seems that the Lord does not design the Elders to be corrupted much by them for nearly all the mines that are worth anything falls to the lot of the gentiles." He thought the influx of non-Mormon miners was "most serious" and noticed that there were "200 known apostates in the 14th Ward and five hundred in the 13th Ward."[65]

Connor prepared the *Kate Connor* to transport ores across Great Salt Lake to the Central Pacific at the new town of Corinne. The craft was refitted with new engines from California. Connor also made necessary improvements for passenger comfort on tours across the lake.[66] By late June the *Kate Connor* was ready to make the first of what were supposed to be regular triweekly trips to Salt Lake City and Lake Point.

Throughout the summer and fall of 1870, Connor shipped ore from Rush Valley first by wagon 20 miles to Lake Point and then by the *Kate Connor* across the lake to Corinne. Using the *Kate Connor* to transport ore was not the success Connor had hoped for, however. Eventually he had to resign himself to the reality that the Mormon-built Utah Central Railroad would transport nearly all of the ore from the Salt Lake City area to the transcontinental railroad.

At this point Connor began to take a much larger role in Utah politics. A new party, non-Mormon in point of view, was organized in the Gentile town of Corinne on 16 July 1870. Connor was elected temporary chairman of the new Liberal Party and spent the rest of his life involved in the party's affairs; in fact, he became an icon for it, eventually receiving the title of "the father of the Liberal Party." Thus, as the father of twins—in mining and politics—he became prominent in Utah affairs in the years from 1862 to his death on 16 December 1891; however, his principal child was mining, and his role as its founder in the territory remains his main contribution to the history of his adopted state.[67]

The completion of the Pacific railroads and influx of non-Mormon miners changed LDS attitudes toward mining. English traveler Charles Marshall observed that "the whole Gentile population is mad with the excitement of the gold fever and the Mormons feel the contagion." Although there were only a couple of thousand Gentile miners in Utah at the time, Marshall thought the rush to wealth would bring 20,000 by the end of the year.[68] J. H. Beadle explained that there were perhaps 3,000 Gentiles speculating in mines, 5,000 "hunting" for mines, and 2,000 really "working mines." He wrote, "The Mormons now propose to become a mining people." The change in Mormon attitudes could be seen in the correspondence of men like William Clayton, who wrote to his business partner, H. Starr, about a mine they were developing: "I estimate the mine to be well worth forty millions of dollars, and I should be very unwilling indeed to sell out for less than three million. . . . We have a mountain of mineral of immense value."[69]

Rossiter W. Raymond, special commissioner for collecting mining statistics in the states and territories west of the Rocky Mountains, gave some reasons for the Mormon about-face toward mining. The new railroads had broken the isolation and destroyed Mormon control of trade; the new mining population offered business opportunities and profits: "They [the Mormons] can no longer help themselves if they would"; and "they have to a considerable extent caught the prevailing fever, and are locating and prospecting ledges with truly Gentile zeal."[70]

Another report by John R. Murphy, who, incidentally, relied on an interview with Connor for the history of Utah mining, gave the mineral production for the years 1869–71. Utah Territory produced 16,200 tons of silver and gold ores with a value of $3 million in almost three years. Whereas there had been only 2 mining districts in the territory in 1868, there were now 44, and 18 smelting furnaces had been built at a cost of $200,000. Murphy gave Connor credit for inducing a "large number of his California friends" to erect smelters in the Rush Valley District.[71]

Connor finally gave up his attempt to operate a smelter at Stockton. In 1870 he collaborated with Simons and Company to build a smelter, but in April 1871 he sold it to "a practical and experienced gentleman in the business, F. Wallace," who began to enlarge and improve the facility.[72] William S. Godbe also built a smelter in the Rush Valley District in 1871. The mine operators of the entire Salt Lake area no longer sent their ores out of the territory because it was "cheaper by far to smelt it here," as Clayton explained.[73]

The upsurge in mining activity brought new growth to Stockton, which Froiseth's mining map of Utah identified with the caption, "P. E. Connor, Proprietor." This was no doubt accurate because Connor was still the chief property owner.[74] By the end of the year, Stockton had a population of 300 and boasted 60 houses, one hotel, a post office, several saloons, one store, and an assay office. J. H. Beadle, on a visit in October, described the town as "a little dull just now" while awaiting the sale of some of the mines to an English company. Noticing a new hotel, he asked its name and was informed that it was the "Lop Ear House." Upon entering the establishment, he learned that it was really the La Pierre House and concluded, "They say it is all the same in French."[75]

Despite Connor's continuing optimism about his Rush Valley mines in 1871, the chief interest was in the new camps in Ophir.[76] R.W. Raymond wrote that ore assayed from 500 to 27,000 per ton, turning "the heads of the oldest miners."[77] Connor purchased 1,633 ½ feet in eight mines in the Rush Valley District in 1871 but apparently nothing in any Ophir District claims.[78]

Although Connor maintained his mining interests in the Salt Lake area in 1872, newspaper notices were not as numerous as in previous years. Connor and some partners sold their Lexington Mine in Little Cottonwood Canyon to Samuel G. Phillips of London for $75,000, according to the *Salt Lake Review* of 13 January.[79] But in Rush Valley, the Jacobs Smelter, built during the summer, ceased operations by the end of the year.[80] The extensive sawmill in Soldier's Canyon was kept busy filling large orders for lumber for the Oquirrh mining camps. Connor also supplied the town of Stockton with water "through lumber pipes a distance of over one mile."[81]

The coming of the transcontinental railroad and the infusion of new Gentile blood into the mainstream of Utah life lessened the apprehension and sense of isolation Connor and other non-Mormon residents of the territory experienced. A greater sense of forbearance and more tolerance of Mormon ways took over. This did not mean that Connor had given up his deep-seated abhorrence of polygamy or his desire to destroy the economic and political control of Brigham Young and the Mormon leadership. Connor retained his basic antipathy to anything Mormon and continued political attempts to disfranchise the Saints. His Gentile friends persisted in rallying around him as the point man for most operations against Mormon power, and Connor accepted this attention as befitting his status in the community.

Striking Lodes in the Star Districts and the Law Courts

Meanwhile, Connor's chief mining interests in 1872 centered on the new Star District in the Picacho Range of mountains several miles southwest of Milford, Utah. The area had divided into two parts on 11 November 1871—the North Star and South Star Districts. Eventually there were 581 mining locations in the North Star District and 1,046 in the South Star District. The boom lasted from 1872 to 1875, and Shauntie, Shenandoah City, Elephant City, and South Camp were the chief mining camps.[82]

The Beaver County Recorder's Office shows that Connor was involved, usually with three or four other men, in locating a number of claims in the North Star District. John P. Gallagher was involved in all but one of the claims and was evidently a close associate of Connor in these mining ventures.[83] Connor also owned the Temperance Mine located in a third West Star District, which was described as "a vein of mineral [which] had held out with an increase of mineral."[84] George W. Crouch, a printer for the *Salt Lake Tribune*, gave up his newspaper job to become Connor's agent in the Star districts.[85]

In the Beaver County Court records, there is also a notice of a judgment in default against John P. Gallagher and P. E. Connor of the Flora Mining and Smelting Company. Dated 8 December 1879, the record shows that Gallagher had failed to pay $682.97 on a promissory note and mortgage, and the aggrieved party also asked $100 in attorney fees. The court ordered that the property be sold at auction to satisfy the debt. If the sale price was not sufficient to pay off the claim, a personal judgment was to be rendered against Gallagher for the difference.[86] Such lawsuits over mining property were to haunt Connor until his death.

Connor started other businesses in the North Star District. He and a man named Lawrence laid plans to build a sawmill in Mill Canyon in the Wah Wah Mountains, about 25 miles west of Milford, to furnish lumber for the mining camps in the Star districts and also Pioche, 60 miles away. In April and May of 1872, Connor and others purchased land and water rights in Beaver County.[87] By late August the Connor steam sawmill was in operation with a "fire engine of 30-horse power" turning out 8,000 feet of lumber each day. The mill was "said to be the finest in the Territory of Utah or the State of Nevada." P. L. Shoaff was the sales agent for Connor in Pioche.[88] By fall Connor also was preparing to build a ten-stamp mill near Shenandoah City on the Beaver River.

In the busy North Star District, there was much activity. Connor and John P. Gallagher were developing the copper, silver, and gold deposits of the Silveropolis group of mines, and the veins of ore were of "mammoth proportions, being from 25 to 100 feet in width and extending, in some instances, a mile in length." Connor was still supplying water and lumber to the camps.[89]

Connor was listed on the Third District Court docket in August and September in proceedings he initiated in the case of *P. Edward Connor vs. Leopold Kramer et al.*[90] M. Livingston sued Connor on 7 July 1873 for a balance of $454.55 he claimed Connor owed for merchandise. Connor testified that he had already paid Livingston $2,238.29, which took care of the indebtedness. The goods may well have been for the Eureka Hotel in Stockton that Connor was renovating at the time. There is no record of a judgment, and the case may have been settled out of court.[91]

In another case concerning the Lexington lode in Little Cottonwood Canyon, Connor emerged the victor. On 7 December 1871, he had loaned Thomas L. Moore $4,000 on a promissory note with a mortgage on 1,280 feet or four-fifths of the Lexington Mine as security. When Moore died suddenly, Robert J. Goldring and Albert P. Dewey, administrators of the estate, refused to pay off the promissory note. A judgment issued on 8 November 1873 determined that Connor was entitled to a decree of foreclosure of the mortgage.[92]

Connor maintained business offices in the Connor Building between First and Second South in Salt Lake City in addition to his residence near the Walker House. His office became a central meeting place for his business and political associates. He had himself listed in the Salt Lake City directory for 1874 as "Connor P. E.—capitalist, office Connor's Building, E. T. es. . . . bet. 1 and 2 so."[93]

One of the early meetings held in Connor's office was "for the purpose of discussing the advisability of petitioning Congress for an appropriation to aid in a geological survey of the mineral resources of the Territory." A committee of nine was appointed to draft the memorial to Congress, and Connor's name led the list of signers.[94] The same nine men had already petitioned the Utah Legislature seeking the appointment of a territorial geologist, pointing out that the total production of the Utah mining industry during the previous four years had amounted to $12,557,357. Connor's name was first among the signers, and he seems to have taken the initiative in both actions.[95]

At his mining locations in the North Star District, Connor's only product for 1874 seemed to be lawsuits. Most cases were initiated and completed in 1874, but some suits extended over several years. Much of the litigation revolved around ownership of mining claims.[96]

In the Ophir District, Connor owned the Chloride Point Mine, which had shipped the first carload of ore to California from the Ophir District. A 180-foot tunnel had been dug to the chloride and horn silver ores, but in 1874 the mine was idle. Connor's most important investment in the Ophir District—the Queen of the Hills and Flavilla Mine—was announced in a notice published in the *Salt Lake Tribune* on 25 August 1874: "All persons are warned against purchasing or removing any ores therefrom, as we are the owners of an undivided interest in said mine or mines, and intend to appeal to the proper tribunals for possession of our interests, illegally withheld. P. E. Connor, L. D. Osborn."[97]

The Queen of the Hills and Flavilla was elsewhere described as an "immense mine" with a seven-foot vein of solid ore 450 feet in depth. In one week the owners shipped more than 3,000 sacks of ore. One writer called it the "finest producing mine, so far, that has ever been found here [Dry Canyon]."[98] Even more impressive evidence of the mine's worth came from the announcement that a resident of Tooele, "John Lawson, who sold an interest in the Flavilla mine for $75,000, is building a $10,000 residence for his aged father and mother."[99] It is understandable why Connor and Osborn were warning off intruders.

Concerning the Stockton area, the *Salt Lake Tribune* of 18 June 1874 reported, "The General is interested here in something over one hundred mines and prospects in various stages of development." Among the most prominent of these claims were the California, with a four-foot vein and silver assaying $80 per ton; the Great Central and Silver Queen; and the Last Chance with a value in silver of $2,000. But the Silver King and the Great Basin remained the leaders. The former was continually mentioned in the press with a note in November that it was "now being worked by contract." The ore was rated at 25 ounces in silver and 65 percent lead. The Great Basin was "working four men daily" under foreman B. F. McCarty by the end of the year. The *Tribune* reported that it "will no doubt open into a valuable mine."[100]

When not operating his mines, Connor occasionally found himself in the Third District Court in Salt Lake City. He faced at least two cases in 1874. Amos Woodward and John S. Worthington, on 24 August 1874, sued Connor, Heber P. Kimball, and W. C. Rydalch for failure to pay off a promissory note of $10,000. The money must have been intended for the Salt Lake, Sevier Valley and Pacific Railroad. Judgment was rendered against Connor and the other two defendants on 12 January 1875 in the sum of $12,075, which included interest at 15 percent. The defendants had 60 days to pay the debt, or the court would sell enough of their joint property to satisfy the claim.[101] The second case, *Steven F. Nuckolls vs. P. Ed. Connor, William S. Godbe, Julian F. Carter, and John H. Latey*, was filed on 27 August 1874, again for a promissory note for $500. On 9 September 1874, a judgment was rendered against the defendants in the sum of $583.74.[102]

Apart from his occasional politicking, the general was very busy managing his mining property. On 9 January 1875, he sold 360 feet in the Shamrock lode and 50 feet in the Shenandoah Mine Number One to John P. Gallagher for $2,000 as he continued to divest his holdings in the North Star District. He also sold to Gallagher, for $250, a claim of 50 feet in the Gay Deceiver Mine on 1 March 1875.[103] The *Salt Lake Tribune* of 22 October 1875 thought that Connor's holdings in the Stockton area were "at last on the road to prosperity."[104]

The Ophir Mining District and More Lawsuits

The Ophir Mining District especially attracted Connor. He purchased 500 feet each in the Delaware and Georgia Mines on 14 January. The Queen of the Hills and Flavilla Mine was now producing 500 tons of ore each week with a return of more than 100 ounces of silver per ton. The *Tribune* praised "the lucky owners of this immense and valuable mine."[105] In June it was reported that a Colonel J. W. Johnson had sold a half interest in the Queen of the Hills and Flavilla to the Chicago Silver Mining and Smelting Company for $350,000 in cash, attesting to its value. The paper further noted that 5,000 tons of ore had been extracted in one year.[106]

Connor was involved in several court cases in 1875 and 1876 regarding ownership in interests in mines,[107] and in a case involving Hiram S. Jacobs and the Salt Lake, Sevier Valley and Pacific Railroad. Jacobs owed the railroad more than $30,000, and Connor was awarded $1,348.46 plus costs of $69.50.[108]

Except for short visits, Connor did not spend much time in Salt Lake City and Utah in 1876. His Silver King Mine was being operated for him by a man named Potts, and little mention was made of developments in his other mining property.[109] His visits to Utah were only occasional because he was spending most of his time in Eureka and Oreana, Nevada, hoping to discover some rich ore bodies.

A significant event which must have brought back memories of the 1860s was the death of Brigham Young on 29 August 1877. The two old antagonists had evidently never met, both being quite careful to prevent that from happening. And yet there was a mellowing of the firm positions the two had formerly held with perhaps a mutual feeling that if they did ever meet, they might actually like at least some aspects of each other's character. The Mormon Prophet never forgot the general's generous offer to sign a bond of up to $100,000 for Young because Connor was so strongly opposed to the actions of Judge James B. McKean in his 1872 court case, where the judge indicted Young for lascivious cohabitation but did not allow him bail. There was in Connor's makeup a quality of conciliation for past hostilities determinedly held at the time. Besides, the general was now an accepted member of the Utah business community and could ill afford to continue open warfare with the leader of the Saints. A new Mormon president, John Taylor, now directing church affairs created an opportunity for a new chapter in Connor's relations with Mormon Utah.

Connor's two chief concerns for the next year, 1878, were mining and politics. The Great Basin Mine received the most notice. Only two men were now employed at the Great Basin, under lease to James D. Coursa. The shaft was down 400 feet, and in 1877 the mine had produced 120 tons of ore which assayed at $35 per ton and gave the lessees $4,200. Connor also spent a few days in December examining mining prospects at Alta in Little Cottonwood Canyon east of Salt Lake City. This visit, plus a few to Stockton, summed up his mining efforts in Utah for 1878.[110] His attention was shifting to Nevada again.

Connor made at least two trips east in 1879 and perhaps a third. The first, from 21 January to 19 April, was to raise capital for his mining enterprises. The second, from 11 May to 27 June, to New York and Boston was for the same purpose.[111] His itinerary, as listed in the local newspapers, suggests that his home was on the road during the year.[112]

Connor found some exciting prospects in his mines in Rush Valley. The 1880 annual Utah governor's report announced that "the business of mining has never been more prosperous or more profitable than at the present time."[113] To construct a smelter for the ores, Connor needed water. He obtained five acres of land in Spring Canyon a mile and a half east of Stockton. He used galvanized pipe to conduct water from the springs to the mill. He also located the Silver King Number-Two Mine in the Rush Valley District in December of 1879.[114] His two Silver King mines were inactive most of 1878 but started up again in the fall and continued to operate during 1879.[115]

His chief interest was the Great Basin Mine, solely his property except a small interest. The De Courcy who ran the Eureka Hotel in Stockton may have held a minor share. One report has De Courcy as the one who in 1879 discovered a large body of ore at a depth of 250 feet in the mine. Connor was able to get some Boston capital on one of his trips east and then incorporated the Great Basin Mining and Smelting Company in May 1879, under Connecticut laws, with a capital stock of $2.5 in 100,000 shares. Besides the Great Basin, the company also owned the General Garfield, the Arthur, and the Silver Queen Mines. Connor employed a man named Gove as superintendent at the Great Basin Mine; he ran two shifts, a daylight one of 10 hours and a night one of 9 hours. The miners received three dollars per shift. The concentration works were in the old Jacobs smelter, built in 1872, and while they were "not models of elegance or convenience, . . . they do good work." During the year the Great Basin Mine produced a little more than 3,382 tons of ore with a return of $48,275.33. The ore was valued at 40 percent lead and from 18 to 21 ounces of silver plus varying amounts of gold. As the *Salt Lake Tribune* wrote, the mine "at present is one of the richest in the Territory,"[116] leading observers to comment that General Connor was on his way to becoming an exceedingly wealthy man.

As often happened when news got out about a rich strike, one man, a Leonard S. Osgood, claimed part ownership in the Great Basin and sued Connor. The case, filed in Third District Court on 21 July 1880, was quite convoluted and involved several people. The suit was finally settled out of court on 13 November 1882, the "action having been compromised."

Some interesting information about the operation of the Great Basin came out during the case. Connor denied that he had extracted ores from the mine totaling $50,000 or that he had taken 150 tons per week at a value of $50 per ton. Instead, he testified, the value had been $15 per ton, and prior to 1880 he had never taken more than 75 tons per week. He further said, "There is no considerable quantity of high grade ore" and until there was "a large outlay of capital, said mine cannot be worked at much if any profit, and had hitherto been worked at no profit." Connor could well have been downplaying the value of the mine.[117]

At the end of the year, the various properties associated with the Great Basin were consolidated under the new title of the Honorine Mine, although many still called it by its original name. L. D. Davis actively controlled the daily operations under Connor and had a crew of 55 men. The ore was valued at 25 to 30 ounces of silver per ton, three to six dollars in gold, and 30 percent lead, and was processed by the Brooke Smelter in the Rush Lake District. The stock of the company was listed on the Boston Exchange on 21 August 1880 at five dollars with 625 shares sold that day. The *Western Mining Gazeteer* observed that "the immense ore-body . . . is beyond a doubt . . . apparently inexhaustible," while the *Salt Lake Tribune* gave Connor credit for the development of the mine: "General Connor has worked upon the property for years, has surmounted a thousand difficulties, and it is a pleasure to think he is about to reap the reward due the sagacity, faith, pluck and patience which has enabled him to develop one of the most valuable mining properties in Utah or any other country."[118]

To allow a night shift to operate his reduction works, the Rush Valley Mining and Smelting Company, Connor installed two electric lights in the building at a cost of $1,500. When residents of Salt Lake City marveled at this new miracle, the *Salt Lake Tribune* published a letter of endorsement Connor had written to the Brush Electric Light Company of Stockton, California. In it the general expressed his satisfaction with the "light electric machine" he had purchased from the firm—"the light is brilliant, clear and beautiful, so that the men can work as well by night as by day." Furthermore, "it is cheaper than oil while an infinitely greater quantity of light is furnished. It requires no special attention from the engineer and runs with perfect regularity. With this light the men cannot shirk. I would not be without it."[119] Evidently this was the first installation of electric lights in Utah Territory.

Because the line he had already built did not supply sufficient water for the mine, Connor constructed another in 1880. He filed on more Soldier Canyon water, which came to be called Connor Springs, and laid five miles of four-inch pipe at a cost of $20,000. This new source also supplied water to the town of Stockton.[120]

The entire operation of the Honorine/Great Basin enterprise was so successful that the Utah Western Railroad began to complain. The *Salt Lake Tribune* of

20 October 1880 reported, "General Connor, in shipping material to his mine and ore from it, has of late overtaxed the capacity of the rolling stock of the road." The newspaper concluded, "If there was one more man in that region like General Connor, the railroad would have to go into liquidation because of overwork."[121]

Connor was also involved with other active mines in Rush Valley and invested in more in 1880. He extracted sufficient ore from his Silver King Mine to pay for further development and sinking a deeper shaft. Another property, the Quandary Mine, "principally owned by Gen. Connor," was located near the Great Basin, and its owner was working to connect the two and "making active preparations for a vigorous onslaught into the ore bodies."[122] He also invested in seven mines, four of which he personally located, and he filed on some land in Soldier Canyon so that he could construct the General Connor Tunnel to better access the ores in the Argonaut, Broughton, Jumboldt, and Roxie Mines.[123]

In Salt Lake City, citizens could watch the general in the Fourth of July 1880 celebration, which he supervised as marshal of the day. It "was the first non-Mormon Fourth of July celebration ever attempted in Utah." To emphasize that fact, the parade included 20 boys dressed as miners, other real miners from the various districts, and young men and women representing the 13 colonies and the 38 states and 11 territories. The *Salt Lake Tribune* gave special notice to all the "new generation" of young Mormons who lined the streets in appreciation of the display: "It shows that a leaven is working in our midst that by and by will leaven the whole lump."[124]

In 1881 the chief mining news in the Rush Valley District was the Great Basin Mine and two lawsuits about it. In the first, the plaintiffs claimed the employees of the Great Basin had ousted them from possession of their vein of ore that lay nearby and had, since then, taken out ores worth $50,000 belonging to them. The plaintiffs asked for the money and an injunction to stop the theft of their ore. On 13 November 1883, the suit was dismissed by the Third District Court without costs when all claims for damages were settled by the participants out of court.[125] In the second case, filed 19 July 1881 in the Third District Court, the plaintiff charged that Connor had promised to pay him $3,000 for testimony he had furnished in another case. There seems to be no record of a decision.[126]

The various mining journals of 1881 were high in their praise of the Great Basin Mine as "one of the most valuable properties in the Territory." The mine shaft was at 800 feet but had encountered water at the rate of 1,000 gallons per hour, a "small matter" to dispose of "when desirable," according to one report. The working crew was getting out 60 tons per day with the ore averaging 25 percent silver and 45 percent lead. The Great Basin Company was processing 8 tons per day at its mill near the mine while another company smelter at Stockton worked 25 tons per day. The *Salt Lake Tribune* recognized the "indomitable will" of Superintendent Connor in making a success of the Great Basin: "After expending his tens of thousands, helping every miner that ever came to him, building dwelling houses and hotels, collecting the meager streams that trickled to the foothills and sunk in the arid earth, and doing all that human power could do to breach the wealth hidden within those bleak and forbidding hills, the treasures seemed ever to elude him—still beyond his grasp." The *Tribune* then concluded, "But, though well-nigh exhausted, in endurance as in means, he struggled on, and to-day his faith and fortitude are rewarded by the possession of the Great Basin mine."[127]

Other Connor-owned mines in Rush Valley also continued to show promise. The Silver King had produced "a good deal of ore." Connor had recently incorporated the Quandary, which was connected to the Great Basin by a tunnel, in Boston under the name of the Rush Valley Mining Company. The *Tribune* reported in January that Connor had purchased for $25,000 the Leonore Mine, which had $40,000 worth of ore in sight.[128]

Connor initiated additional projects in Rush Valley in 1881. He laid claim to five acres of land in Spring Canyon, near Stockton, to construct a Quandary Number-Two mill. He was the owner and manager of the General Connor Tunnel and Mining Company, operated a coal mine on the Weber River, and, according to Edward Tullidge, owned more mining property in the territory than any other mining entrepreneur.[129]

At his town of Stockton, Connor installed an electric-lighting system, evidently making it the first town in Utah to receive more light. When the lights first went on, the Mormon citizens of nearby Tooele thought Stockton was ablaze. John Codman, on a visit to Utah, commented that Gentile Stockton "seemed a representation of misery sought for and found." It was in one of the "bleakest spots" that could ever be chosen, where "scarcely a sage brush can show its head," and the town was built on piles of logs except for an "abortive frame-house called a hotel." It was a most "execrable hole."[130] Codman was too harsh in his judgment; Stockton was no worse than any other western mining town.

Connor went east in February 1881 for a month's visit to a number of leading cities. He was in Washington for the inauguration of President James A. Garfield and took in "other great sights." In July he left Salt Lake City "for a month's run to Boston" on matters concerned with the Great Basin and other Rush Valley mining properties. A month earlier he had gone on a short trip to Montana.[131]

The years from 1878 to 1881 found Connor on the edge of political fame and mining fortunes in Nevada and Utah. The promise of success with the Great Basin Mine and the Eureka Tunnel in Nevada kept him busy traveling much of the time. He moved back to Eureka for another two years, but then, like a moth attracted to a flame, he returned to live out his final years in Mormon Zion.

A cloud was hovering on the horizon for Connor as 1882 opened. A disturbing speech by the general that his fight against Mormon domination had impoverished him seemed uncharacteristic. Connor's opposition to Mormon control was not the sole reason for his financial discomfort. His private meetings with eastern financiers had led to loss of ownership of his Great Basin and Quandary Mines and the unsettling knowledge that he was now a hired hand. While he bore the title of managing director, he was still employed by the faceless Boston corporation he had turned to for financial assistance. Connor in 1882 was not the independent owner/operator of his own mines as in the past. Despite continuing reassurances by newspapers like the *Salt Lake Tribune* and his own friends that success was still just around the corner, things had changed.

In an annual report on mining on 1 January 1882, the *Salt Lake Tribune* gave credit for the success of the Great Basin Mine to Connor for "the able manner in which the extensive property has been managed" and commended the strict economy he employed for the Boston company. A little more than three months after this encouraging report, Connor filed a lawsuit on 12 April in Third District Court contending that the Great Basin Mining and Smelting Company owed him more than $5,000 in back pay for expenses. The suit attached the following property of the Boston company by writ: the Great Basin Mine itself, plus the buildings and machinery, mill, boardinghouse, camp house, "out house," spring of water and pipes, 38 lots in Stockton, and all the dwellings, office buildings, and improvements on the lots, even down to the scales, safe, letterpress, office chairs, stool, button molds, assay furnace, and ore hammer. The list detailed the "blood, sweat, and tears" Connor had poured over the years into this property that was now lost to him. It was not a friendly action. On 14 November 1883, the case was settled out of court and the attachment vacated.[132]

The *Utah Commercial* of July 1882 wrote about Connor, the Eureka Tunnel, and the Great Basin Mine in an editorial:

> It is rather hard papers for Gen. Connor to have to assess his Eureka Tunnel when he might have unloaded $100,000 worth of his stock as well as not; and to have to abandon the Great Basin when he might have closed it out at one time at $50 a share. But he has one consolation. He had not in these two cases enriched himself at the expense of the public and particularly of his friends. If they made nothing by joining him in these enterprises, so neither did he. . . . We trust Gen. Connor will yet find a bonanza in his Prospect Mountain (Eureka) Tunnel, and that the fellows who have traced him out of the Great Basin will go broke in it.[133]

After winding up his affairs in Nevada by 1885, Connor was back in familiar surroundings in Salt Lake City. He picked up his life almost where he had left it when he had moved to Eureka, Nevada, in 1877. However, there was a significant difference in Connor's position. Eight years before, he had been the owner of the Great Basin Mine and had had greater financial security. Having lost both that property and his position with the Eureka Tunnel Company, Connor was forced to scrabble for a living among his lesser investments. He was still affluent enough to take care of his own living expenses and those of his family in Redwood City, but times were harder.

Connor's Life and Fortunes Ebb

Information about Connor's activities for 1887, 1888, and 1889 is fleeting. He seemed to be semiretired, although he still took care of his Rush Valley mining property. In late April of 1887, he gave a lease to George Etaugh to continue work on the General Connor Tunnel near Stockton.[134] Meanwhile, his old Great Basin Mine, now called the Honorine, was still the leading mine in the area.[135] Connor had a nephew, John F. Connor, who also had business interests in Stockton, and the two often traveled together between there and Salt Lake City.[136]

In 1890 Connor presumably returned to the only means of income he had—his mines. The Tooele County Recorder's Office has records of lot transactions in Stockton during 1890, although the sums involved do not register any income for Connor. A typical entry has him quit-claiming lots in Stockton to B. C. and Amelia D. Harvey for one dollar on 15 September.[137]

When not involved in politics, Connor was still a mining entrepreneur, although on a smaller scale. The *Tribune* ran several articles in January 1891 describing the development of the General Connor and its group of mines, composed of the General Connor Tunnel and the Roxie, Broughton, Jane, Humboldt, Little Joe, and Argonaut lodes. Connor owned the properties jointly with Joseph Broughton of Walkerville, Montana. The Argonaut Mine was leased to some men in Stockton who were "happy over the prospects." The General Connor Tunnel had penetrated the mountain 985 feet, and ore from the Argonaut, as an example, assayed 35 percent lead, 36.5 ounces of silver, and $15.07 in gold. The *Tribune* believed that the two partners "are richly entitled to a big mine as they have pushed this enterprise with energy for many years."[138]

In one transaction on 14 March 1891, Connor's son-in-law, B. P. Oliver, gained possession of the Chloride Point Mine in the Ophir Mining District and Connor's two-thirds interest in six mines associated with the General Connor Tunnel in an indenture for $2,000. This cash payment no doubt was a great financial help to the general.[139]

Not only was his financial condition worsening but his health deteriorated also in 1891, and on 17 December 1891, the *Tribune* reported his death: "General Patrick Edward Connor, whom patriots have sometimes alluded to as the 'Liberator of Utah' died in his rooms at the Walker House last evening at 7:55 o'clock after a painful illness of three weeks." His funeral was a "Magnificent Demonstration," according to the *Tribune*, despite a heavy snowstorm and seas of mud. In accordance with his wishes, the adjutant general's office, on 17 June 1889, had set aside a burial plot in the Fort Douglas Cemetery. There the old soldier and persistent mining developer was laid to rest.[140]

Patrick Edward Connor died intestate so the probate court for Salt Lake County appointed two administrators of the estate. The first was Maurice J. Connor, the eldest son, appointed on 13 February 1892 after posting a bond of $2,400. Then, on

4 March 1892, Patrick Edward Connor, Jr., was appointed administrator. The change in administrators was probably made because at the time Maurice was living in Montana. On 10 July Patrick, Jr., issued an inventory of P. Edward Connor's property and money holdings.[141]

The inventory showed the following:[142]

1	A one-half interest in Lots 38 & 39 of Block 1 of the Richland Addition, Salt Lake City	$40.00
2	Cash deposit in W. F. & Co's Bank, Salt Lake City	736.14
3	A one-third interest in Fairmont Lode, Wasatch Co.	500.00
4	Holding of 2,000 ft. in Chloride Point Mine, Ophir	2,000.00
5	A two-thirds interest in General Connor Tunnel, consisting of 6 mines—Roxie, Broughton, Jane, Humboldt, Little Joe, and Argonaut Lodes	1,000.00
6	Series of town lots in Stockton, Utah	920.00
7	Tools, fixtures and other personal property in Chloride Point and General Connor Tunnel Mine	40.00
		$5,236.14

As already indicated, the Tooele County mining records show that the general had granted an indenture to his son-in-law, B. P. Oliver, on 14 March 1891 for $2,000 on the Chloride Point Mine and the General Connor series of six mines. Unless Oliver had released his father-in-law from indenture, General Connor's assets amounted to only $3,236.14. It was a meager reward for 25 years of effort as a mining entrepreneur. Brigham Young had been a true prophet when he had written on 11 August 1866, "Connor is out of the service, and is here now as plain 'Pat,' engaged in mining business, which, as Government pap has been withdrawn, will very likely, if he pursue it diligently, break him up financially."[143] Connor's assets of $3,326.14 at his death contrasted sharply with the $351,000 he had claimed as the worth of his property in 1867.

Although General Patrick Edward Connor has been memorialized as the "Father of the Liberal Party," and even by some admirers as the "Liberator of Utah," his chief contribution to Utah Territory and its history is as the "Father of Utah Mining." He and his soldiers created the first general mining districts and started the movement by Gentile leaders in Utah, aided even by some Mormon promoters, to break down efforts by Brigham Young to discourage development of the mineral wealth of Utah. Connor and other Gentile leaders were determined to break what they and others perceived as Mormon political and economic domination of the territory, and the coming of the transcontinental railroad offered the means of bringing into Utah many outsiders eager to exploit the undeveloped mineral riches available in the Mormon kingdom. Connor became the leader in the initial opening of Utah mines and devoted the rest of his life after the Civil War to a personal search for the "mother lode" in Utah; unfortunately, he lacked the business acumen to achieve his dreams.

4

The Stories They Tell

Carma Wadley

Long before mining was an industry on this continent, it was a quest. From the very beginning of European settlement, myths and legends whispered of gold, of silver, of precious gems and metals all there for the taking. Such stories were a powerful pull for daring explorers in search of easy riches.

El Dorado, where the king supposedly coated himself in gold dust once a year in tribute to the gods, was surely just over the next ridge of hills, they said, despite the fact that for more than two centuries, no one—not the Spanish, not the French, not the English—ever found a city made only of gold. Gran Quivira, where gold and silver were supposed to be so plentiful people ate from plates made of the metals and where the king slept beneath trees filled with tiny golden bells, was just a bit farther north, they said—although Coronado went as far north as the American Great Plains always looking but never finding any gold.

Cortez believed in the islands of the Amazon women, somewhere in the South Sea, because, he said, "everybody who has any knowledge and experience of navigation in the Indies is certain that the discovery of the South Sea would lead to discovery of many islands rich in gold, pearls, precious stones, spices and other unknown and wonderful things."[1] The Seven Cities of Cibola, based on a story of a Christian archbishop who supposedly escaped from the Moors by leading his followers to fabulous islands where even the sand was made of gold, sent numerous Spanish conquistadors on numerous wild-goose chases.

Add to this myth-filled background the fact that mining has always relied to a high degree on luck and fancy. Before—and even after—geologists learned to read the language of the rocks, the richest finds were often capriciously made. Besides that, mining was extremely hard, extremely dangerous work where even slight mistakes could make huge differences. Fate must be carefully courted. And the work took place mostly underground, which was a world apart: dark, spooky, with all kinds of connotations related to the underworld. Given all those factors, it is not surprising that a rich body of lore, legend, myth, and superstition has grown up around mining. And Utah has its share of myths, characters, and stories.

Of Things Imagined

Miners were a superstitious lot, adept at reading signs and symbols into the most mundane of life's everyday occurrences. Most of the superstitions revolved around impending bad luck. Miners who dropped their tools, for example, or who had clothes fall out of their lockers had to worry about that being a sign that they themselves were headed for a fall. So strong was the belief, in some cases, that those miners refused to continue working their shift.

Miners also looked for signs in the flickering candles that provided light before electricity came along. If a candle was snuffed out, that was considered a sign that something bad had happened at home. And if it went out three times in a row, that supposedly meant some other man was involved with the miner's wife. A dog howling in the middle of the night meant a miner would be hurt the next day. Black cats that passed in front of someone on his way to work foretold doom. The presence of a woman in the mine was widely considered bad luck. And announcing in advance that you were planning to quit was a sure way to invite disaster.

Of all the superstitions, however, none seemed to be as generally feared as whistling in a mine, which was believed to dispel the "good spirit" of a mine. The *Park Mining Record* of 2 February 1882 told the story of miners who were leaving an abandoned shaft. One of them began to whistle, and although the others begged him to stop, he only whistled faster and louder. Suddenly they heard a loud, rustling noise and were all engulfed in a cave-in. They all escaped—except the man who was whistling.[2]

From a longtime miner in the Alta District came this opinion of whistling:

> It is a well-known fact, and I have never seen it to fail in all of my mining experience since the turn of the century, that something terrible happened almost immediately; and it has been demonstrated to me several times that the ore body would pinch out or be cut off by a fault or become too low grade to mine, the moment anyone whistled underground near where the stoping operations were being carried on; and in more than one instance the mine caved in, or it was hit by a sudden inrush of water. In no instance did I know of a death caused by a whistle underground, but almost everything else happened that could cause bad luck in the mine.[3]

But inadvertent actions were not the only sources of apprehension. Stories abound of miners who heard strange and unusual noises in the dark mine shafts. Often they were noises of work going on in sections where work had stopped. Sometimes they were the sounds of voices—particularly voices that sounded like miners known to be dead.

Miners from England, particularly from the Cornwall area, gave a face to some of the noises they heard, and they brought stories of the "Tommy Knockers" with them to this country. Tommy Knockers were said to be dwarflike creatures that inhabited dark corners of the mine and seemed to exist solely for the purpose of harassing miners: creating distractions, luring them into dangerous passages and such. But occasionally the Tommy Knockers showed a more benevolent side. It was said that if a sleeping miner was awakened by a Tommy Knocker precisely between the hours of midnight and 2:00 AM and followed the sound of the knocking, he would come upon a new strike.

Of Things Unreal

Miners died. While occasional explosions and cave-ins took heavy tolls, more common were the accidents that killed one or two miners. The abrupt and freakish nature of some of these accidents led naturally to stories that some of the miners came back to haunt the places where they died. Stories were told of dead, and sometimes even headless, miners who would board the mine train with their former companions and ride to that day's place of work before disappearing. Some tales talked of phantom mules and horses, sometimes with a rider, sometimes not, that appeared in abandoned tunnels. Others described unseen forces that came to help push a loaded car or accomplish some other difficult task.

Ghosts have supposedly been sighted in a number of mines, but some of the more persistent stories come from the Ontario Mine in Park City, one of the richest silver mines in that district. After a 1902 explosion in the Ontario, sightings of the Lady in White began. Apparently one of the miners killed in the explosion had been married for only two weeks. When his wife heard the news of his death, she went crazy. She put on her wedding dress and ran into the mine, never to be seen again—except in odd circumstances. Another story from the Ontario tells of the Man in the Yellow Slicker. Several different versions of the story have developed, but they all feature a man who was apparently hired to go into the mine, perhaps to spy on the miners, perhaps to prevent other companies from stealing the rich ore. One day he went down and never came back. Since that time a number of miners claim to have seen him and even been helped by him.

The Ontario was shut down in 1982, and for a time after that, a concessionaire gave tours of the mine to tourists. The ghost stories persisted during those tours. Ron Kunz, one of the tour guides in the mid-1990s, claimed he had seen the ghosts. On two separate occasions, he said, he looked down tunnels and saw figures when it later was proved that no one was there. On another occasion, one of the women waiting for the tour train saw the reflection of a man's mustached face in a pane of glass. When she turned around, no one was there. "Believe what you want to," said Kunz. He personally felt it was an honor to have seen such specters and felt they were there to watch out for him.[4]

A ghost of a different sort may haunt the hillsides of Silver Reef, some 18 miles northeast of St. George, once the site of one of the state's most famous silver mines. Various stories are told about the fact that the mine was discovered when a grindstone was broken up and taken to an assayer in an attempt to trick him. After the man found silver in the sandstone, some stories have him being run out of town; others say he was lynched. And some say he mysteriously disappeared—only to be heard from later when the conditions in the night were just right. A particularly detailed version of the story tells it this way:

> Metalliferous Murphy was a Pioche, Nevada, assayer who lived in the days before Silver Reef was even a glint in the eye of a prospector. It is said that he had an unrelated appearance, as though he had picked the wrong arms and legs out of a grab bag. He slouched like a marionette with no strings attached. Some say he had a beard that looked like a bush on fire, and that he smiled with a lot of teeth like an open piano. Others maintain he had a voice like a fire siren calling its mate. Some men have been known to proclaim that he wore a shirt of dizzy plaid.
>
> The story goes that he devoted his early twenties to the establishment of a reputation as a 'reader of rocks.' But then there have always been liars. At any rate he had manners out of a book, and was famed far and wide for his earthly erudition. Seems he claimed to be a graduate of the University of Edinburgh. "Yep," he used to say, "I'm a mineralogist, that's what I am. Member of the Royal Society."
>
> Many men had it in for Murphy, said he was a smooth-talking scalawag. They made a habit of bringing hunks of sandstone to him. Everyone knew that silver was never found in sandstone. Metalliferous found silver in nearly every piece. The town boiled. The end was in sight. No more nonsense would they stand. They mashed up a grindstone and took it to him. If he found silver in this, by gum, they'd practice a little cedar-tree justice.
>
> Murphy got out his test tubes, his bottles, and his magnifying glass. He peered at the sample as though his eyes were microscopes focused on a deadly germ. Suddenly, he did a short St. Vitus dance and threw his cap in the air. "Silver! Silver!" he cried. "Where did this stuff come from? It'll run 500 ounces per ton." All hell broke lose. The news was broadcast without the aid of a microphone. Like a typhoon going through a palm tree, a number of characters more agile than fragile grabbed Metalliferous. He was never heard of again.

But the story told at campfires and whispered about the towns is that "when the sun is shining just right high above the steep north wall of Grand Canyon, a bony cedar tree reaches gnarled branches over the yawning chasm. Plainly visible yet are hatchet marks where a large limb of the tree with a man roped to it was cut off and let fall over the gorge."[5]

Some say Metalliferous Murphy walks the territory still, looking for those who did him wrong. His voice may be heard in the moaning wind. His ungainly shadow may be seen on nights with unearthly moonlight—because it turns out that his report was accurate: even the sandstone around Silver Reef was full of silver.

Of Things Found

In those early days, ore was where you found it. And finding it was often based more on luck and intuition than on any science. Tales abound of prospectors who came across rich finds by quite literally stumbling across them or tossed their picks away in disgust only to have those very picks stick in the ground precisely where a rich vein was located. Mules kicked up rocks that contained pieces of gold. Some miners claimed they could smell the ore. Others looked for signs: certain plants and flowers that supposedly grew where ore was, or mountain slopes where snow melted too quickly. And they put their faith in happenstance. One story involves a sheepherder in Juab County who was riding through a narrow pass near Fish Springs when the stirrup of his saddle happened to break off a piece of rock. The man noticed something in the rock glistening in the sun, and he discovered a whole ledge of ore. Other miners later reported passing the outcrop numerous times without noticing anything.[6]

But of all the unusual methods for finding mines, few figure more prominently in Utah lore than dreams. Probably more of them remained dreams than became mines, but at least a few bore ore. The Amazon Mine in Logan Canyon, a coal mine between Smithfield and Richmond, a mine east of Santaquin, and one operated in Sardine Canyon were all mines supposedly found after the prospector had had a dream. The Humbug Mine in the Tintic District was discovered after Jesse Knight dreamed of a rich body of ore that was located beneath a certain sagebrush; the name came from what one of Knight's associates reportedly said when told of the dream.[7]

In 1893 a man named John Hyrum Koyle told his friends and relatives that he had been shown in dreams the location of rich ore in the mountains above Spanish Fork. He went to Benjamin Charles Woodward, a local man with some mine experience, and told him that the dream had said Woodward was to direct digging for the ore; in 1895 the men went to the designated spot and began the work. Woodward's son, Charles, who was "to do the cooking, sharpen the hand drills and picks and spend what time I could turning the air blower" picks up the story from there:

> Mr. Koyle had indicated that the working at the top of Windlass Shaft Number Seven would pass over what he called a "hog's back" or a triangle in the formation and that in a few feet we would hit a smooth wall, dipping about 80 degrees. . . . Mr. Koyle placed his hand to his body just below his chest and stated that this was the place where he had seen the ore, that we would only go down about as deep as he indicated on his body and would hit the body of soft, low-grade ore.
>
> But, as work progressed, no ore was found.

> Mr. Koyle was so sure that the ore would be found that he had all of us young fellows who were working at the mine carry wood from the hillside and pile it on the high peak about 1,000 feet above the mine. At the moment the ore was found, this wood was to be burned as a signal to the people of the valley that the ore had been encountered at the Dream Mine.

Koyle came to the mine soon after the Windlass Shaft Number Seven had reached the point where the ore was supposed to be and found young Charles sharpening tools as his father had requested. Koyle wanted Charles to go into the mine and operate a windlass, but the boy refused. Koyle fired him on the spot, then went into the mine to tell the senior Woodward that his services were no longer needed, either; a new dream had told Koyle he was to take over the operation.

> We stayed until the next morning, when Father and I left. The night before we left, however, I gathered the boys together, and we climbed to the high peak where we had been carrying and piling wood, and we proceeded to set it on fire. We were not to be cheated out of that experience after so much hard work. The next morning it looked like a herd of sheep coming up from the valley. The people had seen the signal fire and came to see the final success of the Dream Mine.[8]

Of Things Lost

Just because you found gold once, there was no guarantee that it would be there when you came back again—at least, according to a number of legends surrounding Utah's lost mines. Many of the stories may have grown out of reports that early Mormon leader Brigham Young supposedly knew that there were valuable minerals in the mountains surrounding the valley but did not want his people caught up in the fever gold and silver produced.

One story told of a man who was riding in the hills above the town of Enterprise and happened upon a ledge of richly colored rock. He took it to be assayed, and it was confirmed as valuable ore.

> That afternoon a white-bearded old man was seen riding a donkey past the prospector's house. After he had stopped, he was asked to dismount and take dinner with the family. Almost the first words uttered by the old man were to the effect that a rich mine had been found. The prospector, surprised that anyone should know of his good fortune so soon, asked how he knew of the discovery. The ancient one avoided the question and merely said, "The mine is an evil thing, and any attempt to develop it foreshadows only ruin for you and your boys."

> The prospector accepted the statement rather lightly and excused himself to prepare dinner. When he returned, the old man had vanished. The next day he endeavored to relocate the mine, but search as he did, he could find no trace of the gold-bearing ledge. Nor could he find any of the signs by which he had thought to return to the spot. The very configuration of the hills seemed unfamiliar. After much meditation, he came to the conclusion that the stranger was one of the Three Nephites and that the loss of the mine, after all, was for the best.[9]

Other "lost mines" in Utah include the Lost Josephine Mine near Monticello, the Lost Rifle-Sight Mine near Hite, the Lost Jack Write near Moab, and the Bullet Mine in Uintah County.[10] Folklorist Wayland D. Hand collected this lost mine story:

> An old man in eastern Utah watched gophers throw up particles of gold in their mounds. He located a mine on the property in this remote area and proceeded to mine gold for use in the Salt Lake Temple. He always approached his mine from a different direction, so as not to betray its location to the Indians. The Indians, however, found the mine and finally killed him, but there was a big cave-in in the mine, and they stayed away from it because of a superstition that a bad spirit hovered near.[11]

Of all the lost mines in Utah, however, none is more famous than the Lost Rhoades Mine, believed to be located somewhere in the Uinta Mountains. This rich gold mine was supposedly under the guardianship of the Ute tribe and made sacred by the forced labor and sacrifice of past generations of the Utes. Some accounts tell of the Spanish explorers working the mines and treating the Utes so brutally that they revolted, killed their overseers, and buried the entrance to the mine.

Chief Wakara was the keeper of the mine in the mid-19th century. Supposedly he was told in a dream that when the big hats (Americans) came, he was to tell them about the gold. As the story goes, Brigham Young's policy of feeding rather than fighting the Indians gained him the confidence of Wakara (Walker), who told him of a rich deposit of gold and allowed him to send someone to get some of it for the church to use. The man chosen for the task was Thomas Rhoades, one of the early pioneers in the Salt Lake Valley, who had learned to speak the Ute language.

Thomas Rhoades faithfully fulfilled the task for a number of years, but he became ill and was unable to continue, so Brigham Young chose Rhoades's son Caleb, 19, to continue in his father's place in 1855. Robert A. Powell, a nephew of Caleb, wrote this account, based on a speech he heard his uncle give at an Old Folks' Party in Price in 1898 or 1899:

> Before Caleb was permitted to start, he said he had to make a covenant with God, before Brigham Young and Indian Chief Walker, that he would never

tell or show anyone where this gold was located. Then Brigham Young laid his hands on the head of Caleb and blessed him and set him apart the same as his father had been set apart before he was sent after the gold.

The next step was to secure a guide to show him where the gold would be found. Chief Wakara selected a young Indian to act as guide and instructed him to protect Caleb from other Indians with whom they came in contact on the trip. They started with a pack horse and one horse for each man to ride. When they met other Indians, the guide would explain their mission and they went unmolested. The guide showed Caleb the place and they got ore they could carry, after which they returned safely to Salt Lake City. Caleb said when his father recovered from his illness, they, together made a number of trips of gold, and as long as Brigham Young lived the Indians never interfered. However, after Brigham Young's death, the Indians forbade the taking of any more gold from the mine, and what Caleb was able to get after that, he had to take out and sell unbeknownst to the Indians.

Although he said many other things in his remarks, which lasted about one and one-half hours, the thing that impressed me most of all was his closing statement, in which he said: "Until this day I have kept my covenant that I made with God, before Brigham Young and Chief Walker. I have never showed any man or told any man, and I never intend to do so, as long as I live." These closing remarks . . . explained to me why he never would show any man where it was, but as soon as he thought the reservation was to be thrown open for settlement he had a very strong desire to locate it which would have broken the covenant he said he had made. He even went so far as to send some of the ore back to Washington, D.C., and offered to pay the national debt, which was pretty large at the time, if they would fix it in Washington so he could locate the mine, but they turned him down, then tried to find it themselves by organizing the Florence & Raving Mining Company. Although prospecting on the reservation was prohibited, this company prospected and located all the latterite, filsonite and other minerals they could find, before the opening for the public to file on mineral land, but they never found the Rhoades mine and I know there have been not hundreds, but thousands of men hunting for it. No one as yet has found it, and when the Florence & Raving Mining Company was prospecting the country, Caleb was asked if he was not afraid they would find it. He said, "No, they will never find it because it is where they will least look for it."

One thing that has made me feel there was something in the covenant he made, is the fact that a short time before the opening of the reservation, he took sick and only lived a few days. While he was sick my father visited him often and was there to help wait upon him as much as possible. Father said all Caleb wanted to talk about was his mine and if he could only live to locate it. When he realized that he might die, he tried to draw a map of it so that his wife, then an old woman, would find it after he was dead. She spent two summers searching but failed to locate the mine.[12]

The Lost Rhoades Mine has attracted a lot of interest ever since. In 1971 Kerry Ross Boren and Gale Rhoades, two relatives of Caleb, published a book called *Footprints in the Wilderness: A History of the Lost Rhoades Mine*, which detailed the story of the mine and where they noted, "By definition the story of the Lost Mines is not a legend, for there is ample proof that they did in fact exist."[13]

In 1998 Boren and his wife, Lisa Lee, published *The Gold of Carre-Shinob: The Final Chapter in the Mystery of the Lost Rhoades Mines, Seven Lost Cities and Montezuma's Treasure*, including maps, where they link the Rhoades Mine to ancient Aztec treasures and claim to know where the mine, which they believe belongs to the Utes, is located. "The reader should not expect to be given directions to Carre-Shinob," they write; "the primary purpose of this book is to verify the existence of the world's largest gold source. . . . Keep in mind that thousands have lost everything in their quest for the fabulous wealth of the Rhoades mine, and hundreds more have lost their very lives. It should be kept in mind that even if one is fortunate enough to find the treasure of Carre-Shinob, there would be little opportunity to capitalize on the discovery."[14]

But until the mine is actually found, it appears that the final chapter may not have been written after all. Certainly its lore and legends will continue to tantalize folklorists for years to come.

Of Things Living and Legendary

For all the tales that have developed in and around Utah's mining towns, real-life adventures have also contributed to our understanding of life in these often-remote places. For example, Maggie Tolman Porter wrote of her experience growing up near the town of Ophir:

> Our nearest town was Ophir, a beautiful little mining town nestling is a lovely canyon home four or five miles from our town. I can still see in my mind's eye the beautiful autumn colors as we drove up the canyon to do our marketing. I can remember the purple elderberries dipping and nodding in the creek as it gurgled by the roadside, the flashing maples, the wild flowers that grew on every side. I can still remember the ruins of a house by the roadside, a mile or two from the town, Mother would show me the blackened ruins and tell me how the mother and father locked their two little children in the home and then went to town and both got drunk. Their home caught fire and their children were burned to death. I was always saddened by the sight of those ruins.
>
> I saw my first pansy blossoms in that little town. My mother and I had gone to call on a lady who used to buy butter from mother. I walked around

> the house and there was a bed of gorgeous pansies. I just stood spellbound, unable to speak, they were so beautiful. The lady asked, then, if I liked them and I began to cry. I must have been about eight years old at the time. The lady placed her arm around me and told me I could pick every blossom if I wanted to. I did, but wept when each blossom faded. I was a little girl starved for beauty on a bleak, dry ranch where we could scarcely get enough water to drink the last five years we were there.[15]

Water was also a concern for Angus Cannon, who had tried to find ore in Dugway in western Utah. He had some limited success and established 33 claims in the area, containing copper, lead, silver, and zinc. But, as he told a visiting geologist, nothing "rich enough to pay the transportation to a mill, nor to justify the building of a mill here." But beyond that, "the main trouble here is a lack of water. Every drop used for cooking, dish-washing, bathing or laundry, to say nothing of what the animals drink, must be hauled a distance of 12 miles from the river, either in wagons or by muleback."

The geologist stayed at the camp for several days, examining the terrain and studying ore samples. At last it was time to leave. "I can't encourage you about the minerals," he told Cannon. "Low-grade ore is plentiful, but the expense of converting it to metal would be high, but I can do something about the water situation." He looked around, walked a short distance, looked down at the ground, kicked at some of the dirt, and said, "Dig here."

Cannon called over three available men, who went to work. They dug down about eight feet, and suddenly water began seeping into the hole. Cannon thought of the terrible need they had for that precious liquid. "Here was water, right under the camp all the time!" he exclaimed.[16]

La Platta, a little mining town in the mountains above Huntsville, is another place where real doings almost take on the aura of fiction. A sheepherder, who noticed a particular gleam in the rock he was sitting on, found the first ore here in the 1870s. Men swarmed to the area, and during the heydays of the 1890s, La Platta produced much of the ore used in northern Utah. But those days came to an abrupt end—not because the ore ran out (the basin is still rich in minerals) but because of "the fast grinding wheels of the mines and the slow grinding wheels of the law."[17]

It seems that four mines were operating on a single hill where an especially rich vein was located: the Sundown, the Sunup, the Red Jacket, and the Yellow Jacket. They were all working at full speed when they realized they were soon going to run into each other. One after the other, they filed suits and countersuits, trying to shut down the operations of the others, and, as those suits got bogged down in court, miners began to look for work elsewhere. Eventually the town was left to crumble, just another ghost town among many. There are still a few old buildings and mine shafts left. But all of the land around the old town is now privately owned, making access difficult. The town itself "belongs to no one, or someone, or everyone, but the courts have never said who."[18]

Lore and law came together in an altogether different fashion in the story of Raphael Lopez, a celebrated mystery in the Bingham Canyon area. In 1912 a period of labor upheaval and unrest led to a strike by the union, and the company brought in some strikebreakers, many of whom were Mexican. Lopez may have been one of them. It is known that he later obtained a claim for the Utah Apex Mine.

Lopez regularly ran afoul of the law and was twice sentenced to jail. On 21 November 1913, he became the prime suspect in the shooting of a man named Juan Valdez. Some said the men had known each other in Mexico; others claimed that Lopez was angry because Valdez had testified against him in a trial. A wanted man, Lopez left town pursued by a posse. In a shootout at a ranch near Utah Lake, he shot and killed three of his four pursuers. Returning to Bingham, he gathered and stole food, clothing, and ammunition and holed up in the Apex Mine. Efforts to smoke him out and storm the mine were unsuccessful. Another officer was killed, and a posse member wounded. Entrances to the mine were sealed and guards posted, but Lopez had disappeared into the maze of tunnels. On 15 December a thorough search of the mine turned up nothing, and Lopez was never heard from again.[19] In 1921 the Texas Ranger who had killed Bonnie and Clyde claimed in a book that he had killed Lopez near Del Rio, Texas, where, he said, Lopez was leading a band of outlaws.[20] If so, it would have been a fitting end to what many considered Utah's greatest manhunt.

Other real people have taken on legendary personas, but nowhere did life and legend come together in mining any better than the story of Joe Hill. On the night of 10 January 1914, a grocer named John G. Morrison and his 17-year-old son, Arling, were shot and killed during a robbery of their store on the corner of West Temple and Eighth South in Salt Lake City. It was known that Arling had gotten off a shot at the robber, and when a man was treated for "an ugly wound in the breast" two hours later at the home of a Murray physician, Salt Lake police figured they had their man.

That man was variously known as Joseph Hillstrom, Joel Hagglund, and Joe Hill, a Swedish emigrant who had come to Utah in 1913 to work in the Park City mines and was active in the Industrial Workers of the World (IWW) movement. He was convicted on mostly circumstantial evidence: he claimed he had received the wound from a lady friend he refused to name, and he was wearing a suit with the name "Morrison" printed in the arms and legs with indelible ink.[21] Because of Hill's IWW connections, the case attracted wide attention. Many Americans, including Helen Keller and then-president Woodrow Wilson, were concerned that Hill might not be treated fairly because of his prolabor stance.[22] Despite the public outcry, Hill was executed by a firing squad on 19 November 1915—only to become a legendary figure much larger in death than in life.

During his life Hill had had some success writing ballads, workers' anthems, and parodies of popular songs. His most popular was probably the one he penned to the hymn tune "In the Sweet Bye and Bye":

Long-haired preachers come out every night,
Try to tell you what's wrong and what's right,
But when asked about something to eat,
They will answer with voices so sweet:

You will eat, bye and bye,
In that glorious land about the sky;
Work and pray, live on hay,
You'll get pie in the sky when you die. [23]

After his execution, Hill's body was taken to Chicago, where his funeral, on Thanksgiving Day, attracted thousands of people to the West Side Auditorium. According to his wishes, his body was cremated, and his ashes scattered the following May 1 in several different countries. That area did not include Utah because Hill had always said he did not want to be caught dead in Utah. His songs and his cause were taken up by numerous other labor workers. Stories, books, plays and songs were written about him. Among the most famous was one by Earl Robinson and Alfred Hayes:

I dreamed I saw Joe Hill last night,
Alive as you and me;
Says I, "But Joe, you're 10 years dead."
"I never died," says he; "I never died," says he.[24]

Of Things Still to Come

Alta, located in Little Cottonwood Canyon, was the home of rich silver mines such as the Emma, the Prince of Wales, the Flagstaff, and the South Star. At its height it had a population of about 5,000, with more than 100 business buildings. But, with more than 30 saloons, it had a wild and woolly side as well. By the 1870s it also had a rather large cemetery, not only filled with those killed in mining accidents, snowslides, and avalanches but also, it was said, with more than 100 men killed in brawls or murdered. So it is not surprising that Alta is where one of Utah's best mining stories comes from. The story, told by John W. Smith, goes something like this:

> It happened in the 1870s, when the camp was at its peak. In the 10 years following 1865, when silver was discovered, Alta's cemetery had grown by leaps and bounds. The law of those days was the law of the six-shooter, and that law was frequently enforced. To the graveyard just below the town at the mouth of Collins Gulch were carried the remains of those who met death. Some of the deaths were accidental; some deaths were natural; but many occurred in disputes, barroom brawls, or "grudge fights." Whatever the cause, the graveyard grew until it was occupied by more than 150 "tenants."

> Few names were ever carved on the headstones. So many nicknames were given in camp, and so many aliases were used by the men, that their names really meant nothing. Most of the corpses were carried to their shallow graves "as is," with their boots on, and thrown in without benefit of casket, flowers, burial robes, friends, or clergy.
>
> One day a religious mountebank came into camp claiming the power to resurrect the dead. He proposed, for a nominal sum, to bring back to life the 150 corpses in the cemetery. The camp went into a dither. Could he really do it? And if he could, should he? Alarm spread through the camp as the miners considered how many embarrassing family triangles would result from the resurrection. What a number of double crosses would be revealed! And what a multitude of old grudges would be relived!

No, thanks, they told the man. Ah, he said. He had come to like the good people of Alta so much that he would perform his service anyway—unless, of course, they could, um, give him a little something to help him on his way.

> The camp just couldn't afford to take the chance. Better pass the hat, take up a collection for the miracle man, and ask him to leave.
>
> This they did. The prosperous but panicky miners contributed $2,500 for the bold charlatan, and gave it to him on condition that he would leave Alta's graveyard intact and let its dead rest in peace. So today, the 150 bodies (and others added later) are at rest in the mouth of Collins Gulch—still awaiting the resurrection.[25]

Maybe by that time the truths of all these mysteries and legends will come to light as well. In the meantime, they provide fodder for the imagination and delight for the sensibilities—at least for anyone interested in the significant role that mining played in Utah.

Courtesy Utah State Historical Society.

Patrick Edward Connor, the "father of Utah mining," originally came to Utah as a general with the United States Army in 1862.

Courtesy Utah State Historical Society.

Johanna Connor was not only the wife of Patrick Edward Connor but was also listed as a shareholder in his mining properties.

Courtesy Utah State Historical Society.

Soldiers under Connor's command explored for ore as much as they looked for hostile Native Americans; a few years later, most of them became full-time miners, such as these shown in Stockton in the late nineteenth century.

Courtesy Utah State Historical Society.

Connor built a boat, similar to this one, to haul ore on the Great Salt Lake and named it after his daughter, Kate.

Courtesy Utah State Historical Society.

The Dry Canyon area became covered with claims, some of which overlapped, prompting lawsuits and counter claims.

Courtesy Utah State Historical Society.

Stockton, Utah, became the first city in Utah Territory, and one of the first in the nation, to use electric lights.

Courtesy *Deseret Morning News.*

Joseph Hillstrom, better known as Joe Hill, became an American labor organizer, songwriter, and eventually a martyr when he was executed in Salt Lake City for the murder of storekeeper John G. Morrison.

Part II

Some Mineral Industries

5

Saline Minerals

J. Wallace Gwynn

Utah's saline industry is one of the oldest in the state, beginning with the commercial harvesting of salt from Great Salt Lake by the Mormon pioneers in 1847. The salt industry on the lake has continued from that time to the present, marked by the appearance and disappearance of many companies. Today three companies extract salt from the lake. Other products harvested from the lake include magnesium metal, chlorine gas, potassium sulfate, magnesium chloride, and nutritional supplements. A few of the by-products and potential products include ferrous and ferric chloride, calcium chloride, sodium sulfate, lithium carbonate, bromine, and boron.

Crystalline or rock salt has been mined from the Jurassic Arapien shale in central Utah since 1854, from salt outcrops east of Nephi, and since the 1870s, from the vicinity of Redmond. Today one company is mining salt underground in the Redmond area. Potash mining began on the Salduro Salt Marsh (Bonneville Salt Flats) about 1916. Today the company is producing both potash (KCl) and common salt (NaCl) from subsurface brines. Potash and salt are also produced by a solution-mining operation near Moab in southeastern Utah from the Pennsylvanian Paradox formation of the Hermosa group. Assessment and development were done at Sevier Lake, Millard County, aimed at producing salt and eventually potassium sulfate from subsurface brines, but financial problems curtailed the operation. There are several other areas where deep crystalline salts occur in the state but have not been developed. These areas include the Sevier Desert near Delta, the Preuss salt zone in the Wyoming/Utah/Idaho thrust belt, and small deep deposits of shortite [$Na_2Ca_2(CO_3)_3$] and nahcolite [$NaHCO_3$] in the Uinta Basin.

Use of Utah's saline resources began many years before the Mormon settlers entered the Salt Lake Valley in 1847; Native Americans probably used salt from Great Salt Lake, salt springs and seeps, salt outcrops, and possibly other sources such as Sevier Lake. The first nonnatives known to use salt from Great Salt Lake were the mountain men from Ashley's Rocky Mountain Fur Company during the fall of 1825, followed by John C. Fremont during his 1843 exploration of the West. The first commercial development of the state's saline resources was the production of common salt from Great Salt Lake by the Mormon pioneers. Since then, the lake has produced common salt (sodium chloride), potassium chloride, potassium sulfate, sodium sulfate, magnesium chloride, magnesium metal, and chlorine gas.

This chapter reviews the development of Utah's saline resources at Great Salt Lake and the Salduro Salt Marsh (Bonneville Salt Flats), and in northern and central Utah and the Paradox Basin. It also discusses other salt occurrences in the state that are known but have not been tapped or developed. In 2002 the value of the state's saline-mineral production exceeded $148 million and accounted for more than 25 percent of the state's total industrial-minerals production value.

Salt Mining at the Great Salt Lake

Great Salt Lake, a remnant of the freshwater Pleistocene Lake Bonneville, occupies an area of up to 1,472,000 acres (2,300 square miles) at an elevation of 4,212 feet (lake's historic high) in northwestern Utah (figure 1). At an average surface elevation of 4,200 feet above sea level, the lake has an area of 1,034,000 acres (1,616 square miles) and a volume of 15,390,000 acre-feet. The lake contains approximately 4.5 billion tons of dissolved salts with the following approximate composition (on a dry-weight-percentage basis): Na = 32.0, Mg = 2.5, K = 1.8, Ca = 0.2, Cl = 57.4, SO_4 = 6.1. The main body of the lake is divided into two parts, a southern and a northern arm, by the Southern Pacific Railroad's rock-fill causeway. Because the causeway restricts circulation and mixing between the two arms, the southern arm, which receives nearly all of the freshwater inflow, is much less saline than the northern arm.

The Beginnings of an Industry

Within days of their arrival in the Salt Lake Valley on 24 July 1847, the Mormon leader Brigham Young and a group of his associates traveled to Great Salt Lake.[1] They found an abundance of white, coarsely crystalline salt deposited on its shore. A few days later a group of the pioneers traveled to the lake and gathered 125 bushels of coarse white salt. They also boiled four barrels of salt water that yielded one barrel of fine, white, crystalline salt. It is uncertain when the first commercial salt-boiling operations began, but Charley White built and operated a permanent salt-boiling facility (at an unknown location) from the spring of 1850 until about 1860.

In the mid-1860s Great Salt Lake's salt industry received its first real stimulus due to the discovery of silver in the mining camps around Butte, Montana. The chlorination process used to reduce the silver ores required large quantities of salt. Utah had the salt, and by 1870 the railroads provided the transportation; thus, a substantial salt market was born.

Early Process Developments (1860 to 1895)

Two factors played an important role in changing the way the budding salt industries produced salt. First, the lake began a sustained rise from its 4,200-foot

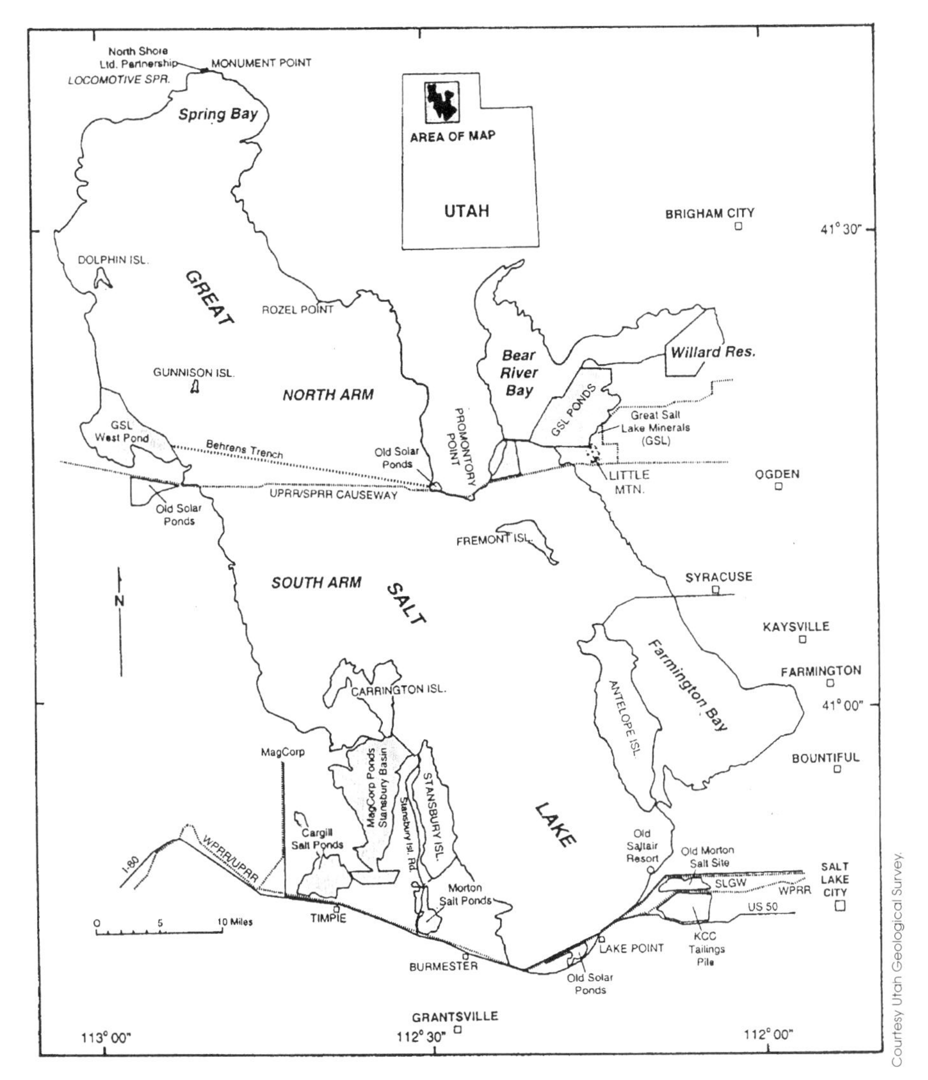

Figure 1. Past and present solar-pond locations and other cultural features around Great Salt Lake, Utah.

elevation in 1860 to its high of nearly 4,212 feet in 1872, and second, there was an increasing demand for better-tasting, higher-purity salt. As the lake level began to rise, the beds of white crystalline salt the pioneers had found along the shore were covered and dissolved, and the lake brine was too diluted to produce salt without additional evaporation. So the salt producers had to build solar-evaporation ponds to produce the salt. They accomplished this by building dikes across the entrances of coves. At first, the dikes were simply constructed of earth taken from the bottom of the ponds. These dikes, however, soon washed away. More durable dikes were then built, consisting of two parallel rows of cottonwood stakes driven into the ground every two feet with a latticework of willows woven between them and backed by bulrushes. The space between the two rows was then filled with earth.

The salt producers depended on the periodic rise of the lake's level to fill their ponds. The periodic rise was due to north-wind-induced tides that raised the water elevation on the south end of the lake as much as 1.5 feet. Since this method of filling the ponds was not reliable, they soon resorted to horse-powered pumps.

The early salt producers found that evaporating a pond of lake water to dryness produced bitter-tasting salt due to its high potassium, magnesium, and sulfate content. The salt also contained mud, sticks, and other debris. To produce a cleaner and better-tasting salt, they developed a process called *fractional crystallization* that used a series of ponds. The first pond settled mud and other debris from the brine. The brine was then moved to a second pond, where solar evaporation concentrated it to the point of sodium-chloride saturation. The concentrated brine was then moved to a third pond, where crystalline salt was precipitated from it onto the pond floor. Precipitation was allowed to continue until the brine concentration reached a density of about 29 degrees Baumé. At that point the remaining brine, called the *bittern*, was drained from the pond back into the lake. This process produced cleaner and better-tasting salt.

Utah salt makers also developed the *split* or a *cleavage-plane* procedure to assist in harvesting salt. At the beginning of each evaporation season, about May or June, a thin layer of very small salt crystals was deposited on the floor of the salt ponds, forming a split between the floor and the large crystals of the annual crop. Two methods were used to make the split, resulting in either a sun split or a mechanical split. A sun split was created by draining the pond until only a small amount of highly concentrated brine covered the floor. A layer of very fine crystals—up to one-eighth inch in size—was precipitated over the large, jagged crystals below. Fresh, highly concentrated brine was then brought into the ponds. The larger crystals of the annual crop grew upon the fine salt layer. A mechanical split resulted from dragging a heavy object, such as a length of rail, across the floor of the pond. This process knocked the edges off the crystals and formed a fine layer of salt that separated the floor from the new crop.

Early Salt Companies, 1880–1915

Early Companies on the South Shores of Great Salt Lake

More than 20 companies produced salt on the shores of Great Salt Lake between 1880 and 1915. The following 6 companies developed into major enterprises on the southern shores.

Jeremy and Company was organized about 1870. The ponds were located at North Point, a portion of the lakeshore three miles north and east of the site of the old Saltair resort. It was one of the first companies to construct artificial evaporation ponds rather than depend on the deposition of salt in small, diked-off bays and inlets. Ponds from 5 to one 100 acres were filled using the natural rise and fall of the lake level that accompanied changes in wind direction and seasons. Government records show that Jeremy and Company harvested more than half of the salt produced in Utah during 1880. With a constant annual production of 10,000 tons, it accounted for only one-sixth of the total lake production in 1890, however. In the spring of 1891, Jeremy and Company reincorporated as Jeremy Salt Company and produced 13,000 tons of salt. The next year, production fell to 5,000 tons. In 1896 the company was sold to Inland Crystal Salt Company.

Inland Salt Company was organized on 21 November 1887 by a group of Mormon entrepreneurs and was the predecessor of the Inter-Mountain Salt Company, Inland Crystal Salt Company, Royal Crystal Salt Company, and Morton Salt Company's Utah branch. In 1888 Inland Salt began constructing its ponds, presumably at the old Morton Salt site (figure 1), which were the first ponds specifically designed to use the fractional crystallization process. In 1888, the first year of operation, 5,000 tons of salt were produced, and by 1890 Inland Salt was producing two-thirds of the 60,000 tons of salt marketed by Utah companies. In April 1891 Inland Salt was sold to buyers from Kansas City for $200,000. In the fall of 1892 money from the sale of Inland Salt was used to develop a resort and a new salt works on the shore of Great Salt Lake. Saltair resort, the Saltair Railway, and Inter-Mountain Salt Company resulted from this investment.

Inland Crystal Salt Company was the result of the reincorporation and renaming of the Inland Salt Company on 1 July 1891. About $50,000 was invested in renovating the old facilities, including installing the world's largest rotary-kiln dryer, coining the brand name "Royal Crystal" which appeared on the table- and dairy-grades of salt, and developing a process for making salt blocks for livestock. As a result of the growing market and renovation of the refining facilities, the total production of the company during 1891 was 90,000 tons or 50,000 tons more than Inland Salt had produced the previous year. Inland Crystal Salt and its successors, including Morton Salt Company, used the name Royal Crystal at least through 1971.

Inter-Mountain Salt Company was organized on 1 October 1892, funded in part by the sale of Inland Salt. Inter-Mountain developed its pond complex directly east of the property owned by Inland Crystal Salt and built a refining mill in Salt Lake City. About four-fifths of the salt sent through the refinery was sold as table salt. Inter-Mountain operated successfully until 2 March 1898, when the plant burned. Stockholders of Inter-Mountain then bought a controlling interest in Inland Crystal Salt Company.

After the merger with Inter-Mountain in 1898, Inland Crystal Salt became the only successful enterprise on the south shore of the lake. However, in 1901 two companies, Diamond Salt Company and Weir Salt Company, attempted to establish salt works in the area.

Diamond Salt Company was located west of the Inland Crystal Salt works and was incorporated on 2 February 1901 with plans of establishing a sanitarium, bathing facility, amusement park, and salt-manufacturing business. Ponds were built north of the Salt Lake and Los Angeles Railway and west of the Inland Crystal Salt works, but the other plans never materialized. The company sold its holdings to E.L. Sheets Company, which in turn was purchased by Inland Crystal Salt in 1915.

Weir Salt Company started its operations about the same time as Diamond Salt entered the industry. Its operations were located on the south shore of the lake near Lake Point, and its ponds were south of the location later occupied by the Hardy plant. Problems plagued the company, and construction of the facility was never completed. Deseret Livestock Company later purchased the property and began construction of its salt works on the same site in 1949.

The Salt Monopoly

The merger of Inter-Mountain and Inland Crystal Salt in 1898 marked the end of an era in Utah's salt industry. Gradually the small companies around the lake failed or were purchased by the merged Inter-Mountain and Inland Crystal Salt. The role of the Church of Jesus Christ of Latter-day Saints in Utah's salt industry grew steadily through the decade prior to 1898 until it became the dominant influence in the emerging monopoly. After the merger, the church-controlled company tried to maintain and strengthen its monopolistic position by purchasing lands suitable for salt operations or buying out competing firms. In 1901, however, Diamond Salt Company and Weir Salt Company constructed facilities on the south shore of the lake, and the Sears Utah Salt Company was established on the east shore of the lake near Syracuse in competition with the monopoly. Unfortunately, Inland Crystal Salt reportedly applied strong-arm tactics, such as intimidation and destruction of property, to control the salt market in addition to eliminating competition through its ability to control the price of salt.

During the first two decades of the twentieth century, the Inland Company maintained its position as the major salt producer in the state. Other small companies produced salt in Utah during this period, but the monopoly was not significantly threatened until Morton Salt Company moved into the area in 1918.

Early Companies on the East and North Shores of Great Salt Lake

George Payne built one of the first salt works on Great Salt Lake's eastern shore in 1880. The ponds were located on the south side of the Syracuse road, southwest of the town. William W. Galbraith later bought up shore land near Payne's property that apparently included it and organized the Syracuse Salt Company. He constructed 90 acres of salt ponds. Lake brine was moved to the ponds by three steam-engine-powered pumps. Most of the salt was shipped directly to the Montana silver mills without processing to be used in refining silver. Syracuse Salt produced 12,000 tons the first year and 2,000 tons the next year. There are no records that salt was produced after 1888.

By 1887 William Galbraith had sold more than half his acreage to the Adams and Kiesel Salt Company, which was incorporated on 17 May 1886. Adams and Kiesel built not only a salt works but also the Syracuse Resort and were involved in cutting and selling cedar posts. The company apparently did not own an extensive refining operation and simply shipped crude salt to the silver mills. From 1889 to 1892, production records indicate that Adams and Kiesel Salt sold 15,000 to 20,000 tons of salt a year. On 19 April 1899, William B. Clarke of Kansas City, Missouri, purchased the company for $30,000. There is no indication that Clarke ever produced any salt.

The Deseret Salt Company, incorporated on 9 October 1883, was located on Farmington Bay, southwest of the town of Farmington. A steam-engine-driven pump lifted the water from the lake into its ponds. Deseret Salt produced crude salt for the silver-mill market. There is no reference to the company after 1892.

The Gwilliam Brothers Salt Company (established about 1890) was located about two miles north and west of the Syracuse Salt Company. Gwilliam Brothers changed the company name to Solar Crystal Salt Company in 1892. It produced salt during the 1890s, but exact production records are missing. In 1901 George W. Gwilliam re-incorporated the company, but how long it was able to operate in competition with Inland Crystal Salt is not a matter of record.

The Sears Utah Salt Company located its operation between Syracuse and Kaysville. Sears Utah Salt, along with several other companies, existed in 1903, but how long it lasted after that is unknown.

For a short time, a salt company owned by A. H. Nelson operated near Brigham City. The company did not report its production in 1891 and discontinued its operation in 1892. The Nelson Company is typical of dozens of small companies that entered the salt business to take advantage of the market for crude salt.

Salt Companies from 1916 to 1970

Salt Companies on the South Shores of Great Salt Lake

Inland Crystal Salt lacked any serious competition in Utah's salt industry until Morton Salt leased a potash plant at Burmester (north of Grantsville) in 1918 and established a foothold. In 1923 Morton Salt purchased controlling interest in the Inland Company from the Church of Jesus Christ of Latter-day Saints. By 1927 the remaining stock was acquired, and Inland Crystal Salt was reincorporated as a wholly owned subsidiary under the name of Royal Crystal Salt Company. Morton Salt produced salt from its plant at Burmester and also from its subsidiary plant at Saltair until 1933, when production and refining facilities were combined at Saltair. Although both companies operated from the same plant, Royal Crystal Salt maintained its separate identity until 1958. When antitrust litigation brought against the company three years earlier resulted in the dissolution of Royal Crystal Salt, the plant subsequently became Morton Salt Company.

Until 1923 the Burmester plant employed manual labor to harvest salt. Machines were subsequently used after experiments proved their feasibility. Ed Cassidy introduced mechanized salt harvesting when he brought his farm tractor to the ponds to replace horses in pulling the salt plows. Machines had not previously been used because of the fear that their heavy weight would break through the thin salt floor of the evaporation ponds into the underlying mud. Following Cassidy's successful trial, however, the company purchased Fordson tractors to plow the salt. These tractors were discontinued in 1936 in favor of eight new machines called "Hootin' Nannys," which were Fordson tractors with a three-quarter-ton-capacity scoop in front. In 1949 the Hootin' Nanny was replaced by another machine called a "Jackrabbit," which used an air-cooled engine and could hold more salt. The Jackrabbit was replaced during the late 1950s by a machine called the "Scoop-mobile," which was larger and had a hydraulic scoop bucket in front. The Scoop-mobile was replaced in 1964 by a revolutionary new machine called the "Palmer-Richards Salt Harvesting Combine," which picked up salt as it moved down a preformed *windrow* (salt piled in a long, straight row) and dumped it into a truck moving alongside it.

Solar Salt Company and Its Predecessors

Growth of the Saltair facility enabled Morton Salt to retain a dominant position in the Intermountain market. Its near monopoly faced a temporary competitive threat from several new plants around the lake during the late 1930s and early 1940s; however, none of the new companies lasted more than three or four years. It was not until the late 1940s and early 1950s that a strong, competitive challenge arose when the Lake Crystal Salt Company, Deseret Livestock Company, and Stansbury Salt Company gained footholds in the salt business and retained them. These new companies benefited from being organized at a time when the salt market was expanding. In the 1960s the market increased 50 percent and doubled again in the 1970s. Although Morton Salt now shared the market with three other companies, the company's production still increased.

Crystal White Salt Company was organized in 1938 to produce salt for the California market. The new company selected a site six miles from Grantsville on the mud flats south of Stansbury Island. It chose the west side of the lake because of the purity and concentration of the lake brine, the accessibility to the railroad, and the geography of the salt flats. Construction began in late summer of 1938, and by the following year, there were 35 acres of ponds. The system was in full operation by June 1939, but the company went out of business in 1941, and the property lay dormant until the newly incorporated Stansbury Salt Company reactivated it.

The Stansbury Salt Company was formed when the properties of the defunct Crystal White Salt Company were sold at a sheriff's auction on 21 June 1948 to Mrs. Mary Godbe Gibbs for $2,500. Her husband, Lorin W. Gibbs, and others then organized Stansbury Salt. A refinery was built in 1950 on the north side of the intersection of U. S. Highways 40 and 50 and the Stansbury Island road. The plant was originally designed to produce 10,000 tons per year. By operating 24 hours a day the plant increased production to 40,000 tons in 1956.

Growth of Stansbury Salt required more capital than was available, and thus limited progress occurred until representatives of Hooker Electro Chemical Company and Pen Salt Chemical Company contacted it in 1954. These two companies organized the Chemical Salt Production Company and engaged Stansbury Salt as agent to build a large, salt-evaporating complex adjacent to the property. In 1956 the two chemical companies merged with Stansbury Salt, and the new combination was incorporated under the name of Solar Salt Company.

Solar Salt was initially incorporated with the intention of shipping salt to the two parent chemical firms in the Pacific Northwest. The new operation was completed in 1960. The ponds were designed to produce 160,000 to 200,000 tons of salt annually, but no more than 50,000 tons per year were shipped to the chemical plants. In 1967 National Bulk Carriers, which needed its own source of salt, purchased Solar Salt's facilities.

Nearly 50 years lapsed between Weir's abortive attempt to enter the salt industry at Lake Point in 1901 and the time when the Deseret Livestock Company reactivated the site in the spring of 1949. In late 1952 or early 1953, David Freed and David Robinson purchased Deseret Livestock, including the Lake Point salt works. In 1955 they offered the salt works to Council McDaniel, a former executive in a West Coast salt firm, for $300,000. McDaniel incorporated under the name Deseret Salt Company and operated the salt works until the latter part of 1958, when he sold it to Leslie Salt Company. In 1961 the Federal Trade Commission charged Leslie, the largest salt producer on the West Coast, with creating a monopoly. The proceeding was settled through a divestiture order requiring Leslie to sell its Utah holdings. On 2 November 1965, Hardy Salt Company of St. Louis, Missouri, purchased Leslie's Lake Point plant.

The pond system constructed by Deseret Livestock consisted of a series of 15 ponds laid out on high ground south of the mill. Later, 10 more ponds were constructed on the relicted lands between the mill and the Western Pacific Railroad. Subsequent owners have added ponds in both locations. Porous soil underlying the floor of the ponds had been a problem for salt producers at Lake Point since 1901.

Constructing seal trenches around the ponds finally solved this problem. Freshwater springs within the ponds have also presented problems.

Salt Companies on the East Shore of Great Salt Lake

Three sites around the east and north portions of the lake have been used for salt production: Spring Bay on the extreme north end of the lake, Promontory Point, and the mud flats west of Syracuse. From the turn of the century until 1939, there was no significant activity on the eastern shore of the lake. Inland Crystal Salt and its successors, Morton Salt and Royal Crystal Salt, were firmly established on the south shore, providing the market with the diversified products it required and jealously guarding their position in the industry. Inland or Morton owned much of the shore land around the lake suitable for salt production and made it very difficult for new companies to locate.

In 1939 the lake stabilized at the bottom of a 15-year declining cycle, exposing large areas of relicted land. The available shore land encouraged C. J. Call to organize the Ritz Salt Company. After constructing an access road and about 55 acres of evaporation ponds, Ritz Salt harvested about 5,000 to 6,000 tons of salt that fall. Plans were made to construct a salt refinery, but it was never built. Call sold his holdings to Morton Salt in 1941 after the company threatened him with a lawsuit for trespassing.

Call moved to the eastern tip of Promontory Point and built a few ponds between the lake's edge and the tracks of the Southern Pacific Railroad. There is no record of production from this site, and the rising lake would have inundated Call's ponds by 1950.

A.T. Smith, owner of Smith Canning Company in Clearfield, became interested in producing salt for his cannery and selling it on the regular market and formed the Solar Salt Company about 1940. Ponds were located a mile below the meander line, straight west of Syracuse. The meander line is a surveyed boundary around the lake below which is state property. A canal was dug from the ponds to the lake, and a gasoline-powered pump lifted the water from the canal into the ponds. Dikes were constructed with wooden shoring filled with clay. During 1940 3,000 tons of salt were harvested. A small mill consisted of a rotary kiln, screens, and a roller for crushing the salt into smaller particles. In 1945 rising waters washed away the dikes and dissolved 20,000 tons of salt. In 1949 the mill was dismantled and used to construct Deseret Livestock's salt plant at Lake Point.

Salt Companies on the Northern Shore of Great Salt Lake

In addition to the common problems encountered by salt makers in other locations around the lake, the north shore has never been considered a prime area for salt operations due to its remote location. It did, however, offer some promise shortly after the transcontinental railroad was completed near the north shore in 1869.

The Housel and Hopkins Salt Company constructed its ponds east of Locomotive Springs on the shore of Spring Bay and was reportedly operating during 1871. No other records of this company exist.

Organization of the Quaker Crystal Salt Company in 1939 resulted from a severe earthquake that affected Monument Point on the northern shore of the lake. Three warm springs began to flow, containing from 11 to 15 percent salt. Analysis of the brine indicated its chemical composition was ideal for cheese making. In addition to the springs, water from the lake was also used to make salt. A mill was built in 1939 or 1940 and consisted of an oil-burning, rotary-kiln dryer and screens to separate the salt into four different grades. The mill had a capacity of 2.5 tons per hour. The pond dikes were washed away in 1948 or 1949 and again in 1952. In 1965 fire destroyed the mill.

Lake Crystal Salt Company was incorporated in 1947. Production and refining operations were located at Promontory Point, and sales and storage facilities were in Ogden. Lake Crystal Salt Company had an advantage not shared by competing salt companies on the south arm of the lake. The highly concentrated brine in the north arm of the lake did not require concentration ponds, and all 300 acres of ponds were devoted to harvesting. Mill construction began in 1947 and consisted of a rotary-kiln dryer, screens, and rollers that produced a variety of grades for the market.

Salt Companies from 1970 to the Present

After the Hardy Salt Company of St. Louis purchased the Leslie Salt Company's Lake Point plant in 1965, it subsequently sold the operation to Lakepoint Salt Company in 1977. Lakepoint Salt consisted of a group of local investors and the former local management of the American Salt Company. The existing Hardy plant was shut down for several months while it was extensively refurbished to return it to profitable operation.[2]

In 1982 Domtar, a Canadian company, purchased Lakepoint Salt, retaining the existing management. In about 1983 the entire pond system was flooded when the Union Pacific Railroad installed a culvert through the railroad causeway to equalize the water on both sides of the causeway (Great Salt Lake was beginning to rise at this time). In 1984, with its ponds flooded, Domtar made an agreement to purchase salt harvested from AMAX Corporation's Stansbury Basin ponds and operated for a year under this arrangement. Then Domtar sold the Lake Point operations to AMAX, which changed the name to Sol-Aire Salt and Chemical Company.

In 1986 Diamond Crystal Salt Company from Michigan proposed a partnership with AMAX to build a large salt complex at Timpie, but negotiations were put on hold when AMAX's northern solar-pond dike broke, flooding the entire pond system. When AMAX decided to abandon the flooded pond system in 1987, Diamond Crystal made an agreement with AMAX to purchase the Sol-Aire salt plant at Lake Point and land suitable for a plant site at Timpie. During 1987–88 Sol-Aire did not have any productive solar ponds and purchased salt from Kaiser Chemicals near Wendover to process at Lake Point. During the same time, however, Diamond Crystal constructed an east-west dike through AMAX's Stansbury Basin evaporation pond, providing access to large quantities of salt on the bottom. This salt was taken for washing to Diamond's new plant at Timpie and then trucked to Lake Point for further processing.[3]

In 1989 AKZO, a large European company with salt operations in the United States, purchased Diamond Crystal, and Sol-Aire's name was changed to AKZO Salt of Utah. From 1991 through 1994, AKZO installed a small drying operation, a bulk rail-loading facility, and other improvements that completed the current salt-producing facility at Timpie. The old Lake Point plant was closed and partially reclaimed. In 1995 Nobel (a Swedish Corporation) merged with AKZO to become AKZO Nobel. Finally, in 1997 Cargill Salt, which for a time in the early 1960s operated the Lake Point plant under the name of Leslie Salt, purchased all the AKZO Nobel salt-producing facilities in the United States. Since the purchase, Cargill has invested heavily in the Timpie facility (figure 1) to bring its salt-drying and processing capacity up to the level of the salt-producing solar ponds.[4]

Morton Salt continued to operate successfully on the south shore of the lake until 1982, when heavy rain fell just south of the solar ponds, causing a flood in the C-7 canal that ran between the ponds and Kennecott's tailings. The flood washed over the crystallizing ponds, wiping out dikes and dissolving much of the salt laid down in previous years. It also flowed through the stockpile area, dissolving much of that salt, too. In addition, the rising lake diluted the lake brine, prompting construction of an additional 670 acres of concentrator ponds and elevation of the brine-supply pumps. Because of these problems, Morton was unable to produce enough salt and had to purchase supplemental salt from the AMAX ponds in the Stansbury Basin, located west of Stansbury Island.[5]

In 1991, when Kennecott Utah Copper needed acreage to expand its tailings pond, Morton's adjacent ponding area was a logical choice for acquisition. Kennecott bought the North American Salt Company plant at the south end of Stansbury Island and traded it to Morton Salt for its Saltair plant (figure 1). This trade enabled Morton Salt to continue producing salt from Great Salt Lake. Morton Salt moved some of the old equipment from the Saltair site to the new facility northwest of Grantsville, upgraded the old equipment in the North American mill, built some new facilities, repaired old solar-pond dikes, and added new dikes. At the end of 2000, Morton Salt and its 140 employees were well positioned to continue providing quality salt to a large area of the United States.[6]

After the Solar Salt Company (located on the southern end of Stansbury Island) was purchased by National Bulk Carriers in 1967, the North American Salt Company entered the Utah salt market. In 1972 North American Salt, owned by D. George Harris and Associates, purchased Solar Salt and became one of the largest producers in Utah.[7] North American Salt operated at the south end of Stansbury Island until 1991, when Kennecott purchased the entire facility for its trade with

Morton Salt. As a result of the trade, North American's management moved north to the Great Salt Lake Minerals and Chemicals Corporation's salt plant, also owned by D. George Harris and Associates (figure 1). The Great Salt Lake Minerals plant had been producing salt since about 1970. North American Salt was purchased by IMC Global on 1 April 1998 and renamed IMC Salt, Inc.[8]

On 28 November 2001, IMC Global sold its salt unit (IMC Salt, Inc.) to Apollo Management, LP. This new business entity is now known as Compass Mineral Group with headquarters in Overland Park, Kansas.[9] As part of this transaction, IMC Salt changed its name back to North American Salt Company.[10]

During World War II, National Lead Industries began to develop technology to produce magnesium by operating a government-owned plant at Lucky, Ohio. The company gained additional metals-production expertise in 1951 with the formation of a jointly owned company, Titanium Metals Corporation of America (TIMET), at Henderson, Nevada.

During the early 1960s, National Lead began investigating the possibility of producing magnesium metal and searching for sources of magnesium. The company joined with Hogle-Kearns, a Utah investment firm, and Kerr McGee, a diversified chemical company, in a venture to assess Great Salt Lake's potential for producing magnesium metal. The partners sought rights from the State of Utah to develop a solar-pond system in Stansbury Basin.[11]

During 1965 and 1966, National Lead conducted pilot operations. Scale-model solar ponds were constructed at Burmester, and a pilot manufacturing plant, designed to produce electrolytic-cell feed, was built near Lake Point. Magnesium chloride from this plant was trucked 450 miles and fed into a prototype cell at TIMET in Henderson, Nevada. Construction of the integrated magnesium-production facility began in 1970 with Ralph M. Parsons as the general contractor. The plant was located 10 miles north of Interstate 80 on the west side of Stansbury Basin (figures 1 and 2). In the summer of 1972, magnesium production began. Process difficulties made it necessary to shut down operations completely in 1975 to do some reengineering, with the help of Norsk Hydro, a magnesium producer in Norway. During the mid-1970s National Lead changed its name to NL Industries. In 1980 NL Industries sold the magnesium operation to AMAX Inc., a diversified mining and natural-resource company.[12]

Shortly after the transfer of ownership to AMAX, the lake began to rise. On 7 June 1986, in spite of AMAX's efforts to raise and fortify its dikes, a storm breached the main dike separating the solar ponds from the lake, completely flooding the ponds. During the flooding of the1980s, AMAX continued producing at a reduced rate, using concentrated magnesium chloride brines purchased and trucked in from Reilly Chemical near Wendover, Utah, and Leslie Salt, located near San Francisco. An alternative solar-ponding site was identified near Knolls. Ponds were constructed, and brine was obtained from the Newfoundland pond, part of Utah's West Desert Pumping Project (figure 2). After concentrating the brine at Knolls, it was moved 41 miles to the magnesium plant by pipeline. The West Desert Pumping Project

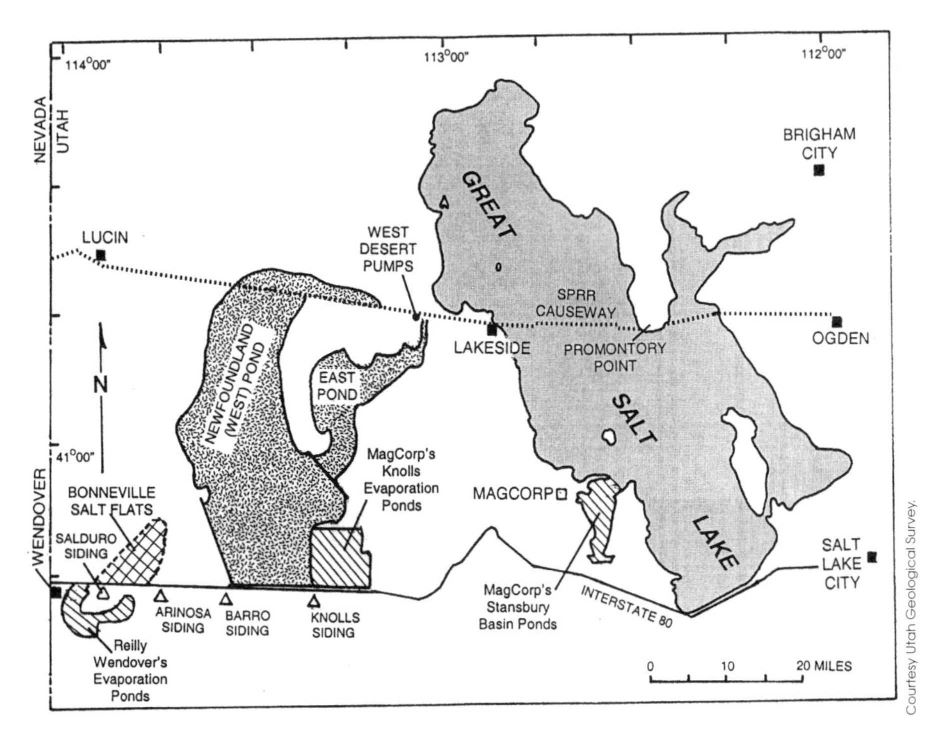

Figure 2. Components of the West Desert Pumping Project, MagCorp's Stansbury Basin and Knolls evaporation ponds, Reilly Wendover's evaporation ponds, sidings on the Union Pacific Railroad (paralleling Interstate 80), and other cultural features.

operated from April 1987 through June 1989 and then shut down.[13] Brine was available to AMAX until mid-1990, when the supply dried up.[14]

In 1989 AMAX sold the magnesium facility to Renco Inc., a privately held company in New York. The magnesium operation was renamed Magnesium Corporation of America (MagCorp). In the late 1980s and early 1990s, the level of Great Salt Lake receded, and by 1992 it had dropped to a level that allowed MagCorp to begin recommissioning the ponds in the Stansbury Basin. In 1995 the first brine harvest from the Stansbury Basin ponds was brought into the plant.[15]

The raw material used to manufacture magnesium metal is concentrated magnesium-chloride brine. Preparing the concentrated brine to feed to the electrolytic cells entails removing unwanted impurities and preventing further concentration. The product is then reduced to a gray $MgCl_2$ powder, melted in the reactor, and fed to the electrolytic cells. Here the molten magnesium chloride is separated into magnesium metal and chlorine gas. Besides magnesium metal (and magnesium alloys) and chlorine gas, other products include calcium chloride, ferrous chloride, and ferric chloride.[16]

In 1965 Lithium Corporation of America (Lithcoa), based in North Carolina, began investigating the production of lithium from Great Salt Lake (figure 1). When it became apparent it was not economical to extract lithium alone, the corporation joined with a German potash company, Salzdetfurth, which had expertise in producing potassium fertilizers (potash). From 1968 to 1998, the potash operation was known as Great Salt Lake Minerals and Chemicals Corporation or GSL. In April 1998 GSL was acquired by IMC Global and renamed IMC Kalium Ogden Corporation.[17] On 28 November 2001, IMC Global sold IMC Kalium Ogden Corporation to Apollo Management, LP. This new business entity is now known as Compass Minerals Group. As part of this transaction, IMC Kalium Ogden Corporation changed its name back to Great Salt Lake Minerals Corporation (GSL).[18]

By 1968 construction of a 12,000-acre solar-pond complex had begun, the production of lithium was placed on hold, and the production of potassium sulfate became the primary goal. To increase potash production, the solar-pond complex was first expanded during 1970 and 1971 to 19,000 acres; a second large expansion in 1991 brought the total pond area to 35,000 acres. A third expansion in 1998–99 increased the area to 40,000 acres. The main body of GSL's solar ponds and the processing plant are located between Little Mountain, about 15 miles west of Ogden, and Promontory Point.[19]

With ever-increasing world-market demands for potassium, GSL constructed solar ponds on the west side of the lake. The brines are transported to the east side using an eastward-sloping, 21-mile-long, underwater canal called the Behren's Trench to a pump station on Promontory Point.

GSL currently produces sulfate of potash (SOP) that is an excellent fertilizer. In addition to being shipped domestically, large quantities of SOP are also exported to the Pacific Rim. Magnesium chloride is also produced in both liquid and solid form. Anhydrous sodium sulfate, also known as salt cake, is another product that can be processed from the lake brine.

In the 1960s Hartley Anderson started selling Great Salt Lake water as a nutritional supplement. The business he started has evolved so that today Mineral Resources International (MRI) and its sister company, North Shore Limited Partnership (North Shore), create both liquid and powder mineral supplements from Great Salt Lake brine. North Shore produces concentrated brines in its 20-acre solar-evaporation complex at the north end of the lake (figure 1). MRI processes this brine into final nutritional supplements at its facilities just west of Ogden.[20]

Development of the Bonneville Salt Flats Potash Resources

The Bonneville Salt Flats, also a remnant of Pleistocene Lake Bonneville, is a salt-crust-covered portion of the Great Salt Lake Desert, located west of the lake in the northwestern portion of Utah (figure 3). The salt flats cover about 70 square miles, and the Great Salt Lake Desert about 4,000 square miles. The salt flats are normally dry during the summer months but are usually covered by water during the winter and early spring. Beneath the surface of both the salt flats and much of the desert, the

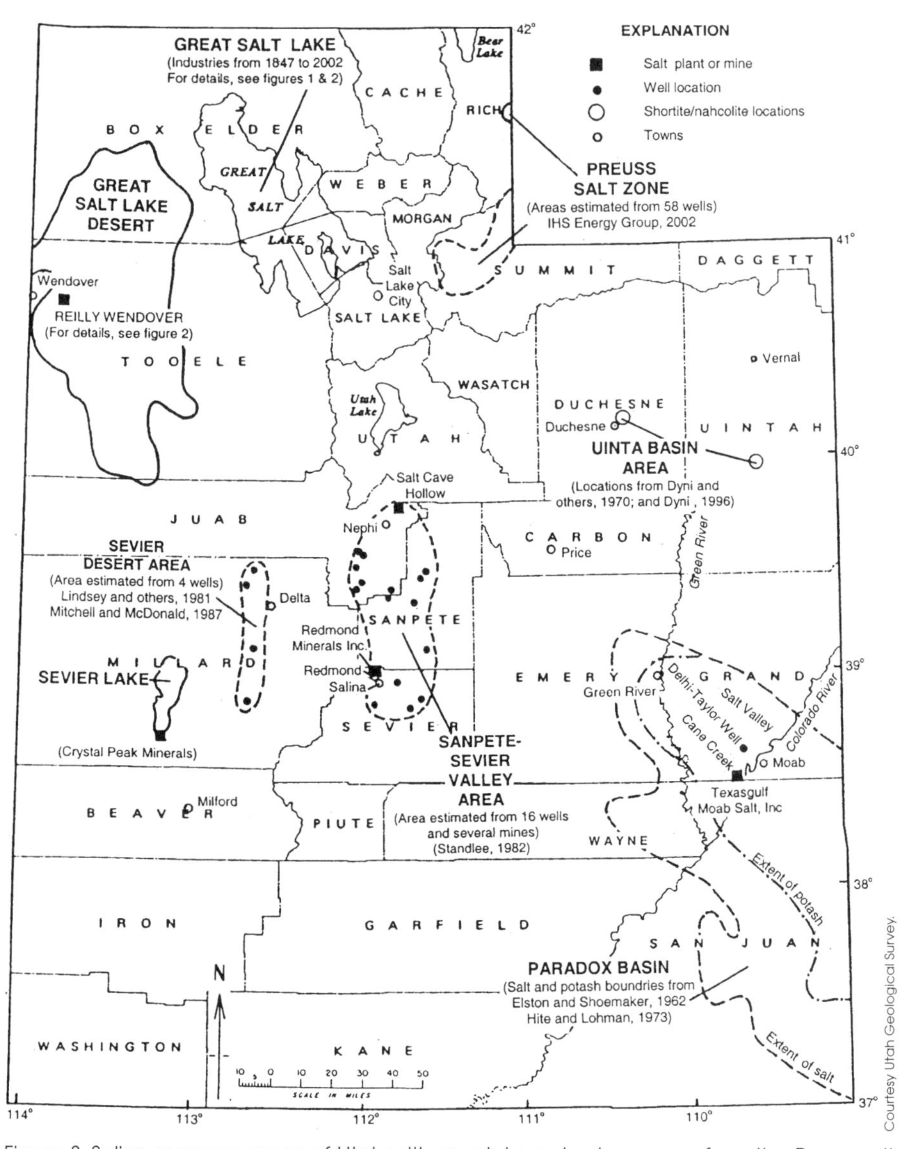

Courtesy Utah Geological Survey.

Figure 3. Saline resource areas of Utah with special emphasis on ones from the Preuss salt zone, Uinta Basin, Sevier Desert, Sanpete-Sevier Valley, Paradox Basin, and Sevier Lake areas. Detail for the Great Salt Lake and Bonneville Salt Flats appears in figures 1 and 2.

sediments are saturated with brines of varying salinities. Though the chemical composition of the brine varies from place to place, the following analysis (in dry-weight percentage) represents a typical brine breakdown: Na = 34.4, Mg = 1.6, K = 2.2, Ca = 0.6, Cl = 58.3, SO_4 = 2.9.

This section, originally published in Gwynn (1996)[21] and more recently in Gwynn (2002),[22] outlines the developmental history of the Bonneville Salt Flats from recognition of their salt and potash potential in 1906 or 1907 by Western Pacific Railroad engineers to the production of potash by Reilly Wendover today.

The first 10 to 12 miles of Interstate 80 east of Wendover, Utah, traverse the seemingly endless, flat, white, salt-covered expanse of the Bonneville Salt Flats, known in the early 1900s as the Salduro Salt Marsh. The salt flats and surrounding Great Salt Lake Desert are remnants of the bed of an ancient, large, cyclic lake, whose latest cycle, Lake Bonneville, occurred from about 32,000 to 14,000 years ago. Lake Bonneville was more than 1,000 feet deep and covered an area of 20,000 square miles in western Utah plus small portions of southern Idaho and eastern Nevada. Though the water of Lake Bonneville was relatively fresh, it contained small amounts of dissolved salt, including chlorides and sulfates of sodium, potassium, and magnesium. These dissolved salts precipitated on the surface of the Salduro Salt Marsh as the lake evaporated and are the source material of potash.

The Salduro Salt Marsh played an important role in the early development of a potash industry in the United States. Potash is mostly used as a fertilizer; the chemical industry utilizes lesser amounts. Prior to World War I (1914–18), Germany supplied nearly all of the growing demands for agricultural and industrial potash in the United States. The blockade of Germany by Great Britain during the war, however, cut these supplies off, and the U.S. had to find alternative sources of potash. By the end of the war, at least 128 American plants were producing potassium compounds from kelp, wood ashes, lake brines, alunite, cement dust, sugar-beet waste, blast-furnace dust, and other sources.

As early as 1906 or 1907, the engineers building the Western Pacific Railroad across the western Utah desert brought the existence of the salt beds of the Salduro Salt Marsh to national attention. The salt beds were soon covered by mining claims, and almost immediately the claim owners organized the Montello Salt Company, headquartered in Ogden. After several years of unprofitable attempts to produce salt, the claims were leased to the Capel Salt Company of Salt Lake City. Capel erected a small mill near the Salduro siding, about 10 miles east of Wendover, and produced and sold common salt for a short time.

In about 1916 Capel merged into or was transferred to the potash enterprise of the Solvay Process Company. During the war years, Solvay investigated many saline deposits and in 1917 began to extract potash from the subsurface brines of the Salduro Salt Marsh. The operation was constructed on the south side of the Western Pacific Railroad at the Salduro siding. There, in the center of the marsh salt-beds concentric, circular canals were dug into the salt and underlying muds. Salty water or brine flowed into these canals, where it concentrated through solar evaporation. The most concentrated brines were continually pumped inward, over the dikes separating the outer from the inner concentric canals. Potassium-bearing salts precipitated from the highly concentrated brine within the innermost canal. From there the salts were harvested and processed to produce potassium chloride.

Production of potassium chloride began in 1917, and at the end of 1918, Solvay transferred its interest in the potash operation to the Utah-Salduro Potash Company (USPC). By 1920 USPC was the largest single producer of potash in the United States. In 1921 the plant suddenly closed, mainly due to the fall in high wartime potash prices and the reorganization of the Solvay Process Company. After that time USPC restricted its operations to producing common salt.

In 1919 the Bonneville Potash Corporation (BPC), formed by J. L. Silsbee of Salt Lake City, erected a potash plant at Wendover, Utah, near the Utah/Nevada border (figure 2). From 1920 to 1936, BPC unsuccessfully attempted to produce potash commercially through the solar evaporation of brines. In 1936 a new operating company, Bonneville Limited, formed and built a new plant to recover potassium chloride by flotation from solar-precipitated salts. The first potash from this new plant was shipped in 1938. By 1939 Bonneville Limited was successfully producing potash and went on to become a significant, long-term potash supplier. Since that time, the operation has survived several ownership changes and now operates under the name of Reilly Wendover.

A third company, Chloride Products Incorporated (CPI), formed in 1921 by Frank Cook and a group of California capitalists, also attempted to produce potash from the brines of the Salduro Salt Marsh. CPI constructed canals, evaporation ponds, and a small processing plant near Arinosa, a few miles east of the USPC operation. CPI's developmental work is recorded through at least 1925; after that no further information about it is available.

The development of the potash industry on the Salduro Salt Marsh faced many challenges. The postwar decrease in potash prices made it increasingly difficult for the Utah companies to compete with other domestic and foreign suppliers. The surface conditions on the marsh were another critical factor. During spring the surface was normally covered with water, hindering development work, and wave action frequently destroyed dikes and filled the brine-collection ditches with sediment. Also heavy equipment frequently broke through the salt crust and sank into the underlying mud, necessitating the invention and use of special wide-metal wheels on the equipment. Hot, dry summers and cold winters, accentuated by the ever-present wind, made working conditions on the marsh unpleasant.

Early production of potash from the brines of the Salduro Salt Marsh by Solvay and the USPC played an important and sometimes singular role in supplying the United States with fertilizer during the latter part of World War I. In spite of domestic competition, international competition, and other economical, logistical, and climatological obstacles, the potash industry on the Bonneville Salt Flats survives.

Today Reilly Wendover, the potash industry's lone survivor on Utah's western desert, is an important contributor to potash production in the United States and the economic base of both Tooele County and the State of Utah.

Salt Mining in Central Utah

Salt deposits underlie the Sanpete-Sevier Valley area and probably account for the many local, salt-related names such as Salt Creek, Little Salt Creek, Salt Spring Creek, and Salina.[23] The salt is contained in the Jurassic Arapien shale, which consists of calcareous mudstone, siltstone, sandstone, limestone, and evaporates, including halite and gypsum. The Arapien shale is complexly deformed and shows signs of intense compression. Its thickness is uncertain because of the deformation but is estimated to range from about 4,000 to as much as 13,000 feet.[24]

The journal of the Escalante-Dominguez expedition into Utah in 1776 mentions the salt deposits in the mountains of Nephi, including a comment that the Indians mined their salt from that location. It was not until the Mormon pioneers moved to the area that the deposits were commercially exploited, however. Timothy B. Foote was one of the first white men to mine the salt deposits. In 1854 he built a toll road to his salt works and charged 25 cents for every wagonload of lumber or wood hauled over it. Four years later David Salisbury, Richard Jenkins, and Thomas Booth began mining salt from a cave they located about seven miles north and east of Nephi, probably in Salt Cave Hollow (figure 3). The trio was able to crush, boil, and dry about 500 pounds of salt each day, selling it for six cents a pound. Because of Indian trouble in the area, the plant moved to Nephi in 1862.

After the Eureka mining district opened, thousands of tons of salt were mined and hauled in horse- or mule-drawn wagons to the Shoebridge mill to use in refining ore. Livestock consumed most of the rock salt from Nephi; however, small amounts were boiled down and sold as table salt. Salt was mined intermittently from Nephi until about 1942, but there are no reports of activity at this site after 1943. In fact, any information about salt production from Nephi after 1897 is rare.

A corporation made up of Nephi residents organized the Nebo Salt Manufacturing Company in 1892 with the intention of producing a pure salt from the waters of a spring in Salt Creek Canyon.[25] The company brought water 350 feet from the spring to a 20-by24-foot building constructed to house the boilers, evaporators, and other salt-making machines. The brine was converted into table salt, dairy salt, and packing salt. Analysis of a random sample showed the salt to be 99.172 percent pure NaCl. Nebo Salt operated for about four years before Inter-Mountain Salt Company purchased it in 1897 to eliminate competition.

Like the Nephi area, salt in the Redmond area is concentrated in the Jurassic Arapien shale. Where the Arapien salt is mined near Redmond (figure 3), it appears as near-vertical, much-contorted beds interleaved with reddish brown, calcareous mudstones.[26]

In the late 1800s, the first settlers in Redmond noticed that the Indians camped north of town for a few days at least a couple of times each year. One time after the Native Americans had left, a couple of the settlers rode to the location and found that the Indians had been chipping salt from an outcrop. During the 1870s pioneer locators of the salt formation took a stock interest in the Sevier Valley Salt Company and produced salt for more than 30 years. The Gunnison Valley Salt Company acquired the property in 1909 and mined salt until 1926, when it was sold to the Great Western Salt Company. Six or eight other companies or individuals produced salt from the formation east of Redmond from the 1870s through the 1890s, including I.N. Parker.[27]

In the late 1800s, William P. Poulson, who had worked in the mines in Denmark, came to America. He eventually found his way west, where he worked for a time in the Park City silver mines. Poulson heard of the salt deposits near Redmond, married, and moved to the area. He opened a salt pit east of town near the I.N. Parker claim and formed the Great Western Salt Company. The salt was poor quality, and he soon abandoned the operation. In 1902, after a great deal of prospecting, Poulson opened a new mine on the west side of Arapien Valley with his sons—Francis, John, Milo, and Albert—in a location containing a higher grade of salt. He retained the Great Western Salt Company name. Poulson's sons later bought him out and changed the name to Poulson Brothers Salt Company.

Albert Poulson left his brothers in 1920 and operated the Inland Crystal Salt Company mine (exact location unknown), owned by the Church of Jesus Christ of Latter-day Saints, under his name on a royalty basis. He also purchased adjoining property from L. Jacob of Salina. Inland Crystal sold its mine to Royal Crystal, which in turn sold it to Morton. In 1952 Albert Poulson purchased the mine and operated it under the name of Albert Poulson and Sons until his death in 1980. After Albert's death, his son Willis sold the mine to LaMar and Milo Bosshardt.

John Poulson bought out the remaining two brothers of the Poulson Brothers Salt Company and continued the business with his sons: Blaine, Jewel, and Wallace. The sons later acquired the company from their father and managed it until 1965, when they sold it to Redmond Clay and Salt Company, owned by LaMar and Milo Bosshardt.

Early methods of surface or open-pit mining used a hand drill and a double jack or sledgehammer. The *churn drill,* a long steel rod with a bit on the end, came somewhat later. To operate the churn drill, a person sat down, put the drill between his knees, and began to churn (drop, lift, twist, and drop, etc.). Water was periodically poured down the deepening hole, and when a soupy mixture of cuttings and water formed, it was scooped out with a long spoon. Augers, first turned by hand, were later mechanized. An air-driven hammer, operated by a man sitting on a board pushing the reciprocating drill into the salt with his feet, was really an improvement. The air hammers were later mounted on jacklegs that supported the hammer, then on track-mounted devices. Now large hydraulic drills can auger out a 10-foot hole in three minutes. Early blasting was done with black powder, followed by dynamite, and now ammonium nitrate.

Horse-drawn wagons were first used to haul the rock salt. Chunks of salt too large to lift were rolled up boards and into the wagons. Each wagon hauled about two tons. Ten-wheel trucks replaced wagons, and now semitrucks do the hauling. For many years the salt was mined by open-pit methods, but since the early 1970s, underground mining techniques have replaced them. Large modern equipment, front-end loaders, and hauling trucks have replaced smaller pieces of equipment.

The mined salt was hauled to the mill, where it was crushed and passed through coarse, medium, and fine mesh screens. In some cases the salt was mixed with various chemicals (probably mineral supplements) to give the customer a choice of grades. By about 1968 the two mines in Redmond were producing about 10,000 tons of salt annually, which was shipped as far away as Arizona, Nebraska, and Canada. Most of the salt was used for road deicing with small amounts for livestock.

In 1959 LaMar and Milo Bosshardt went into the salt business under the partnership of Redmond Clay and Salt Company. By 1965 they had installed the first mixer, the first dryer, and the first sewing machine for paper bags in Sevier County. They also purchased the Poulson Brothers operation the same year. In 1966 the Bosshardts installed equipment to make a fourth screen size of salt. In 1968 they built a mill at the mine so all the salt crushing and loading of bulk salt could be done there rather than hauling it to town. By 1969 they became the first local salt company to deliver with a semitruck.

In 1976 the Bosshardts formed American Orsa Inc. as the food division of Redmond Clay and Salt to sell table salt and food-grade clay under the names Orsa Salt and Orsa Clay. In 1980 they purchased Albert Poulson and Sons to become the only operating salt company in the valley for the first time this century. In 1988 the name Orsa Salt was changed to RealSalt, and Orsa Clay became Redmond Clay.

In 1990 the Bosshardts built a new warehouse at the salt mine but continued to operate the old warehouse in town until 1993, when full operation moved to the mine. In 1997 the food division was absorbed into the parent company, and the company name changed to Redmond Minerals Inc. In 1999 the ownership of Redmond Minerals transferred from the Bosshardt family to Rhett Roberts, the present CEO.

The salt and potash deposits of the Paradox Basin of southeastern Utah lie in the Pennsylvanian Paradox formation of the Hermosa group. The area underlain by salt includes parts of Emery, Wayne, Garfield, Grand, and San Juan Counties in Utah (figure 3) and Montrose, San Miguel, Dolores, and Montezuma Counties in Colorado. The area underlain by potash is somewhat smaller, located in the northeastern portion of the basin, and includes parts of Emery, Grand, and San Juan Counties in Utah and the previously named counties in Colorado. The salt and potash in the Paradox Basin were deposited in a series of evaporite cycles. At least 29 salt beds or cycles are recognized, separated from each other by marker beds consisting of variable combinations of anhydrite, shale, mudstone, siltstone, and dolomite. Not all areas of the potash zone are underlain by all 29 salt beds. Eighteen of these cycles are known to contain potash. The thickest sequences of salt occur in 14 northwest-trending, salt-cored anticlines within the northeastern, potash-bearing portion of the basin.[28] In the Cane Creek area, salt 1 (the top cycle) is absent, and the potash section mined is zone 5 (K5), near the top of salt 5.[29]

The existence of potash in the Paradox Basin has been known for many years. The Crescent Eagle well, drilled in Salt Valley in 1925, reported significant signs of potash. In the 1940s Salt Valley experienced active exploration for potash. However, reports at the time indicated that, although potash was present, development should not be undertaken. In 1952 the Delhi-Taylor Oil Company started drilling in the Seven Mile area, about eight miles northwest of Moab and subsequently outlined a large ore (potash) deposit (figure 3). The Cane Creek deposit was recognized in 1956, and tests were so favorable that the Delhi-Taylor Company drilled eight test holes. Texas Gulf Sulphur Company (TGS) optioned the property on 15 April 1960 and, during the next four months, completed nine confirmation and development wells. The drilling program outlined a large, high-grade deposit of sylvinite ore (a mixture of halite and sylvite). Additional pilot holes were drilled during 1960.[30]

On 23 February 1961, TGS began construction of the Cane Creek Mine's head frame.[31] The other surface facilities were constructed from 1961 through 1963. TGS started construction of the underground portion of the Cane Creek Mine in 1962 (figure 3). The potash zone lay 3,000 feet below the land surface. Two parallel shafts were sunk to a depth of 2,789 feet, and then an incline was driven nearly 4,000 feet into potash bed K5.[32] On 13 January 1965, a boxcar was filled with the first load of potash from the Cane Creek Mine and delivered to market on a 36-mile-long Denver and Rio Grande Western rail spur built for the plant. TGS announced that the average thickness of the developed Cane Creek potash ore body was 11 feet, and the grade was between 25 and 30 percent K_2O.[33] The operation produced 439,999 tons in 1969, and by 1970 TGS had cut some 340 miles of headings in a room-and-pillar configuration.

TGS was plagued with problems from the start. The mine contained explosive methane, it was hot, and instead of being level and flat, the ore layer was distorted into undulating sections.[34] On 28 August 1963, during the development of the incline, a methane explosion killed 18 workers.[35] Because of the problems associated with underground mining, management started discussing the possibility of converting to solution mining. Extensive research produced a full report in August 1969, which convinced management that solution mining was, in fact, a practical and economic plan. The board of directors approved the project in July 1970, and the unique program was launched.[36] The facility was converted to solution mining between 1970 and 1972 by abandoning the mine, drilling wells into the mine from the surface, and constructing 420 acres of solar-evaporation ponds. The production record for the solution mine was set in 1974 at 261,002 tons of potash. In 1972 TGS became known as Texasgulf, Incorporated.

To operate the solution mine, water is pumped from the Colorado River, which flows adjacent to the plant. The river water runs across the surface salt tailings to assist reclamation before it is injected into the mine as brine. The brine flows through the mine, dissolving potash from the original pillars and walls and salt from formations above the potash. The brine is then pumped back to the surface and into one of 23 evaporation ponds. Solar evaporation leaves an 8- to 12-inch layer of sylvinite crystals in each pond. Each pond is harvested once per year during the period from September through May. The ponds are not harvested during the summer months to maximize evaporation.

The sylvinite is harvested from the ponds with 25 two-ton scraper loaders. How deep the blades of the scrapers cut is controlled by laser to protect the Hypalon pond liners. The harvested sylvinite is pumped as slurry 3.5 miles to the plant, where the sylvite (KCl) is separated from the salt (NaCl) by flotation. The refined sylvite and the salt are then dried, screened, and stored in large warehouses. Each of the two warehouses has a capacity of about 100,000 tons.

Both standard and granular potash are produced. The standard product reaches industrial and animal-feed markets in the Intermountain area, while the granular goes into agricultural products in the Intermountain region and on the West Coast.

Three grades of salt are produced: fine, medium, and coarse. The fine salt is used primarily for animal feed and hide processing. The medium is sold primarily as feedstock for chlor-alkali plants and other chemical products. Coarse salt is used primarily for water softening and ice control.

A high-capacity bagging and palletizing operation packages about 50,000 tons per year of both salt and potash. Shipments from the facility are via the Southern Pacific Railroad and truck. The plant (located in sec. 24, T. 26 S., R. 20 E., SLBM) is just 30 miles south of Interstate 70, which provides highway access to both the West Coast and the Midwest.

Texasgulf was the original owner of the mine. Moab Salt, Inc., was formed in January 1988 as a joint-venture partner with Texasgulf to market salt while Texasgulf continued to market the potash. Texasgulf bought out Moab Salt in January 1990 to become sole owner once again. Texasgulf was sold to the Potash Corporation of Saskatchewan (PCS) in April 1995, and PCS sold this operation to Intrepid Mining LLC in Denver in February 2000.

Other Saline Resources in the State

Sevier Lake

Sevier Lake, or playa located in central Millard County (figure 3), is a remnant of Pleistocene Lake Bonneville. Though its surface is sometimes covered by water during the spring runoff, most of the year it is dry. Just below the surface, however, muds remain saturated with brine to considerable depth. The 200-square-mile lake lies at the terminus of the Sevier River. The river and its many tributaries drain an area of about 16,200 square miles. A typical analysis of the subsurface brine (in dry-weight percentage) is as follows: Na = 23.4, Mg = 1.9, K = 1.7, Ca = 0.2, Cl = 58.7, SO_4 = 14.1.

Beginning in 1978, Crystal Peak Minerals Corporation started a multiphase, saline-resource-assessment program at Sevier Lake. The program included collecting and multiparameter analyzing brine and sediment samples, determining brine-salt-phase chemical relationships, conducting a gravity survey, characterizing climatological patterns, and determining local hydrologic conditions.[37]

Encouraged by the findings of the assessment program, Crystal Peak moved into resource development. This work included constructing dikes to isolate and protect a 3,000-acre solar-evaporation pond complex and excavating a six- to seven-mile-long, north-trending, brine-collection canal. Roads, a load-out facility, and an operations center were constructed, and salt floors were deposited in the evaporation ponds. Work started on a salt-processing plant and associated facilities. While Crystal Peak planned initially to produce only sodium chloride, future plans included processing potassium sulfate and other potentially profitable saline products once the operation was financially stable. In about 1991 funding for the project terminated, and the project was abandoned.

Preuss Salt Zone in the Wyoming/Utah/Idaho Thrust Belt

A thick salt zone lies at the base of the Jurassic Preuss sandstone throughout much of the Wyoming/Utah/Idaho thrust belt (figure 3).[38] Salt in this formation was mined in the 1920s from outcrops in Crow and Stump Creeks, Idaho,[39] and the extent of the salt was delineated during oil-well drilling in the area.[40] The deposit covers an area of 10,000 to 12,000 square miles, extending 180 miles north and south from Jackson, Wyoming, to the Uinta Mountains of Utah and 60 miles east and west from the Tip Top and LaBarge oil fields in Sublette County, Wyoming, to just west of Bear Lake in northeastern Utah and southeastern Idaho. It is an area of major north-south-trending thrust faults, with eastward displacements of the thrust sheets totaling possibly 60 miles. Folding associated with and generally parallel to the strike of the faults has created traps for many of the oil and gas fields of the area.[41]

The depth to the top of the Preuss salt in Utah varies from a minimum of 38 feet to a maximum of more than 5,000 feet. Maher describes the salt zone in the Pineview area as consisting "primarily of salt and anhydrite with light gray to gray, moderately soft, micaceous shale."[42] The exact extent and distribution of the Preuss salt zone beyond the wells that have penetrated it are not known, and no commercial development of the salt resource within the state has occurred.

Sevier Desert Area

Salt (age and stratigraphic unit uncertain) has been encountered in at least four wells drilled within the Sevier Desert of western Utah in a north-south-trending deposit under the town of Delta (figure 3). The shallowest evaporites encountered to date are at a depth of 2,550 feet in the Argonaut Energy Number-One Federal well (sec. 23, T. 15 S, R. 7 W., SLBM), and the deepest lie at about 8,790 feet in the ARCO Oil and Gas Company Number-One Pavant Butte well (sec. 35, T. 19 S., R. 7 W. SLBM). The salt is 5,152 feet thick in the Argonaut Energy well.[43] No development of the salt has taken place, but the area has been considered for a gas-storage site.

Uinta Basin Area

Saline minerals are found in the Eocene Green River and Uinta formations of the Uinta Basin in eastern Utah (Duchesne and Uintah Counties), mainly within the saline facies of the Uinta formation. The main salt minerals are shortite [$Na_2Ca_2(CO_3)_3$] and nahcolite [$NaHCO_3$]. The saline minerals within the Uinta Basin are not being developed (circles on figure 3). They are, however, valuable in increasing understanding of the complex nature of the Green River formation and the conditions under which it developed.[44]

Oil-Well Brines

The Uinta Basin (Duchesne and Uintah Counties) and the Paradox Basin (Grand and San Juan Counties) produce most of the state's hydrocarbons (figure 3). Sodium chloride brine is often a by-product of oil and gas production. Magnesium, potassium, and sulfate are minor constituents in most cases. The brines with the highest total dissolved solids occur in the Pennsylvanian and Mississippian reservoirs of the Paradox Basin. With few exceptions, these brines are little used, and their proper disposal becomes a costly liability to the producers. An evaporation test on brines from the Lisbon oil field (T. 30 S., R. 24 E., SLBM) by J.G. Gwynn in1989 revealed that potentially valuable products such as sodium, potassium, and magnesium chlorides can be produced from them. [45] Commercial production of salts or other products from these brines requires economic and environmental evaluations of the extraction process and a market study of the final salts and/or concentrated brines.

6

Coal Industry

Allan Kent Powell

Throughout the history of Utah, coal has been a critical resource for the development of the state's economy. Initially coal was used to heat homes and buildings, as fuel for the steam locomotives, and when the coal was of high enough quality to produce coke, as fuel for smelting metals mined from Utah's mountains. Later, when natural gas replaced ash- and soot-producing coal as a source of heat, railroads converted to diesel fuel to power their locomotives, and the principal market for Utah coal became local coal-fired power plants and markets outside the United States.

In addition to water and land, coal was one resource that Brigham Young and other early leaders considered essential to develop a viable and self-sufficient economy in the Mormon homeland. Considerable effort was expended to open mines as part of the Iron Mission when coal was discovered near Parowan and Cedar City in 1851.[1] The attempt proved unsuccessful, but in 1854 a new source of coal in the Sanpitch Mountains on the western slope of Sanpete Valley, about 120 miles south of Salt Lake City, offered renewed hope. Brigham Young visited the outcrops in 1855, and several wagonloads of coal were shipped to Salt Lake City that year. Several families of Welsh coal miners were sent to mine the coal, and the name of the settlement was changed from Coalbed to Wales. However, Wales did not become the mining center that its name promised.

Closer to Salt Lake City, outcrops of coal were found on Chalk Creek in 1859, and mining operations began. The settlement of Chalk Creek was renamed Coalville in hopes that this location 45 miles east of the Salt Lake Valley would become a regional coal center. Several coal mines opened in the area, and the completion of the Union Pacific Railroad through Echo Canyon and down the Weber River in 1869 offered a much easier transportation system than horse-drawn wagons. While coal continued to be mined in the Coalville area until the midtwentieth century, it was Carbon County, not Coalville, that became the new center for Utah's coal-mining industry.

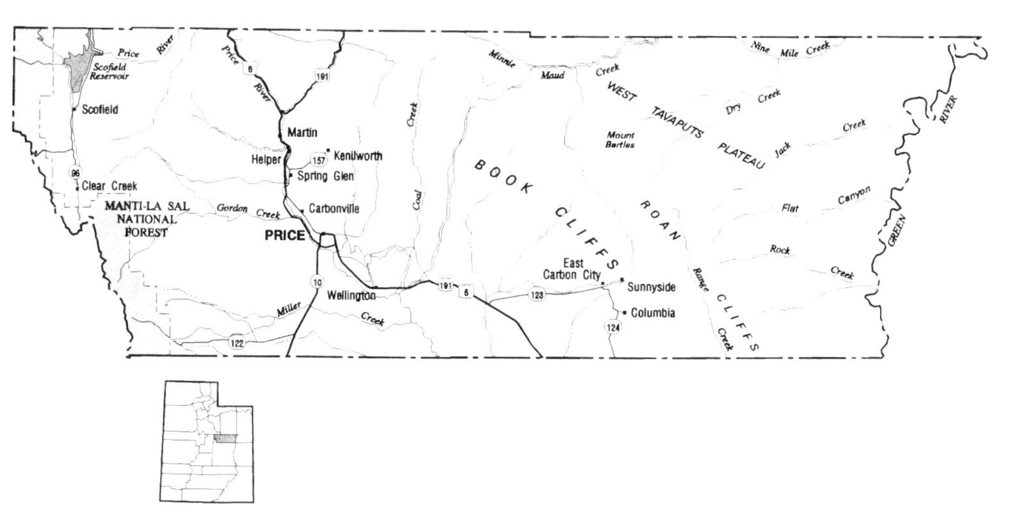

Carbon County has been the center of Utah's coal industry for more than a century.

Carbon County

Carbon County and coal are synonymous and have been ever since the county was created in 1894. The 27th of Utah's 29 counties established and the last before statehood in 1896, Carbon County was carved out of the northern end of Emery County. Castle Valley, which encompasses both Carbon and Emery Counties, was first settled in the mid-1870s as settlers from Sanpete Valley to the west and Utah Valley to the northwest crossed the coal-laden Wasatch Plateau to establish farms along the streams flowing through the valley on their way to the Green River—Price River, Huntington River, Cottonwood Creek, and Ferron Creek. However, with the completion of the Denver and Rio Grande Western Railroad through the northern end of Castle Valley in 1883 and the expansion of coal-mining operations along its route, the towns of Price and Helper became economic centers for the region, while the communities of Huntington, Castle Dale, and Ferron to the south languished as agricultural villages tucked between the mountainous Wasatch Plateau to the west and the expansive, largely unexplored wilderness of the San Rafael Swell to the east. Residents and businessmen along the Price River felt that their interests were not being served and tired of the 30- or 40-mile wagon or horseback ride from Price and Helper to the county seat in Castle Dale. Initially they worked to have the county seat relocated to Price, but when their efforts failed, petitions were circulated to establish a separate county. The Utah territorial legislature passed a law, and on 4 March 1894, territorial governor Caleb B. West signed the bill creating Carbon County.[2]

At the same time as settlers moved across the mountains to establish farms along the Castle Valley streams, fledgling coal companies organized to pursue opening recently discovered coalfields in the high mountain valleys. The Fairview Coal Mining and Coke Company was organized in May 1874 and opened a mine in Coal Canyon near the head of Huntington Canyon. The company also built 11 coke ovens and named the fledgling mining camp Connellsville after the large Pennsylvania coking center that the Utah venture hoped to emulate. Coking efforts were carried out for about three years. Wagons hauled coke to Springville, but transportation costs and the questionable quality of the coke doomed Utah's Connellsville to a short existence.

Eastward nine miles from Connellsville in Pleasant Valley—another high mountain valley of the Wasatch Plateau—the first of several major Carbon County coal-mining locations opened when the Pleasant Valley Coal Company, a local business headed by Miland O. Packard of Springville, began mining coal from its Winter Quarters properties in 1877. That same year the Pleasant Valley Railroad Company was organized and began construction of a narrow-gauge line that was completed from Springville to Winter Quarters in October 1879.

The decision to connect Salt Lake City and Denver with a railroad line running through western Colorado and eastern Utah had a major impact on the development of coal in eastern Utah. As Denver and Rio Grande Western crews surveyed the route along the northern rim of Castle Valley and up Price Canyon to Soldier Summit, a company geologist located rich coal seams in what would become Sunnyside in the northeast corner of Castle Valley and at Castle Gate, a few miles up Price Canyon. In 1882 the railroad company began to acquire the Pleasant Valley Coal Company and Pleasant Valley Railroad.

The arrival of the Denver and Rio Grande Western ended the monopoly that the Union Pacific had held on the Utah coal trade since the completion of the transcontinental railroad in 1869.[3] The Union Pacific attempted to meet the competition head on through the Utah Central Coal Company, which opened a mine in Pleasant Valley in 1882. The mine operated until 1884, when the mine tipple caught fire and the fire spread to the mine itself, killing John McLean and his son. The mine was permanently sealed after attempts to extinguish the fire were unsuccessful. A second mine opened in 1885, and in 1890 the Union Pacific Coal Company took control of the former Utah Central coal mines. The Union Pacific had planned to construct a railroad to its mining operations in Pleasant Valley; however, the line was never built, and the excessive freight rates the rival Denver and Rio Grande Western line charged the Union Pacific to haul coal forced the railroad to cease coal mining in eastern Utah in 1897.

Once under the control of the Denver and Rio Grande Western, the Pleasant Valley Coal Company expanded its Winter Quarters operations near Scofield and opened a mine at Castle Gate in 1888. Several coke ovens were built at Castle Gate in 1889. They proved successful, and by 1896 104 coke ovens had been constructed at Castle Gate, and coke production totaled 20,448 tons that year.[4] By 1901 an additional 100 coke ovens were operating at Castle Gate, and production amounted to approximately 5,000 tons a month.[5]

In the meantime, the Denver and Rio Grande Western organized a second coal company—the Utah Fuel Company—to open a mine at Clear Creek, four miles south of Scofield, and began developing the rich Sunnyside coal deposits. A company town was established at Sunnyside, and subsidiary companies—the Magnolia Trading

Company and the Wasatch Store Company—were organized to manage the saloon and mercantile trade.[6] At Sunnyside 200 coke ovens were constructed in 1900, and by 1907 there were an additional 450 coke ovens, increasing production from 50,620 tons in 1900 to 324,692 tons in 1907.

The Panic of 1907 impacted coal and coke production significantly. Carbon County coal production declined from 1,816,133 tons in 1907 to 1,606,853 tons in 1908, while coke production dropped to only 134,195 tons in 1908, a reduction of 190,500 tons from the previous year.

Beginning in 1904, Utah Fuel became involved in a series of legal battles and suits over the company's illegal acquisition of public coal lands. Agents of the company hired individuals to file on desirable coal property as agricultural land and then transfer title to the coal company. The litigation lasted for five years, and in the end Utah Fuel was found guilty, fined, required to pay for coal extracted from land fraudulently acquired, and forced to return some of the property and forfeit entry fees it had paid for the lands.[7]

The coal fraud cases against Utah Fuel and the promise of lucrative markets in Salt Lake City and other parts of Utah encouraged independent coal companies to organize and open mines in locations within Carbon County outside of the Pleasant Valley, Castle Gate, and Sunnyside areas controlled by Utah Fuel. The first of these independents was the Independent Coal and Coke Company, which opened a mine at Kenilworth a few miles northeast of Helper in 1906. Although the company built a railroad line to its mine, it was still dependent on the Denver and Rio Grande Western to haul coal to the Salt Lake market.

Two other companies, the Consolidated Fuel Company and the Castle Valley Fuel Company, opened another major area about 15 miles southwest of Price at Hiawatha, Black Hawk, and Mohrland beginning in 1908.

Mining began in the Spring Canyon area just west of Helper after 1912 when Jesse Knight, a Mormon businessman from Provo and mine developer in the Tintic Mining District, acquired 1,600 acres of land in the canyon at what became known as Storrs. A second mine in Spring Canyon, the Standardville, developed by the Standard Coal Company, began mining coal in 1913. Three other mines opened in Spring Canyon: the Latuda Mine, owned by the Liberty Fuel Company, in 1914, and mines at Peerless and Rains in 1916.

Also in 1916 Ogden businessman W.H.Wattis opened the Wattis Mine, located a few miles north of Hiawatha. Across the valley, the Columbia Mine opened three miles southeast of Sunnyside in 1922 and became an important factor in the production of cast iron at the Columbia Steel Company's plant at Irontown, just south of Provo in Utah County.

The last important coal-mining area opened in Carbon County was Gordon Creek, located south of Spring Canyon and approximately halfway between the county's two principal towns—Price and Helper. Work began in 1923 when the Sweet Coal Company, National Coal Company, and Consumers Mutual Coal Company developed a series of mines.

The rapid expansion of mining in eastern Utah during the 1890s made it the state's largest coal-producing area, led to the creation of Carbon County in 1894, saw the establishment of coal-company towns around the mines, and brought the first of several waves of immigrant miners to Carbon County when Finnish miners moved into the Scofield-Winter Quarters-Clear Creek area and Italian miners were recruited for the mines at Castle Gate and Sunnyside. They worked side by side with English and Welsh miners. There were also American-born miners, some of whom had farms in the surrounding valleys.

Such a group of miners entered the Winter Quarters Mine on 1 May 1900, when an explosion ripped through the mine killing 200 miners. The Winter Quarters or Scofield Mine disaster was, in terms of the number of miners killed, the most tragic coal-mine disaster in the United States to that time.[8] The explosion received national and international attention and raised the question, "What can be done to make coal mines safer for the miners?" One practice immediately adopted was already in place in the Castle Gate mines—to shoot off explosive charges only when the miners were outside the mine. Many contended that the unsafe practices of taking in excessive amounts of blasting powder and discharging the explosives while the men were inside the mine that had led to the explosion. Later, in 1924, the Castle Gate Mine was the site of an explosion nearly as disastrous as the one at Winter Quarters. One hundred seventy-two miners were killed at Castle Gate when inspectors unknowingly touched off a pocket of methane gas that ignited an excessive amount of coal dust inside the mine. Fifty Greek miners died in the explosion.[9]

While these two disasters still rank among the most tragic in American coal mining history, hundreds of other coal miners died from falling rocks and other accidents.[10] J. Eldon Dorman, a young coal-company doctor at the coal camp named Consumers, recalled the tensions that were typical throughout Utah and other coal camps:

> I had a mine telephone in my office and one at home. If an accident happened, I was frequently requested to meet the more seriously injured at the portal. It was quite an uphill walk, so I started driving my car. No road existed, but I could drive almost to the mine entrance. The entrance stood in view of the entire town, however, and every woman in camp recognized my car. Soon, each was standing on her front or back porch wringing her hands on her apron and wondering if her husband or son was hurt or killed. I only drove up there twice. The ever-present fear of death in mine accidents, explosions or afterdamp already haunted the families enough.[11]

Immigrant Populations

Carbon County coal miners came from a variety of backgrounds. Some were Mormon farmers from Castle, Sanpete, and Utah valleys. They traveled to the mining camps

in the fall after harvesting crops and took up residence in boardinghouses, tents, and the cabins of relatives or townspeople who had permanent homes in the coal camps. In the spring and summer when coal production fell off, they returned to their farms and another season of agricultural work. Some came as Mormon converts from the coal mines of England and Wales, where they had entered the mines as young boys and become skilled miners before their departure for Utah. Others came from older coal-mining areas of the United States in the East, Midwest, and such western states as Colorado, Wyoming, and Montana.

However, it was the immigrants from Finland, Italy, Greece, Austria, the Balkans, China, Japan, and other nations that brought a unique character to the Carbon County coal mines and towns, creating a legacy of diversity long before Americans at large began to celebrate its virtues. The first of these groups was the Chinese. They arrived in the Scofield area during the 1880s and helped open the Clear Creek mines. Their excellent pick work became legendary. However, as the flames of anti-Chinese sentiment were fanned by the Knights of Labor and other groups and individuals, the Chinese were forced out of the coal mines and moved on to other locations.

During the 1890s Finnish miners arrived at Winter Quarters, Clear Creek, and Scofield, where they became an important part of the workforce. They built amusement halls and a public sauna, and they adapted well to the long winters and short days of the high, mountainous Pleasant Valley. The size of the Finnish population is reflected in the tragic statistic that of the 200 miners killed by the Winter Quarters explosion, 62 were Finnish. Among these were the six sons and three grandsons of 70-year-old Abe Louma, who had traveled with his 65-year-old wife from Finland to join their seven sons and their families in Scofield just three months before the disaster.[12] In time most of the Finns left, though some stayed, and their descendants are still among the handful of year-round residents of Pleasant Valley.

While the Finnish immigrants were attracted to the Scofield-Winter Quarters-Clear Creek communities of Pleasant Valley, Italian immigrants found more employment opportunities in the Castle Gate and Sunnyside districts. By the mid-1890s, a few years after the Castle Gate Mine opened in 1888, Italians were recruited by labor agents in the East and made their way on the Denver and Rio Grande Western to Castle Gate. When the Sunnyside Mine opened in 1900, Carbon County was already a well-known destination for Italian immigrants traveling to the West.[13] By the end of 1903, Castle Gate was known as an Italian mine because the majority of the nearly 500 miners there were Italian. At the other camps, they were a substantial minority, as reported by the *Eastern Utah Advocate* in the following statistics: "At Sunnyside there are 358 English, 246 Italians, 222 Austrians; at Clear Creek, 128 Finns, 172 Italians, 95 English speaking; at Winter Quarters, 181 English speaking, 126 Finns, 74 Italians and a few others."[14] The Italians tended to divide into two groups—northern and southern, even establishing and joining fraternal organizations based on this split.[15]

If Castle Gate was the portal for Italian immigrants to enter Carbon County, Sunnyside served the same purpose for Austrian miners. While there were some

German-speaking coal miners in the county, the majority of those called "Austrian" were in reality Serbian, Croatian, and Slovenian, immigrants from lands that were part of the huge Austro-Hungarian Empire, which extended south from German-speaking Austria into the Balkans. Like the Italians who had come a few years before them and the Greeks who came a few years later, the Austrians were primarily rural peasants.[16] They did not remain only in Sunnyside but moved into other camps within the county and in the 1920s and 1930s made up a substantial part of the population in Spring Canyon and Gordon Creek.

By 1905 Carbon County caught the attention of another group of European immigrants, the Greeks, who began a large-scale exodus from their homeland after the currant crop failed in 1907, and labor and steamship agents began recruiting young Greek workers with promises of wealth and riches in the promised land of America. Leonidas G. Skliris, a Greek immigrant from Sparta who established headquarters at 507 West 200 South in Salt Lake City, was responsible for many Greeks coming to Utah. As a padrone or labor agent for the Denver and Rio Grande Western and Western Pacific Railroads, the Utah Copper Company, and coal mines in Carbon County, Skliris recruited Greek workers, receiving commissions from steamship companies for fares purchased by the immigrants in Greece. After their arrival, immigrants were obligated to pay Skliris for the jobs he had obtained for them.[17]

Greek immigrants constituted the largest of the ethnic groups in Carbon County. Their arrival in the county followed in the aftermath of the 1903–4 coal strike and as new coal mines opened, beginning with Kenilworth in 1906. Greek coffee houses sprang up in Price, Helper, and many of the coal-mining camps to provide a social center for the new immigrants.

Coming first as young men intent upon working for a few years to earn money for their sisters' dowry, to pay off family debts, and return to Greece to purchase farms or acquire small businesses, some of the immigrants sent to Greece for wives or "picture brides" and began to establish families and sink roots into a new homeland. The dedication of the Greek Orthodox Church of the Assumption in Price on 15 August 1916, attended by Greek miners from all over the county, forged a permanent bond between their new home in America and the Greek homeland.[18]

The story of Japanese coal miners in the area is less known. They came first to fill jobs on railroad gangs. Many were recruited in California by Edward Daigoro Hashimoto, who established headquarters in 1902 at 163 West South Temple in Salt Lake City's Japanese Town.[19] Japanese boardinghouses and amusement halls were built in Sunnyside, Kenilworth, and other coal camps in the county.

The operation and expansion of Carbon County's coal industry required an army of miners recruited among the longtime residents of the area and from the ranks of newly arrived immigrants from around the world. Following the practice in other mining and industrial centers in the United States, local coal operators fostered diversity among their miners to maintain control and help check labor unrest. The fact that the young immigrant miners spoke a dozen or so different languages hampered communication and fostered mistrust. This strategy proved successful in many circumstances but not always, and the story of the miners' struggle to redress grievances and secure recognition by coal companies for their labor unions is the sequel to a strong current of ethnic hostility that was eventually extinguished through their common interests and collective action.

Miners' Unions Organize

The first effort at union organization in the coalfields occurred in the 1880s, when the Knights of Labor established a local in Pleasant Valley. During the winter of 1883, Pleasant Valley miners went on strike in the first recorded labor dispute in the eastern Utah coalfields. Most of the miners were members of the Church of Jesus Christ of Latter-day Saints, and at the request of their stake president, Abraham O. Smoot, who traveled from his home in Provo to meet with them, they called off the strike and returned to work.[20]

In the aftermath of the Winter Quarters Mine disaster, miners in Pleasant Valley went out on strike in January 1901.[21] The loss of many veteran coal miners and their replacement by inexperienced miners who thought the wages did not compensate for the hard work and danger led miners to demand increases of approximately 10 percent over the average pay of $2.50 and a recognition of the right of miners to appoint their own check weighman to insure that they received proper credit for the coal they mined.

The Pleasant Valley Coal Company insisted that the coal miners were paid as much as their counterparts in Wyoming, who were obligated to work 10 hours a day instead of the 8 provided for in the Utah State Constitution, adopted in 1896 when Utah became a state. The company asserted that profits were very low because of the already high wages. Agreeing to the general idea of the check weighman, company officials nevertheless insisted that before they could collect money from individual miners to pay for that person, the miners would have to give written permission for the deductions. The coal companies did not want to make easy what many officials considered the first step toward a coal miners' union.

The unrest began with miners at Winter Quarters, who voted on 13 January 1901 to go out on strike. They appointed committees to oversee the strike, meet incoming trains to discourage potential strikebreakers, and visit the miners in Clear Creek, Castle Gate, and Sunnyside to try to persuade them to join the strike. They were successful in nearby Clear Creek, where miners walked out two weeks later on 29 January. The strikers were less successful in Castle Gate, where, after a week's effort, the miners voted not to join the strike. In the aftermath, at least 20 Castle Gate miners were discharged for speaking out in favor of the strike.

Confined to the Scofield-Winter Quarters-Clear Creek area in the northwest corner of Carbon County, the strike dragged on for six weeks until it was called off on 20 February 1901. Miners who wanted to return to work were required to sign yellow-dog contracts renouncing union membership, promising not to join a union if an attempt was made to organize one, indicating that they were satisfied with Pleasant Valley Coal Company wages and methods of doing business, and stating that if any grievances arose, they preferred to settle them directly with company officials rather than through union officers.[22]

The coal operators had a valid concern about Utah miners' union sympathies. In November 1903 miners throughout the county joined together in a strike that lasted more than a year.[23] Unlike the 1901 strike, the 1903 strike began in the eastern end of the county at Sunnyside and spread west to Castle Gate and finally Pleasant Valley. Furthermore, the strike was tied to one by Colorado coal miners earlier that fall, when the United Mine Workers of America (UMWA), following its successful anthracite strike of 1902 in Pennsylvania, undertook an organizing campaign among western coal miners. Where the 1901 strike had been a local effort without any UMWA involvement, the 1903–4 strike had strong support from outside the area as labor organizers were sent to Utah to lead it. Where the 1901 strike had involved English-speaking miners, supported by Finnish miners, and had not been successful in recruiting Italian miners at Castle Gate, in 1903 the UMWA saw its greatest success in recruiting Italian miners in Utah, though English-speaking and other immigrant miners, including the Finns and Slavs, joined the strike as well.

The threat of violence and the power of the coal companies brought the Utah National Guard to Carbon County to protect imported strikebreakers and those men who refused to join the strike. The guard also kept a close eye on strikers who had been evicted from their homes in the coal camps by company guards and had taken up residence in tent colonies established by the union. In April 1904 the legendary labor organizer Mother Jones arrived in Utah and spent a month in the state working for the strike. Her residency in a tent colony north of Helper at the mouth of Price Canyon led to a confrontation between local law-enforcement officers and the strikers that resulted in the arrest and incarceration of 120 Italian strikers in a "bullpen" in Price.[24]

The miners demanded a 10 percent pay raise, appointment of their own check weighman, an end to unfair practices at the company stores, adherence to Utah's eight-hour-day law for miners, and recognition of the UMWA. The strike quickly focused on the last issue because coal operators were adamant in their refusal to allow the UMWA to enter Utah coalfields.

Convinced that only through union recognition could any hard-won concessions be maintained and coal-company abuses curtailed, the strikers clung tenaciously to their demand. The coal companies countered with a barrage of measures designed to win public support. The agitation was described as a "strike by foreigners," who had no appreciation for American democracy and the opportunities it gave them. There were also claims that "real" miners could earn a decent living through their own skills and hard work and that the "old-time" miners did not support the strike. Labor organizers were branded as radicals and outside agitators, whose political philosophy was grounded in socialism or anarchy.

The strikers did win the support of the Utah Federation of Labor and some sympathy from national guardsmen sent to Carbon County; however, state officials were clearly distressed by the threat of high coal prices for heating fuel and the disruption that the union threatened to bring to Utah. Moving ever closer to an antilabor, pro-right-to-work position, some leaders of the Church of Jesus Christ of Latter-day Saints offered no support for the striking miners, a good number of whom were also members of the LDS faith, but assisted in recruiting strikebreakers among the farmers and residents of surrounding communities.

In the end the strike was defeated. Many of the striking Italian miners did not return to the coal mines but moved away or took up other occupations in Helper and other locations in the county. The overextended UMWA failed to honor its promises of ongoing support for the strikers, leaving bitter feelings against the union that lasted for nearly two decades. Greek immigrants, who were brought to the county as strikebreakers, in turn became the next immigrant group to be involved in labor disputes.

It did not take Greek miners long to begin to protest the injustices they experienced in Utah coal mines. In March 1907 recently imported Greek and Slavic miners at Winter Quarters walked out, claiming the coal company was cheating them at the weighing scales. The following year George Demetrakopoulos, a labor agent employed by Leonidas Skliris, was shot and killed by a fellow Greek, Steve Flemetis, who had been dismissed by the Utah Fuel Company at the instigation of the agent.[25] Two other deaths occurred at Kenilworth in February 1911, when Greek miners, also protesting cheating at the weighing scales and the discharge of several of their countrymen, walked out on strike. Armed with rifles and pistols, they took up positions in the rocks on the hills surrounding the mining camp. Shots were exchanged, leaving Thomas Jackson, a company guard, and Steve Kolasakis, a Greek striker, dead. The Greek miners returned to work, though tension in the coal camps continued as threats were made and other acts of intimidation committed.[26]

Utah coal miners kept abreast of developments and conflicts in other parts of the country, including neighboring Colorado, where two women and 11 children died in a strikers' tent colony at Ludlow on 20 April 1914.[27]

World War I provided a new opportunity for the UMWA in Utah: the demand for coal led to the expansion of existing mines and opening of new ones while United States Department of Labor policies reflected a much more benevolent policy toward organized labor in an effort to keep workers on the job and clear of disruptive labor conflicts. Union organizers returned to Utah, and UMWA locals were organized in a number of the coal camps, though often in secret and under the real threat that local officers would be dismissed and forced to move from company-owned housing. Utah miners did not join the November 1919 nationwide coal-miners strike; however, soldiers from Fort Douglas, Utah, and Camp Kearny, California, were sent to Carbon County to guard railroad bridges and mine tipples and insure that the shipment of Utah coal would not be curtailed.[28]

Tensions continued in the post-World War I years, and, with contracts due to expire on 1 April 1922, the UMWA prepared for another nationwide strike. This time Carbon County miners were full participants, joining with 650,000 other miners in the largest coal strike in American history.[29]

In the aftermath of World War I and its high demand for coal, by 1922 Carbon County miners saw wage reductions of approximately 30 percent as the rates for mined coal fell from 79 to 55 cents per ton. In addition, long-standing grievances, including coal companies' cheating on the weighing scales, high rents for company housing, mandatory trade at company stores, favoritism, and other abuses, compelled nearly three-fourths of Utah's coal miners to join the strike.

Tensions increased as company guards evicted strikers from company-owned houses and patrolled streets and roads to intercept organizers who sought to persuade undecided miners to join them. Violence flared as strikebreakers were brought into Carbon County under the protection of company guards. Shots were fired in Scofield, Kenilworth, and Standardville and on the Spring Canyon road, where a Greek immigrant striker, John Tenas, was killed by one of the coal-company guards. A month after the Tenas shooting, Arthur Webb, a company guard, was killed on 14 June 1922 when the strikers opened fire on a trainload of strikebreakers as it emerged from a tunnel near the mouth of Spring Canyon. In the aftermath of the Webb shooting, martial law was declared, and the Utah National Guard was sent to the coalfields to restore order and prevent further violence. Eight Greek strikers were arrested and tried for the murder of Arthur Webb. Five were found guilty and given prison sentences.

The nationwide strike lasted from 1 April until 16 August 1922, when John L. Lewis, president of the UMWA, signed an agreement with a majority of the nation's coal operators and ordered his men back to work. In Utah, however, miners did not return to work until September, when mine owners agreed to restore the previous wages but not before raising the price of coal to consumers. The agreement was only a partial victory as the coal operators steadfastly refused to negotiate with union leaders or recognize the UMWA as the legitimate representative of Utah coal miners.

After 1922 the UMWA struggled to maintain a presence in Carbon County. A strong antiunion sentiment emerged in the aftermath of the strike, reflected in the Utah Legislature passing a Right-to-Work Act in 1923 that guaranteed workers the right to employment without union affiliation even though a majority of the shop or industry workers had become organized. Local UMWA leaders such as Frank Bonacci were blacklisted, and industrial spies were hired to keep watch on clandestine organizing activities.[30] In 1928 a 20 percent wage cut was implemented, and the following year, a few months before the Wall Street stock-market crash of 1929 propelled the nation into the Great Depression, UMWA President John L. Lewis ordered an end to further organizing activities in Carbon County.

The Great Depression had a serious impact on the coal industry. Miners suffered layoffs and reduced hours. Carbon County coal miners supported the election of Franklin D. Roosevelt in 1932 with his promise of a New Deal for the American people. For coal miners and other workers, one of the first benefits of the New Deal was the passage of the National Industrial Recovery Act (NIRA) in June 1933 and its all-important Section 7A—often called labor's Emancipation Proclamation because it recognized the right to organize.

While the NIRA was making its way through Congress, developments were already under way in Carbon County that led to a struggle between two unions—the National Miners Union (NMU) and the UMWA—to represent Utah's coal miners. The NMU, considered to be more militant, more radical, and communist directed, was formed in 1928 by discharged members of the UMWA who believed they had been sold out by John L. Lewis and the union during the 1927–28 strike in Pennsylvania.

Early in 1933 the NMU sent organizers to Carbon County and at a rally in Helper on 28 May 1933 announced that organization was under way with the selection of miners to work in their respective areas with union organizers; the establishment of a local headquarters at Millerich Hall in Spring Glen, a small community just south of Helper; the organization of a women's auxiliary and youth section; and the publication of a weekly newspaper, the *Carbon County Miner.*

When UMWA president John L. Lewis learned of the NMU campaign in Utah, he immediately sent back his own organizers, and throughout the summer of 1933, the two unions battled for the allegiance of the miners. In August 1933 the NMU struck against several coal companies, and the county seemed on the verge of civil war when local officials declared martial law and the UMWA joined with coal companies in a temporary and distrustful alliance to put down the strike.[31]

With the backing of the Roosevelt administration's NIRA legislation, the UMWA recognized that the situation in Carbon County presented the first real opportunity in its 43-year history to secure coal-company recognition of the union and that many miners were reluctant to support a communist-led labor union, so they worked quickly and effectively to establish locals in the Carbon County coal mines. On 8 November 1933, representatives of District 22 of the UMWA, headquartered in Rock Springs, Wyoming, but with jurisdiction over the Utah coalfields, met with officials of the Utah Coal Producers and Operators Association at the Newhouse Hotel in Salt Lake City to sign a joint contract which conformed to the provisions of the National Bituminous Code of the NIRA. Embodied within the agreement was recognition of the UMWA by Utah coal operators, and, as they signed the agreement, the nonunion era in Utah's coal-mining history came to an end.

Throughout the rest of the twentieth century, the UMWA played a leading role in county affairs. In the early 1950s, District 22 headquarters moved from Rock Springs to Price, where union affairs were conducted out of rented office space on Main Street until a modern building was constructed in 1976.

From Coal Camps to Towns

For Carbon County coal miners, the union experience was one thing, and life in the coal camps was another. That life was a paradoxical mix of rich social activity and opportunity with some unique restrictions imposed by the sometimes-benevolent, sometimes-oppressive watch of the coal company. James B. Allen writes of this ambivalence in company towns, where "company ownership and control was used to oppress employee-residents both economically and politically. 'I owe my soul to the company store' is a familiar expression which flashes into the minds of many whenever the term 'company town' is mentioned."[32] Exploitation of workers was a constant complaint and ever-present grievance when miners went on strike. Miners balked at requirements that they trade at the company store, where they had to purchase goods at exorbitant prices using the company scrip they were paid and credit they were offered. "Under such a system, . . . the employee was little more than a serf, tied to company property not only through the need for work but also through his perpetual debt."[33]

Company houses were provided, and occupants were under the careful supervision of company officials to ensure that houses and property were properly maintained. In the early years, individuals were permitted to build houses on company property; however, it was clear that this right was temporary and that permission could be withdrawn and residents evicted from the houses they had constructed at the discretion of the coal company. Thus, during labor disputes, one of the first actions taken by coal-company officials was expelling strikers from company-owned houses and others occupied by strikers on coal-company property.

Yet, as Allen continues, "While there is apparently much substance to these charges, it is also obvious that this is not the whole picture. On the contrary, owners of many company towns actually had the interests of their employees at heart in the cooperation of company houses, company stores, and other economic activities."[34]

There were 24 different company towns built in the Carbon County coalfields between the 1880s and 1942. These towns lay in four different areas of the region. The first one, Winter Quarters, was built during the mid-1880s and sat at the mouth of a canyon that opened into Pleasant Valley about a mile west of Scofield. A second company town in the Pleasant Valley area, Clear Creek, was established later, in 1899, in the mountains about four miles south of Scofield.

The railroad town of Helper served as the hub for the second area, which can be described as the north end of Castle Valley near the mouth of Price Canyon. North of Helper was Castle Gate, opened in 1888. East of Helper, the Kenilworth Mine opened in 1906. West of Helper, in the Spring Canyon area, were the towns of Spring Canyon, Standardville, Latuda, Rains, Mutual, and National. South of Helper in Gordon Creek were Coal City, Sweets, and Peerless. South of Price, on the east slope of the Wasatch Plateau, were the camps of Hiawatha, Blackhawk, Wattis, and

Mohrland. East of Price, Sunnyside was established in 1899, followed by Columbia and Dragerton, the last company town to be built in the county in 1942.[35]

Company towns existed first to provide accommodations that gave miners and company officials relatively easy access to the mines and tipples. Company stores offered the essentials—food, clothing, mining tools, and supplies—often at rates higher than those charged by stores in Price, Helper, and Scofield, but with ready access and easy credit that were convenient for the isolated miners. The company towns also tried to maintain some control over the miners, ensuring that malcontents, political agitators, and union organizers did not have easy access to company property and employees. In time most company towns integrated the principles of welfare capitalism by providing recreational and entertainment opportunities, educational facilities, and medical care in a nicely landscaped, well-maintained environment.

However, the early company towns had a long way to go as few facilities were provided—usually a boardinghouse and temporary tent accommodations that offered shelter but not much more. Sometimes miners built their own houses on land leased from the company. After this initial phase, company towns became more permanent by constructing uniform, four-room, wood-frame cottages that included two bedrooms, a kitchen, and a living room. Company officials and doctors resided in more elaborate houses, sometimes two stories high, that were built in the most desirable location in the camp—a section usually referred to as "silk-stocking row." Other sections often reflected the nationality of the dominant residents: Jap Town, Greek Town, Bohunk Town, Wop Town were commonly used names.[36]

At times this ethnic division carried over into public facilities—usually among the Japanese, who in some camps had their own bathhouse, boardinghouses, and amusement hall. Some company towns allowed the operation of a coffeehouse that functioned as the center of Greek life. When Greek coffeehouses were not allowed on company land, they were built on private property near the towns.

The heart of company towns was usually in the center and included the mine office, doctor's office, store, post office, school, and amusement hall. In Hiawatha a stone jailhouse was built near the mine office and amusement hall. Also in Hiawatha the company built two churches—one for members of the Church of Jesus Christ of Latter-day Saints and another used by other religious denominations. Amusement halls included basic facilities for dances, plays, movies, and public meetings. Others, including the Hiawatha Amusement Hall, were outfitted with "bowling alleys, pool and billiard tables, card tables, lodge rooms, a dance floor, reading rooms and rooms where the women of the community may entertain and hold their parties."[37]

If the amusement hall nurtured the social and recreation life of the camp, the company doctor provided health care that almost always equaled and often exceeded that in nonmining Utah towns. Company doctors were expected to meet every need for any member of the family. A company doctor had to be able to set and treat broken and crushed bones, severed limbs, and cuts and lacerations; perform tonsillectomies and appendectomies; extract teeth; diagnose communicable diseases and enforce quarantines; deliver babies; give physical examinations; treat venereal disease; patch up those injured in barroom brawls and shootings; and handle any other medical emergency that might come along.[38]

Company towns offered inexpensive housing, a sense of community, recreational opportunities, and much more. Yet all this came at a price. It was always clear that the coal company was the landlord and could fire miners and evict residents at will. Conformity with company rules was expected and usually demanded. Consequently, many coal-camp residents came to resent the restrictions and the always-present, if not always-spoken, fear of retaliation for misconduct.

Beginning in the 1930s, several factors led to the demise of coal-company towns. As more and more miners acquired private automobiles and could commute to the mines from surrounding towns, the need to live within walking distance evaporated. Recognition of the UMWA by Utah coal operators led to demands from the union that miners not be required to live in company-owned housing as part of their employment. Finally, coal companies discovered that with miners providing housing on their own, the surplus of coal miners after World War II, the difficulties of being landlords, and the opportunities to sell company houses to residents or sell and move them to other locations in the area, it was prudent to end the company-town era in Utah's coal-mining history. In a sense the demise began only a few years after the last company town—Dragerton—was built in 1942. Nevertheless, it was a process that stretched across the decades of the 1950s and 1960s and into the 1970s.

Growth in the Late Twentieth Century

The last quarter of the twentieth century brought other changes. The construction of coal-fired power plants in Emery and Millard Counties saw a significant increase in coal production. By 2001 electric utilities were consuming 22.3 million tons or 83 percent of the 27.02 million tons mined in Utah that year.[39] Utah coal mines supplied more than 27 million tons of coal in 1996 (27,071,000 tons) and 2001 (27,024,000 tons), marking the two highest production years in Utah's history. Moreover, the output came from the fewest number of miners since the earliest days of Utah's pioneering efforts at coal mining in the 1870s. In 1996 2,077 employees mined 27.07 million tons of coal for a productivity rate of 5.91 tons per miner hour. In 2000 the productivity rate jumped to 6.91 tons per miner hour with 1,672 employees mining 26.92 million tons of coal. Compared with a productivity rate of 2.05 tons per miner hour less than two decades earlier, when 4,296 employees had mined 16.91 million tons of coal in 1982, it was clear that fewer and fewer miners were producing more and more coal. State officials maintained that "Utah's high productivity is largely credited to excellent management, a capable engineering and geological staff, a high degree of mechanization, and a highly skilled workforce."[40] The high degree of mechanization is obvious from one statistic for 2001, when eight long-wall panels accounted for 21.5

million tons, or 79 percent of the total production, while 18 continuous-miner sections produced 5.5 million tons; these two methods accounted for virtually all of the 27.02 million tons mined in 2001.[41]

Accompanying this expanded production by fewer miners was a shift from the traditional Carbon County coal mines to ones in Emery and Sevier Counties. According to the Utah Department of Natural Resources, "during the 1960s and 1970s Carbon County was the leading producer, with Emery County second, and Sevier County producing small amounts. During the 1980s, coal production from Carbon and Emery Counties was roughly equal, but by the 1990s Emery County became the leading producer. In 1999, Sevier County moved past Carbon County into second place in coal production."[42]

With the ever-increasing demand for energy and the untapped coal resources in Utah, coal mining will continue to play an important role in Utah's economy into the foreseeable future, though technology will enable fewer and fewer miners to produce the coal. It seems that the days of mine disasters killing hundreds of men are over—as are the days of bitter labor disputes that engulfed whole communities. Coal-camp life is a thing of the past, but the influx of immigrant miners and their families with their vibrant heritage and cultures remains preserved in the churches they built, the festivals that commemorate their ethnic heritage, the oral histories that have been collected, the books and articles that have been written about their experience, and the artifacts they left, which are now exhibited in such places as the Helper Mining and Railroad Museum.

7

Uranium Boom

Raye C. Ringholz

Some 200 million years ago, as a vast prehistoric sea drained from what is now the Colorado Plateau, a network of residual marshes and rivers deposited accumulations of mud, salt, sand, and gypsum across the arid land. The surface warped and tilted; raging winds and flash floods ripped huge trees from the hillsides and carried them to flatlands, where they were left to petrify among the conglomerates; and massive dinosaurs left their three-toed footprints on the ground. Then, after another cycle of inundation, upheaval, and drainage, deepening sands formed multicolored layers to create an imprint of time.

Wind, water, frost, and heat did their work on the landscape to fashion a fantasy of carved red-rock cliffs and chasms, where ancient peoples left their marks in pictographs and petroglyphs. Much later immigrants and explorers fought their way through this desolate country, but despite the hardships of travel, many marveled at the beauty.

John Wesley Powell wrote,

> When thinking of these rocks, we must not conceive of piles of boulders or heaps of fragments, but a whole land of naked rock, with giant forms carved on it: cathedral-shaped buttes, towering hundreds or thousands of feet; cliffs that cannot be scaled, and canyon walls that shrink the river into insignificance, with vast, hollow domes and tall pinnacles and shafts set on the verge overhead; and all highly colored—buff, gray, red, brown, and chocolate—never lichened, never moss-covered; but bare, and often polished.[1]

Early Uses and Discoveries

Early Navajo and Ute Indians, who pounded the reddish and canary-colored rock into powder to mix with animal fat for war paint, were the first to find a practical use for the stone. Colorado gold miners in later years left huge tailing piles of waste

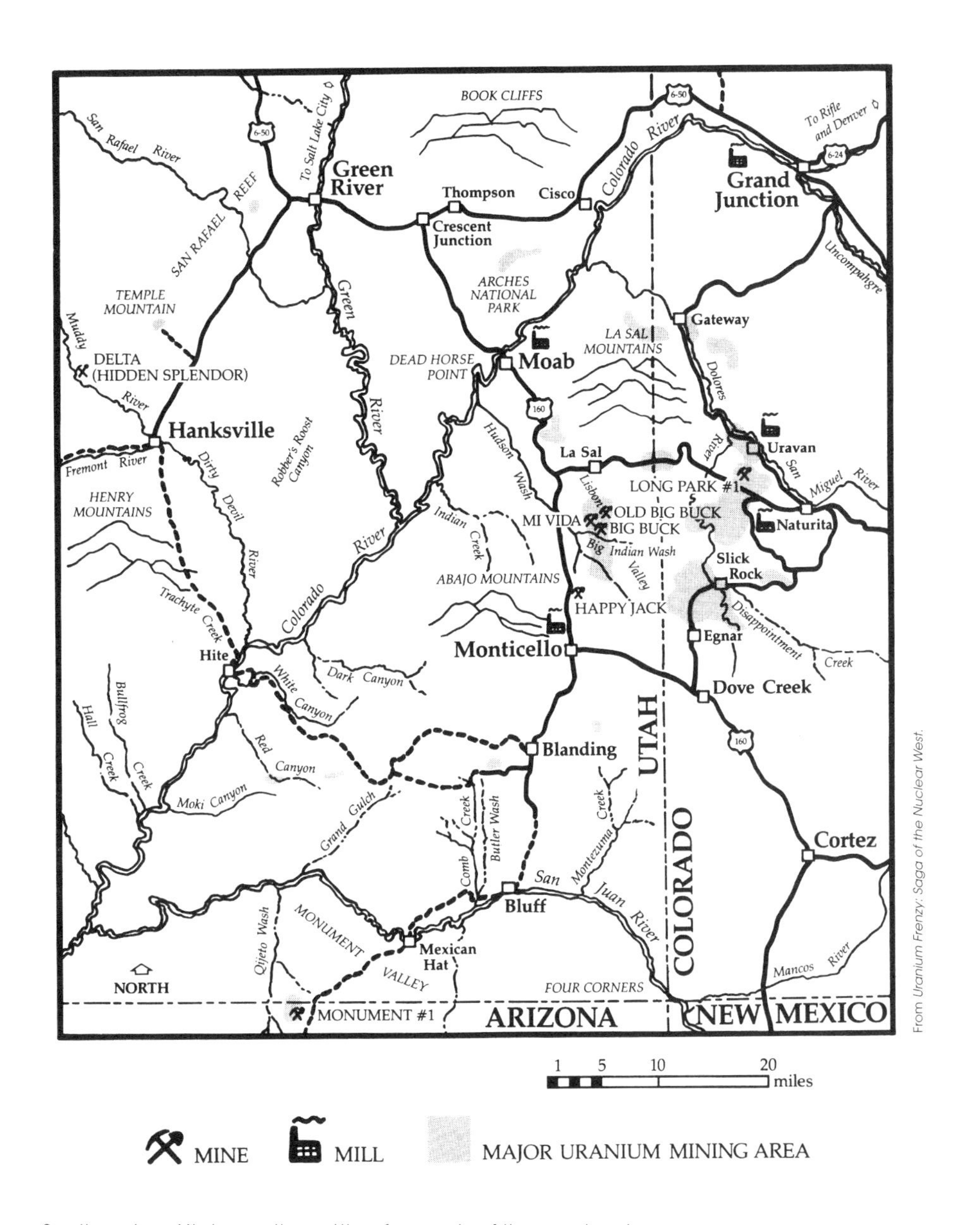

Southeastern Utah was the setting for much of the uranium boom.

saturated with an irksome black, tarlike substance that stuck to their tools. No one suspected that anything of value lay within the bothersome refuse—the carnotite and pitchblende that was discarded.

It wasn't until 1789 that pitchblende was analyzed. Working with ore samples from mines in Saxony, M. H. Klaproth discovered uranium, the heaviest element known at that time. Almost 100 years later, Henri Becqueral made further tests and found that uranium was highly radioactive and could be used to expose photographic plates. Two years later, in 1898, Pierre and Marie Curie and G. Bemont isolated the "miracle element" radium from pitchblende they had obtained from the Erz Mountains in Germany and Czechoslovakia.

The world was agog at this promising cure for cancer, but after a few years, the European mines were depleted, and the Curies had to look elsewhere for a supply. Their attention turned to the colorful sandstone desert of the Colorado Plateau and the pesky pitchblende that had annoyed Colorado gold seekers. The Curies ordered their first shipment of the ore in 1913, and, when another French chemist discovered a new source of radium in yellow carnotite identical to that found in Utah and Colorado, the orders increased, and hordes of ore seekers stampeded into the southwestern backcountry.

Prospecting for radium was a relatively simple matter. Four times the land had been swallowed by prehistoric seas, then drained by upheavals and ruptured by faults that left massive plateaus faced with sheer cliffs and deep, river-cut canyons. The shifting and warping exposed the colorful veins, which were easy to recognize. Dark, sticky pitchblende was found in the Shinarump conglomerate laid down in the Upper Triassic period. Carnotite occurred in the Salt Wash member of the Morrison formation of the Upper Jurassic period. Within a few hours, prospectors could stake claims on surface outcrops and petrified logs protruding from the riverbeds. In March 1912 alone, 10 claims were staked.

Finding the ore was easy, but mining and transporting it was more difficult. Miners worked the rock with single-jack, four-pound hand hammers and drill steel. Ore was mucked into wheelbarrows, then sacked and carried by burros to a stockpile, where it awaited horse-drawn wagons. As roads in the remote area were virtually nonexistent, it was a long, bumpy ride to the railhead, where the carnotite could be shipped to France.[2]

For several years the Colorado Plateau became the Curies' primary source for carnotite, but the heyday was short lived. The outbreak of World War I brought mining to a halt, and discovery of high-grade ores in the Belgian Congo dealt the final deathblow in 1922.

The radium era was over, but a new bonanza lay waiting to be discovered in the carnotite waste.

In 1929 William E. Ford, professor of mineralogy in the Sheffield Scientific School of Yale University wrote of uranium, "The chief interest in the mineral lies in the fact that it is the principal source of radium. Experiments have been made looking toward its use in steel. In the form of various compounds it has a limited use in the coloring of glass and porcelain, in photography and as chemical reagents."[3]

In the early 1930s, eastern glass manufacturers, discovering vanadium, a waste product of carnotite, to be an excellent coloring agent, contacted Moab resident Howard Balsley, an ore buyer for Vitro Chemical Company, to procure materials for them. Among the prospectors Balsley contacted were the Shumway brothers—Arah, Seth, Harris, and Lee—from neighboring San Juan County. The Shumways located claims near Cottonwood Canyon and Montezuma Creek, where they mined numerous "high grade trees," some 60 feet long and 4 feet in diameter that "looked like giant crayons stuck in the rock."[4]

With the Shumways, the Redd family, and other prospectors finding new deposits and sifting though abandoned radium dumps, Balsley shipped 50 50-ton carloads of low-grade vanadium ore to the porcelain plants within the next few years.

Then chemists discovered that vanadium oxide had another important use. Adding the high-grade ore and iron increased the tensile strength, elasticity, and durability of molten steel. As industrialization burgeoned in the United States and Europe, the demand for vanadium to manufacture automobiles, locomotives, and heavy machinery increased. Rumblings of the Second World War exacerbated the need as vanadium steel became a mainstay of armor plate on warships and planes.

Prior to World War I, the valuable ores were obtained from Peruvian mines and roscoelite deposits in Colorado, but by about 1934 those mines had played out. Once again attention focused on the Colorado Plateau. Large concentrations of vanadium in carnotite ore had increased the difficulty in extracting radium and been considered a useless nuisance, and as a result, they had been relegated to the tailing piles. But when manufacturers started demanding new sources of vanadium, the abandoned leftovers became valuable, and a second mining rush erupted as the refuse was scoured for the reddish rock. Prospectors returned again to salvage petrified logs and chip at ore faces like those in the Salt Valley anticline and Polar Mesa of Grand County and the White Fawn or Balsley's Yellow Circle claims southeast of Moab. When the surface ore was depleted about 1938, prospectors started burrowing into hillsides only reachable by roads they had scratched out of the brush for their pickup trucks. They labored long, dusty hours over 35-pound jackhammers joined by long hoses to compressors at the mine face.

Stockpiles of vanadium called for mills to process the ore. Vanadium Corporation of America (VCA) and the U.S. Vanadium Corporation, a subsidiary of Union Carbide, joined the search and constructed several reduction plants in Colorado. In 1937 Frank Garbutt and H.J. Kimmerle built Utah's first vanadium mill southwest of Blanding, Utah, and in the early 1940s, VCA operated a facility at nearby Monticello.

But few people knew that vanadium was not the only product being recovered in the mills. By then a top-secret government project was under way on the desolate plateau. In August 1939 physicists Albert Einstein and Leo Szilard wrote a letter to President Franklin D. Roosevelt warning him that recent scientific data revealed the possibility of constructing "extremely powerful bombs" from uranium. The letter also noted indications that German scientists were experimenting with uranium.[5] When Hitler brutally attacked Poland a few weeks later, Roosevelt recognized the terrible possibility of a Nazi nuclear attack. He immediately appointed an Advisory Committee of Uranium to coordinate and accelerate atomic research.

Successful Nuclear Reactions and the Hunt for Uranium

On 2 December 1942, scientists working beneath the stands of Stagg Field at the University of Chicago created a nuclear chain reaction. Roosevelt promptly authorized construction of production plants to build a bomb. The project was placed under the Army Corps of Engineers as the Manhattan Engineering Project. As uranium would be the bomb's triggering element, procurement of an ample supply was imperative.

But at that time the product could only be found in mines in the Belgian Congo and Canada. Even those sources were nearing depletion. Discovery of a domestic source was critical. Knowing that a small stockpile of uranium existed in the old carnotite tailings, Washington turned its attention to the Four Corners area, where Utah, Colorado, Arizona, and New Mexico meet.

A covert mission quietly directed government engineers into Grand Junction, Colorado, where they converted an abandoned lumberyard cabin in the willows behind cemetery hill into a makeshift headquarters. Commanding officer Lieutenant Phil Leahy knew nothing of the project's actual scope or purpose. His orders were to convert vanadium mills in Durango, Rifle, Naturita, and Uravan, Colorado, and Monticello, Utah, to extract every possible bit of uranium from the old tailings. He also oversaw construction of an experimental uranium-processing mill at the Manhattan Project compound. The resultant sludge (which they called "mineral X") was shipped from a little-used railroad track beside the offices to Oak Ridge, Tennessee; Hanford, Washington; and Los Alamos, New Mexico, for some unknown use.

Army geologists were dispatched to the area to look for undiscovered deposits. They pioneered a new method of prospecting by using metal-detection devices called Geiger counters to facilitate mapping anomalies promising possible deposits of the precious mineral. Civilian prospectors were also encouraged to continue selling their ore to the mills for its vanadium content. They were not told of the vital need for uranium and were not paid for that element recovered from their claims. The top-priority product went to the Manhattan Project free and unheralded. (Congress later compensated the miners for the uranium they sold.) On 6 August 1945, a 20-kiloton uranium bomb exploded on Hiroshima, Japan. Only three days later, a 21-kiloton plutonium device devastated Nagasaki. The closely held government secret was out.

Mixed emotions of relief and horror engulfed the country with the end of World War II. The nuclear age was greeted with wonder: promises of peaceful uses such as environmentally clean power, gas-free vehicles, preservation of meat, distillation of seawater—and the necessity for continued experimentation and development of atomic weapons for national security. All of these things demanded a domestic supply of uranium.

In 1946 Congress disbanded the army's Manhattan Engineering Project and passed the Atomic Energy Act, which established the civilian Atomic Energy Commission (AEC). That summer the president also approved a testing series for the new bombs on the tiny Pacific atolls of Bikini and Enewetak in the Marshall Islands.

The AEC was given complete control of procuring and processing uranium for the tests. The agency's resources were virtually nonexistent, however. By that time only one of five vanadium mills operating during the war was in production, and a handful of independent miners, approximately 55, operated vanadium/uranium mines. The feds initiated an undercover program to locate all potential sources of uranium on the Colorado Plateau.[6]

In September 1946 W. P. Huleatt, Scott W. Hazen, Jr., and William M. Traver, Jr., submitted to the U.S. Department of the Interior and Bureau of Mines a report that outlined promising geologic areas: "The Thompsons area is situated on the Salt Valley anticline in Grand County, Utah. The ore deposits are found in the lower unit of the Morrison formation, which ranges up to 200 feet in thickness. Occurrences of ore in 'rolls' or lenses striking east to northeast are common, and most of the ore mined is of the higher grade than average for the vanadium region as a whole."

The report went on to say that in the Polar Mesa area, a small plateau near the Colorado state line northeast of the La Sal Mountains, "The Morrison formation is exposed around the rim of the mesa, which is capped by Dakota sandstone. . . . The deposits occur in a sandstone layer 20 to 60 feet thick and 200 feet above the base of the Morrison. As a result of erosion, a large part of the ore horizon lies at relatively shallow depths and lends itself readily to surface drilling."

It also noted that the Yellow Circle group, about 20 miles southeast of Moab, has deposits "at the top of the lower unit of the Morrison formation. . . . The thin layers of the lower sandstone, which contains most of the ore, thicken at places to form northeast-trending rolls."

The report concluded with a negative assessment of the Happy Jack Mine, which was located 10 miles north of Monticello in San Juan County. "The ore body being mined at the time was nearing exhaustion," they wrote, "and the operators were following a thin seam in anticipation of developing more ore."[7]

The operators were Joe Cooper and Fletcher and Grant Bronson, who had paid $500 for the promising copper deposit in White Canyon near the Colorado River. The Happy Jack claim had been high-graded for copper during the 1930s, and the partners thought there was still a good chance of turning a profit on it. They built a road into the backcountry, mined and hauled the ore by hand, and sent it to the American Smelting and Refining Company in Salt Lake City. "About two weeks later," Grant Bronson subsequently wrote, "we received a letter telling us the ore assayed 11 percent copper but not to send any more as it was radioactive."[8]

The disappointed partners decided it would not pay to process the low-grade copper, so they tried to sell the mine. Fortunately for them, they didn't find a buyer before the feds sent out their urgent call for uranium. With the newly invented Geiger counter, they found a highly radioactive drift along the Shinarump formation. In 1948 VCA built a sampling plant and pilot mill at White Canyon near Hite and bought mill feed from the Happy Jack and a few other independent producers. Between 1949 and 1953, the AEC purchased 128,145 pounds of uranium from the White Canyon mill at an average cost of $13.93 per pound.[9]

In the early 1950s, the Hite mill closed, and the Happy Jack began shipping ore to the AEC mill at Monticello. Cooper and the Bronsons succeeded in blending high- and low-grade ores to bring every shipment up to standard. By the middle of 1957, the partners had blocked out 433,000 tons of ore with an average grade of 0.34 percent uranium, earning them more than $25 million.

Other so-called doghole miners suddenly found their humble diggings to be worth a fortune. The Redd family staked claims for copper and vanadium in their Blue Lizard property in 1943. It was an unsuccessful venture until they drifted a shaft 700 feet into the mountains and hit pitchblende in the Shinarump formation just when the AEC launched its drive for uranium. The family grossed $6 million when the Continental Uranium Company bought the Blue Lizard in 1955.

Pratt Seegmiller, operator of a souvenir shop by the Big Rock Candy Mountain, had always been a rock hound and spent many hours scouring the area for mineral specimens to sell to tourists. When he learned that the AEC was buying uranium, he remembered some bright yellow material he had collected some years before, but he couldn't recall where he had seen it. All he could visualize was an outcrop of carnotite near an old copper claim.

In April 1948 it came to him. He hiked to the location and staked two claims. The Freedom and Prospector Mines, later leased to VCA, became two of Utah's richest uranium deposits, and Seegmiller collected enough royalties "to support our family in a style to which we were not accustomed."[10]

Seegmiller's revelation came at an opportune time because the AEC had purchased the Monticello mill and renovated it with a new extraction process and a capacity of 100 tons per day. The improved operation removed 93 percent of the uranium and 85 percent of the vanadium from the sludge.

At the same time, the agency initiated a program offering miners higher prices for uranium/vanadium ore. Vanadium was priced at 31 cents a pound, while uranium ores brought a bonus of 50 cents per pound, plus the price of hauling ore to the mill. To encourage prospecting even more, the AEC cut several roads into the barren landscape to alleviate transportation problems.

News spread fast, and scores of outsiders started joining the farmers and sheep ranchers seeking the ore. A group of high school students staked 40 claims and sold them for $15,000. Blanton W. Burford, from Texas, hit an eight-foot vein of high-grade uranium on Rattlesnake Mountain near Moab while bulldozing a road to his claims. He and his partners reaped a profit of $4 million.

An article in a 1949 *Engineering and Mining Journal* added fuel to the fire. Associate editor A.W. Knoerr raised the question, "Can Uranium Mining Pay?" His answer was a resounding "yes!" But he added a few caveats as a warning to inexperienced prospectors.

"Uranium deposits of the Colorado Plateau are distributed over an area of approximately 200 miles in diameter," he wrote.

> Steep canyons of the San Miguel and Dolores Rivers and their tributaries cut through the heart of the Plateau which ranges in elevation from 5,000 to 7,500 ft. Secondary roads over which most of the ore has to be hauled are only fair. Trucks sink to the hub-caps in wet weather. Haulage is further complicated by the fact that most mines are situated on the rims of canyons. Mine supplies and ores have to be hauled over steep rough access roads from mines to secondary roads in the canyons, or over long round-about routes on plateaus. . . . Water to operate drills and cool compressors often has to be hauled many miles.

Knoerr also warned that, while the Morrison carnotite deposits promise greater uranium content, the ore bodies tend to follow vaguely defined channels, with varying thickness and veins that often "swing up or down violently and pinch out." Carnotite deposits of 10,000 to 20,000 tons would be considered large. "Given average ores and favorable mining costs, a modest profit is possible, possibly in the order of $5 to $7 per ton," he concluded.[11]

Aspiring prospectors tended to downplay Knoerr's discouraging assessment. Rather, they focused on an additional "golden ring" proffered by the AEC. In addition to a 10-year guaranteed minimum price for uranium, the agency added a $10,000 bonus for each separate discovery and production of high-grade ore. What's more, the government was the only buyer and was desperate for the product. The reason? The Soviet Union had spread an "iron curtain" over eastern Europe and in late August 1949 had tested its first fission bomb. Consequently, a new global struggle was developing into a cold war with the two powerful nations locked in a nuclear-arms race. The loss of China to Mao Tse-tung's forces and outbreak of the Korean War exacerbated the threats.

Escalation of the cold war made development of a bigger, stronger nuclear bomb a crucial priority in Washington. A search for a closer, more economical testing site commenced. Frenchman's Flat, 75 miles northwest of Las Vegas, Nevada, was selected. At dawn on 27 January 1951, a one-kiloton bomb, dubbed Shot Able, was dropped from an Air Force B-50D for the first time on native soil. The secret was out. "Vegans `Atom-ized,'" the Las Vegas *Review-Journal* reported. "The super solar light lighted the sky so brilliantly that residents of southern Utah, scores of miles away, saw the flash."[12]

Shot Able was the first of hundreds of above- and below-ground tests at the site. Public revelation about the program stimulated interest in the AEC's uranium-procurement efforts, and ore seekers responded in droves. "More uranium was mined in the Colorado Plateau in 1951 than in any previous year," Moab's *Times-Independent* headlined. "In November 1950, 145 claims were staked; February 1951 tallied 600."[13] The newspaper also recorded that J.W. Gramlich received the first AEC bonus payment of $9,672 for 2,763 pounds of uranium oxide from his Morning Star and Evening Star claims on Lion Creek.

Charlie Steen's Explorations

Charles Augustus Steen was another fortune hunter who answered the call. Leaving his three toddlers and pregnant wife, M.L., with his mother in Texas while he ventured alone into the Utah desert was not a practical move for the 28-year-old, but Steen's financial situation was dismal, and he was a dreamer. He was barely making ends meet as a carpenter, but he held a B.A. degree in geology from the Texas School of Mines and Metallurgy and considered himself more qualified for the hunt than the scores of inexperienced schoolteachers, salesmen, and sheepherders who rallied to the government's challenge.

Steen was an experienced mineralogist. He had conducted field research in Peru for the Socony Vacuum Oil Company and been a geologist for Standard Oil Company of Indiana (until he was fired and blacklisted from the industry for refusing to complete necessary paperwork). He had studied geologic features of the Colorado Plateau and the area's uranium mining history and researched articles describing new prospecting techniques developed by the AEC. He had analyzed locations of existing uranium mines and charted a prospecting area for himself in a triangle from Moab, Utah, to Dove Creek, Colorado, and Grand Junction to the north.

But Steen failed to follow the advice offered in the *Engineering and Mining Journal* and other expert sources. Knoerr advised against independent operators and suggested group ventures that could pool exploratory equipment, trucks, and other necessities. Steen headed out alone, with a second-hand jeep and a broken-down drill rig. Experienced prospectors touted the use of Geiger counters. Steen could not afford to buy one. Besides, he didn't need a radiation detector; he could base his explorations on what he had learned as a geologist.

In a pamphlet, R.P. Fischer, head geologist at the Grand Junction AEC complex, listed a few practical clues:

1. Ore deposits occur in the thickest parts of the favorable members of the Morrison formation. Beds of sandstone less than 20 ft. thick rarely contain valuable deposits.

2 Deposits often follow vaguely-defined trend-lines conforming to stream channels in the beds. Once a trend becomes apparent, additional deposits may often be found by projection of drilling or exploration work along these lines.
3 Sandstones in the thicker central parts of beds are dominantly medium-grained, while those in the less-favorable thinner parts of the beds are dominantly fine-grained.
4 The color of ore-bearing sandstone is dominantly pale to light yellow-brown and speckled with limonite stains, while sandstone which is whitish to pale yellow-brown or that which is reddish-brown contains few deposits.
5 Interbedded mudstones are normally red. Near the ore deposits, however, the mudstones within the ore-bearing sandstone are altered to grey, also the upper few inches to few feet of the mudstone beneath the ore-bearing sandstone member.[14]

Knoerr also quoted other engineers and mine operators who pinpointed uranium ores 250 to 300 feet above the Slick Rock member of the Morrison formation and claimed the outcrops were easily identifiable and could be traced for miles along the canyon walls. "If you can detect the slightest sign of canary yellow, you probably have a good shipping ore," they said.[15]

But Charlie Steen had his own ideas. He decided to concentrate his search in areas government geologists had investigated and labeled worthless. He would use his battered drill to probe deeper than the Morrison river channels on the surface of canyon rims. It was his theory that uranium collected deep underground like reservoirs of oil, then leeched up into the Morrison or Shinarump strata. He wanted to find downward-sloping, or anticlinal, structures behind existing claims that had produced small amounts of uranium. He planned to concentrate on the geologically older Shinarump conglomerate, rather than the more popular Morrison formation.

In 1950 Steen arrived in Dove Creek, Colorado, with a $1,000 loan from his mother, Rose, in his pocket and a ramshackle trailer for a home. A few months later, he got Bill McCormick, owner of the Dove Creek Mercantile Store, to grubstake his search and sent for his wife, three sons, and new baby to join him. They moved to Yellow Cat Wash in southeastern Utah and linked the trailer to an 8-by-16-foot shack.

While Steen doggedly rattled along scratchy trails in his jeep and trudged up and down jagged cliffs to detect signs of ore, he was beset with bad luck. His drill pump froze and burst twice. Bits broke. His family struggled through a hard winter. People laughed at "that nutty Texan" who persisted in looking for uranium where none existed, and, worse yet, without a Geiger counter. "Steen's folly," they called his endeavors.

But with spring came a change of fortune. One day he happened upon Dan Hayes, a prospector working some claims in the Big Indian Wash area on the western slope of the Lisbon anticline. In 1916 there had been some unsuccessful attempts at mining small uranium deposits in what was known as the Adams group. Hayes, with partners Jim Bentley and W.Y. Brewer, had restaked 15 claims there in 1948 and renamed them the Big Buck. The earliest uranium workings in that district, the mine produced about 3,700 tons of ore averaging .36 percent uranium within five years. The owners shipped the ore to the Manhattan Project.

This was just what Steen was looking for. If low-grade ore outcropped on the southwest side of the Lisbon anticline, he figured high-grade ore would be found by drilling down the dip. Hayes gave him permission to check the terrain behind the Big Buck outcrops, and Steen staked 12 claims. He filled out location notices with Spanish names—Mujer Sin Verguenza, Mi Corazon, Besame Mucho, Pisco, Fundadoro, Te Quiero, Linda Mujer, Mi Amorcita, Ann, Bacardi, Mi Alma, and Mi Vida ("My Life").

Bursting with his exciting news, Steen rushed back to his family at Yellow Cat Wash only to discover that Rose had suffered a heart attack in Tucson. By the time she recovered, her medical bills had wiped out any possibility of a further grubstake. The Steens moved back to Arizona, and Charlie spent a tedious year hammering nails again.

By 30 April 1952, having accumulated some small savings, Steen could stand it no longer. He sold the trailer house for $350, piled his belongings in a two-wheel cart and on top of the jeep, and moved his family to Cisco, Utah, a bedraggled railroad whistle-stop west of Grand Junction.

"We arrived looking like `The Grapes of Wrath,'" Steen remembered, "but we were in fine spirits and full of hope. We rented a tarpaper shack for $15 a month and stoked the stove with scavengered railroad coal."[16] A few lucky breaks helped the struggling family. Some friends he had written sent checks to help him out, and his mother gave him the profits from the sale of her house and furniture and volunteered to help him set up camp at the claim site.

Steen lost no time. He paid a bulldozer operator to scrape a four-mile track from the county road to Big Indian Wash, then he and Rose hauled supplies the 40 miles from Moab to their makeshift cookshack and bunkhouse. He borrowed a secondhand diamond drill from Bill McCormick in exchange for 49 percent of the action—if any. Then he violated all proven practices for hunting uranium. "The damn fool placed it 1,700 feet back from the rim in a location that didn't have a single indication of surface ore," the locals guffawed. "Steen's folly" for sure.

Steen was convinced of his theory. He bored through the Wingate formation that sandwiched itself several layers below the Morrison and just above the Chinle and Shinarump strata. He knew the AEC had tested the Morrison in the area as negative, so he would target the Shinarump, which had produced a few recent discoveries. Entering through the Wingate would lessen the drilling distance to carnotite, hopefully exposing the ore face before the patched-up drill gave out. No such luck. The pump broke. Drill bits wore out. Equipment repairs and supplies ate up the little money he had. Everything depended on this one hole, an impractical position for an experienced prospector.

> He started drilling on the day before Independence Day, 1952. It was hot, in the nineties. The desert burned. Charlie and Rose stood beside the spindly derrick. It looked like a toy in the vast, empty canyon.
>
> "Well, here goes," Charlie said and he fired the motor.
>
> There was a whirring sound. The drill revolved. Grinding, gnawing noises filled the silence as the bit screwed itself into solid rock. The bit, encrusted with diamonds, the hardest material known to man, chewed through the petrified detritus. Inch by inch it probed geologic strata seeking the horizon of ore. Smoke and dust spewed from the deflector as it burrowed.
>
> Charlie sweated over the hole. He began pulling up cores. He laid the foot-long cylinders of sandstone in a row according to their order of occurrence. If he hit something, he would know just how deep it was by its position in the line.
>
> He examined the sample for traces of uranium. There wasn't a sign of the bright yellow ore. He would recognize carnotite if he saw it.
>
> On July 6th the weather changed. Dark thunderheads gathered to cool the temperatures to just under 80 degrees. By the time the afternoon rains came, the bore reached 72 feet. The cores turned dirty grey. He continued eating into the formation until he had pulled out fourteen feet of the stuff. It was foreign to him. He tossed it on the ground with the other rock lengths and continued drilling. His calculations indicated paydirt at about 200 feet. He still had a way to go.
>
> The summer heat returned. One blistering day seemed to meld into another. The old drill started groaning and smoking, but Charlie persisted. He got down to 197 feet. Then it happened. The stem spun free. There was a loud, whirring noise.
>
> "Jesus Christ!" he yelled, as he dove to switch off the engine.
>
> He knew by the sound that the pipe had broken and most of it was stuck in the hole. He had no way to retrieve it. No more money.[17]

On 18 July a dejected Charlie Steen returned to Cisco. He stopped at Buddy Cowger's service station and casually put one of his disappointing cores under Cowger's Geiger counter. The needle swung clear off the scale. Steen had not recognized the gray samples from his Mi Vida claim as pitchblende. But nobody believed him. It was common knowledge that the AEC had pronounced the Big Indian area barren of ore. Everyone thought that Steen had salted the Mi Vida because the strike had occurred in the Chinle formation, a shallower horizon than usual and a layer not then known as a uranium producer. To make matters worse, Bill McCormick revealed that he had a silent partner in his 49 percent interest, and his associate claimed the discovery was worthless. They wanted out.

Further exploration of the strike and drilling would be an expensive proposition. Charlie and Rose had no more money to put into the venture. Attempts to raise funds in Salt Lake City, Grand Junction. and Denver proved fruitless. Then two employers he had known in Houston saved the day. William T. Hudson and Dan O'Laurie bought out McCormick's partnership for $15,000, and O'Laurie loaned Steen $30,000 to start work. It sounded like a fortune. Still, it was not enough to cover the accepted practice of drilling a number of additional holes to prove the field. Howard Balsley stressed the importance of blocking an ore body and estimating its tonnage. "Uranium ore is just about the most erratic and inconsistent mineral I know of," he said in a speech before the American Mining Congress. "Unless a uranium property has been drilled and the ore bodies proved and delimited, one may have a wonderful face of ore today and witness its complete disappearance tomorrow. There is no assurance whatever, so far as I know, of the continuity of an uranium pocket or roll."[18]

Steen knew it was a gamble, but he decided to put all of the funds into a single hole. He started a six-by-eight-foot timbered shaft 30 feet southeast of the borehole on 4 October. Then he contacted Mitch Melich, a Moab lawyer, to form his corporation, Utex Exploration Company.

Work on the mine was tedious and slow. Steen and his small crew had to drill through layers and layers of rock, blast out the hole, muck out debris with electric shovels, then timber the sides of the hole to keep them from caving in. The work was made all the harder by the uncertainty of success.

Then on 1 December, his 31st birthday, he got the present of a lifetime. The drills reached the 68-foot level and bottomed into 14 feet of gray, primary pitchblende. A later radiometric assay indicated 0.34 to 5.0 percent uranium. It was the biggest strike in the history of the Colorado Plateau. The public furor over Charlie Steen's discovery was just what the AEC needed to get its domestic uranium program rolling. Its office was flooded with hopeful ore hunters seeking hot tips on how to prospect and where to look.

Vernon Pick Arrives

Then news of another successful prospector surfaced. Vernon Pick lost his auto-repair shop in Minnesota to a fire in 1951 and was headed for California to look for work in an aircraft factory when he caught uranium fever. He was in a trailer park in Colorado Springs when he heard about the AEC's incentives for new high-grade discoveries and decided to go to Grand Junction to investigate. Pick knew nothing about geology or prospecting but figured he could learn from the free how-to pamphlets and maps the AEC furnished would-be prospectors. He wrangled an interview with one of the agency chiefs, who advised him to try the little-explored Dirty Devil River country of the San Rafael Swell in southeastern Utah. He decided to spend a few hundred dollars of his insurance money on camping gear and prospecting equipment and set out for Hanksville, Utah, to give it a try.

The dusty little town was not new to Pick. He had been there 16 years earlier, so he looked up former acquaintances and teamed up to go prospecting with June Marsing, a local cowpuncher. But after two months of squabbles over expenses for parts and repairs on Marsing's four-wheel-drive vehicle, the pair split up, and Pick set out alone. Leaving his pickup truck at the end of the road, he backpacked for eight months into the trackless desert to places like Little Wild Horse Mesa, Circle Cliff, and Poison Springs. He was plagued with mishaps. He suffered arsenic poisoning from polluted water and was bitten by a scorpion, tracked by cougars, and startled by rattlesnakes. The weather vacillated between blistering heat and drenching electrical storms. Finally, in the spring of 1952, he came out of the wilds and spent all but his last $300 on a scintillometer, a radiation-detection device more accurate than a Geiger counter.

> On June 21, he camped on the banks of the Muddy River that flowed through a desolate area northwest of Hanksville. The water was high from winter runoff and his hike along the turbulent stream exhausted him. Frequent fords across the waist-high torrent were necessary. Stones cut his feet and his wet boots raised blisters. To make matters worse, his drinking water was polluted by the bloated carcass of a cow festering in the weeds.
>
> Sick and discouraged, he began to prepare for the next day's search, and picked up his scintillometer to adjust the dial calibration. The needle was stuck at the high register. It wouldn't return to normal. He wondered if the batteries were weak. But if they were, why would they register on high? With an electrician's curiosity, he started to walk with the instrument.
>
> "It would indicate and then drop off as though the batteries were low," he later recalled. "I walked into a gully and the reading was still higher. I climbed and the higher I went, the higher it read. I was traveling pretty fast now. It was 500 feet to the top of the ledge. I had never seen the counter like that. It was clear off scale. I couldn't tell what was where until it dawned on me that the whole damned ledge I had been walking on, looking for an ore body, was all ore. Nothing but high grade ore!"
>
> He jammed his pick into the rock. It was the Shinarump conglomerate, riddled with canary yellow carnotite. You could see it.[19]

Elation overcame exhaustion as Pick scrambled to pile stone markers for his claims. He used the Greek alphabet for names: Alpha, Beta, Gamma, and Delta One, Two, Three, and Four. The next morning he started back to civilization. The river was still in flood, so Pick decided to make a raft from an old driftwood log tied with his belt, bootlaces, and scintillometer strap. He piled his supplies under a tarpaulin, then straddled the awkward craft and tried to steer it downstream with a stick. By the time he reached his truck, the raft had ricocheted against so many canyon walls and boulders that he was battered and bruised and had lost most of his food. Fortunately, two AEC men who were investigating his unidentified parked pickup truck gave the weary prospector a hot meal.

Pick named his bonanza the Delta Mine. In the first year, he blocked out close to 300,000 tons of ore valued at $40 each. Monthly production averaged 1,500 tons of uranium. Coming on the heels of Steen's discovery, the strike furnished the fuel the AEC needed to spur domestic production of uranium. National magazines featured the two folk heroes, and thousands of fortune seekers, most of whom knew nothing about hunting minerals, converged on the southeastern desert. Moab became the Uranium Capital of the World. Traffic through town increased by 300 percent. Motels filled, and strangers rented rooms in private homes or camped in their cars. Facilities that had serviced a population of 1,200 were stretched to the limit. Water lines were overtaxed. Electricity failed frequently.

Stories about Steen fanned the uranium frenzy. Readers from afar were titillated by reports of Charlie flying in his Cessna 195 to get a better view on his onboard television, gilding his prospecting boots, lavishly hosting celebrities in his hilltop mansion, and treating the entire community to extravagant "discovery parties" in the huge new airport hangar.

Less colorful, but still newsworthy, Vernon Pick's ordeals made hot news for *Life* magazine and the *St. Louis Post-Dispatch*. The publicity backfired, however, when locals claimed his story was grossly exaggerated and filed lawsuits asserting that his strike was the result of a tip-off by AEC geologists. Pick was later cleared by FBI and AEC investigations, but he found developing his claim frustrating. The mine was on a high escarpment in a wild, remote area that was difficult to reach. It was expensive and grueling to haul drilling equipment to the site, and ore had to be packed out by horses. When internationally known financier Floyd B. Odlum offered $9 million and a PBY airplane for the property, Pick sold. Odlum renamed the mine Hidden Splendor.

Radiation Health Issues Arise

While Pick and Steen were making history, a little-noticed drama was playing out in offices of the U.S. Public Health Service in Salt Lake City. The proliferation of vanadium mining prior to World War II had turned the attention of medical researchers to literature dating back to the early 1500s. They found reports of unprecedented deaths of miners working in the Erz Mountains of Germany and Czechoslovakia. The mines, exploited for silver, cobalt, bismuth, nickel, and arsenic, also contained pitchblende, which had earlier been the origin of uranium for dyes and later became the first source of radium for the Curies. For centuries the illness that claimed miners was known as *bergkranheit*, or "mountain disease." Hundreds of underground workers in the prime of life were suddenly stricken by a wasting illness that usually killed them within a year. The annual death rate totaled 1 percent of the total workforce, and 70 percent of the deceased succumbed to lung cancer. By the early twentieth century, epidemiologists suspected exposure to radioactive materials was the cause of the unusually high mortality rate.

Washington's unprecedented demand for a domestic stockpile of uranium and the surge of mining on the Colorado Plateau following the Steen and Pick discoveries revived the interest of medical researchers in the "European experience." One of the first to warn of possible radiation danger to American miners was Duncan Holaday. During World War II, Holaday, an occupational-health specialist for the U.S. Public Health Service, inspected government-owned, contractor-operated arsenals for toxic gases, dusts, and other hazards. Following the war, he participated in cleanup operations for ships that had been anchored in range of nuclear-bomb tests on Bikini. Upon his return to Washington, his expertise in radiation-related industrial hygiene put him in much demand as a consultant.

In 1949 Ralph Batie, chief of health and safety for the Colorado Raw Materials Division of the AEC, contacted Holaday to alert him about potential radiation problems in the fledgling uranium-mining industry. Batie had previously sought help from Dr. Merril Eisenbud, head of the AEC Health and Safety Laboratory in New York, and prominent radiologist Dr. Bernie Wolf. The two doctors had distinguished themselves by eliminating chronic lung disease and other debilitating health problems in the beryllium industry by pinpointing toxic dust as the cause and proposing better ventilation and hygiene practices as the cure. After visiting the Colorado Plateau and reviewing Batie's findings, the two specialists suggested implementing similar procedures to the ones prescribed for beryllium plants in uranium mines and mills. Upon their return to New York, they sent Batie radiation-measuring equipment to help him establish a firm database for further study.

Holaday was interested in Batie's preliminary findings and agreed to go west for a look. He was deeply troubled by the results. The AEC had adopted a standard of 10 picocuries per liter of air for all of its uranium-refining plants and laboratories but had not yet published permissible levels for airborne radioactive dust or gases in privately owned mines and mills. A *curie* is a unit of radioactivity undergoing 3.7×10^{10} disintegrations per second. A *picocurie* is one-trillionth of a curie.

"At some of the more dusty operations in the mills, the concentration of airborne alpha emitters [charged particles emitted from the nucleus of some atoms] was several thousand times as high as those that would be permitted in other AEC installations," Batie told Holaday. "In one of the extensive mines, the concentration of radon gas was over ten times the permissible level which is used in other industries. The readings indicated the probability that severe internal radiation hazards existed in many operations."[20]

Eisenbud had been convinced that, once measurements were taken in the mines and mills, the AEC would write established radiation standards into uranium-procurement contracts similar to those they had imposed on the beryllium industry. If the rigid requirements for ventilation and hygiene were not met, federal funds would be withheld. The AEC denied such responsibility. It referenced the Atomic Energy

Act of 1946, which stated that safety and health matters were not under AEC jurisdiction until the source material was removed from its place of origin. The agency was not responsible until the ore landed in the mills. What happened in the mines was a matter for the states involved to address.

But the state industrial hygienists had no expertise on mine radiation. They had no measuring instruments. They didn't know what data needed to be collected or how to collect it. They had no methods of testing or evaluating any findings. More importantly, they had no authority to impose control measures on privately owned mills and mines without a legislative mandate.

After Holaday had toured Batie's laboratory and reviewed all of the evidence, he returned to Washington with a mission. "This situation is one that requires immediate investigation, in order to determine the exact extent of the hazards to which workers are subjected and to institute corrective measures," he wrote in a memo to Henry Doyle, regional representative of the U.S. Public Health Service, on 11 April 1949. "At present some 700 workers are involved and the number may shortly rise to several thousand."[21] He concluded by urging a preliminary study of the uranium industry in an attempt to rectify hazards that had been permitted to develop.

Four mines in Monument Valley, primarily manned by Navajos, were selected for testing. Holaday was shocked by the results. None of the workings had change rooms, toilets, showers, or drinking water. Employees were not given preemployment physical examinations or a medical program. The nearest health facility was 40 miles away over roads little more than trails. There was a first-aid kit but no qualified attendant. Samples of radon measured "4,750 times the maximum allowable concentration."[22]

Holaday enlisted the support of other health scientists and petitioned the U.S. surgeon general for permission to conduct a more extensive study of the uranium-mining industry that would result in imposition of adequate radiation controls. Their request was granted. In 1950 Holaday was transferred to Salt Lake City to head the environmental phase of the research.

A comprehensive uranium-mining health and safety study was initiated that July. Holaday and his small crew of engineers went underground in scores of mines, filling vacuum flasks with the trapped air and collecting dust with hand-operated pumps. At the same time, a team of doctors and laboratory technicians gave physicals to miners, testing blood and taking X-rays in a mobile clinic. Mining companies allowed these activities only on the condition that workers would not be alarmed. Nothing was to be said about radon dangers, and any medical problems discovered could not be revealed. Miners could only be advised to contact personal physicians.

Upon completion of the field surveys in October, Holaday and his colleagues were more alarmed than ever. Most of the mines depended upon inadequate, natural ventilation. The Public Health Service had concluded that a maximum concentration of 100 picocuries of radon per liter of air were safe in a mine, but a single sample taken at the face of VCA's Prospector Number-One Mine at Marysvale, Utah, showed 26,900 picocuries per liter of air.

Holaday shuddered when he thought of the consequences of only 1,000 picocuries per liter of air registered in the German and Czechoslovakian mines. He couldn't understand why, when the AEC already paid miners a 50-cent-per-pound development allowance, it failed to pay a few cents more to force companies to cut churn-drill holes and buy fans and gasoline motors to run for better ventilation. A similar situation in the beryllium industry had worked.

By 1952 Holaday felt he had enough information to suggest remedies for the radiation problem. With Public Health Service colleagues Dr. Wilford D. David and Henry Doyle, he wrote an interim report for the Uranium Study Advisory Committee. The report presented methods for correcting potentially harmful conditions by using respirators, wet drilling, and other dust-suppression techniques; improving ventilation; encouraging better personal hygiene of workers; and implementing periodic medical exams and histories. Two thousand copies of the report were distributed to AEC offices, mining companies, and state bureaus of mines. No follow-up meetings were scheduled for discussion or review. The paper gathered dust in agency files.

A year later Holaday, claiming the problem had been sufficiently defined and the solution methods developed, announced that there was nothing more he could do and concluded his environmental-research program. Implementation was now a legal or legislative matter.

But the medical-examination program was another issue. It had been a hit-and-miss project with physicals given on the basis of chance. Follow-up exams needed to produce definite data were virtually impossible to conduct in the rugged terrain on workers who moved from job to job as opportunities arose.

Then a 51-year-old uranium miner from Nucla, Colorado, died of lung cancer, and the possibility of radiation-induced illness became a reality. An autopsy revealed 34.4 picocuries per liter of polonium-210 (radio lead, the final product of disintegrating radon daughters) in his urine. The fast-growing, inoperable, "oat cell" carcinoma was the same evidenced by deceased miners in Germany and Czechoslovakia. It appeared that the epidemic had started.

The National Cancer Institute initiated an epidemiological study directed by the Public Health Service field station in Salt Lake City in cooperation with the AEC and the Colorado Health Department. Mobile clinics were sent into the backcountry to perform physical examinations on an extensive cohort of miners, whose cases would be followed for an indefinite number of years to monitor their health. In addition, a survey of inactive and retired miners who might have quit their jobs due to health reasons was included. But the information had to remain classified. Once again the mining companies agreed to the program only with the stipulation that their workers must not be alarmed

While medical teams focused on uranium mines and mills, persons living downwind of the Nevada Test Site became aware of another radioactive threat. On 19 May 1953, the 32-kiloton nuclear bomb Shot Harry was detonated at the site. Dr. Arthur F. Bruhn and 30 students in his geology class at Dixie Junior College witnessed the predawn blast from a mountaintop halfway between St. George, Utah, and Las Vegas. The youngsters had all initialed letters from the AEC affirming that there would be "absolutely no danger" involved in this wonderful opportunity of "furthering their education."

The explosion, known as "Dirty Harry," blew with such force that the rocks and soil beneath it, the tower supporting it, and the cab container and casing holding it were vaporized. A violent updraft sucked loose debris into a huge, mushrooming cloud that drifted east to spread over Lake Mead and north to St. George and Cedar City. The air on the mountaintop where Bruhn's class was watching grew thick with radioactive dust that reddened the students' skin and burned their eyes. In a small Mormon town on the Utah/Nevada border, unsuspecting children frolicked in storms of unseasonable "snow" that fell from the springtime sky. Within an area 40 miles north and 160 miles east of Frenchman's Flat, 1,420 lambing ewes and 2,970 new lambs died from the effects of Harry's fallout.

Despite government protestations that the tests posed no danger to humans or the environment, "downwinders" in Southern Utah began to lose hair, suffer stillbirths, fall ill, or succumb to Hodgkin's disease, leukemia, brain tumors, lung cancer, and other radiation-induced illnesses. By the late 1970s, there were 20 victims and 14 deaths in St. George alone.

The numbers multiplied as the latent effects of chronic exposure to radiation intensified over the years. Still, hundreds of above-and underground detonations exploded in the desert until President George Bush entered into a unilateral moratorium on nuclear-weapons testing in 1992. By then, after tedious court battles and legislative forays, Congress had passed a Radiation Exposure Compensation Act to provide some financial relief to uranium miners, mill workers, and their families. The law was finally revised and expanded in 2000 to include test-site workers, ore transporters, and downwinders with additional illnesses covered. President George W. Bush remedied an embarrassing funding shortfall the next year.

Uranium Fortunes and Wall Street

Although the AEC was aware of the covert environmental and medical-health studies being conducted by the Public Health Service, the agency continued to step up its campaign to amass huge stockpiles of uranium. Besides bonuses of $1,000 for new discoveries of high-grade ore, the only legal customer for the ore paid a guaranteed price of up to $50 per ton for .3 percent uranium, threw in six cents per ton haulage allowance, and continued to construct roads, build processing mills, and furnish the services of professional geologists.

By August 1953, six months after operations began, Steen's Mi Vida had produced $1 million worth of ore assaying up to 87 percent uranium. That December he and his former partner Bill McCormick paid Dan Hayes, Joe and Don Adams, and Edward Saul $50,000 for an option to buy the Big Buck claims for $2 million. Vernon Pick had also enjoyed a successful first year. He had blocked out almost 3,000 tons of ore with an estimated value of $40 a ton. The Delta Mine's monthly average was 1,500 tons of uranium.

Run-of-the-mill doghole miners viewed Steen and Pick as proof that anyone could hit it big, but the majority had little more than barren holes in the ground. Those with promising properties lacked the thousands of dollars needed to prove their claims with exploratory drilling. Big mining companies were not interested in the little, independent prospectors.

Then a seedy old promoter named J. Walters, Jr., got an idea. If Uncle Sam was willing to pay big for uranium, and the market was guaranteed, but the majority of miners were unable to finance proving their claims, why not raise the capital by issuing over-the-counter penny stocks? The cheap issues would be affordable to almost anyone. Three million shares at one cent each added up to $30,000, enough to finance most of the Colorado Plateau ventures.

Walters knew his innovative idea would work because Utah's securities laws were virtually nonexistent. It was a simple matter to get a broker's license. There were no educational requirements nor examinations to pass before filling out a simple form and paying a $25 fee. A person was made an agent "if the commission shall find that the applicant is of good repute."[23]

A 1925 statute had created a Utah Securities Commission that allowed three methods of underwriting new corporate enterprises. An intrastate offering permitted Utahns needing a relatively small amount of money to raise funds by sale only to state residents. No formal SEC filing was necessary. A "short-form" filing with the SEC under Regulation A could be done if the amount was $300,000 or less, including some sales out of state. Interstate sales exceeding $300,000 required a full SEC registration.

When Walters read about Steen and Pick, he rushed to Moab, bought some claims, and registered them as Uranium Oil and Trading Company. Then he went to Salt Lake City to sell his idea. He convinced Jack Coombs, a young Salt Lake City stockbroker, that his penny-stock promotion would fly and got Coombs to agree to raise $50,000 for his venture. Coombs, in turn, enlisted Frank Whitney to help him sell the over-the-counter stocks from his coffee shop in the Continental Bank building. Frank, and his brother Dick, raised the entire $50,000 at a penny a share in one week, then sold the coffee shop and opened their own brokerage house. Some months later Walters launched Alladin Uranium Company. The Whitney brothers promptly sold $43,000 worth of stock.

Then Salt Lakers Dewey Anderson and Hal Cameron approached the brothers about claims in Big Indian Wash that they had purchased from Shorty Larsen and Bud Nielson in Huntington, Utah. The Swedish farmers had given Steen a few hundred dollars when he was raising money to sink a borehole. When Charlie hit ore, he repaid the pair by advising them to stake a couple of claims "right by the Mi Vida." Sensing a sure bet, the Whitneys enthusiastically joined forces with Anderson and Cameron and enlisted a board of directors to form Federal Uranium Company. On 30 December 1953, they offered 7.2 million one-cent shares of the new stock.

Almost overnight Salt Lake City became "the Wall Street of uranium." People from all walks of life, many of whom knew nothing about geology or prospecting, surged into the backcountry staking claims whether there was any sign of uranium ore or not. Promoters and self-proclaimed stockbrokers followed with penny offerings for quickly formed companies with exotic names like Absaraka, Apache, Arrow, Atomic, Black Jack, Jolly Jack, Lucky Strike, and King Midas. The AEC fanned the fire with one million feet of exploratory drilling per year, construction of 993 access roads, payment of $3,725,000 in bonuses for new discoveries, and distribution of maps and geologic information.

By the fall of 1953, big money had joined the race. Salt Lakers Zeke Dumke and A. Payne Kibbe paid Steen $65,000, with an overriding royalty of 12.5 percent and a guaranteed contract with his Moab Drilling Company, to buy 10 promising claims near the Mi Vida. They named their project Lisbon Uranium Company. On 15 January 1954, 1,079,000 shares of the stock were offered at 20 cents a share. Ten days later Steen and Bill McCormick optioned the Big Buck claims to Ray and Ralph Bowman for $50,000, 750,000 one-cent shares, and a drilling contract. The resultant Standard Uranium Company was the first offering fully registered with the SEC. On the first day, 1,430,000 shares of common stock at $1.25 per share sold out.

In April Kibbe, who had optioned three additional claims from Nielsen and Larsen in the Lisbon deal, learned that Floyd Odlum was looking for uranium properties. His Hidden Splendor Mine had pinched out and become known as Odlum's Hidden Blunder. Kibbe proposed that Odlum invest in Lisbon so they could drill the Swedes' claims. Odlum accepted, and the very mention of his name shot Lisbon stock from 40 cents to 95 cents in three days.

Salt Lake City went wild. Where there had been 20 stockbrokers in 1953, there were 112 traders by the next spring. They operated out of dingy "bucket shops" and elegant, walnut-paneled offices. They sold anything with the word "uranium" in the title. When a brand-new offering named Timco Uranium announced five strikes of commercial-grade ore, its stock soared 50 times over the subscription price of a few weeks before. Cromer Brokerage Company traded 6,729,000 three-cent shares of Apache in one week. Then newscaster Walter Winchell sent things over the top. During his broadcast on Sunday, 23 May 1954, he announced that Federal Uranium had hit 0.4 percent uranium-oxide ore. The next day a record-breaking 7 million shares of uranium stock were sold on the Salt Lake Stock Exchange.

By the mid-1950s there were 600 producing companies shipping uranium ore from the Colorado Plateau. Ore production doubled every 18 months. There were more than 8,000 workers in the industry. Yet the number of processing mills to transform raw rock into yellowcake was not keeping up with the trend.

Steen decided to build a plant of his own. He obtained permission from the AEC to construct the uranium reduction mill (URECO) in Moab, then talked Floyd Odlum into purchasing 30 percent of the stock through his Atlas Company's wholly owned subsidiary, Hidden Splendor Mining Company. URECO was set to process 1,500 tons of ore per day that would be delivered from the Mi Vida, Hidden Splendor, Lisbon, Mountain Mesa Uranium Company, LaSal Mining and Development, Hecla Mining Company, and Radorock Incorporated mines. The mill utilized the radically new, resin-in-pulp metallurgical process developed in AEC pilot plants that had not yet been tried in commercial operations. Most existing plants were ball mills that used steel balls to crush ore placed in large, rotating cylinders. The ore was pulverized into fine sand, then mixed with salt and dropped through a screen into the roaster, a pipelike apparatus that was heated to 1,800 degrees. Finally, the ore was dropped into tanks of water and sulfuric acid.

A major component of the resin-in-pulp process was tiny, tapioca-sized beads of yellow resin. Raw ore was crushed, then acid-leached into an unfiltered slurry. Then specially sized, ion-exchange resins in wire-mesh baskets were slowly shaken in a flowing stream of uranium-bearing pulp. When the beads were fully coated, the uranium was stripped from the resins in a chemical bath.

URECO started operating on 4 October 1956. By then the uranium industry was experiencing a change. Small independent mining operations were merging or being absorbed by corporate giants like U.S. Vanadium, VCA, Anaconda, Homestake Mining Company, National Lead, Vitro, Kerr-McGee Oil Industries, Climax Molybdenum, New Jersey Zinc, and the Santa Fe Railroad. The anticipated "atomic age" of nuclear-power plants was not developing as quickly as expected, and with 15,000 tons of concentrate stockpiled, the federal government had enough to satisfy all its military and electrical needs.

The Boom Breaks

Between 1948 and 1956, the AEC had received 1,524,000 tons of ore. Concerned about possible overproduction, the commission announced a stretch-out in the federal-buying schedule. Allan E. Jones, the new manager at Grand Junction, reported a revised guaranteed government market for 500 tons of concentrate per year from any one operator at a flat price of eight dollars a pound through 1966. The former policy had set rates for ores and negotiated prices of concentrate. The agency would also only buy ore from established contracts and make no new commitments. When the United States and former Soviet Union agreed on a moratorium for nuclear-weapons tests on 31 October 1958, Jones revised his policy, limiting amounts of concentrate to be purchased by specifying that they come only from ore reserves developed before 1958. The announcement virtually took the wind out of prospecting activity.[24]

The penny-stock craze suffered similar woes. The buying public was losing interest in penny stocks as the industry moved from inexpensive and appealing one-man operations to big business. And the Utah Securities Commission's imposition of more stringent controls and fees on broker/dealers was followed by the National

Association of Securities Dealers' indictments of traders for infractions of SEC regulations. Eighteen firms had their licenses revoked due to unsound financial reports. The fabled Wall Street of uranium closed as abruptly as it had started.

The flame rekindled in 1966 with an AEC proclamation that it would no longer be the sole buyer of uranium ores. From 1951 to 1965, the commission had purchased 56,623 tons of concentrate from Utah operators alone at a cost of $509,084,000. Now, with uranium on the free market, commercial activity shifted into gear. Fifteen nuclear-power plants were already operating in the United States, 8 more were under construction, and 20 were in the planning stages.

Prospectors returned to the Colorado Plateau, but things were different this time. The AEC was no longer purchasing ore, and their buying stations, haulage payments, and bonuses were a thing of the past. Uranium exploration had become a sophisticated business. Ore was deeper, and sensitive detection instruments were needed to find it. Deep drilling was expensive, and it was estimated that locating a deposit, developing a mine, and building a new processing plant would take eight years and cost $20 million. Giant industries like Westinghouse and General Electric were initiating drilling programs, not the independent doghole miner of the past. Rio Algom Mines, Ltd., of Canada drilled a large ore body on the Lisbon fault and built a $23 million mine/mill complex near La Sal.

Still the dreamers came. The Grand County recorder tabulated 5,810 location notices filed in 1967, and in the chill, usually slow month of December, 259 claims were staked. Forty-five affidavits of the required annual labor on a property were submitted. In neighboring San Juan County, 19,521 claims were filed that year, 539 of them in October alone.

The money men on Salt Lake's Wall Street found a new way to get back into the uranium game. They dug out SEC registrations for worthless, defunct uranium corporations and used the corporate-structure shells to launch new development companies that sought financing by issuing stock. In this way they avoided the time-consuming and expensive process of starting up a new corporation and filing with the SEC and state securities office. Penny stocks revived, and people dug in their attics for useless uranium stock certificates that were suddenly worth something. Soon approximately 120 stocks were listed, some retaining former names for new enterprises, others changing. Arrow Uranium became Controlled Metals, Strategic Metals became Greenwich Pharmaceutical, and Trans-Western Uranium changed to Flying Diamond. But what started as a legitimate effort to expedite formation of new companies, quickly deteriorated into a "fast-buck" scheme. It didn't take long for the SEC to intercede and put a stop to the practice.

By the early 1970s, it appeared that the unique uranium boom was coming to an end. Exploration and drilling activities tapered off, and Utah's uranium-processing mills closed. Construction of nuclear-power plants fell below predictions after disasters such as the one at Three Mile Island. Evidence of radiation health hazards triggered the Uranium Mill Tailings Radiation Control Act of 1978 that ordered cleanup of waste dumps, a problem exacerbated in later years by the issue of disposing of spent fuel rods from nuclear-power plants.

The uranium story was an exciting chapter in the annals of Utah mining. The famous Lisbon Valley mines produced more than 10,000 tons of uranium from 1953 though 1982. In the same period, 1,000 to 10,000 tons were removed from the La Sal, Dry Valley, Monument Valley, Green River, White Canyon, and San Rafael areas. Up to 1,000 tons came from the Paradox Basin, Thompson, Montezuma Canyon, Cottonwood Wash, the Henry Mountains, Cane Creek, Seven Mile Canyon, Marysvale, and Topaz Mountain and lesser amounts from the Circle Cliffs, Silver Reef, Indian Creek, the Wah Wah Mountains, and North San Rafael.[25]

8

Beryllium Mining

Debra Wagner

Brush Resources Inc., formerly Brush Wellman Inc., mines beryllium-bearing ore from the company's Topaz Mining Properties. Discovery of beryllium ore occurred in 1959. The find created much excitement within the mining industry. As time unfolded, these deposits proved to be a valuable resource, opening the door for the beryllium industry to grow. This ore source permitted the company to become fully integrated with a controlled supply and a production capability for all major commercial beryllium products.[1]

The first pit opened in 1968, and the mill near Lynndyl began operation in 1969. Since that time, the open-pit mining operations have been continuously active. The value of this project was demonstrated dramatically in the growth of the industry that followed. Many new and varied products came to the market as a result of the Utah venture.

The Topaz Mining Properties are located in Juab County, Utah, and are approximately 47 miles west/northwest of the company's mill near Delta. Access is by Highway 174 west from U.S. Highway 6. The company owns fee-simple title to the surface of the mine property.

Geology

The mining properties are located in the Spor Mountain/Topaz Mountain area in western Juab County. This region has been a commercial source of uranium, fluorspar, and beryllium. The beryllium district lies on the west and southwest slopes of Spor Mountain. Bertrandite, a hydrous beryllium silicate ($Be_4Si_2O_7(OH)_2$), is the ore mined. Until 1969 the beryllium industry in the United States had been dependent upon imported beryl ore for its only source. Beginning in 1969, the company's extraction plant near Lynndyl has been in constant production, using bertrandite ore feed from the mining properties. Beryllium is classified as a "strategic metal" by the United States Department of Defense.

The Spor Mountain area is part of the Thomas Mountains/Tintic Mountains subdivision of the Basin and Range physiographic province. The mining properties are made up chiefly of westward-tilted and intricately faulted Paleozoic sedimentary rocks that have been locally intruded by volcanic rocks of Tertiary age. Flows and tuffs of Tertiary age also overlie the Paleozoic rocks, creating pronounced angular unconformity. The area is extensively faulted. Most of the faults trend northeast/southwest and have displacements ranging from 50 to 800 feet. They have played a major role as conduits for the beryllium-mineralized solutions.

Tertiary volcanic rocks of the Spor Mountain formation consist of two informal members: the beryllium tuff and an overlying porphyritic rhyolite. The formation dates at 21 million years (Lower Miocene). The two members occur together in most places and are restricted to the vicinity of Spor Mountain. The porphyritic rhyolite member breaks out as flows, domes, and small plugs.

The beryllium tuff rests unconformably on older volcanic rocks of Tertiary age and sedimentary rocks of Paleozoic age. It is an important stratigraphic unit because all beryllium production in the district originates from it (see figure 1). Mining operations within the beryllium tuff by the company have encountered many variations in particle size and composition of the ore zone. Hydrothermal (epithermal) fluids have partially altered the beryllium tuff deposits to a fine-grained mixture of montmorillonite-kaolinite clay, potassium feldspar, silica minerals, and fluorite. Distinctive zones of argillic and feldspathic alteration enclose the actual deposits. The bertrandite ore mineral of beryllium is submicroscopic, disseminated in the tuff, and concentrated in fluorite nodules.

Many authors have published information on beryllium mineralization in the tuff, including several publications by David A. Lindsey.

Land Use

Before mining, the land was used for grazing primarily winter and spring sheep and wildlife habitat. Currently some sheep and cattle still graze on the mine property. The wildlife in the area is confined to small mammals, birds, and antelope, which range there throughout the year.

Public safety is provided in compliance with the company's policies as well as Mine Safety and Health Administration (MSHA) rules. The mining properties are on private land. Unescorted public access is limited to county roads that travel through the property. Signs notify visitors to register at the mine camp when entering the area. No unescorted access is permitted in either existing or proposed mining areas.

Mine Operations

The unique bertrandite ore bodies within the company's mining properties are geographically and geologically separated. They occur as stratiform tuff deposits of widely varying thickness and inconsistent grade that dip steeply underneath massive rhyolite flows. Ongoing mining operations have provided mill feed continuously since 1969.

The method for removing the rock overburden from the ore is known as "open-pit prestripping." Traditionally an earth-moving contractor removed the rock from the open-pit area to expose a three- to five-year supply of bertrandite ore. This method was used exclusively from 1968 through 1997, when the last prestripping occurred. Overburden was placed adjacent to the stripping area according to the approved mining and reclamation plan at the time.

The ore is mined using a modified bench system, where the bench generally follows the ore body's strike and migrates down-dip as mining advances. The beryllium mineralization in the host tuff is visually indistinguishable from unmineralized material, widely disseminated, and relatively low grade. These characteristics require a unique, highly sophisticated approach to determine the beryllium grade and control the ore. The ore is sampled extensively, mapped meticulously, and dressed and lifted to the stockpile with the utmost care. All engineering and mining efforts revolve around the ability to detect the beryllium with a neutron-activated beryllium analyzer (the berylometer). The laboratory berylometer assays the drilling samples to enable detailed mine planning, and the field (portable) berylometer determines the exact cutoff point in mining.

The ore is lifted from predetermined areas within the open pit and placed on a designed stockpile pad. During stockpile construction, the ore is carefully spread into relatively thin and intermingling layers. This method creates a fairly homogeneous blend that is acceptable for mill feed.

In compliance with the Utah Mined Land Reclamation Act of 1975, the company filed a complete Notice of Intention and Mining and Reclamation Plan with the Division of Oil, Gas and Mining in March 1977. The division granted tentative approval for this plan later in 1977. The company submitted revised plans to the division in 1981 and 1988.

Topsoil and overburden are handled carefully to ensure the best possible use of the soil. Sufficient soils are salvaged at each site to topsoil the dumps and backfilled pits. These sites can then be seeded to provide a vegetative cover similar to the native plants in the area.

The company designed a test-plot program to evaluate varying topsoil thicknesses and fertilizer rates for revegetating future dumps. Another test-plot program evaluated the use of growth media other than topsoil for revegetating dump tops. Results from these studies indicate that existing topsoil and growth media substantially increase the area available for revegetation. During this same time, the company began using innovative techniques in dump construction, seedbed preparation, and reseeding. The company was subsequently awarded the division's 2000 Earth Day Award for its efforts.

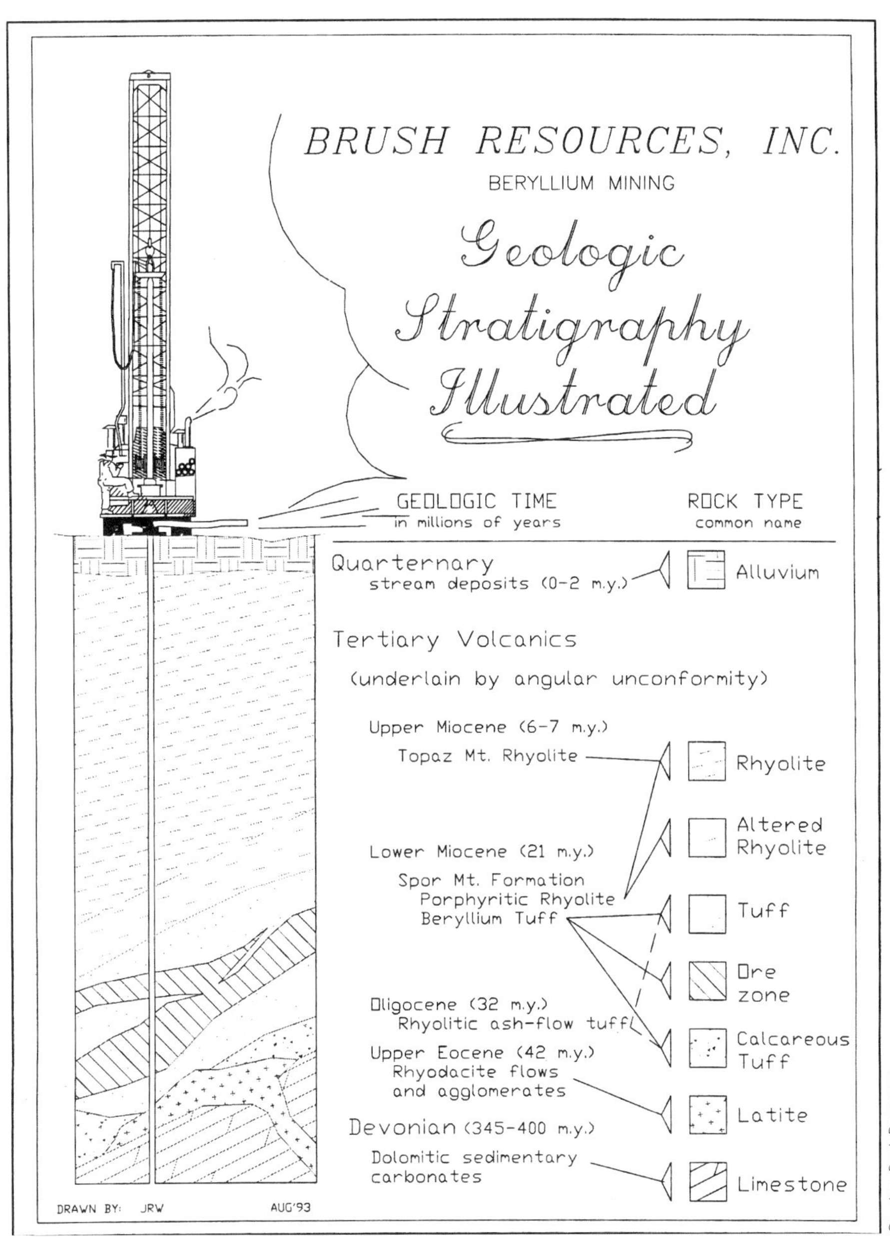

Courtesy Brush Resources.

Milling Operations

The company's Delta mill is capable of processing two types of ore—bertrandite ore from the mine and imported beryl ore. Different processes are used to solubilize the two ore feeds. The resulting aqueous steams, both of which contain beryllium in a common form, are blended together. An organic solvent then extracts the beryllium from the blended stream.

Beryllium Products

Beryllium metal offers an unmatched combination of physical and mechanical benefits. At one-third less density than aluminum, beryllium is one of the lightest structural materials available, yet pound-for-pound it offers nearly seven times the stability of steel. Beryllium has a very high heat-absorbing capacity and is an excellent thermal conductor. Dimensional stability is outstanding over a wide range of temperatures. Standard machines and methods can handle the metal. Beryllium is the preferred material for many complex, high-performance parts for aerospace structures, military systems, medical components, and audio and computer systems.

Beryllia ceramic, the oxide form of beryllium, offers a combination of performance features unmatched by any other ceramic material, such as excellent thermal conductivity, electrical insulating properties, and a low dielectric constant. Because beryllia ceramic dissipates heat, many designers of electronics use it to remove potentially damaging heat from dense circuitry. Beryllia ceramic is an ideal choice for a variety of products used in wireless telecommunications and the automotive industry and is often part of power circuits for motion control.

Beryllium alloys (copper and nickel based) offer many advantages, including high electrical and thermal conductivity, corrosion resistance, durability against stress and wear, extra strength and hardness, and good formability. Beryllium alloys are preferred for a wide variety of both consumer and industrial products. They are highly used for computers, telecommunications, automotive electronics, energy systems, appliances, plastic molds, and other thermal-management devices.

Courtesy of L. Tom Perry Special Collections, Harold B. Lee Library, Brigham Young University.

The mining camp home of the John Westenskow family in Sunnyside had a horseshoe hung with the open end up, to hold good luck, over the door and a box swing for the baby.

Courtesy Utah State Historical Society.

Miners load coal onto an ore car for the Lion Coal Company. The size of the rocks and the tight area where the miners are working give some indications of the difficulty of coal mining. Photo by William H. Shipler (9 March 1921).

Courtesy L. Tom Perry Special Collections, Harold B. Lee Library, Brigham Young University.

These miners spent long hours digging with picks and shovels to extract coal from deep inside Utah's mountains.

Courtesy L. Tom Perry Special Collections, Harold B. Lee Library, Brigham Young University.

In 1892 miners relied on horsepower to work this mine in Huntington Canyon.

Courtesy *Deseret Morning News* from copy donated by Robert Edwards. Glass plate in L. Tom Perry Special Collections, Harold B. Lee Library, Brigham Young University.

On 1 May 1900 coal gas ignited in the Winter Quarters Number-Four Mine at Scofield, killing more than two hundred miners. The blast knocked cars off their tracks outside of the mine.

Courtesy *Deseret Morning News*. Glass plate in L. Tom Perry Special Collections, Harold B. Lee Library, Brigham Young University.

One of the 107 women widowed and 4 of the 268 children left fatherless by the Scofield mine disaster sit in the foreground of the burial services in Scofield Cemetery. Photo by George Edward Anderson.

Courtesy Utah State Historical Society.

This Castle Valley Coal Company store and office building was located at Mohrland. Photo by William H. Shipler (taken 25 April 1911).

Courtesy Utah State Historical Society.

The interior of the Castle Valley Coal Company store shows the range of products for sale. Photo by William H. Shipler (22 February 1911).

Courtesy Utah State Historical Society.

The Castle Valley Coal Company tram waits with its load at the top of the hill. Photo by William H. Shipler (22 February 1911).

Courtesy Utah State Historical Society.

Officials of the Independent Coal and Coke Company pose in front of their office. Photo by William H. Shipler (15 September 1910).

Courtesy Utah State Historical Society.

The Independent Coal and Coke Company built housing for the miners who worked for them. Photo by William H. Shipler (1 December 1907).

Courtesy Utah State Historical Society.

The Independent Coal and Coke Company hotel stood out in Kenilworth, the company town. Photo by William H. Shipler (15 September 1910).

Courtesy Utah State Historical Society.

The store, market, and bakery were part of Independent Coal and Coke Company operations. Photo by William H. Shipler (15 September 1910).

Courtesy Utah State Historical Society.

Miners line up along the side of a mountain at the Lion Coal Company in Wattis. Photo by William H. Shipler (9 March 1921).

Courtesy Utah State Historical Society.

Loaded coal cars descend on the Kenilworth Independent Coal and Coke Company tram to the unloading station at the bottom of the hill. Photo by William H. Shipler (15 September 1910).

Courtesy Museum of Western Colorado.

A prospector checks the radiation reading on his Geiger counter.

Courtesy Western Mining and Railroad Museum.

Charlie Steen and companions examine ore samples underground.

Courtesy Sheldon A. Wimpfen.

Vernon Pick turned to prospecting after his auto-repair shop burned down.

Courtesy Duncan Holaday.

Duncan Holaday and Dr. W. F. Bale inspect a uranium mine.

Courtesy U.S. Department of Energy.

Shot Harry, part of Operation Upshot-Knothole and known as "Dirty Harry," was a 32-kiloton, weapons-related device fired from a tower.

Courtesy Special Collections, Gerald R. Sherratt Library, Southern Utah University.

Lindsay Hill Pit near Iron Springs is now used as the Iron County landfill. Photo by York Jones.

Courtesy Special Collections, Gerald R. Sherratt Library, Southern Utah University.

Blowout Pit (1961) on the south side of Iron Mountain, mined by Utah Construction from 1947 to 1968, is now locally known as the "toilet bowl" because it is half filled with turquoise-colored water. Photo by York Jones.

Frame scaffolding remained in 1996 at the Jennie Mine in the Gold Springs District on the Utah/Nevada border. Photo by Janet Seegmiller.

Courtesy Special Collections, Gerald R. Sherratt Library, Southern Utah University.

Iron County Coal Company reopened the Corry Mine on Lone Tree Mountain east of Cedar City in 1913. Photo by R. D. Adams.

Enos Wall recognized the possibilities of mining for copper in Bingham Canyon and organized the Utah Copper Company.

Courtesy Kennecott Utah Copper.

Courtesy Kennecott Utah Copper.

Daniel Cowen Jackling realized that mining in Bingham could become productive by removing large quantities of overburden.

Courtesy Utah State Historical Society.

The town of Bingham in 1903. In the center of the picture is the mountain that was known as Copper Hill. That mountain no longer exists; today the Bingham Canyon open-pit copper mine has taken its place.

Courtesy Kennecott Utah Copper.

This early smelter, under construction near the Great Salt Lake, was completed in 1908 by American Smelting and Refining Company.

Courtesy Kennecott Utah Copper.

Social opportunities in early Bingham Canyon were limited. But Bingham's White Elephant Saloon offered some alternatives. This 1907 photo shows players at the faro table, and the stacks of chips indicate the stakes were high.

Courtesy Kennecott Utah Copper.

Early equipment for moving copper ore included Shay steam engines, used between 1900 and about 1912. Here a pair chugs up an incline in 1907. Miners said a man could walk up the hill faster than the engines moved, and the trip downhill with loaded cars was very scary.

Courtesy Kennecott Utah Copper.

Steam shovels began operating in Bingham Canyon in 1906. This Marion shovel awaits the arrival of ore cars on one of the early mine terraces.

Courtesy Kennecott Utah Copper.

Getting around in Magna in the early days was not easy. This 1915 photo of Main Street shows at least one grocery store, two automobiles, and the rough and rutted road they had to travel.

Courtesy Kennecott Utah Copper.

Bingham Mine operators used a smaller engine for shorter hauls. Called a Dinkey, it weighed about 20 tons. This 1916 photo shows a Dinkey moving wooden ore cars into position for loading. The cars could carry about 12 tons of copper ore.

Courtesy Kennecott Utah Copper.

The town of Bingham appears in the lower left of this 1919 photo, as the now-terraced Copper Hill is being slowly mined away.

The town of Carr Fork was located above Bingham. This 1925 photo shows its railroad viaduct, which in its day was considered an engineering and construction marvel.

In 1947 as electric power took over, this Mallet "110" engine was the last of Kennecott Utah Copper's steam-powered locomotives.

Courtesy Kennecott Utah Copper.

In 1940 an ore train crosses Dry Fork railroad viaduct in Bingham Canyon.

Courtesy Kennecott Utah Copper.

Environmental concerns and the Clean Air Act of 1970 prompted modernization of Kennecott's smelter, including this huge stack, completed in 1978, which stands 1,215 feet high.

Courtesy Kennecott Utah Copper.

This 1944 photo shows construction of a new giant power plant for Kennecott, ultimately rated at 175,000 kilowatt-hours. Modernization saw electric locomotives pulling huge trains of cars, electric shovels moving 16 tons of rock in one bite, and efficient electric processing centers.

Courtesy Kennecott Utah Copper.

By the mid-1980s rail haulage in the Kennecott mine was phased out and replaced by huge-capacity diesel-electric trucks. This ore truck could carry 255 tons.

Courtesy Kennecott Utah Copper.

In 1954 the State of Utah observed Kennecott Utah Copper's 50th anniversary. On August 14, Daniel C. Jackling's 85th birthday, the state placed this bronze statue in the Utah State Capitol Building rotunda.

Courtesy Kennecott Utah Copper.

This recent photo shows the vast topographical changes due to the Kennecott Utah Copper mine and its milling operations. The original Copper Hill, where mining began, along with a huge surrounding area, is today the world's largest open-pit mine, more than three-quarters of a mile deep and two-and-a-half miles wide. At the lower center of the photo is Kennecott's Copperton concentrator. The open pit is so large it is easily spotted by astronauts.

Courtesy Kennecott Utah Copper.

During World War II, many women took over jobs traditionally held by men who left for the armed services. These women are working in Kennecott's mines and smelters.

Courtesy Kennecott Utah Copper.

Part III

Major Mining Regions

9

Iron County

Janet Seegmiller

Mining has come full circle in Iron County. Explorers in the early 1850s found vast reserves of iron and coal, leading to the settlement of the area. Deposits of silver, gold, lead, fluorspar, and gypsum were discovered later. These finds produced headlines in local newspapers, but only iron and coal mining prospered, and that occurred almost 100 years after settlement. At the turn of the twenty-first century, little mining remains. Despite the mineral resources available, it is currently easier and cheaper to mine elsewhere.

Iron Mining and Manufacturing

Without question Iron County is accurately named.[1] Within its borders lie the richest and most accessible iron-ore bodies in the western United States.[2] The mining district is 3 miles wide and 23 miles long, occupying only 69 of the county's 3,300 square miles. However, economically and historically, its impact has been greater than any other natural resource.

The iron-ore bodies were created in Tertiary times, when igneous intrusions of molten rock, or magma, pushed up toward the Earth's surface, forming bulges between layers of a blue gray Jurassic limestone called the Homestake formation. The magma hardened into quartz monzonite. Iron-rich emanations, as liquid or gas, followed the perimeter of the igneous rock up, creating deposits of magnetite and hematite, iron ores. Iron ore occurs in cracks and fissures in the quartz, indicating that the iron deposit took place after the intrusive igneous rock had cooled to a solid state. Replacement ore is characteristically a blend of hematite and magnetite iron, whereas the ore in fissure deposits is usually high-grade magnetite.[3]

Geologists believe the intrusions were formed under a surface cover that was 2,000 to possibly 8,000 feet thick. Subsequent erosion has partially stripped the cover from three prominent laccoliths or domes, namely Iron Mountain, Granite Mountain, and Three Peaks, exposing the monzonite cores. The iron ore exists on the tops

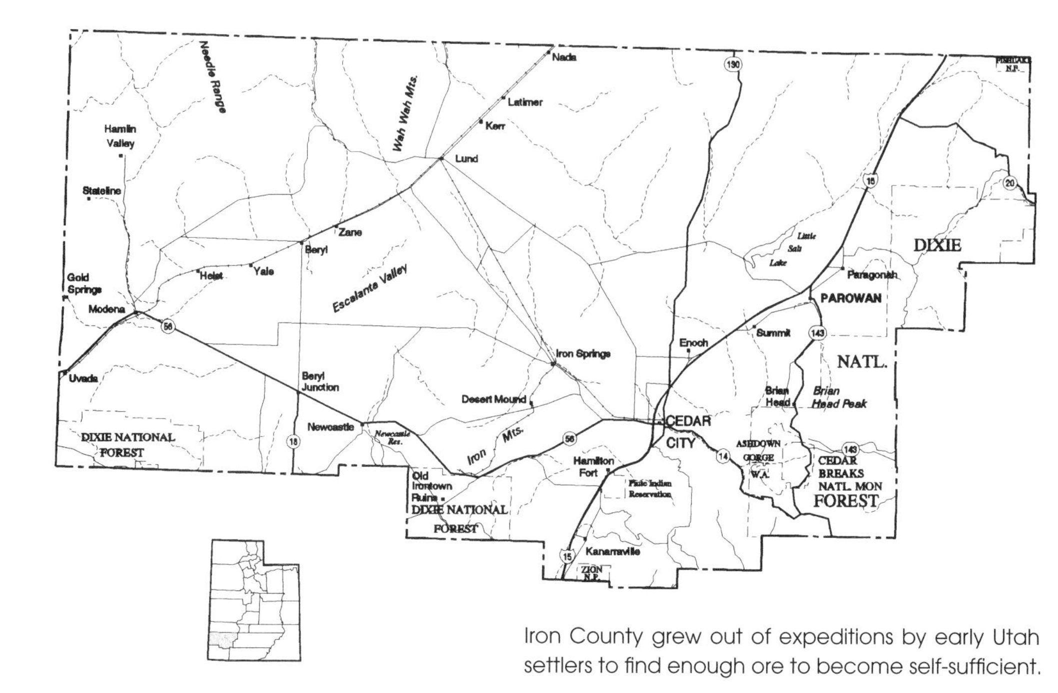

Iron County grew out of expeditions by early Utah settlers to find enough ore to become self-sufficient.

and around the flanks of these laccolithic humps down to the floors of the surrounding valleys. Some of the ore bodies have been exposed, totally or partially eroded, and then reburied with sediment during the intervening thousands of years.[4]

The rich iron deposits of southern Utah attracted the earliest interest as they lay along a section of the Old Spanish Trail, first used in its entirety in about 1830 for commerce, slave transportation, and emigration from Santa Fe to Los Angeles. In July 1847 Mormon pioneers escaping religious persecution in Illinois and Missouri arrived in the Salt Lake Valley under the direction of Brigham Young and immediately began exploring the Great Basin. They were especially eager to establish a corridor of Mormon colonies from Salt Lake to southern California. In October 1849 apostle Charles C. Rich, on his way to select a settlement site in San Bernardino, recorded the discovery of iron ore near Iron Springs, 25 miles southwest of the Little Salt Lake (now Parowan) Valley. Traveling companion Addison Pratt, a missionary bound for the Society Islands, also wrote about the ore, noting its abundance.[5]

In 1849 discovery of gold in California was less important to Brigham Young than Pratt's report of "immense quantities of rich iron ore" near the Little Salt Lake. Self-sufficiency was Young's goal for the pioneers in the Great Basin. Gold and silver mines he could live without because they would attract non-Mormons to the area, but iron was essential for homes, farms, factories, and transportation. Therefore, a colony in what was to be Iron County was a priority.

In November 1849 Young placed apostle Parley P. Pratt at the head of the Southern Exploring Expedition with the commission to explore the area and evaluate potential locations for settlement. Pratt recommended establishing an outpost on the Little Salt Lake and also spoke highly of Cedar Valley, "a large body of good land on the Southwest borders" with "a hill of the richest Iron ore" rising in the middle of "thousands of acres of cedar contributing an almost inexhaustible supply of fuel." [6]

In 1850, apostle George A. Smith was called to lead a group of iron missionaries—119 men, 30 women, and 18 children in 101 wagons—on an expedition to create a community that would sustain a full-fledged ironworks in the middle of the desert. After building a fort in early 1851, precursor to the city of Parowan, near Center Creek and planting spring crops, the pioneers discovered coal on the banks of the Little Muddy 20 miles to the southwest and renamed it Coal Creek. Exploring farther, the settlers found "several veins" of coal "varying from 3 inches to 3 feet in thickness" five miles up the canyon. [7]

With the discovery of significant coal deposits in Coal Creek Canyon, the iron missionaries decided to build the ironworks on the banks of Coal Creek some 5 miles from the promising new fuel source and 10 miles from the iron-ore deposits. In November 1851 Henry Lunt and two companies of men moved to Coal Creek to establish a settlement, soon named Cedar City, and began preparing to produce iron. In May 1852 Burr Frost successfully smelted iron ore in his blacksmith's forge and produced a small amount of iron that was shaped into nails. To further the venture, Brigham Young organized an iron-manufacturing company with Richard Harrison as superintendent, Henry Lunt, clerk, Thomas Bladen, engineer, and David B. Adams, furnace operator.

One month later Henry Lunt reported "a considerable amount of work has been done," toward making iron:

> The fire bricks for the furnace were ready for laying and have proved to be of the best quality. The timber which was needed for the framing of the machinery was hauled from a canyon five miles south of Coal Creek. . . . an extensive blacksmith's shop was completed. The iron work for the machinery was progressing rapidly. The iron for this purpose was mostly obtained by taking tires from their wagons, expecting to replace them from the Iron Works, as the settlers felt very hopeful that they, in a few weeks, would have iron of their own manufacture. . . . brethren from Parowan and [Cedar City] have been working out their taxes in making a road up Coal Creek Canyon during the past week. The road is made within one mile of the coal.[8]

However, differences divided the men, and leadership roles were not clear. With so many tasks demanding immediate attention in the new frontier communities, the pioneers clashed over whether to devote limited manpower first to farming chores or iron production. Furthermore, Mormon converts from the British Isles with experience in iron manufacture had understandable difficulty adapting their skills to conditions in southern Utah, and disputes arose over matters of procedure, such as how to test the iron ore.[9] Erastus Snow and Franklin D. Richards would write later, "[W]e found a Scotch party, a Welch party, an English party, and an American party, and we turned Iron Masters and undertook to put all these parties through the furnace, and run out a party of Saints for building up the Kingdom of God."[10]

The first test of their blast furnace happened on 29–30 September 1852. One hundred loads of coal were hauled and coked, loads of dry pitch pine were brought to mix with the coke, limestone was hauled to charge the furnace, and tons of iron ore were crushed with sledge hammers and brought to the furnace. Men worked all day and night charging the furnace and tending it. At six o'clock in the morning, they crowded in front of the furnace while an ironworker tapped it and a molten stream ran out. Instantly pent-up anxiety broke loose in a spontaneous cry of joy. On the spot five men were chosen to carry samples of the pig iron to Brigham Young in Salt Lake City to display at the October church conference. Although the metal had a peculiar appearance, they had made the first iron west of the Mississippi.[11]

During the conference, Brigham Young called for 100 families to strengthen the Cedar City colony. George A. Smith preached an "iron" sermon, promising that plows and kettles, guns and wagon wheels, sawmill cranks, nails, and door trimmings would all be made from Cedar City iron.[12] Ironworkers, coal miners, blacksmiths, and farmers were recruited, and by November and December, they began arriving at the ironworks, bolstering the spirits of the earlier settlers.

The enterprise desperately needed capital. At Brigham Young's request, apostles Erastus Snow and Franklin D. Richards organized the Deseret Iron Company in England on 28 April 1852. Snow and Richards obtained subscriptions totaling 4,000 British pounds ($19,360) from wealthy church members in the British Empire, and they visited ironworks in England, Wales, Ireland, and Scotland to obtain more information on making iron. Upon their arrival in Cedar City in November 1852, they bought out the pioneer iron company for $2,866, absorbing many of the original workers into the new organization. This was probably the first foreign-owned mining operation in the Intermountain West.[13] With the establishment of the Deseret Iron Company, some ironworkers felt the missionary aspect of the venture was finished and they had no further obligation to stay, even though leaders counseled otherwise. A number of iron missionaries returned to the northern settlements; others went to California.

Between September 1852 and September 1853, through experimenting and great effort, the furnace produced about 20 tons of pig iron from approximately 50 tons of ore, even though iron masters were still not satisfied with the quality. From the pig iron, a variety of skillets, andirons, kettles, wheels, and other goods were successfully cast in sand moldings. Henry Lunt took the iron products to Salt Lake, where President Young displayed them at the April 1853 General Conference as the first cast iron made by the Latter-day Saints.

The infant industry suffered several setbacks. On 3 September 1853, a mountain cloudburst sent a flood down Cedar Canyon which swept away dams, bridges, and the road to the coal mines. Three feet of water inundated the ironworks, which lay east of the creek, leaving 10 inches of mud inside the furnace and buildings. Hundreds of bushels of charcoal, lumber, and wood were carried away. The flood, coupled with the Walker Indian war, forced Brigham Young to shut down the ironworks temporarily and put its employees to work on the adobe fort encircling Cedar City.

After rebuilding and repairing the dam and other breaches along the creek, the Deseret Iron Company started its first trial on 9 January 1854, but the weather turned bitterly cold, freezing the creek and stopping the waterwheel and air blast, thus shutting down the furnace. Company leaders decided to build a new furnace west of Coal Creek with a larger waterwheel and air cylinder, and they set it on a rock foundation. The impressive new stone-masonry structure, called the Noble Furnace by the pioneers, was completed in September 1854 and had a sandstone lining. Although the furnace seemed to work well, the output was disappointing. Different charges were used to compensate for the minerals that did not combine well to make good pig iron. By April output was improved, and a total of 10 tons of "good iron" was produced, including 1,700 pounds during one 24-hour period. From this run the company made a bell, which is the only known casting of Deseret Iron still in existence.[14]

The company still lacked the capital to accomplish its assignment, and workers suffered greatly without suitable clothing, bedding, and other common comforts for themselves and their families. Following his spring visit in May 1855, Brigham Young concluded that "the brethren have done as well as men could possibly do, considering their impoverished circumstances, and the inconveniences they have had to labor under. They have probably progressed better than any other people would upon the face of the earth."[15]

Many problems threatened to close Deseret Iron during 1855 and 1856, including a food shortage which left both humans and animals near starvation. Some workers packed up their families and moved away. During the winter, severe weather cut off the supply of coal and froze the creek and waterwheel. Summer drought stopped the waterwheel again, and there were more mechanical breakdowns. A May 1856 announcement in the *Deseret News* asked for 150 more workers and 50 additional wagon teams to supply the furnace with fuel and ore.

In the spring of 1857, Brigham Young's steam engine from his sugar mill in Salt Lake City was sent to remedy the inconsistent water flow from the creek. By July it was in place and working, but the workforce continued to drop, and the iron company struggled. The men spent their time repairing broken machinery and making improvements to the furnace. They had barely restarted the ironworks in late summer when orders came to suspend all production and turn attention to harvesting grain due to the approach of federal troops under Colonel Albert Sidney Johnston. Amid the turmoil created by the army's approach, men from the area were involved in the tragic attack on a wagon train of Arkansas emigrants at Mountain Meadows in September 1857. The aftermath of this event discouraged and disquieted many in the county. The exodus from Cedar City continued.

Iron making resumed the next spring despite the remaining specter of war. Members of the Iron District Militia were sent to scour the mountains for a refuge site if Johnston's army forced the Saints to flee their homes. In April 16 men and teams went to the mines near Las Vegas to bring back lead ore to make into bullets.[16] A new furnace was finished in April, and new trials began. However, the lining of the furnace gave way in May, September, and October.

During the summer, Johnston's army entered the territory and established Camp Floyd southwest of Salt Lake Valley, bringing with it a large inventory of iron in the form of wagons and weapons. In October 1858 Brigham Young ordered Isaac Haight to close down the ironworks. His letter said, "Put everything in as good a condition for preservation as possible and let it rest. Such fruitless exertions to make Iron seem to be exhausting not only the patience, but the vital energies, and power of the settlement."[17]

Within five months, the population of Cedar City dropped by two-thirds.[18] Those with means left. Those who had given their wagon wheels to the ironworks remained behind. Ironworkers had little to show for their labor. What little pay they received had come in goods from the company store. Two who remained, Joseph Walker and John Pidding Jones, made the last run of the ironworks in 1860 by melting down seven wagonloads of federal cannonballs and then casting flatirons, dog irons, molasses rolls, sawmill and gristmill irons, grates, and other implements. In 1861 Erastus Snow took all the removable assets, including machinery, to the new colony at St. George.[19]

The ironworks failed to live up to expectations for reasons the settlers could neither understand nor control. Location of the works required time-consuming transportation of both coal and iron. Inadequate and erratic flow of Coal Creek did not permit consistent furnace operation. Local clay and sandstone proved to be poor firing material for lining the furnace. Coal mined from Cedar and Right Hand Canyons slacked too quickly, which reduced the efficiency of the furnace, and contained too much sulfur, which produced brittle iron. Lack of adequate financial backing kept the company from paying its workers; they and their families often were barefoot or hungry. Furthermore, the unknown chemical makeup of raw materials caused metallurgical problems in smelting the ore. Experience could not overcome lack of scientific knowledge. Disagreements on technical aspects of iron making had a lasting effect on the morale and efficiency of the Iron Mission.[20]

The reward for most of the ironworkers was knowing they had done what was asked of them by their leaders. They had not come to get rich but to establish an industry to help every community in the Great Basin. Perhaps the most remarkable aspect of the Iron Mission was the perseverance of the people in continually trying to make iron and build their community while living in abject poverty.

In the years after 1860, John P. Jones supplied some badly needed iron implements by building a small cupola furnace along Coal Creek and utilizing a large waterwheel which was shared with a cabinet shop and flour mill. Later at Johnson's Fort (now Enoch), he built a blacksmith shop, where he made tools, and eventually

another cupola furnace, a foundry, and a coke and charcoal oven. Utilizing scrap iron, Jones painstakingly molded stove and fireplace grates, skillets and irons, horseshoes, and horseshoe nails. The largest single casting was a 500-pound hammer to drive piles for the dam being constructed on the Virgin River by the St. George and Washington Irrigation Company. Cogwheels, shafts, tracks, wheels, pulleys and rollers were made for sawmills, the molasses mill in Washington County, and mines at Silver Reef and in Lincoln County, Nevada. Mine operators were the best cash customers of John Pidding Jones and Sons Iron Company. Their furnace operated on and off for nearly 20 years.[21]

With high hopes a second phase of iron mining and manufacturing began in Iron County in July 1868. A company was formed by several of southern Utah's more successful businessmen: Ebenezer Hanks, Peter Shirts, Chapman Duncan, Seth M. Blair, and Homer Duncan. They called it the Union Iron Works or Pinto Iron Works. Their location was at Little Pinto Creek, some 23 miles southwest of Cedar City, close to ore fields at the south end of Iron Mountain. Dr. T. L. Scheuner, a Swiss metallurgist, was superintendent. A smelting furnace, a beehive charcoal oven, and a number of buildings to support the ironworks were built along the creek at a place now known as Old Irontown but officially named Iron City.[22] By 1870 a fair-sized settlement, complete with shops, homes, a post office, and farms, was in place. The 1870 census shows 89 persons in Iron City: 27 males and 24 females over 12 and 38 children in 19 households.

The furnace soon was producing 800 pounds of good-quality iron every eight hours around the clock. The challenge, however, was not iron production but selling the product. Seth Blair wrote to the *Deseret News* to plead for a foundry somewhere in Utah to buy their cast iron and turn it into steel or wrought iron. Between 1868 and 1871, large supplies of iron ore and needed materials were gathered for continuous furnace use, and the operation produced many machinery parts and household implements. However, because it lacked capital and laborers, when the materials ran out, production ceased, and the furnace shut down. The largest project sent pig iron to Salt Lake City, where it was cast into 12 oxen for the St. George Temple baptismal font.

To infuse more capital into the venture, Union Iron Works was taken over by the Great Western Iron Mining and Manufacturing Company, with Thomas Taylor, a Salt Lake businessman, and his son-in-law, John C. Cutler, as major stockholders.[23] Cutler later became governor of Utah. Litigation prevented Taylor from developing his properties for some time. Despite continual legal problems, however, Taylor added the land holdings of Ebenezer Hanks to his own on 8 January 1881.[24]

A committee of the Zion's Board of Trade from Salt Lake City organized another company, the Utah Iron Manufacturing Company, in August 1881. It obtained properties at Iron Springs from Thomas Taylor, Henry Lunt, and the LDS Church, which had secured patents on coal and iron reserves in 1880. Utah Iron's plan to construct a 150-mile railroad from the mines to the Utah Southern Railroad terminus at Juab was not implemented. Its mining claims were challenged, and it spent the next three years in litigation, trying to document its best claims.[25]

Meanwhile, the LDS Church's First Presidency and Board of Trade attempted to develop uncontested claims at Pinto and Iron Springs by forming the Iron Manufacturing Company of Utah (IMCU). Many shares went to Thomas Taylor in return for his properties. IMCU obtained coal claims in Cedar Canyon and purchased Great Western Iron's plant at Iron City, including the blast furnace, machine shop, engine house, pattern shop, foundry, store, schoolhouse, and residences. Increased capital was essential, and church leaders were determined that it should come from within Utah, rather than lose control by selling stock to eastern or Gentile (non-Mormon) interests. Church members were encouraged to subscribe to IMCU stock, but most of the subscriptions were promises of labor and material rather than cash.

President John Taylor received permission from LDS church members at the April 1884 General Conference to put church funds into the ironworks, which enabled IMCU to buy the Pioche and Bullionville Railroad, a narrow-gauge line with 20 miles of rails, two locomotives, 25 cars, a roundhouse, and other equipment. Tracks were to be laid between the coal mines in Cedar Canyon, iron deposits at Iron Mountain, and the furnaces at Old Irontown. The road was to transport itself by repeatedly extending the rails in front of the engine, moving the engine and cars onto the rails, removing the rails from behind, and again placing them in front. However, the method proved too time consuming, and the railroad equipment was finally transported by oxcart and wagon from Jack Rabbit, Nevada, to Cedar City, a distance of 80 miles. Though some grading for the railroad was done, no tracks were ever laid.

Mormon church leaders, forced into hiding to avoid prosecution for polygamy, were unable to pursue development of the iron company. Thomas Taylor claimed to have opportunities to sell the company's properties but felt that he was prohibited by the stalling actions of George Cannon, who held the mortgage papers after President John Taylor's death.[26] In 1886 a Cedar City "observer" wrote the *Salt Lake Herald*, "Today, as far as the iron industry is concerned, we are quiet as a church yard, and nothing left to remind us of our past hopes and great anticipations, but the roadbed . . . a few pair of railroad car wheels, a portion of a locomotive and tender, and a few hundred feet of rails, all of which seems to be quietly laid away, at least until times brighten up."[27]

After 1872, when federal mining law established rules for claim location, annual assessment work, and patenting procedures, hundreds of prospectors covered the Iron County mineral belt. From 23 recorded claims in 1880 in the Pinto and Iron Spring Mining Districts, the number grew to more than 100 in 1900 and more than 1,000 patented and unpatented mining claims in 1922, most of them worthless. Patenting was done by those hoping to interest investment capital to develop the area or by companies, such as Colorado Fuel and Iron Company (CF&I), which purchased and patented many claims for its steel plant near Pueblo, Colorado, the first in the western United States.

Between 1899 and 1923, people who earnestly believed that Iron County would yet become a great iron-producing area waited for the right combination of demand and capital. Events that augured well for the eventual development of an iron industry included completion of the Los Angeles to Salt Lake City branch of the Union Pacific Railroad, increased demand for iron and steel products on the West Coast as population increased, and the opening of several small-scale steel plants in California.[28]

Columbia Steel Company, a leader in California's steel industry, became interested in the county's iron-ore deposits because of the close proximity of large coal deposits in Carbon County. A 1922 feasibility study showed "there exists a body of coal and iron ore in Utah with other raw materials necessary for the production of pig iron and that are available at a comparatively low cost and can be assembled at some point near Salt Lake at as low a figure as any other similar materials are assembled in other parts of the United States."[29] L. F. Rains, president of Carbon Fuel Company, had already purchased or located iron-ore claims on the north side of Granite Mountain that were ultimately sold to the new Columbia Steel, merging the California facilities with the Utah iron and coal properties. Columbia Steel began building a blast furnace south of Provo near Springville, equidistant from coal and iron sources. Limestone was available nearby, as well as an abundance of water at Utah Lake. The site became Ironton.

While the furnace was under construction, the coal mines being readied, and iron mines opening in Iron County, the Union Pacific built a branch railroad from the main line at Lund through Iron Springs Gap to Cedar City. The tracks were brought to Cedar City in less than three months from April to June 1923. The Milner spur was also constructed to the Pioche Mine about a mile south of Iron Springs Gap. By April 1924 coal and iron ore were being shipped to the Ironton plant. On 30 April 1924, the furnace was charged and blown in, and three days later 150 tons of pig iron were on their way to the Pacific Coast. Iron County mines were alive again. The commencement of Columbia Steel Works was celebrated at Utah Steel Day on 13 June 1924, and the old iron bell cast in 1855 in Cedar City was exhibited at the celebration.

Columbia Steel first mined its own Pioche and Vermillion ore bodies using a modified *glory hole* system. A drift or tunnel was driven under the ore body, and a series of raises or shafts were dug to the surface. At the top of each shaft, a heavy grid or screen was installed, covering the shaft opening. Blasting or jackhammers broke up the ore; then heavy draft-horse teams pulling a scraper conveyed it over the grids. The ore passing the screen fell down the shaft into five-ton pit cars on a narrow-gauge rail system. The cars were hauled outside the mine to a crushing and screening plant, where the sized ore was loaded into railroad cars for shipping. There were enough five-ton cars in the mine to transport 1,000 tons of ore daily. However, the system was slow, inefficient, dangerous, and expensive.

Within a year Columbia determined that the chemistry of the ore was not exactly right for the blast furnace. Investigations showed that the ore at Desert Mound was better, and in May 1925 Columbia Steel contracted with Archibald Milner and Brothers, principals in the Utah Iron Ore Corporation, for 1.5 million tons of ore from Desert Mound to be furnished to Ironton at a minimum rate of 500 tons per day. Utah Iron built a three-and-one-half-mile branch-railroad line to Desert Mound and its 527 acres of patented ore in the Iron Springs District. Milner's reserves were estimated at 15 million tons within a depth of 100 feet.

Utah Iron mined by open-pit blasting. The ore was loaded by a single, small steam shovel onto cars on a narrow-gauge rail system and transported to a processing plant, where it was crushed, screened, and shipped to Ironton. Eventually the grade out of the open pit became too steep for the railroad, and dump trucks replaced it. The steam shovel could handle 300 tons per eight-hour shift, and so a second (and sometimes third) shift was added to meet the contract of 500 tons per day. (By comparison in the 1960s, any one of the five crushing plants in the district could produce 500 tons per hour.)

From 1924 to 1936, Utah Iron mined 2.4 million net tons of iron ore with 1.5 million tons supplied to Columbia Steel, 778,350 tons sent to the CF&I furnace in Pueblo, and 134,000 tons sold for flux to various foundries and smelters. Desert Mound went out of production in 1936.

Iron Springs was at its height between 1924 and 1936. A post office was located in the branch store of the Cedar Mercantile Company, and the school board moved a schoolhouse to the town from Yale. The school operated from 1924 to 1930 with Leslie Green, Geneva Heaton, Kate Isom, and Grace Bates as teachers. About 40 men were employed at the Pioche Mine. Community baseball was a favorite pastime, and the team from Iron Springs played other community teams and ones from the CCC camp during the 1930s.

The purchase of Columbia Steel by United States Steel Corporation (U.S. Steel, later USX) in 1929 significantly impacted iron mining in southern Utah. U.S. Steel acquired Columbia's properties in Utah and California and set up its own mining operations under its subsidiary, Columbia Iron Mining Company, which operated in Iron County from 1935 to 1985. Due to the depletion of suitable grades of ore at Desert Mound, Columbia moved its mining operations to Iron Mountain. Twelve miles of track were laid to extend the railroad from Desert Mound to the south side of Iron Mountain. A new crushing plant started construction in August 1935 and went into operation nine months later.

Open-pit mining began at the Black Hawk outcrop and utilized the first electrically powered shovel brought to Iron County, plus an electric-powered churn drill to make the blast holes and two specially built 24-ton Mack trucks. Unfortunately, the hard, dense magnetite from Black Hawk proved more rugged than the new equipment, causing many breakdowns. However, Black Hawk ore had more iron than other available ores. Since it improved furnace performance, demand increased. Ore shipments to Ironton rose from 175,000 tons per year in 1936 to nearly 300,000 tons in 1941.

Prior to the Japanese attack on Pearl Harbor, United States defense plans included locating an inland steel mill somewhere in the West as a precaution against possible closure of the Panama Canal. A site at Orem, Utah, was selected because it was safe from possible air attack and was equal distance from the major naval bases. In addition, there were good transportation systems and plentiful sources of coal, iron ore, and water within a reasonable distance. The Geneva mill required four times as much iron ore as the Ironton plant. Between 1940 and Geneva's opening in 1944, the number of mine workers in Iron County increased from 21 to more than 300. Shifts and work hours multiplied to provide iron ore for Geneva, Ironton, and other plants supplying steel for the war effort.

A second major steel operation was added when CF&I contracted in 1943 with Utah Construction Company (UCC) to build a loading plant and open a mine on the Duncan claim, about a mile southwest of Columbia Steel's mines. The first six-month contract launched a 40-year operation in Iron County for UCC. It worked the Duncan Mine, then the deep and spectacular Blowout Mine, and finally the large Comstock ore body on the northeast side of Iron Mountain. In 1944 UCC purchased and leased property on both sides of the railroad near Granite Mountain and Three Peaks. A crushing plant and loading facility were constructed on the south side of the tracks in Iron Springs Gap. Iron ore mined by UCC was sold on the open market, primarily to Kaiser Steel Company at Fontana, California.

No labor union represented iron miners in the county until 1943, when United Steelworkers of America representatives met with Columbia Steel miners to organize a local in the mining district. This union was part of the Congress of Industrial Organizations (CIO) labor group. When UCC began its mining activities in the county, its workers were members of the Construction Trade Unions, part of the rival labor organization, the American Federation of Labor (AFL). They earned higher wages than Columbia Steel workers, which resulted in contention over wages and contracts at the Columbia Steel mines for a number of years.

The Geneva mill operated for 21 months as a defense plant and stopped production on 3 September 1945, just weeks after Japan surrendered. When the Geneva plant and its wartime facilities were first offered for sale in 1946, companies were not interested. Political pressure by President Truman and Utah's congressional delegation finally elicited six bids. The one from U.S. Steel was the most favorable. The company spent $47.5 million for the plant, which cost about $200 million to build. U.S. Steel then spent about $17 million converting the plant to a peacetime operation. The purchase of Geneva was of tremendous importance to Iron County. The local newspaper editor commented, "Completion of the sale assures the peace-time operation of this great war developed plant and brings to the west its greatest chance of industrial development. . . . And since the Geneva plant is dependent upon the ore from Iron County mines to feed its blast furnaces, Cedar City immediately takes its place as an important cog in the industrial development of the West, and will benefit tremendously."[30]

Ore requirements for Geneva's blast furnaces mandated an increase in mining production as well as a more evenly blended furnace feed. Geneva Steel and Columbia Iron Mining decided to blend low-grade ores of 40 percent iron with higher-grade ones. A blending facility was built at Geneva to allow use of the lower-grade ores, a step in conserving available resources and fully using the iron as it was mined.[31]

In 1949 Columbia Iron Mining reopened the Desert Mound ore body and the Short Line deposit next to it. Columbia also contracted with UCC to remove some three million cubic yards of overburden from the planned pits. In response to an edict from the local power company that mining equipment could operate only at night because Escalante Valley farmers needed all the daytime power for irrigation, Columbia built its own power plant with three diesel generators at Iron Mountain. Company-owned power lines were strung from the plant to Desert Mound, helping assure Geneva Steel an uninterrupted and adequate supply of iron ore.

During the 1940s more than 17.4 million net tons of ore were mined in Iron County, five times more than during the previous 87 years. The 1950s proved to be the largest production decade in history. Combined shipments from UCC and Columbia Iron Mining exceeded 41.85 million tons. More than 600 people were employed in mining, and the county benefited from high wages and a mine-oriented tax base.

During the 1950s the unions were strong, and strikes every three or four years by the United Steelworkers of America hurt the local economy and threatened the existence of Geneva Steel. In 1949, as part of a national strike, 165 local workers were out for 6 weeks. In 1952, 220 local workers went on strike for 10 weeks, extending their strike past the national settlement to resolve a local pay issue. In 1956, 241 members of the local union struck from July to November. By a domino effect, some UCC and railroad workers also were laid off during these times. On occasion separate railroad-union strikes resulted in curtailed production and layoffs of mine employees. Local businesses felt the consequences, which sometimes persisted for months after the strike settlement as families recovered from the loss of income. Some businessmen resented the high wages paid miners because they caused discontent among their own workers.

Settlements may have appeared worthwhile, but in the end they hastened the downfall of the American steel industry. Hourly wages in the steel industry were the highest in the country in the 1950s and 1960s. However, by the 1960s Japan, operating the most modern steel plants in the world, could ship iron ore from Utah, fabricate steel products, ship them back to San Francisco, and still undersell U.S. Steel.[32] Thus, during the decade of greatest demand and production, the seeds of decline were sown.

Two other issues affected U.S. Steel's long-range plans in Iron County. Development of the 100-million-ton Rex ore body was held up by contested mining claims, requiring years of litigation and demands for extensive royalty payments. The other issue was additional local taxation, deemed unfair by U.S. Steel. Some residents were concerned that ore reserves were being exhausted without sufficient return to the county. In 1949 Iron County attorney Durham Morris drafted legislation, introduced by State Senator L. N. Marsden and Representative E. Ray Lyman in the state legislature, that was designed to increase tax revenues from mining. In 1950 the legislature approved the "net-proceeds tax," to be levied on all iron-ore shipments. Mining companies' protests resulted in some modification, but the companies still faced a large tax increase after 1951. The value of the iron mines in the state, all located in Iron County, was set at $24,177,127 for 1951 tax purposes, which accounted for 67 percent of the total county tax value. By comparison the tax value of the same properties in 1950 had been $3,737,415, just 23 percent of the county's total value.[33]

A lengthy lawsuit ensued with U.S. Steel arguing that Iron County and the state were wrongfully collecting taxes. Fifth District Judge Will L. Hoyt upheld the Utah State Tax Commission's right to set a value on ore for tax purposes when the ore was sold under contract between two subsidiaries of the same parent company, in this case Columbia Iron Mining and Geneva Steel. The court's decision was important to the county and impacted taxes collected in 1949 and subsequently. Thereafter, mining companies paid a net-proceeds tax on ore mining, a mine-occupation tax to the state, taxes on patented mining claims, taxes on equipment and fixed assets, and state and federal corporate income tax, as well as large royalty payments to patented claim owners. In response U.S. Steel drastically curtailed mining operations in Iron County in 1962 and opened the Atlantic City Mine in Wyoming. Tonnage shipped was cut in half to 2 million tons each year.[34]

During the 1960s and 1970s, UCC mined and improved ore using a $1.3-million ore-beneficiation mill built in 1961, which concentrated low-grade ores. The mill used a washing-flotation and magnetic-separation system. Beneficiation allowed more complete utilization of ore reserves, saved railroad freight by not shipping waste rock, and furnished better iron ore to the blast furnaces. UCC also developed a 500-ton, mobile, dry-magnetic separation unit to upgrade ore found in the alluvium fans surrounding major ore bodies. UCC shipped concentrate to Geneva Steel and cement plants in Utah, Idaho, and the Pacific Northwest, where the ore was used to give special properties to cement. The mill also concentrated low-grade ores from other companies in the area.[35]

UCC mined its own properties as well as Blowout (1947–68),[36] Comstock (1954–81), Queen of the West (1956–67), and Mountain Lion (1970–81), owned by CF&I. Purchases of new and heavier equipment in 1975 indicated company commitment to improved production, automation, and safety. Four 75-ton trucks, a new rotary drill rig, and a mammoth 10-cubic-yard electric shovel weighing more than 400 tons and costing a million dollars, were major purchases. During the 1970s Utah International (formerly UCC) was the largest mining operator in Iron County, with a workforce numbering 180 in the winter and 230 in the summer. Utah International and its AFL unions somehow handled contract negotiations without major strikes, so they were only idle when the railroad or steelworkers' unions shut down the industry.

However, seemingly uncontrollable situations, including labor slowdowns, state and federal regulatory-agency demands (EPA and OSHA), and rising freight costs on the railroads increased operating costs and eventually took a toll on the iron industry. Even as it became evident in 1971 that the industry could not compete with foreign steel, steelworkers negotiated a settlement promising wage increases of 31 percent over three years. Plant modernization was needed, but instead USX (formerly U.S. Steel) began closing older facilities. Geneva was nearly closed in 1979–80 by the Environmental Protection Agency, then temporarily closed its doors in 1986 during labor-contract negotiations, and permanently shut down a year later when no labor agreement was reached. USX stopped mining and shipping from Iron County in 1980. Beginning in 1984, its mine facilities in Utah and Wyoming were closed and dismantled. The CF&I mill at Pueblo and the Fontana mill in California likewise closed and were dismantled. Some blamed inept management; others censured the unions for the steel industry's troubles.

In January 1981 Utah International (now known as BHP-Utah International) closed down its mining operations in Iron County but continued shipping from its 1.5-million-ton stockpiles at Iron Springs for four more years. According to operations manager York F. Jones, "We had just priced ourselves out of the market."[37] The beneficiation mill and alluvium concentrator and the crushing and loading facility stood idle for five years and were then dismantled. BHP-Utah International's employees were terminated or transferred to other corporation facilities.

During the 1950s and 1960s, iron-mining industries and their associated operations and services (railroad, electric power, and other utilities) paid approximately $923,000 in county taxes, or 60 to 70 percent of the tax bill for Iron County. By 1975 their share had decreased to about 37 percent ($894,000), but iron mining was still the major industry, the major employer, and the major tax payer in the county.[38] The demise of mining came in the 1980s and adversely affected the Union Pacific Railroad and all other county businesses. The blow of losing this major tax payer, plus its related high-paying jobs and increased property values, seemed insurmountable in the early 1980s. The county's iron empire appeared dead.

A partial resuscitation began in August 1987, when Basic Manufacturing and Technology of Utah purchased the idle Geneva plant from USX and its ore reserves in Iron County. The next year it bought the iron-ore crushing and loading plant at the Comstock Mine from BHP-Utah International, followed by CF&I's property, including the Comstock pit and other reserves, in 1989. Iron ore from stockpiles and ore reserves at the Comstock and other mines supplied the reopened Geneva Steel mill. Gilbert Development of Cedar City shipped ore from the Comstock, Mountain Lion, Excelsior, Chesapeake, and Burke pits to Geneva, where the blast-furnace burden used about 60 percent iron pellets from Minnesota and 40 percent raw ore from Iron County. More than 800,000 tons were shipped annually in 1989 and 1990, but less than 175,000 tons were shipped in 1994. Taxes paid to the county between 1987 and 1994 ranged from $5,000 to $40,000.[39]

In February 1999, faced with increased competition from imported steel and a slowing domestic economy, Geneva Steel became one of at least 25 American steel producers to file for Chapter 11 bankruptcy between 1998 and 2001. Buoyed by a $110 million loan from Citicorp USA, the company emerged from bankruptcy in January 2001 as Geneva Steel LLC, only to refile in January 2002. When additional efforts to reopen the plant as a minimill failed to materialize, the company began planning to remediate the property. On 28 July 2004, a bankruptcy judge approved the sale of 62 acres at the former mill site for the construction of a 534-megawatt, natural-gas-fired power plant, which will be owned and operated by PacificCorp.[40]

Remaining iron ore in the Pinto and Iron Springs Districts is estimated at more than 200 million tons. The undeveloped Rex ore body has in excess of 150 million tons and is considered the single richest accessible iron-ore body in the western United States. However, mining these reserves requires the investment of millions of dollars to strip overburden and build facilities to produce a marketable product. Iron County has recovered tax losses by building an economy based on other manufacturing industries and Southern Utah University. Still, vast valuable ore reserves wait for a time when the price is right to mine iron once again.

As this book goes to press, the first mining of iron ore in ten years appears imminent. Palladon Venture Ltd. and Luxor Capital Partners have closed a secured long term loan in the principal amount of $12,750,000 to refinance the acquisition of the Comstock/Mountain Lion Iron Project in Iron County, Utah. Palladon has contracted with Gilbert Construction of Cedar City to begin mining and ore will be shipped before the end of 2006.[41]

Coal and Coal Mining

Immense coal beds are prominent features of the Cretaceous formations of southwestern Utah. East of the Hurricane cliffs, four major coalfields exist on the Markagunt and Paunsagunt Plateaus in the Colorado Plateau province. They are the Kolob (which covers southeastern Iron, northeastern Washington, and northwestern Kane Counties), Kanab, Kaiparowits, and Henry Mountains fields. There is also one minor coal-bearing area, the New Harmony field, on the border between Iron and Washington Counties. Mines were opened near New Harmony early in the twentieth century but have been abandoned for many years.

Coal in Iron County occurs in practically inexhaustible quantities, but it has not been commercially mined to the extent of reserves in other Utah counties. On fresh surfaces in the mines, the coal is deep black, moderately glistening, and slick to the touch—qualities soon lost after exposure to air. Analysis shows the coal of the Kolob field burns as hot as that mined in Carbon County, and its moisture and ash content are also about the same. However, the sulfur content is higher at 6 to 7 percent. Iron County coal is suitable for heating and cooking and was used for almost 20 years to generate electric power. Nevertheless, its high sulfur content makes it unsuitable for blacksmithing, iron making, and other metallurgical processes, a fact the early settlers did not understand.[42]

The fortuitous discovery of stone coal in the stream called Little Muddy or Cottonwood Creek in the spring of 1851 and Peter Shirts's location of two veins of coal up Cedar Canyon in April 1851 led the Iron Mission leaders to locate iron manufacturing on the banks of the stream, now called Coal Creek, 10 miles from the iron-ore deposits and about 5 miles from the coal. The chemical composition of the coal, however, hindered, rather than helped, iron manufacture.

Coal came by pack and wagon to fire the blast furnace in September 1852. It was mined at an outcrop five miles up the canyon at the Walker Mine, near the mouth of Maple Canyon, and at other canyon sites as miners continually sought better coal for the ironworks.[43] The Jones-Bulloch (later Macfarlane, and then Koal Kreek) Mine, eight miles up the canyon south of Coal Creek, was the first one of any size.[44] The Leyson Mine in Right Hand Canyon opened in 1854. Nearby, the first coke ovens in the region were built. When charcoal from cedar trees replaced coke in the blast furnaces, coal mining languished. Typically coal-mining operations lasted a few years, yielded a few hundred hard-won tons of coal, and were then abandoned.[45]

Mining activity spread in the 1880s. Andrew Corry opened a mine on Lone Tree Mountain in 1885. Although this mine had reportedly the "best coal" in southern Utah, it required a long haul in good weather and was totally inaccessible during the winter. A mine above Kanarraville (near the later Graff Kleen Koal Mine) supplied coke to the stamping mills at Silver Reef during the 1880s. Ovens were built nearby, and coke was hauled by team down the front of Kanarra Mountain and on to Silver Reef. A number of small mines south of Graff Point operated by P. Arnold Graff, Jesse Williams, and others supplied domestic coal for Kanarraville.

From 1890 to 1915, coal production was sporadic. Methods were slow and inefficient. No production was recorded for some years, and 524 and 575 tons were reported in 1898 and 1899.[46] Small mines worked by hand tried to meet the needs of local residents, schools, and businesses that burned coal for heat.

William C. Adams, who spent the better part of 30 years working in local mines alongside his father, described mining "done the hard way" with picks and shovels and handmade cars and wheelbarrows. To open a vein, miners hand-drilled a hole three to four feet deep in a vein of coal and then used lime and black powder, or a *squid,* to create a small explosion and break out a small part. Miners started in the middle of the vein, widening out to each corner to make a square room. At first candlesticks poked into wooden props provided the only light. Carbide lamps replaced candles and provided better illumination, but lighting remained dangerous until the Utah Industrial Commission required air courses in the mines to vent off explosive gases. The clay streaks through a coal vein were separated out by hand, and then the coal was scooped by shovel into a waiting car. The best miners averaged 85 cars a week, each about 1,000 pounds, but one-third of the load was waste, and the small pieces were screened off before loading. Much more tonnage was mined than the wagons ever brought down the canyon.[47]

Two early but unsuccessful efforts were made at bringing coal to Cedar City by tramway. In 1913 Dr. Earnest F. Green and the Iron County Coal Company reopened the Corry Mine on Lone Tree Mountain above Green's Lake. The *Iron County Record* of 28 November 1913 described Green's mine in enthusiastic detail as a tunnel 45 feet deep into a 10-foot ledge of "fine appearing" coal.[48] Miners at the Corry Mine lived in a two-story frame hotel and boardinghouse.[49] Construction of a tram began in February 1918 but was never finished because the company collapsed in the post-World War I financial depression.[50] Ten years later a tramway a little more than a mile long was built from a mine on the north side of Lone Peak to the mouth of the canyon. It operated for a year or two, but the small amount of good coal obtained did not justify continuing it.[51]

County coal production ranged between 1,000 and 3,000 tons per year in the 1920s. During the 1930s production increased, and during the 1940s 6,000 to 8,000 tons were mined annually. Iron County coal was used in Iron, Washington, and Beaver Counties as fuel for households and businesses. The early pattern of small-scale, pick-and-shovel mining operations improved somewhat because of better equipment, especially after 1945.

An interesting story surrounds the Kleen Koal Mine on the western rim of the Kolob terrace. In 1937 Dr. Arnold L. Graff, son of P. Arnold Graff, was trying to locate a certain section corner on Kanarra Mountain and went to the land surveyor general's office to look at the field notes of the original survey. The notes mentioned coke ovens near an old coal mine. Since Graff owned the property where the ovens were supposedly located, he sent two miners to search for the coal deposit. After a few days, Graff; Parley Dalley, a chemistry professor at Branch Agricultural College; William C. Adams; and Albert Marsden, an attorney, drove to the site. When they arrived, they were startled to see the miners picking at a huge face of very high-quality coal. Nearby lay coke that was very hard and apparently in as good condition as when it had been made 50 years earlier for the mills at Silver Reef. Graff had discovered a forgotten mine.[52]

Graff hired William C. Adams to open the mine and called it the Kleen Koal Mine. Twenty to 25 men worked from August to October 1937 to activate the mine before winter. The mine was at 8,500 feet elevation, and a three-quarter-mile cable tramway was built to deliver coal quickly and cheaply during all seasons. The tram, anchored to a ledge near the top of the mountain, ran to a tipple or load/storage site above Red Hill. The tipple site was on a newly constructed road, still known as the Graff Tipple Road. From mining to loading, the coal moved by gravity. Twenty buckets, each carrying 450 pounds of coal, moved on 15,000 feet of cable with the plummet of loaded buckets returning the empty buckets to the mine. Storage bins at the bottom of the tram had a 200-ton capacity.[53]

Guy C. Tucker later leased the mine from Graff. During his years of operation, the tramway was extended two and a half miles to flat land, and new buckets were made, which carried 1,000 pounds of coal each. The loaded buckets traveled so fast that the brakes wore out, so a generator was worked into the cable to slow the tram to about 250 feet per minute. The tram lines were used all winter long. Men drove to the Red Hill tipple, then walked to the mine in the snow along the tram line, which carried groceries and supplies up to the mine. Graff had cabins and a mess house built near the mine. Tucker and his sons operated the Kleen Koal Mine until 1941, when Tucker closed it because the military draft took all his good men.[54] The Old Kanarraville, Davis, and two Williams' Mines also operated on a small scale during the 1930s and 1940s. A road went to these mines, but in the winter bobsleds hauled the coal out.[55]

In 1944 Southern Utah Power began constructing a modern, coal-powered, steam-generating electrical plant with a 2,500-kilowatt capacity one mile up Cedar Canyon. Reed Gardner, manager of the power company, contracted with the Tuckers to supply the coal needs of the power plant. Water for the plant came from nearby Coal Creek. Tucker considered a number of mine sites and finally started up the Tucker Coal Mine on property owned by Kenneth Macfarlane in Right Hand Canyon. When the power plant went online in July of 1945, Tucker supplied 3,000 tons of coal per month, increasing to 5,700 tons in 1947, when an additional 10,000-kilowatt plant began operating. The power plant grew to meet escalating demands for electricity prompted by postwar expansion in iron-ore production and the rapid increase in the use of deep wells as a source of water for agriculture in Cedar Valley and the Newcastle area.

By 1952 coal was supplied by the three largest mines in the area: Koal Kreek and Webster in Cedar Canyon and Tucker in Right Hand Canyon. The operators were Grant and Floyd Tucker and Lewis Webster. Their mines became highly mechanized as the demand for coal increased in the late 1940s. The mines were all underground and used the room-and-pillar method, where rooms were filled with rock waste as mining advanced and the pillars were normally not recovered. In each of these mines, the coal was undercut, loaded by machinery, and moved to the surface by endless belts. At the mine mouth, large pieces of waste were handpicked from the coal, and then it was loaded onto trucks for hauling. Coal which went through a water-cleaning plant built by the Tuckers in 1948 was too wet to use at the power plant but was fine for heating homes, schools, and the college. The Tuckers later built an air-cleaning plant in Cedar Canyon to process coal from all three mines. Their investment was just under $100,000, and air cleaning lowered ash content to 8 percent. Four hundred tons of coal per day could be cleaned, with 200 tons coming from Koal Kreek Mine and the other 200 from the Tucker and Webster mines.[56]

With increased mechanization between 1942 and 1958, average output per man per day went from 13.57 to 18.97 tons. During the 1940s and 1950s, an average of 19 men were employed in coal mining. Thirty-five men worked at mining and hauling and in the cleaning plant in the 1960s. However, the three mines closed in 1965 when the power plant shut down because California Pacific Utilities, owner of Southern Utah Power, could buy cheaper electricity from Glen Canyon Dam.

Estimated coal reserves in the Cedar Mountain quadrangle total nearly 260 million tons. Although 90 percent is in thick beds and reasonably accessible to existing lines of transportation, the coal reserves remain just that in the early twenty-first century. Coal mining is historically important but not economically significant today.[57]

Silver and other Minerals

Silver, gold, lead, fluorspar, and other useful minerals were also produced by igneous intrusions through limestone formations in the western mountain ranges of the Great Basin and have been mined in Iron County.[58] In the early 1870s, an old prospector known simply as "Pike" found placer gold in the low mountains on the Utah/Nevada border, some 20 miles east of Pioche, Nevada.[59] He probably panned some nuggets at the lower end of a wash and worked his way up, led by the distinctive color, until he found a deposit worth mining, which was, thereafter, called "Pike's diggings." The Stateline and Gold Springs Mining Districts formed on either side of the diggings in the Buck/Paradise Mountains, and mines developed where mother lode veins of silver and gold were associated with quartz, pyrite, adularia, and sometimes lead and fluorspar. These mines were worked off and on between the 1890s and the 1930s as ore values rose and fell with the changing economy.

The Stateline Mining District was organized in 1896 in Stateline Canyon, immediately west of Hamlin Valley, about 18 miles northwest of Modena. A mining camp complete with stores, hotels, a school, a doctor, and a newspaper, *The Stateline Oracle*, flourished for several years. The largest mines were the Johnny, Ofer, Big Fourteen, Gold Dome, and Creole. Contractors Joseph Dedrichs and James Burke built a mill in 1902 to handle ore from the mining claims of the Ophir Mining and Milling Company. In 1904 the Ophir was sold to satisfy a judgment won by the mill contractors, who had not been paid.[60] Ore mined at Stateline was taken by wagon to the railroad station at Modena for shipping. An estimated 13,000 ounces of gold and 173,000 ounces of silver were taken from these mines.[61]

In 1903 Stateline claimed to be the gateway to gold-bearing camps of the Gold Springs District, although more people lived on the Nevada side of the district than in Utah. The Nevada mining camp was at Fay in Deerlodge Canyon. The census of 1900 shows 232 in Deerlodge precinct and 118 in Stateline precinct, and in 1903 the *Stateline Oracle* reported 180 miners at Fay and Deerlodge. Stateline probably had more than 200 residents in 1902–3; the town was a mile-long stretch of stone and false-front buildings. Deposits were never mined out, but ore values dropped in 1903–4, and most miners moved to better prospects. Only 35 people were left in the Stateline precinct in 1910.[62]

Women in Stateline usually worked as merchants or ran boardinghouses or hotels, except for Martha Tilley, who was part owner of the Mammoth lode claim. In 1905 she filed a notice of forfeiture against her partner, Henry Bowen, certifying her expenditures of more than $200 in 1903 and 1904 in labor and improvements; he had to pay his portion within 90 days, or the claim became hers.[63]

Other mine operators came from out of state, including J. H. McDonald from New Jersey, George Buel from New York, Joseph Carter from Minnesota, and Zeth Drake from Wisconsin, or from out of Iron County, like William Leamaster, George Rice, and Isaac C. Wolf. However, some Iron County men became involved after 1909, possibly picking up mining properties for delinquent taxes. In the spring of 1909, Samuel A. Higbee, John S. Woodbury, and A. R. Corry of Cedar City were "large owners" in the Big Fourteen Mine, which was reporting assays of 2,650 ounces of silver and $108 per ton of gold. R. J. Bryant, Jr., of Snow and Bryant Company reported progress in building a steam stamp mill during 1911 which would make Stateline a "producer" again, but there is little evidence of great success.[64] In the 1980s an operation in Stateline to recover more precious minerals by leeching was closed because of improper environmental procedures.

Gold Springs Mining District,[65] organized in 1897 or 1898, straddles the Utah/Nevada border about 17 miles northwest of Modena. Prospectors C. A. Short and H. R. Elliott found rich gold outcrops. Their mine, the Jennie, was the largest gold producer both during the early days from 1898 to 1904 and in later operations during the 1930s with reports of 4,000 ounces of gold and 21,000 ounces of silver.[66]

H. T. Johnson of Minneapolis was general manager and part owner of mining properties at Gold Springs in 1917–18. The area mines received electricity from Dixie Power Company and had telephone lines strung from the mine to Modena. In January 1918 thieves cut down and removed from 4,000 to 5,000 pounds of heavy copper wire strung in the hills between Modena and Gold Springs. Iron County Sheriff Alfred Froyd apprehended two men trying to sell the wire in Beaver and Salt Lake.[67]

John Jordan operated the Jennie during the 1930s, when the shaft was extended 300 feet down into the vein. Ore was lifted to the surface for crushing and then moved by gravity into a large mill, where it was further broken down and the precious metals somewhat refined by an oil-flotation process. As a teenager in 1935 and 1936, Blair Maxfield, later professor of geology at Southern Utah University, worked at the Jennie, where his father was superintendent. Total gold production in the district is estimated at 13,000 ounces as a coproduct with silver, with about 9,000 ounces coming from the mines in Iron County (Jennie, Jumbo, and Independence).[68]

In 1893 Henry D. Holt discovered silver ore on the desert west of Shoal Creek. Holt and three other men held three mining claims in common, but when Holt's partners would not help with assessment work, he bought them out for eight cows. Known early as the Holt and later as the Escalante Mine, the site is about four miles southwest of Beryl Junction and seven miles directly north of Enterprise. The ore seemed fairly good to the men working the mine in the 1890s, but the shaft filled with water and subsequently closed. It is said that George A. Holt offered the water to anyone who would pump it out, but nearby dry farmers were not interested.[69] Decades passed with only occasional attempts at mining. Title to the claims somehow passed to Heber J. Grant, president of the Church of Jesus Christ of Latter-day Saints, then to his widow after his death in 1945. She gave the claims to the Enterprise LDS ward, which sold them to help pay for a new meetinghouse. Sam Arentz, a mining engineer and developer from Salt Lake City, bought the claims, and in the late 1970s, sold them to Ranchers Exploration and Development Corporation of Albuquerque, New Mexico. Ranchers Exploration developed the mine in 1980 and operated it until Hecla Mining Company of Wallace, Idaho, bought out the company in 1986.[70]

An economical system of mining was developed in the 1970s and helped make the Escalante Mine profitable. Dewatering the mine was a major concern for the farmers of the valley, who used underground water for irrigation and feared that mining would affect their supply. Ranchers Exploration agreed to compensate farmers if dewatering caused an excessive drop in the water table. With an investment of approximately $30 million, Ranchers Exploration developed the mine and built a mill which removed approximately 1,999 pounds of waste from every ton of ore mined to recover about eight ounces of silver. The mill handled up to 750 tons of ore per day. Once a week a gas-fired furnace melted the precipitates and fluxes from the gray powder concentrate to a liquid at 2,200 degrees Fahrenheit, and then the furnace tipped, and the molten concentrate poured into cast-iron, cone-shaped molds. Silver, heavier than slag, settled to the bottom, leaving the slag on top. After cooling, silver and slag were removed from the mold, and the slag was chipped away. The 50-pound silver "button" was shipped to a commercial refinery for final processing to 99.9 percent pure silver.[71] General Manager Ed Hahne of Cedar City headed the operation that employed 115 workers from the surrounding communities during the mid-1980s.

The mine was dewatered by pumping at an average rate of 19,500 gallons per minute. Ranchers Exploration made adjustments to the system of water disposal in 1981 and 1982 and successfully discharged three-fourths of the mine water onto a farm which it had purchased for recharging the aquifer; one-fourth of the water recharged Shoal Creek.[72] Approximately 300,000 tons of ore were milled and 2.3 million ounces of silver produced each year from 1982 to 1990, when the ore body was mined out. The last smelting occurred in 1991. Recovery of approximately 25 million ounces of silver makes the Escalante Mine unquestionably the second most successful one in Iron County.[72]

Almost all other precious-metal mining occurred sporadically between 1890 and 1940—a period when mining excitement ran high, but results proved disappointing. There are a number of other mining areas in Iron County history, notably in the Indian Peak Range (Arrowhead Mine, Skougard Mine, and Cougar Spar Mine) and on the west side of the Antelope Range (Bullion Canyon and Chloride Canyon Mines). Discovery of ore at the Arrowhead occurred sometime in the 1890s, and the mine was worked off and on. The ore was chiefly lead and zinc, with a little silver and gold. As with most other new mining ventures, developers claimed it would be "one of the very best mines in Utah."[74]

Bullion and Chloride Canyons lie on the west side of the Antelope Range between Silver Peak and Antelope Springs. In 1903 and 1904, Bullion Canyon was thoroughly prospected, and ore containing lead, copper, and some silver was assayed. Most of the claims were filed by local men who had formed mining companies that picked away at a variety of claims without finances to develop the mines properly.[74]

In 1910 interest turned to Chloride Canyon, where local businessmen joined with George Ray of Chicago and Ronald B. Rankin of St. Louis to form the Standard Consolidated Mining Company, capitalized at $250,000. The treasury stock was underwritten by Rand and Rankin, representing the Interstate Venture Company, who contracted for $40,000 of the stock. With ample working capital, 10 men with equipment and supplies began mining in 1910 with expectations of finding the rich mother lode of lead and silver at the 200-foot level, but they were disappointed.[76] In October 1918 the Copper Zone Company reported a "phenomenal" strike in the face of a 600-foot tunnel tapping the ore 300 feet below the surface at Bullion Canyon. Expectation for rich silver, lead, copper, and gold production remained unrealized, however.[77]

Fluorspar, a fluoride of calcium, came into commercial production in Utah during World War II to meet demand created by Geneva's open-hearth steel furnace, where it was used as a fluxing agent. It was mined in the Wah Wah Mountains at the Cougar Spar Mine and near Mountain Springs, where Otto and Lou Fife mined for five or six years. Lead and zinc mines were profitable until President Truman vetoed a bill to extend mineral bonuses. His action shut down 500 lead and zinc mines, and Fife's mine was one of them. Fluorspar is found in larger quantities in Beaver, Millard, and Juab Counties.

The Cedar Mountain quadrangle contains enormous quantities of gypsum. This useful industrial mineral is well exposed along Utah Highway 14 in Cedar Canyon. In the early days, gypsum was used locally to make small amounts of plaster for homes. Mammoth Plaster and Cement Company quarried gypsum commercially in Cedar Canyon in 1923. At first raw gypsum was quarried, crushed, trucked to the railroad cars, and shipped to Los Angeles to fill a contract with the Blue Diamond Material Company for 400 tons a day. However, the high-grade material specified in the contract was buried under a top layer heavily contaminated with impurities which could only be removed at great expense. Lehi W. Jones, acting for the company's board of directors, went to Los Angeles and arranged to break the contract, saving even greater financial disaster. Company officers still wanted to build a mill, but investors were unwilling to put more money into the venture. Many people from Cedar City had invested heavily in the company, some putting in thousands of dollars, and dreams of financial wealth ended when the quarrying stopped and the crusher and bins were abandoned. In 1937 Cedar Plaster Company installed a plaster mill in the old gristmill at the mouth of Cedar Canyon. Obtaining gypsum from deposits owned by Samuel F. Leigh and Emil Roundy, the mill shipped a "fairly large amount of high quality plaster" until production ceased at the beginning of World War II.[78]

Gypsum has many uses in manufacturing cement, plaster, and wallboard. It also serves as a filler in paint and paper, a conditioner for alkaline soil, and a stabilizer for the ammonia present in manure. It is a valuable and versatile raw material, present in quantity in the Cedar City area, but it has never been of great economic benefit to the community.[79]

During the 1950s several Utah residents began searching for uranium in Iron County. Geiger counters produced readings of varying levels in many areas of the county, but there was no uranium; prospectors were actually picking up readings of nuclear fallout, which covered the county following the Nevada atomic tests. In the 1970s a Colorado firm proposed mining aluminum-producing alunite in the Wah Wah Mountains of Iron and Beaver Counties, but the proposal came to nothing. In the 1980s geologists searched for oil beneath the Escalante Desert, once again sparking headlines but producing no results. Mining and mineral news always made for bold headlines in the local newspapers, and occasionally actual discoveries brought excitement and employment, but no fortunes were made in Iron County's silver, gold, lead, or fluorspar mines.

10

Bingham Canyon

Bruce D. Whitehead
Robert E. Rampton

Mining Begins in Bingham Canyon

The first mining claim in the Utah Territory was filed on 7 September 1863 after the discovery of mineral-bearing ore in Bingham Canyon. Articles of formation of the West Mountain Quartz Mining District were approved on 17 September, and the following December, the first mining district was established. Various historical accounts offer slightly different versions of the discovery of ore in Bingham Canyon. One story attributes it to two early Mormon pioneers, Thomas and Sanford Bingham, who were the first to use the canyon for cattle grazing and to whom it owes its name.[1] Another says the ore was found during a picnic in Bingham Canyon.

Both of those stories were denied by W. W. Gardner, son of Archibald Gardner, a noted mill builder in the Salt Lake Valley, who was involved almost from the beginning in Bingham Canyon mining operations.[2] Gardner says the story of that first mining claim begins when his father started getting timber for his mill at West Jordan from the West (Oquirrh) Mountains in Bingham Canyon. George B. Ogilvie, who with his brother, Alex, ran cattle in the mouth of Bingham Canyon, was sent out to cut timber and "noticed some stones which one of the logs had loosened in the dragway. . . . He showed it to other loggers and they were all impressed that it might be a piece of valuable mineral." Since none of them knew what it was, they took it to Gardner.

"Archibald Gardner knew that General [Patrick E.] Connor and some of his soldiers had had some experience while in California in mining.[3] Archibald Gardner advised the men to take the stone to Fort Douglas and show it to General Connor and ask him for advice. Ogilvie and others did as suggested and asked him to examine the sample, and if it had value, tell them what they should do about it." After

Locations of the early towns and camps in Bingham Canyon. All are now gone, absorbed by the modern Kennecott Utah Copper open-pit mine.

Courtesy Kennecott Utah Copper.

examining the rocks, two groups of about 12 men each—the Bingham Canyon loggers and some of the officers from Fort Douglas—met at the Jordan Ward House with Bishop Archibald Gardner, who was appointed recorder for the meeting.[4] During that meeting, the Jordan Silver Mining Company was formed, and each member of the group was given one share, except for Ogilvie, who was given two. In December 1863 Connor led the organization of the West Mountain Quartz Mining District in the Oquirrh Mountains, which included the Jordan lode of the Jordan Silver Mining Company.[5] Thus, the stage was set for what was to become one of the most amazing mining ventures the world has ever known.

In the spring of 1864, several companies of Connor's Volunteers were officially ordered to prospect in various promising locations. Numerous outcrops of copper and other nonferrous metals were located in Bingham Canyon and elsewhere, but none of these were worked extensively or successfully because of the lack of transportation and difficulties in smelting.[6] Since the railroad hadn't yet arrived in Utah Territory, mining supplies were excessively expensive: a shovel cost $2.50, and a keg of powder sold for $100.[7]

Regardless of the costs and the inconvenience of primitive mining methods, interest in Bingham Canyon was fired up by the discovery of gold and silver deposits. Not only were Connor's men prospecting the area, but also, as the word spread, men from around the West converged upon Bingham Canyon. Many were soon discouraged. It was hard work and involved lode mining with drilling tunnels, hauling the ore out, and then transporting it to the few crude mills by wagon.

"Had it not been for the discovery of gold in 1864, the camp might have been abandoned until a later date due to the decline in lode mining. However, placer gold was discovered in the gravels there and many miners became actively engaged in panning gold along the various streams. Gold mining could go on due to the fact that the profits from gold were greater from smaller amounts than could be obtained from the metals."[8] Placer mining, which had already proved its worth during the California gold rush, required simply that streambed gravels be washed in moving water to remove the lighter gravel, leaving the heavier gold at the bottom. In Bingham,

> the most productive and richest placer was that of Clay's bar, from which $100,000 in gold had been taken out by 1868. This bar consisted of a shaft that had to be abandoned due to excessive water encountered below a depth of 120 feet. Dan W. Heaton suggested that even though the shaft had to be abandoned, the old tailings, if re-worked, might produce a fair wage. The tailings were thrown into the sluice boxes and the result was wages ranging from $7 to $15 per day for the three men who owned the bar.
>
> One day while Ben Clay was cleaning out the sluice boxes, throwing the larger cobbles high upon a dump, one piece rolled off the shovel. He picked it up and discovered it to be a gold nugget weighing eight ounces.

> This was said to be the largest nugget ever washed out in the state and was valued at $128.65.
>
> The Clay boys marketed their "dust" in Salt Lake City, where their frequent comings with such quantities of gold soon attracted attention and quite a few men followed them to the "diggings."[9]

Most of these miners used traditional sluicing, but, again taking their cue from the California gold-rush miners, soon replaced the old style with ground sluices, larger boxes, and a small hydraulic plant. But the high costs of mining supplies and equipment didn't prevent the development of small underground mines and camps up and down Bingham Canyon and Carr Fork, the two main canyons that straddled what became known as "the Hill," a mountain that later became the focal point of the largest mining enterprise in the nation.

The more important early mines in the two canyons were the Galena, the Kingston, the Julia Dean, and Silver Hill. By 1868 there were perhaps 100 people living in primitive cabins in Bingham Canyon. Two years later the U.S. Census Bureau noted that 275 were living in the canyon, and by this time the miners had recovered more than $2 million in gold from the gravels. By 1871 Bingham Canyon's growth had made it large enough to be organized as a voting precinct in Salt Lake County.[10]

Entrepreneurs and the Railroads Arrive

Initially the ore was hauled out of the mines through the canyons and gulches by wagon trains; however, with the completion of the transcontinental railroad in 1869, the mining profile of Bingham Canyon and Carr Fork changed dramatically. Heavy equipment became more readily available and set the stage in 1873 for construction of a branch line, the Bingham and Camp Floyd Railroad, to Bingham Canyon. In the winter of 1873, the narrow-gauge Copper Belt Railroad was completed, bringing a new level of efficiency to the operations. While its capabilities were limited, the shipment of ores from Bingham Canyon mines greatly increased, consequently enlarging profits. Gear-driven Shay engines provided the power for moving the copper ore. They were powerful and built to haul ore over steep grades. But they also were slow. At four miles an hour, a person could walk as fast as they traveled. It was still "a fight to get up the hill and then one to keep from falling down."[11] Ore by rail was successful enough that by 1901 the Copper Belt had been extended into the upper part of Bingham Canyon.

> There was a ten-year burst of silver and lead mining in the canyon. Several million dollars worth of lead-silver ore was smelted at a Bingham smelter, at Salt Lake Valley smelters, or at smelters in San Francisco, Baltimore or Wales. Some copper was produced as a by-product of these efforts, but the low-grade copper-bearing ores were not as easily smelted as Bingham's

lead-silver ores. During the heydays of Utah's silver fever of the 1870s and early 1880s, the production of copper never exceeded a million pounds a year. The rich finds of copper that have dominated Utah's minerals industry in the past century were not discovered and worked until the 1890s.[12]

Probably the first person to realize that the huge deposits of low-grade copper ores at Bingham Canyon held possibilities for riches was "Colonel" Enos A. Wall, a Pennsylvania lawyer turned mining entrepreneur. He visited Bingham Canyon in 1887, and his attention was drawn to signs of copper mineralization in Carr Fork. A spring on the hillside left green stains on the rocks as the water meandered down. Wall had a close look at the ridge rocks and obtained an ore sample from an abandoned mine. His sample assayed at 2.4 percent copper. Further investigation showed that a large part of the ground surrounding the exposed mineral deposits had been abandoned and was therefore subject to refiling. He immediately staked two claims and located other deposits he envisioned with possibilities.[13]

"By 1900, the spirited Colonel owned all or part of 19 claims covering an area of 200 acres. He recognized the marginal nature of the property and even agreed to allow the local supervisor to use the dumps on his property for road making. Local residents disparagingly referred to his claims as 'Wall-rock.'"[14] Even though short of money for development, Wall managed to keep up the assessment and development work. Up to 1900 he had spent some $20,000 and driven 3,250 feet of tunnels into the hillside, "following fractures and veinlets in the hope of finding larger masses of rich ore."[15]

During this same period, Samuel Newhouse and Thomas Weir formed the Highland Boy Company as a gold-mining venture. The fact that they erected a cyanide mill in Carr Fork for processing oxidized gold ore reveals that they had not the remotest idea of developing copper deposits. "Their operation was in business only a short time when the workings revealed vast zones of copper sulphide. The copper present in the ore interfered with the recovery of the gold and caused an exceptionally large consumption of cyanide. The mill proved to be a dismal failure and while the venture seemed to be a loss, it marked the beginning of a new era for Bingham."[16]

Newhouse was the money promoter for Highland Boy, and while on a trip to Denver to secure backing, he received a telegram from his partner, Weir, that ore containing 15 percent copper had been discovered. The pair soon decided to build a "modern copper smeltery," with its associated crusher and concentrator and a day-by-day capacity of 250 tons, to process the ore from their mine. The construction contract was awarded in September 1898, and the following spring the facility was ready for business. The plant was the first to be erected in Utah primarily to reduce copper ores.[17] With massive quantities of moderately rich and low-grade ores, the Highland Boy Mine became the largest sulfide copper producer in Utah and one of the largest in the West.

The widely publicized success achieved by Newhouse and Weir in developing the Highland Boy Mine into a profitable copper mining and smelting

> venture soon brought their enterprise under the covetous eye of a "Standard Oil Company syndicate," headed by William Rockefeller [brother of John D.] and Henry H. Rogers [an associate of John D. in Standard Oil]. The "Standard Oil crowd," as it was called, purchased control of the Utah Consolidated Gold Mines, Ltd., from Newhouse and Weir, for a reported $12 million. A new corporation, the Utah Consolidated Mining Company, including the absorption of the productive Highland Boy mine, resulted.[18]

The gold/copper-mining industry was rapidly changing with many companies forming, claims consolidating, financiers buying and selling companies, claims and properties trading back and forth—all jockeying for positions of power. Some of the early mines and companies were the Highland Boy Company and the Old Jordan Silver Mining Company in 1893, Utah Consolidated Mining Company in 1896, Boston Consolidated Mining Company in 1897, Yampa in 1897, and U.S. Mining Company in 1899.[19] The need for housing led to the creation of the mining towns of the era, including Leadmine, Frogtown, Verona Gulch, Winamuck, Freeman Gulch, Heaston Heights, Markham Gulch, Copper Heights, Carr Fork, Apex, Phoenix, Frisco, Silver Shield, Niagara, Commercial, Telegraph, Terrace Heights, Copperfield, Greek Camp, Japanese Camp, and Dinkeyville.

The rapid change from gold to copper and the resulting success prompted the formation of numerous companies and exchange of shares of existing companies, all with visions of financial success. Among them were the Bingham Consolidated Mining and Smelting Company and the United States Mining Company (later to become the United States Smelting, Refining and Mining Company). Those two concerns, plus Utah Consolidated and the American Smelting and Refining Company (ASARCO), built large smelters. ASARCO came on the industrial scene in early 1899, soon becoming the dominant nonferrous smelting firm in Utah.[20] In early 1904 there were three large copper smelters at the south end of the Salt Lake Valley in Murray and Midvale, and ASARCO had a large lead smelter in Murray.

Local farmers near the valley smelters, however, claimed their crops suffered severe damage from the sulfur dioxide gases from the venting stacks. They held several meetings in late 1904 and early 1905 and subsequently filed a suit in the United States District Court of Utah. "A lengthy trial resulted in a verdict against the four smelter companies which perpetually enjoined them from the future roasting or smelting of sulphide ores carrying over ten per centum of sulphur, at their locations so as to discharge said sulphur into the atmosphere in the form of gas or acid, or from further discharging into the atmosphere of arsenic in any form."[21]

That court action brought an end to Utah's booming sulfide copper industry and ordered the closure of at least two copper-smelting plants in the Salt Lake Valley. Just how closely the court order was enforced is not clear, however. In just a matter of months, United States Smelting, after close experimenting, developed a method of running all the smoke from its lead and copper smelters through bags. That process successfully filtered sufficient sulfur, and other undesirables, from the stack emissions to satisfy the critics.[22]

In 1909 District Court Judge John A. Marshall issued a permanent decree which gave permission to resume smelting ores. An obviously biased account of this controversy appeared in the Bingham press:

> The farmers of the Salt Lake Valley now seem as anxious to have the smelters remain as they were a little while ago. Just as soon as they found their bluff was taken in earnest, there was much hurrying to and fro across the valley, and the prospects of seeing the easy money they had been getting from the smelters, fade away, did not appeal to the gentleman with the bunch of whiskers on his chin [presumably Uncle Sam]. For the last four years the smelters have paid in damage suits, in and out of court, a half million dollars annually.[23]

New Investors Arrive with New Techniques

Meanwhile, new investors and promoters came to Bingham Canyon. Notable was Joseph R. De Lamar, a former sea captain who developed an interest in mining properties in Georgia, Colorado, Idaho, Nevada, California, and eventually in Utah. He purchased the Brickyard Mine in Mercur from Wall in 1894 and, following some investigations, saw the value of Bingham Canyon's porphyry copper deposits. De Lamar's Mercur operation used the services of a young mining engineer, Robert C. Gemmell, as supervisor of ore sampling, and of Daniel C. Jackling, a metallurgist in charge of milling tests.[24] De Lamar's mining and milling tests proved to be more than just encouraging. During the heated negotiations while forming the mining company, Gemmell resigned and went to Mexico, but Jackling stayed and later proved to be the ultimate driving force in the evolution of Utah's copper mining and refining industries. His theory that the porphyry copper deposits in Bingham Canyon had enormous potential and could be profitably mined and milled by using large-scale, low-cost mining methods proved to be correct.

Daniel Cowan Jackling was an example of the American dream, working his way from a poor, orphaned Missouri farm boy to an internationally recognized and admired mining engineer and industrialist.

> Jackling was one of those rare individuals with dogged determination and foresight—gifts that eventually launched the Bingham project. Those traits were honed by a hard childhood. Daniel C. Jackling was born in 1869 at Hudson, a small town in western Missouri, where his father engaged in trading and forwarding on the old Santa Fe Trail. Before Jackling was a year old his father died and his mother lived only a year longer. Jackling was placed

in the care of his mother's sister. Jackling and his aunt migrated from one farm to another, from Missouri to Arkansas, to Illinois, and back to Missouri. Finally landing in Sedalia, Missouri, and at age sixteen, Jackling managed to complete the eighth grade, then took a job as a teamster in a freight business owned by his uncle.[25]

By 19 Jackling had determined that he wanted to go to school and become a teacher, so he enrolled in the state Normal School, at the same time working on his uncle's farm "where he observed some engineers using a transit to lay out a building site." He changed career directions and enrolled in the Missouri School of Mines in 1889. He supported his education by spending his vacations working as an assistant to a railroad survey party and an assistant to the professor of chemistry and metallurgy. He completed the four-year course in only three and was awarded a bachelor's degree in the science of metallurgy.[26]

After a brief stint at a smelter in Kansas City, Missouri, Jackling struck out for Cripple Creek, Colorado. The story goes that he borrowed enough money for a second-class railroad ticket and got as far as Divide, the end of the line, without money for stage fare to Cripple Creek, then a mining boomtown. He talked a fellow passenger going to Cripple Creek into taking his baggage along, then walked the 18 miles through ice and snow and arrived at the camp with only three dollars in his pocket. Several days later he obtained a job in an assayer's office.[27] After working at several camps as a miner, assayer, mill hand, and metallurgist, Jackling took a job as metallurgical superintendent of the Golden Gate mill at Mercur, Utah. It was there that he met De Lamar and was subsequently assigned, along with Gemmell, to examine the copper property at Bingham, where De Lamar held an option on the properties that would become the Utah Copper Company.[28] Their report to De Lamar was to have far-reaching effects.

In 1901 Jackling returned to Colorado, where he was hired as a consulting engineer by Charles M. MacNeill and Spencer Penrose, owners of a controlling interest in the United States Reduction and Refining Company. His interest in the Utah porphyry coppers never waned, and his previous report to De Lamar on the Bingham Canyon potential caught the interest of MacNeill and Penrose, so he returned to Utah to negotiate for the property. But negotiations with the property owners proved difficult, and offers were submitted and resubmitted. Finally, Enos Wall, a major company holder, agreed to convey to Hartwig Cohen, Jackling's associate, two-thirds of his interest, but Wall kept the remainder to be part of the development and retained the right to nominate one member of the governing board. De Lamar was tired of holding what he felt was a frozen asset and was therefore willing to sell Jackling and his backers his own quarter interest for $125,000. His deal with De Lamar successfully completed, and Cohen's option from Wall safely in hand, Jackling returned to Colorado Springs. Taking a copy of the Jackling-Gemmell report with him, he visited Charles MacNeill to sell him on the new venture. He had, Jackling told MacNeill, "without exception, the greatest opportunity in the world and . . . he just had to get in on it."[29]

Skeptical at first, MacNeill agreed to a professional inspection of the property and sent F. H. Minard to Utah. Minard's investigation and report verified the estimates on the tonnage and grade of ore, but he "rather praised the property with faint damns." While he estimated that the workings disclosed nine million tons of copper-bearing rock, he pointed out "certain physical difficulties" and questioned Jackling's cost estimates. "Minard's final recommendation was that a 200- or 300-ton plant be erected to make extended experiments covering a period of at least a year, and this only on the condition that they would be able to acquire an interest in the property for the construction of the plant *without any payment whatever*."[30]

"On June 1, 1903, MacNeill, Spencer and R. A. F. Penrose accompanied Jackling to Salt Lake City, to inspect personally the property. They drove to the mine and walked over the property, at the conclusion of which 'Dick' Penrose said to MacNeill that he thought they should go ahead. That evening Jackling gave a dinner at the Knutsford Hotel to commemorate the occasion. The dinner is said to have cost Jackling his last $100."[31] Thus, on 4 June 1903, the Utah Copper Company was born and duly incorporated under the laws of Colorado with a nominal capital of $500,000 in one-dollar shares. MacNeill and Penrose took 250,000 shares, and their friends paid $250,000 in cash for the others.[32] That same month Jackling was given the go-ahead to build a 300-ton experimental concentrator, and a lease was firmed up for the surface rights on 20 acres in lower Bingham Canyon, along with the rights to dump tailings. Utah Copper paid $250 a month for the rights, which it would terminate upon abandoning the mill. Company executives began to focus on specific development objectives. They determined what measures to take to mine the porphyry copper ore and how to get it to the mills despite little experience in the kinds of undertakings required to mine, mill, and make a profit.

As mining increased, the 1901 Copper Belt Railroad was not enough to meet the demands of new mills in Magna and Garfield. The High Line extension of the Denver and Rio Grande Western Railroad was built to transport the huge volume of ore produced at the Utah Copper Mine to its mill in Magna. Although the High Line set records in ore transport, clearing as much as 6,000 tons of ore daily, it faltered because of a winding, circuitous route. In the spring of 1908, Utah Copper engineers surveyed a more direct line. The next year mine production hit well over 71 million pounds, and construction began on the Bingham and Garfield Railway in 1910. Declared "one of the most marvelous little roads in the world," the tracks traversed the foothills of the Oquirrh Mountains for 20 miles from a connection with the San Pedro, Los Angeles & Salt Lake Railroad at Garfield.[33]

The railroads, however, solved only half the problem. The old system of *caving* by blasting the hillsides and letting the results fall and break up into smaller pieces was too expensive and slow and did not produce enough usable ore to be profitable. Jackling did not give up his vision that, given sufficient quantities, the low-grade copper ore underneath the cap could move from mine to mill at a profit, however. His assays of Copper Hill showed about 2 percent copper content. That yielded approximately 39 pounds of copper from each ton of ore, which was not acceptable to the mining engineers of that day. Today the average tests show about .6 percent copper content in the huge open-pit mine, yielding 12 to 13 pounds of copper per ton, all at an acceptable profit.

Open-Pit Mining in Bingham Canyon

Jackling knew that in the iron mines of Minnesota, steam shovels had significantly reduced the per-ton cost of mining and thought there was no reason that the methods could not be adapted to the Utah Copper properties. In early 1906 Jackling, now general manager of Utah Copper, appointed his old associate, Robert Gemmell, as general superintendent. In April of that year, both men visited the Minnesota Mesabi Iron Range to study how steam shovels mined the iron ore. The steam shovels could strip away 70 feet of oxidized *cap,* or overburden, to prepare for terrace mining the higher-grade ore beneath. There the two men met William J. Olcott, a distinguished engineer in the iron-mining industry and a classmate of Gemmell's from the University of Michigan. Olcott recommended that J. D. Schilling take charge of any steam-shovel operation. Five months later in August 1906, Utah Copper put its first steam shovel into service, just two months after Boston Consolidated had activated its huge machines.[34]

The first steam-shovel equipment included two Marion shovels, one Vulcan shovel, four small Davenport locomotives, and six-yard wooden dump cars. These commenced the job of stripping the overburden from the hillside at the rate of about 100,000 tons per month, or the equivalent of nearly one acre of ground every 30 days. Thus, the mining technique was effectively married to rail transportation, and by June 1907 the shovels had moved about 700,000 cubic yards of capping, exposing nearly six acres of ore. Eighteen months later the shovels had stripped more than 3.2 million cubic yards.[35]

But the Utah copper industry's picture still was not complete, and difficult decisions remained. Concentrators and mills still had to be financed, strategically located, and built to handle the ever-increasing tonnages of copper ores. As early as the gold/silver/lead heydays of the mid-1800s, a concentrator and leaching works had been erected, and in 1900 the Bingham Gold and Copper Company built a smelter at Bingham Junction. The following year Highland Boy enlarged its plant and added a concentrator at the mine.

In 1904 Utah Copper constructed a mill at Copperton, followed in 1906 by the company's 6,000-ton mill at Garfield. Profitable mining of the Bingham Canyon and Carr Fork properties had been plagued by fits and starts from the beginning, but with the advent of steam-shovel mining, expansion of the railroads, and mill and concentrator construction, the industry rapidly moved ahead. The real secret to the area's development was money. The mining companies, particularly Utah Copper and Boston Consolidated, caught the attention of eastern entrepreneurs.[36]

If Jackling's thesis of mining and processing huge tonnages of lower-grade copper ore was to work, it was obvious that some consolidation had to take place. In 1907 Jackling moved to take control of the Arizona property of the Ray Consolidated Copper Company, then added the Chino porphyry copper deposits in New Mexico. The Guggenheim interests brought the Nevada Consolidated Copper Company, located near Ely, into the fold as the fourth addition to the Jackling organization. As these other western copper properties were developed, their mining, transportation, milling, and concentrator facilities were engineered and modeled after the installations serving the needs of Utah Copper.[37]

The Newhouse interests in charge of Boston Consolidated shared Jackling's vision of dealing in huge tonnages; they also shared the same mountain of ore deposits: Boston Consolidated had the upper portion of the hill, and Utah Copper held the bottom. Merger negotiations were attempted but seemed to be ending in failure and not without some bitterness. Enos Wall, still a minority power in copper development, viewed the proposed merger as a conquest of Boston Consolidated and a violation of common decency and moral ethics. Lending some credence to the conquest theory was Jackling's candid appraisal: "Sooner or later, I knew that we would have to take them, or they would have to take us."[38] Finally, on 25 January 1910, "the merger of the Utah copper companies was consummated before the end of the day, on the basis of two-and-one-half shares of Boston Consolidated for one share of Utah Copper."[39] The merger of the two Bingham Canyon copper giants set the stage for unprecedented growth of the survivor, Utah Copper Company, and the copper industry for the next five decades.[40]

While mining rock containing only 39 pounds of copper per ton was viewed with skepticism by most competent engineers, Jackling foresaw the economic advantage of mining copper at a production scale hitherto unheard of:

> He had the courage to back his convictions to the limit; the personality to induce capitalists to finance the enterprise because of their confidence in him; and he had the resourcefulness to devise the methods and create the organization for bringing his elaborate plans to fruition. . . . Concurrent with his development of Utah Copper, he brought into prominence porphyry copper properties in Arizona and New Mexico, the most famous of which were the Ray Consolidated and Chino mines, as well as properties in Nevada. As "Father of Porphyry Mining" Daniel C. Jackling would guide Utah Copper Company for the next thirty-two years.[41]

The movement of so much topsoil made dramatic changes in the Bingham Canyon topography. What once was Copper Hill, mined at the top by Boston

Consolidated and lower levels by Utah Copper, slowly disappeared. That source of copper ore finally became the world's largest open-pit mine as it succumbed to modern mining, concentrating, and refining technologies.

Immigrant Populations, Social Changes, and Labor Unions

The social changes near the mines over the years were perhaps even more dramatic than those to the physical landscape. They were driven by immigrants from many countries converging on Bingham Canyon to work the mines. The early years saw the immigrants heading for the canyon for employment opportunities. In the six years after the first gold strike in 1863, 276 new residents arrived in Bingham Canyon, mostly Irish who had fled their native potato famine. But the Irish resented the growing number of English immigrant workers, whom they called Cousin Jacks, and it wasn't long before the Irish began to leave the canyon. In 1880 the U.S. Census Bureau in the 10th census described the Bingham Canyon population by origin: America (includes American-born children of immigrants, mostly British and Scandinavian), 452; British Isles, 170; Scandinavia, 83; Ireland, 51; Italy, 35; China, 32; Canada, 22; Finland, 19; Germany, 17; Prussia, France, Nova Scotia, 2 each; Greece, Austria, Africa, Holland, and Portugal, 1 each.[42]

During the next two decades, the ethnic profile of Bingham changed dramatically with Finns, Swedes, Italians, Slovenes, Croatians, Serbs, Greeks, and Armenians coming to live. Japanese and Korean laborers, added to resident Chinese, expanded the Asian colony. In 1912 the Utah Bureau of Immigration, Labor, and Statistics reported to the U.S. Department of Commerce the following: Greeks, 1,210; northern Italians, 402; southern Italians, 237; Austrians, 564; Japanese, 254; Finns, 217; English, 161; Bulgarians, 60; Swedes, 59; Irish, 52; and Germans, 23.[43] It was obvious that English-speaking workers were leaving mining for other opportunities, and new immigrants quickly took their places. Settlement names often reflected the origin of their residents, such as Frogtown, with its French Canadian workers. Copperfield had adjacent towns of Greek Camp, Japanese Camp, and Dinkeyville, named for the tiny old railroad engines. Highland Boy and Phoenix were home to some 1,200 southern Slavs and Italians, and Carr Fork was mostly Finns, with some Swedes, Norwegians, and Irish. The town of Bingham itself was the center for the canyon's old Anglo families. Nonetheless, an estimated 65 percent of Bingham Canyon's population was foreign born.[44]

With the rapid growth in mining came equally swift changes in the society surrounding the mine. In 1900 the first rural free delivery mail route outside of a farming district in the United States was established. Businesses in support of the miners and other workers—grocery stores, transportation systems, clothiers, boardinghouses, hotels, restaurants, and a host of other enterprises—came into being. By 1900 the population of Bingham Canyon was tabulated at about 3,000, and Bingham's Main Street had 30 saloons.

Labor unions were another of the social changes evolving from the developing copper industry. After the merger of Utah Copper and Boston Consolidated in 1910, the first labor dispute occurred.

> The years prior to World War I were times of labor ferment. The International Workers of the World (IWW), commonly called Wobblies, was moving into industrial communities (especially in Montana mines) spreading the doctrine of workers' rights. Bingham did not escape its attention. In addition the Western Federation of Labor (WFL) was trying to establish itself as the representative of the miners.
>
> Coupled with these movements was a growing resentment among the miners against the system of labor recruiting then in effect. Certain labor agents arranged with the company to supply workers in any desired number at any time. Those agents collected an initial hiring fee from the workers plus monthly dues as long as the jobs lasted. No one could be hired without going through the agents.[45]

Company executives and political figures denied that this padrone system existed, however.

The leading agent was Leonidas G. Skliris, who provided workers under a padrone system for Utah Copper, the Western Pacific Railroad, the Denver and Rio Grande Western Railroad, and the Carbon County coal mines at Castle Gate, Hiawatha, Sunnyside, and Scofield.[46] His power earned him the title of "Czar of the Greeks"; he had contacts with labor agents in Idaho, Wyoming, Colorado, Nevada, and California. He "could, within minutes of a telephone call, have men on a train traveling to a destination where they would be hired as workers or strikebreakers."[47]

With his power as the labor agent, Skliris lived high. He resided in the newly completed Hotel Utah in Salt Lake City and traveled extensively throughout the country, making new alliances and recruiting workers. His influence was greatest among the Greek population because he spoke the language. He charged $50, $25, whatever the market would allow, to place a worker and an additional monthly fee for the worker to keep the job. His success was in great measure due to the large number of Greek immigrant workers. Few spoke English, so Skliris spoke for them in negotiating employment and pay. Naturally they gravitated to the jobs, the mines, and the businesses where their native language was used.[48] Additionally he coerced workers into trading exclusively with the Pan Hellenic Grocery Store and threatened them with discharge if they didn't or if they were not spending enough.

The padrone system affected other immigrants, especially the Japanese. Skliris's Greek workers resented the Japanese because they were paid more. Asians mostly were "bank men," who were lowered by ropes and manually swung picks to loosen the ore in the steep banks, a dangerous occupation. The difference in pay was a major divisive factor in the mines.[49] As the resentments and conflicts, both socially and on the job, boiled to the surface and the unions gained favor with some of the workers, a strike seemed imminent. Charles Moyer, president of the WFL, came to Bingham from Colorado to head the bargaining with the companies.

The workers felt the company was abetting the unfair system and deliberately trying to get as much work done as possible using cheap labor. Some workers were paid as little as $1.75 a shift, and it did not matter how the labor was furnished. Many of the workers joined in a union-sponsored strike against the company in 1912. As labor unrest grew, men were imported from other parts of the United States as well as foreign lands, primarily as strikebreakers, when workers voted in favor of this first strike in 1912, which turned violent. The Utah National Guard and a large contingent of deputy sheriffs were called in to keep order. The strikers in turn armed themselves, and for a time the Bingham atmosphere was very tense. With the deputies, company guards, strikebreakers, 50 sharpshooters from the Utah National Guard, and many of the strikers armed, lives were lost. Many families became frightened and moved out of Bingham; as a result, commerce ground to a halt. Armed groups of both strikers and strikebreakers intimidated residents, who were afraid to leave their homes. Ultimately some 1,200 miners were involved in the walkout. Many of the Mexican strikebreakers remained in Bingham.[50]

WFL's President Moyer and company officials were inflexible in their positions, and no meetings ever took place. Moyer forwarded the union demands to the mining companies, and the next day the United States Smelting, Refining and Mining Company informed him that it would not "accede to demand." On 17 September about 1,000 miners assembled at the Bingham Theater to take a strike vote. At that point even Moyer urged caution and pointed out that they might not be able to win with a strike. "When the men were asked whether they favored the strike, they leaped to their feet and broke into cheers. When the negative vote was called for, not a single miner rose to his feet. The vote was unanimous," announced G. E. Locke, the union secretary.[51] The strike would commence at seven o'clock the next morning and would affect 4,800 miners at Bingham's principal mines: Utah Copper; United States Smelting, Refining and Mining; Utah Consolidated; Utah Apex; Bingham New Haven, and Ohio Copper. The workers exited the theater firing shots into the air. As the Salt Lake smelters were drawn into the strike, the affected workers totaled some 9,000.[52]

Word of the strike vote spread rapidly, and anarchy reigned. As a mob of strikers terrorized the town, Salt Lake County Sheriff Joseph C. Sharp and his chief deputy, Axel Steele, mobilized 25 deputies, armed them with several thousand rounds of ammunition, and dispatched them to Bingham aboard a special train. By order of the mayor, all saloons were closed on 18 September and drugstores were forbidden to sell alcohol on penalty of revocation of their licenses.[53]

With both the company and the strikers refusing to budge, the workers took matters into their own hands. Miners, mostly the Greeks who wanted Skliris immediately fired as agent, bought firearms and ammunition from Salt Lake–area hardware and sporting-goods stores. They then positioned themselves at strategic points on the mountainsides. "With 800 foreign strikers armed with rifles and revolvers strongly entrenched in the precipitous mountain ledges across the canyon from the Utah Copper Mine, raking the mine workings with a hail of lead at every attempt of railroad employees or deputy sheriffs to enter the grounds, the strike situation . . . reached its initial crisis."[54] Moyer finally admitted that the union could not handle the Greeks.[55]

Governor William Spry issued an ultimatum for the strikers to leave the hillsides and mines and went to the Bingham Theater, expecting to meet with them. The strikers ignored his order until a bearded priest in black robes with a priest's black hat on his head walked up Main Street and then up the mountainside.

> Their warlike spirit, subdued temporarily by a lone Priest from the Greek Church, Father Vasilios Lambrides, who exhorted them in the name of their religion to refrain from further violence and defiance of the law, the army of strikers on the mountainside commanding the works of the Utah Copper Company, voluntarily descended from their stronghold yesterday afternoon. The little father dressed in flowing clerical robes with a glittering cross of gold upon his breast, went among the militant strikers like the spirit of peace and brought "the truce of God." Everywhere, guns were laid aside for him and hats were doffed in respectful salute.[56]

Soon after, nearly all the men left their positions to go listen to Governor Spry.

Bedlam still prevailed, and some miners sympathetic to the companies were urged to go to work when operations started up again. Then, in early October, Jackling; his assistant, Gemmell; and R. H. Channing, president of Utah Copper, accompanied Sheriff Sharp on an inspection of the Bingham properties. Following that meeting, Jackling told reporters from the *Salt Lake Tribune* that "the company's property is there and the company desires to operate it. The company wishes to begin work as soon as possible and the company is looking for men."[57]

> On the morning of 8 October the Highland Boy whistle sounded at six, again at seven, and finally at eight o'clock and a group of miners reported for work at the mine, accompanied by a dozen deputies. Close behind them trooped between 100 and 200 irate strikers who tried to persuade them not to go to work. According to *The Tribune*, [one of the strikers] became "officious" and was "tapped on the head with the butt of a rifle." He then went away and molested the men no more. The fifty miners passed onto company property and work was resumed at the Highland Boy Mine. The following day, under the surveillance of a hundred deputies posted on adjacent hillsides, Utah Copper commenced work with a skeleton crew of 150 men using one steam shovel and a single locomotive.[58]

By late October Utah Copper had 1,500 workers on the job, along with 15 steam shovels, and by the end of the month, it was moving 11,000 tons a day, more than half the usual tonnage.

One major festering sore was resolved. Skliris, the Czar of the Greeks, returned from a trip to Idaho and Colorado, still fervently denying that the padrone system of hiring workers existed. He even offered a $5,000 reward to anyone who could prove that it did. Nothing came of his offer, and on 22 September 1912, Skliris "resigned."[59]

It took the companies five months to return to normal operations. The actual expense of fighting the strike was slight compared to the loss of profit caused by the walkout. Utah Copper estimated its loss at $1.25 million, and the other mining companies posted similar figures.[60] The efforts of unions at mines all over the West were affected by the failure of the 1912 Bingham strike and the damaged prestige of the WFL, but the labor movement at Bingham was not dead. The next year the radical IWW again cast a covetous eye on the mine and mill workers in Bingham's copper operations. Between 1914 and 1916, the IWW was agitating all over the United States and, in due course, found a few followers in Bingham. The IWW committee attempted to stir up a strike among the Bingham workers and circulated handbills with six demands, which the companies regarded as outlandish and simply would not consider. Few workers were in sympathy with that union, and even fewer were members. Attempts to organize strikes at two properties came to naught. Labor unions were not a force in Bingham for many years and did not gain a foothold on "the Copper" until 1944, when the first collective bargaining agreement on wages and working conditions was negotiated.[61]

The strike did bring about some improvement in general working conditions. In a few years, the companies granted many of the workers' demands, such as showers, change rooms, and company insurance plans. In 1913 the company created a training manual for workers and published it in English, Serbo-Croation, Greek, Italian, and Japanese.[62] Working conditions also improved through the actions of state agencies and legislation favorable to workers generally. In 1917 the Utah Industrial Commission was formed, the office of the state mine inspector was established, and safety regulations were put in place. That same year the Utah Legislature enacted the Workers Compensation Act.[63]

With labor strife fading into the background for the time being, Utah Copper moved forward with Jackling's theory that mining and processing huge tonnages of lower-grade copper ore could be profitable. But that couldn't be done without some heavy financial backing. Companies were consolidated, ownership shares were traded back and forth, and major financiers took notice of the potential.

Kennecott Enters

Kennecott Utah Copper's genesis came in the early 1900s, when Dr. Robert Kennicott was commissioned to head an expedition to run a telegraph line across Alaska for

Western Union Telegraph Company. Because of his knowledge of Alaska, he had considerable influence on the exploration and development of that territory. The Kennicott Mining District was named after him. A mine was located and developed in the new district, but a clerical error changed the spelling from Kennicott to Kennecott, resulting in the eventual name of the Kennecott Mines Company. In 1908 New York's Guggenheim financial interests acquired the Alaska claims and developed and operated the mines under the new Kennecott name.

In a financial consolidation move, Guggenheim decided to bring all its copper interests together for stock-sale purposes, and on 29 April 1915, Kennecott Copper Corporation was incorporated under New York law as a holding company consolidating the properties throughout the world.[64] Nine months later Kennecott Copper Corporation acquired shares totaling 25 percent of Utah Copper from the Guggenheim Exploration Company. During the next eight years, by stock purchase and exchange, Kennecott obtained 77 percent of Utah Copper stock, giving it undisputed control over the company by 1923.[65] Under this single leadership, open-pit mining with removal of heavy tonnages could proceed using increasingly heavy equipment.

After 17 years the original steam shovels had been phased out, and by 1919 the new open-pit operation was using 21 new behemoths, mounted on rails, each scooping up seven tons of material with each bite of its three-and-a-half-yard dippers. Due to continuing improvements in the milling and processing plants, the appetite for even larger ore tonnages grew. More-efficient electric shovels replaced the remaining steam shovels. The use of electric power gravitated to the railroads, and by 1928 mining operations had become increasingly modern with the introduction of electric locomotives.[66] Use of electric power at the mine was nothing new. As early as 1901, the Telluride Power Company built a 44,000-volt main line from Cedar Valley to Bingham, and a branch of the main line from the Provo generating station to Mercur. That power, produced by Telluride's five generating stations—one as far away as Grace, Idaho, 160 miles to the north—operated an air compressor, saws, and other equipment.[67]

Disasters and City Development

Profound changes were also occurring in Bingham Canyon, fueled by factors only peripherally related to industrial evolution. The early days saw divisiveness among the residents, many of whom were recent immigrants. Language and culture at times were major barriers, but over the years, the actions of several individuals and a series of natural disasters—fires, floods, and snowslides—welded residents of the communities together in caring for the dead and their families, the injured, and the homeless. Fires were numerous over the years, primarily due to the wooden construction of the majority of houses. Fire danger was intensified by the crowded conditions in the canyon, where houses were actually stacked upon each other and packed closely together. Fighting the fires was extremely difficult due to the lack of water and proper equipment. Additionally many of the homes were located on terraced hillsides, where it was impossible to pump enough water and fire trucks could not go.

That problem became evident on 8 November 1880, when a fire—some believe caused by arson—started in a building by Reed's store and quickly spread to 31 other buildings.[68] In 1895 two fires, one in July and the other in August, again spread quickly, but the August fire had far more serious consequences. The origin is not known, but the houses in the narrow confines of the canyon were excellent fuel. Lacking proper equipment, the bucket brigades worked heroically but were stonewalled by the blaze. As the fire raced through the canyon, 45 homes were lost, valued at about $200,000. Fortunately, no one died, although many people were homeless for a time. Concerned citizens provided food and shelter. Bingham rebuilt and in November 1903 organized a fire department. By December the apparatus committee reported it was advertising for equipment, and on 23 January 1904, A. L. Heaston and J. Bryant were elected Bingham's first fire chiefs. A month later, on 28 February, the town of Bingham incorporated.[69]

The organization in place to combat disasters had plenty to do during the next few years. In 1924 a brisk wind whipped a blaze through Bingham, and a main water valve ruptured, making fire fighting essentially impossible. Six families were made homeless; two volunteer firemen, Tommy Price and Harold Anderson, were killed; and Leonard Gust was injured by a falling wall.[70] Another fire on 10 February 1925 razed 20 buildings, but even worse was a huge snowslide in 1926. Heavy snow had fallen for several hours, and early that morning the snow at the top of Sap Gulch started moving. The earth trembled, a terrible crash was heard, and the gigantic avalanche descended on Highland Boy. "Seventeen dwellings and a three-story boarding house were swept from their foundations and buried beneath the on-rushing thousands of tons of rock, snow, ice and other debris." The lives of 39 people were lost amid the tangled wreckage. Work was suspended at the Utah Apex and Utah Delaware Mines, and some 200 men began the task of removing both the living and the dead from the ruins. Some 200 residents were affected by the slide. At the time that avalanche was said to be most disastrous in the history of Utah.[71]

Six years later, on the morning of 11 August 1930, down the canyon at Markham Gulch, weather struck again, this time with a very heavy rainstorm. The fury lasted for about half an hour and soon reached the flood stage. Thick muddy streams roared down the mountainside, in through the rear doors and windows of homes, and out the front doors and windows. Cars, debris, and other objects were swept down the main road by a torrential stream of muddy water. Nearly every home on the west side of the main canyon was damaged by mud from the immense waste dumps higher up as well as rushing water. Then landslides followed the floods. Though the rainstorm was brief, the floods lasted from 11:30 AM to 2:00 PM. They demolished 20 homes and damaged more than 100 buildings. The loss was estimated at $400,000.[72]

Both those disasters, however, paled in comparison to the Highland Boy fire of 1932, labeled one of most devastating in the town's history. It started in an old theater where, it was rumored, some children were playing with celluloid film and set it afire to watch it burn. Within minutes the town was an inferno. Wooden houses on the terraced hillsides quickly burned to the ground, and people could save only a few belongings. That fire raged for three hours, fought by local, Murray, and Salt Lake County fire departments. Everything on both sides of the narrow canyon for one-third of a mile was reduced to smoking rubble.

Again, the residents mobilized, and all available space was turned into sleeping quarters. A relief station was set up in the Highland Boy Community House, and local organizations brought all possible aid. Fortunately, no lives were lost, although some victims were severely burned. They were treated by a local doctor, who set up a first-aid station at the scene.[73]

While these disasters helped solidify the feeling of community, several individuals also worked to unite the Bingham Canyon residents. The lives of three remarkable women provide outstanding examples of that effort. Georgia Lathouris Mageras (affectionately known as Magerou, the genitive form of Mageras) was a midwife who arrived in Magna from Greece in 1909. She delivered the babies of miners' wives and earned the reputation of never losing a mother or child in her long years of practice. Her services were sought by Greek, Italian, Austrian, and Slavic workers and their families, who preferred her to the company doctors. Finally, she found herself doing what was called "practicing medicine without a license." She then began assisting doctors with deliveries, rather than taking charge herself. When babies arrived before the doctor, he had only to sign the birth certificate. Her impact was profound because she insisted on cleanliness, not only for herself in the deliveries but for the mothers and their babies.

Additionally Magerou treated many people outside of childbirth with folk-medicine remedies she had learned while growing up in Greece. Her herbal treatments were widely acclaimed, and workers traveled long distances to be treated. She earned the respect of not only the residents of Snaketown, Ragtown, and all of Bingham Canyon but also the company and community doctors.[74]

Alta Miller, who was born in Bingham in 1904 and eventually taught in the same school she had attended, recalls the various ethnic groups in the community as sources of fun and inspiration:

> I remember the wonderful celebrations we had; I especially remember Columbus Day. All the Italian people in Bingham would get together. They would dress in the old Italian costume of Columbus and they would make three ships—the *Nina*, the *Pinta*, and the *Santa Maria*. Then they would have all the children excused from school. We would line up. Each child had two flags, and we would hold one in each hand—one Italian and one American flag. We would follow the three ships up the main street. The bands would play, the people would wave their flags, and everybody in Bingham came out and lined the sidewalks. It was wonderful! In the afternoon

> they had all kinds of races—foot races, three-legged races, and horse races. Then the Italian people would serve refreshments. We all looked forward to Columbus Day.
>
> We had other celebration days. The Japanese people had Kite Day and Doll Day. One thing about the people in Bingham, everybody supported everybody else. I feel there was no prejudice among the people of Bingham at that time. There were many different nationalities, especially Spanish, Greek, Austrian, Italian, quite a few English, lots of Scandinavians—about 18 different nationalities. Many of them were young men who came to work and later brought their families. They built homes, many of them up on the side of the mountain just like a crow's nest. Many of the single men lived in boarding houses.[75]

Many Bingham residents think much of the unity that Miller recalls came from the work of Ada Duhigg, who arrived in 1932 and became affectionately known as the "angel of Bingham Canyon." A native of Iowa, she prepared herself for a teaching career, but early in her life, she felt a spiritual call. She enrolled in the Kansas City National Training School for Christian Workers in 1931. A year later she graduated as a Methodist deaconess and joined a group of some 600 deaconesses who would be called to serve in 500 communities. Her assignment by the Women's Home Missionary Society was superintendent of the Highland Boy Community House (HBCH) in Bingham Canyon, Utah. "The building had been dedicated only five years earlier as a missionary project by the Methodist Church, and a neighborhood house for all races and creeds."[76] Under Duhigg's guidance, the community house served as a place of worship, a gathering place for fun and parties, and a learning center with a library of some 1,500 books and guided-learning programs for those who knew English only as a second language. The gatherings included leadership training and meetings for Cub Scouts, Boy Scouts, Brownies, Girl Scouts, and Campfire Girls. Training was provided in first aid, Red Cross skills, and nursing. There were well-baby clinics and even tonsillectomy clinics that included the impromptu use of the kitchen as an operating room. Bingham Canyon residents observed, "It's not surprising that there was very little juvenile delinquency in Highland Boy. Between school and the multitude of activities offered at HBCH, there was little energy left over for mischief." Teachers and mission-house people worked with families. Duhigg was the intermediary between children at school and their parents and did much to unite the immigrants with the native Americans and make Bingham a closely knit community.[77]

World War I and the Great Depression

That community, like all others in the nation, felt the impact of World War I. The outbreak of the war in Europe saw a worldwide slump in the copper market, and Utah Copper was forced to curtail operations by half. In an effort to cut the flow of strategic metals to Germany, Great Britain in 1914 placed copper on the conditional contraband list. However, the slump was short lived, and the next year the market rebounded due to the increase in wartime demand for copper. This resulted in a production increase to 33 percent above normal. "During World War I, Utah Copper was second only to Montana's Anaconda as a source of newly mined copper. In 1916 company profits rose to an all-time high of $33.7 million on production of 93,800 tons of copper."[78]

At the close of the war, however, the copper market slumped again, and operations were curtailed drastically. The Magna mill closed in 1919, and the Arthur mill shut down in 1921. The following year the postwar demand for copper began to increase, and both mills were reopened. During the shutdown and as the plants went back on line, they were dramatically overhauled and made much more efficient. Froth flotation units were installed, and the copper recovery gradually went up from the 1917 average of about 61 percent of capacity to 73 percent in 1918; it continued to rise to 81 percent in 1923 as both mills fully utilized the floatation process. By 1926 the capacity of the mills had increased to 50,000 tons per day.[79] The never-ending process of developing technology continued, and by 1963 the two plants had a combined capacity of 90,000 tons per day.

Innovation was not confined to Utah Copper's mills. In the late 1920s, the entire system was electrified. At the end of the decade, 41 electric locomotives were in service. "The modernization of the mining equipment and the initiation of better handling techniques enabled the company to move its 232 millionth cubic yard of material from the Bingham mine in April 1935. By this time the company had moved as much earth as had been moved in the construction of the Panama Canal."[80]

The volatile copper market started another downward slide with the Great Depression of the 1930s, and Utah Copper Company's Bingham operations were again curtailed. The Arthur mill closed, and the Magna mill went on a reduced operating schedule. "The company staggered the employment to allow the greatest number of employees to be retained—giving them approximately one-half of full-time employment. Production continued to decline, however, due to the meager demand for copper, reaching a low point in 1933 when operations were only one-fifth of normal capacity."[81] The market was not much better the following year because domestic consumption of copper increased by only 5 percent.[82]

Despite the Depression-driven low market, Utah Copper continued with innovations to make copper mining, haulage, and recovery more economical and profitable. In the face of the 1929 Depression, the company built a precipitate plant directly at the mouth of Bingham Canyon. By the next year, the Bingham and Garfield Railroad was operating more than 148 miles of rails, including sidings, tracks, and switchbacks in and near the mine. The face of Bingham Canyon was changing. The once dominant Copper Hill was transformed into an open-pit mining operation to compensate for the decreasing concentration of copper in the ores with increased tonnages. The fleet of electric shovels, the haul trains, and the mills and concentrators were constantly upgraded. Utah Copper appeared to be the beneficiary of a slowly recovering metal market.

The mine operated at varying rates of production in 1935, and the ore was treated entirely at the Magna mill. The Arthur mill continued to lie idle but was reconditioned to ensure its readiness to be put in service on short notice. The plan of rotating employment continued during the latter part of the year for all employees. The slowly increasing demand for copper finally afforded about 84 percent of full monthly time.[83]

It was not an easy time, however, because the Depression and its effects on the workforce and unemployment were major factors in Bingham Canyon. As the company slowly increased the number of workers, resentment again boiled over in the canyon, apparently because some workers were recruited from other communities in the Salt Lake Valley with a detrimental effect on local business. The Town Board complained to Utah Copper, "These employees make their living here and the majority of them never spend one cent in Bingham. It is believed that if the mines will employ strictly local labor—men who reside in the canyon—local business houses and property owners will benefit, and the town will attain a degree of prosperity that is its just due."[84] It is not known that this resolution had any effect on mining company employment practices.

Within a few years, the company rebounded. A major financial milestone occurred on 10 November 1936. "Kennecott acquired all the property and assets which had been formerly owned by Utah Copper Company. This acquisition completely unified the Utah properties with Kennecott. Your Corporation had previously owned approximately ninety-nine percent of the Utah stock."[85] The corporation's annual report for 1936, reflecting the final consolidation with Utah Copper, listed the new divisions: Nevada Consolidated Copper Corporation; the Nevada, Arizona, and New Mexico operations were listed as subsidiaries, along with the Nevada Northern Railway Company; the Ray and Gila Valley Railroad Company; the Alaska Division; the Braden Copper Company, a Chilean operation; the Chase Brass and Copper Company, Inc.; and the Kennecott Wire and Cable Company.[86] Kennecott merged all its Utah properties into the parent company as a wholly owned subsidiary.

That final unification reflected Jackling's efforts during the previous two decades in bringing together major copper entities in the West. The company from that period until Jackling's retirement on 1 October 1942 was referred to as the Jackling organization.[87] Then in 1947 Kennecott dissolved the Utah Copper Company and officially organized the Utah Copper Division as an operating division of Kennecott Copper Corporation.[88] The 1915 acquisitions by Kennecott had set the stage for decades of growth and the development of new mining, milling, and concentrating techniques.

Utah Copper did not confine its efforts just to copper over the years. While copper, gold, silver, and lead were important by-products of refining almost from the beginning, the recovery of molybdenum caught management's eye as early as 1898, when its presence was noticed. At first the graphitelike substance on the water used in copper concentrating was considered simply another "waste" product, not worth any investigation because of its low density. But the development of higher-strength and corrosion-resistant steels following World War I commanded some attention for molybdenum.

By 1935 molybdenite (MoS_3) proved sufficiently interesting that a workable process was developed, successful enough that a test plant was installed at the Magna mill. "By mid-1936, the operation had proved so successful that the recovery of molybdenite began on a commercial scale. Production increased rapidly as the entire milling operation was converted to the recovery of the metal. By 1938, the Utah Copper Company was the world's second largest producer of molybdenite."[89] By the early 1960s, the company was still in second place in world production, and the annual output of this important alloying metal had a value in excess of $30 million.

The investigation of additional minerals did not stop there, however. Utah Copper embarked on a complete spectrographic analysis of copper ore, copper concentrate, and molybdenite concentrate. The analysis showed the presence of some 38 elements in addition to the usual components.[90] The early development of molybdenite-recovery technology, which served the new high-strength and superhard steel markets, gave Kennecott a leadership position as molybdenum found its way into other metal-enhanced steel alloys, electrodes in electrically fired furnaces and forehearths, nuclear-energy applications, and the manufacture of aircraft and missile parts.

Impact of World War II

The onset of World War II impelled a number of changes. Kennecott had recognized unions as the employee-bargaining representatives as early as 1938, but the first collective bargaining agreement covering wages and working conditions was not negotiated until 1944. That same year Utah Power and Light built a plant to make Kennecott independent of outside sources of electrical power. But wartime mobilization took its toll on Kennecott's operations, as well as on the town of Bingham. Some workers abandoned mining work, relocating to other areas for jobs in the defense industry. The shortage of mine and mill workers initiated a government order for draft boards to defer some skilled and nonskilled mine and smelter workers so they could stay on the job. The shortage even stimulated importation of Puerto Ricans to work on six-month contracts.

"Despite the labor shortage, Bingham mines—subsidized by the government and assured of no labor unrest by a freeze on union grievance committees—produced a staggering amount of metal during the war years. From 1941 through 1944, the district produced over three billion pounds of copper—more than one-half of all the copper mined in the United States. Bingham's contribution to the defeat of the Axis powers was enormous." Germany capitulated on 8 May 1945, and Bingham rejoiced. Four months later, on Tuesday, 14 August 1945, Japan surrendered. Within minutes of the announcement, the entire town of Bingham shut down for a two-day holiday and, apparently, much revelry.[91]

World War II brought about not only subsidiary operational changes but far-reaching executive organizational ones as well. Workers going into military service and others leaving Bingham Canyon to take jobs in the defense industries had to be replaced to keep up with the demands. Throughout the entire industry, women in mining were unusual, even in clerical and secretarial jobs. Utah Copper though created its own versions of Rosie the Riveter.[92] During the war years, it was not uncommon to see women laying railroad tracks, working on maintenance gangs, running the railroad switches, operating overhead cranes, staffing the parts and supply shops, operating lathes, supervising the vats in floatation mills, and repairing turbines in maintenance shops.

Postwar Production

With the end of World War II, all America felt a universal sense of relief and joy that the uncertainties of war had ended. But the transition from a wartime, all-out military effort with its hang-the-costs mentality to a peacetime economy with a sense that business and the entire nation had to move in a direction where sound practices prevailed was not easy. The economic and social driving forces were headed in many, and at times new, directions. Wartime restrictions, price controls, and business decisions imposed by the federal government were gradually lifted. One of these restrictions—the freeze on union grievance committees that assured no labor unrest—terminated early. That left the door open for organized labor to flex its muscles, long dormant since the first collective bargaining agreement had been forged in 1944 with the International Union of Mine, Mill, and Smelter Workers as the recognized bargaining agent. Unionized workers went on general strike in 1946.

> Although Kennecott's copper business reached maturity after World War II, it was far from stable. The three deepest production slumps were the result of strikes. The 1946 strike lasted 152 days. In 1959, workers were out 144 days and the strike, which ran into 1960, totaled 171 days. Again a strike of 259 days spanned two calendar years, 1967 and 1968, with 170 of those days in the low production year of 1967. Strikes were a regular occurrence at contract renewal time. Over the 34-year period from 1946 through 1980, the workers represented by labor unions at Utah Copper were on strike an average of 29 days per year, almost one month out of 12.[93]

Union members saw this as progress, but management did not. In his 1945 annual report, E. T. Stannard, president of Kennecott Copper, told stockholders that during the reporting year, Kennecott operations had largely been at a standstill because of strikes. He complained that laws and executive and administrative rulings "have been manifestly favorable to labor. The injurious strikes now so widespread are due in large measure to this increase in power without any corresponding accountability."[94]

In spite of general labor unrest and the up-and-down gyrations in the demand for copper, Kennecott continued to refine and expand existing processing technology and search for new methods of producing high grades of copper at less cost. Undoubtedly, the first major move in this direction during the national recovery from war mobilization was the design and construction of the first mine rail-haulage tunnel, a 4,650-foot tube built at an elevation of 6,040 feet. It eliminated hauling ore from the bottom of the pit to the top and replaced the old Bingham and Garfield Railroad line. The tunnel provided safer and faster movement by longer trains. It was completed in 1948, and electric locomotives replaced the steam ones for hauling ore to the mills.

The tunnel and railroad line improved the economics enough that a year later construction began on a second mine-haulage tunnel, 7,000 feet long and at 5,840 feet elevation. That project was coupled with the 1950 start-up of an electrolytic refinery at the town of Garfield to produce copper cathodes, gold and silver bars, and commercial-grade selenium. Tunnel haulage was so attractive that in 1958 construction began on yet a third mine rail-haulage tunnel, this one 18,000 feet in length at 5,490 feet elevation. Expansion and consolidation continued a year later with Kennecott buying out ASARCO's interest in the Garfield smelter. The general updating required that the company's power plant expand to 175,000-kilowatt capacity by 1960.

The rapid pace of technological research and development following World War II sparked a renewed interest in some of those exotic minerals present in the Bingham deposits. In 1951 Kennecott established a new metallurgical research laboratory at the University of Utah. The objective was to improve the recovery of copper, gold, and molybdenite and facilitate the recovery of metals outside the Kennecott production inventory. That research brought enhanced recovery of molybdenum, as well as developing technology to recover platinum, palladium, tellurium, selenium, rhenium, and nickel sulfate—all virtually unknown to engineers three decades earlier. In addition, the research laboratory was the impetus for expanding research by Kennecott's four western mining divisions—Utah Copper, Nevada Mines, Ray Mines, and Chino Mines.

In 1954 the National Society of the Sons of Utah Pioneers presented a statue of Colonel Daniel C. Jackling to the State of Utah. The heroic statue, standing nine feet tall, was carved by Dr. Avard Fairbanks, world-famous Utah sculptor, and weighs 2,000 pounds; 86 percent of the bronze casting is copper from the Bingham Mine. It was placed in the rotunda of the Utah State Capitol Building and dedicated on 14 August, Jackling's 85th birthday. The date also marked the 50th anniversary of the first shipment of copper ore to the Copperton mill. Jackling was too ill to attend the ceremonies, but government, business, educational, religious, and civic leaders did. Utah Governor J. Bracken Lee and Nicholas G. Morgan, Sr., president of the National Society of the Sons of Utah Pioneers, paid tribute.[95]

Jackling died on 14 March 1956 at his home in Woodside, San Mateo County, California. He was 86. The *New York Times* proclaimed, "Mr. Jackling's first major accomplishment came . . . in 1896, when his successful experiments on metallurgical processes of low-grade ores were largely responsible for the Consolidated Mercury Gold Mines in Utah. This was followed in a few years by his big copper discovery and its first exploration at Bingham, Utah, where other companies had refused to operate because of the low grades of the ores they encountered."[96]

Throughout the 1950s, the profile of Bingham Canyon was rapidly changing, with copper mining, processing, and recovery systems constantly updated. The old Carr Fork bridge, built in 1910 and 1911 to connect the rail system from the mine to the Magna and Arthur mills, was targeted for demolition. The bridge was 690 feet high and 190 feet long. It had remained essentially the same since the day of its completion. In its day it was considered an engineering and construction marvel, but it had to be torn down to make way for the expansion of the open-pit mine.

The decade of the 1960s saw even more intense modernization of existing facilities and construction of new ones. Declining ore grades as the Bingham open-pit mine advanced down into the earth drove much of this effort. In its simplest terms, more ore had to be mined and moved to concentrate, smelt, and refine the same amount of copper. In like fashion, more overburden had to be blasted and removed, increasing the stripping ratio, to get at the recoverable copper.[97] The result was a $100 million expansion program, begun in 1963 and designed to increase the Utah Copper Division's production capacity by 100,000 tons a year. One of the first steps was to start using diesel electric trucks to replace the slower and more cumbersome train hauling the mine waste.

The multimillion-dollar expansion announcement caused considerable concern among employees about whether they would fit into the new operation. Would it require people with new skills? What would happen to people with outdated skills? Would the pay scales stay the same? J. P. O'Keefe, general manager of the Utah Copper Division, addressed those concerns in a statement to employees in the company publication:

> New occupations will be created. Further, some of these new occupations will require standards of physical and mental coordination not necessary in many present occupations. For example, the drivers of the big haulage trucks will have to have more than the usual driver's license; they must be able to pass physical and aptitude tests which demonstrate that the driver will not be a safety hazard to himself and to others.
>
> Another good example is the smelter. When the process changes have been completed, the plant and many of the jobs will be entirely different than now. Just as different, in fact, as if a new smelter had been constructed.
>
> Many other examples could be cited.
>
> However, let me assure you that wherever practical present employees will be trained for the new jobs. This is not the easy way. In fact, this

> intention will, by itself, create problems. Fortunately, these problems can be solved, provided we work toward a smooth transition in a spirit of understanding and cooperation. It is impossible in this brief message to deal with specific jobs, but let me again assure you that we will spare no effort to see that you fit into the new Utah Copper Division.[98]

A major step in the expansion was the opening of a cone precipitate plant at Bingham in 1965. The plant increased the efficiency of leaching solutions, specially treated water pumped over waste dumps and permitted to percolate to the bottom of the heap. The precipitation plant removed the copper from the solution and then recycled the solution. The new plant was designed using cones developed by the Kennecott Research Center at the University of Utah, opening the door for many new approaches to copper recovery.[99]

The following year the company opened a molybdenite oxide plant, creating a new market for recovering the vital steel-alloying metal, previously bypassed in the copper-production processes. Molybdenum joined the family of gold and silver as important revenue-producing by-products of the Utah Copper Division.

Over the years change came about as necessary and usually without major disruptions in the overall copper environment. But through the 1950s, change accelerated, and as the world's largest open-pit mine expanded, both downward and outward, something had to give. The town of Bingham was being squeezed out. In 1959 Kennecott made a bid for whatever private property remained in the canyon. By 1971 the once-colorful town of Bingham, the center of commerce for the mining community, had ceased to exist. Today Bingham has been absorbed into the open-pit mine; homes, businesses, and streets are gone, but the legends live on.[100]

Last Quarter of the Twentieth Century

Kennecott's ore-processing facilities were not immune to change, and the driving force was the necessity to reduce costs. Both the Arthur and Magna mills were hopelessly outdated in spite of continuing redesign and updating. Those facilities had originally been built in the early 1900s, and the escalating costs of maintenance were killing profitability. In 1988 the crushing and grinding operations were discontinued, although flotation facilities remained in limited use. With the construction and expansion of the Copperton concentrator, the old Arthur mill was shut down and demolished. The Magna operation followed with everything now centralized at the modern Copperton concentrator. The old 15-mile rail line from the mine to the Magna and Arthur mills was replaced by a 5-mile ore conveyor from mine to concentrator.

Rapid changes affected not only Utah's copper production but Kennecott Utah Copper's infrastructure as well. New environmental regulations dictated nonproductive capital expenditures and increased operating costs. To comply with the Clean Air Act of 1970, Utah Copper Division's $300 million expansion included new smelting technology that saw construction of Noranda smelting furnaces in 1977 and a new, 1,200-foot stack at the Garfield smelter in 1978.

> In addition to compliance costs, uncertainties surrounding the emerging environmental agenda interfered with long-term planning for the copper operations. Two oil crises and the soaring inflation of the 1970s further exacerbated Kennecott's problems by escalating copper production costs at an alarming rate. Costs of materials and supplies consumed by the copper operations rose rapidly. In addition, inflation rapidly boosted labor costs as a result of contracts that provided automatic wage increases linked to the consumer price index. By 1980, Kennecott's copper properties had become high-cost operations.[101]

In addition, 1980 marked the beginning of a worldwide copper recession. The acquisition of Kennecott Copper Corporation by Standard Oil of Ohio (SOHIO) in 1981 was the first step in significant corporate restructuring. Plagued by recessions and difficult labor negotiations, operations at the Bingham Canyon Mine shut down in 1986. Labor negotiations produced new agreements the following year, and all Kennecott Utah Copper operations resumed.

That same year British Petroleum (BP), a London-based firm, took control of SOHIO, making Kennecott a part of BP Minerals America. With new leadership and financing, the company in 1988 announced a new $400 million modernization program under the leadership of a new president, G. Frank Joklik. He had moved up from being head of the new Kennecott Minerals Company's exploration, technology, and planning section. He was well grounded in exploration and large-project development, both in his native Australia and the United States, but had little experience in operations. Rather than proving a hindrance, however, this situation proved to be a plus. He was not encumbered by the traditional perspectives that had long constrained innovation in Kennecott's operations. He brought a fresh approach to Salt Lake City as he formulated a new strategic plan for Kennecott's minerals business.[102]

With new energies and realistic approaches to the economics of operations efficiency, a revitalized Kennecott Utah Copper moved into 1988 with completion of a peripheral tailing-discharge system at the tailings pond near Magna and the opening of modernized facilities at Bingham and Copperton. The new production process included an in-pit crusher and conveying system, and three grinding lines in the Copperton concentrator, increasing production to 85,000 tons per day, some 13 percent above the original design capacity. Utah's copper industry was becoming even more efficient.

Along with the general overhaul of copper processing with new facilities, new corporate changes occurred. BP had long been anxious to get out of the mineral business and in January 1989 agreed to sell most of its worldwide holdings—BP Minerals America, Kennecott's parent, included—to Rio Tinto Zinc (RTZ) Corporation, one of the world's largest mining enterprises, also based in London. Unlike BP, RTZ's major business was, and still is, mining. This gave Kennecott a new and solid operational support base, something neither SOHIO nor BP had provided.[103]

The struggles associated with the 1980 copper recession and frequent corporate ownership changes eliminated most of the internal pressures against major expansions and upgrades. In 1989 RTZ (now Rio Tinto PLC) authorized a $227 million expansion program to increase production by adding a fourth grinding line at the Copperton concentrator, making it one of the world's largest. Two years later the line was completed ahead of schedule and under budget and increased Kennecott's annual production capacity to 125,000 tons per day.

"In the early 1990s Kennecott Utah Copper, employing 2,400 people, produced approximately 300,000 tons of copper annually plus significant quantities of molybdenum, silver and gold. In 1993 Kennecott started construction of a new smelter and modernized refinery at the company's Utah Copper operations at a projected cost of $880 million—the largest private investment ever undertaken in Utah."[104] That project was completed in 1995, making the new copper smelter the largest and cleanest in the world, capturing 99.8 percent of the sulfur contained in the copper concentrates.[105]

Since 1991 Kennecott Utah Copper has cleaned up more than 25 million tons of waste materials. Most of that waste was generated by mining operations that predated Kennecott's ownership of the mine. Another 20 million tons of clean material has been returned to the sites. Kennecott has spent more than $350 million on reclamation projects.

Bingham Canyon Today

Modernizing facilities, reducing costs where possible, and, perhaps most importantly, training employees in new skills where required have combined to continually improve the company's profitability, while at the same time protecting its workforce. In 2003 the International Society of Mine Safety Professionals recognized the Bingham Canyon Mine employees and contractors for working two million man-hours without a lost-time injury.[106]

Kennecott's open-pit operation has been a major tourist and visitor attraction for many decades. Utahns, out-of-state visitors, and especially schoolchildren are fascinated by a special overlook area of the open-pit mine and watch the blasting, mining, and loading operations. In 1992 the company completed construction and dedicated the new Bingham Canyon Mine Visitors Center. It opens each spring in April and remains open until October 31, depending on weather conditions. The Visitors Center attracts an average of over 100,000 visitors each year. It was an especially active drawing card during the 2002 Winter Olympics. The minimal entrance fees are the source of major donations to more than 100 Utah-based community charities and nonprofit organizations. These contributions are administered by the trustees of the Kennecott Utah Copper Charitable Foundation. The foundation has annually donated more than $100,000 to organizations providing assistance to the poor and needy, the disabled, the elderly, youth groups, transplant patients, health and nursing-care organizations, and other important community-based charities.[107]

Since 1903, Kennecott Utah Copper has been one of the world's largest mining families. That workforce, which has included grandfathers, fathers, sons, mothers, grandsons, aunts and uncles, brothers and sisters, and cousins, forever changed the way ore is mined and processed. Employees trace family roots to the southern and eastern Europeans, Latin Americans, and Asians who first began making mining history in Utah.

The Bingham Canyon Mine has been the financial resource that has fed more families, educated more people, and contributed more jobs than any other nongovernmental business in Utah. As with any natural-resource industry, the stock of material is not infinite. Nonetheless, Kennecott Utah Copper plans to extend the open-pit operations for as long as economically feasible. Drill tests show copper deposits extend to a depth of 100 feet above sea level or almost a mile deeper than the present pit, and exploration reveals that the recoverable copper deposits can take the open pit about 650 feet deeper than it is now. Current projections are that the open-pit mine can be extended and produce profitably until the year 2028 with other options, including underground incursions, possible after that. Mining in Bingham Canyon still looks forward to a long life.

11

Silver Reef and Southwestern Utah's Shifting Frontier

W. Paul Reeve

In a series of letters to the *Salt Lake Tribune* in the fall of 1875 and spring of 1876, prospector William Tecumseh Barbee announced a "singular discovery": silver had been found in sandstone. "The country is wild with excitement," Barbee exclaimed and noted that southern Utah's "sandstone country beats all the boys," especially the "sheets of silver which are exposed all over the different reefs. . . . This is the most unfavorable looking country for mines that I have ever seen in my varied mining experience," Barbee explained, "but, as the mines are here, what are the rock sharps going to do about it?" According to Barbee, the new discoveries were so promising that "even our Mormon brethren are getting chloride and horn silver on the brain."[1]

No doubt spurred on by Barbee's favorable dispatches, a flood of miners, merchants, and speculators soon poured into Washington County in the southwestern corner of Utah Territory. Before long, a raucous mining camp sprang to life at what later became Silver Reef. Over the next two decades, silver mining proved a powerful force in the development of Washington County. The Silver Reef mineral rush generated intense enthusiasm and produced marketable ore until just past the turn of the twentieth century.

Although Silver Reef merits most of the attention due Washington County mining, extractive activity there both pre- and postdated that camp's boom years. In fact, a series of mining events during the 1860s was responsible for reducing the size of not only Washington County but the territory of Utah as a whole. In the early twentieth century, hot on the heels of Silver Reef's decline, the county also experienced a short-lived oil boom. Even though the county's extractive activities proved ephemeral, they nonetheless spawned economic bursts which spilled over to benefit the region's long-term LDS settlers. Mining also added splashes of contrasting color to the region's religiously motivated agricultural history.

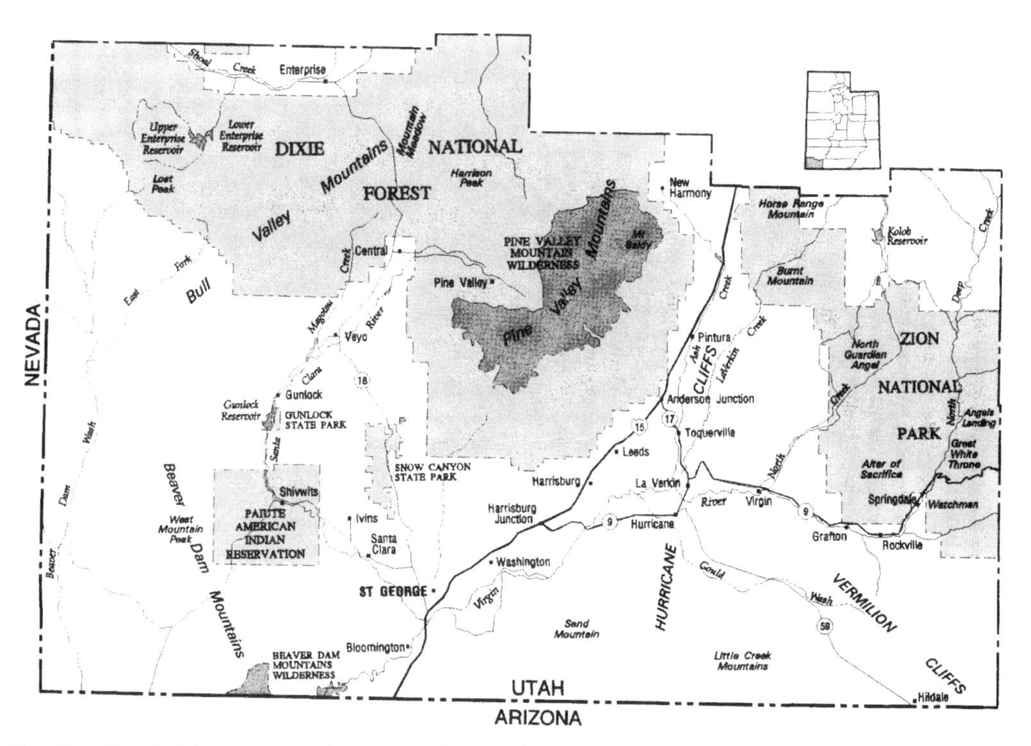

The Silver Reef mining area was located just west of Leeds in Washington County in the southwestern corner of Utah.

Early Mining in Washington County

Washington County's mining history had a rather obscure beginning in the fall of 1863, more than a decade before the rush to Silver Reef. When it started, the site of a very rich silver deposit lay hidden from Anglo-American prospectors in remote southwestern Utah Territory. Before it ended, however, the Utah/Nevada border would shift to enfold that site within the Silver State. Although Nevada laid claim to the story's conclusion, its origin belongs to Utah and in many ways provides an essential backdrop to the saga that developed later at Silver Reef.

Initially only local Southern Paiute bands were aware of an outcrop of ore at what would eventually become Pioche, in present-day southeastern Nevada. In fact, what Pioche residents came to call Meadow Valley Street in their town actually was a worn Paiute trail, leading to the ore.[2] A leader of a Paiute band whom Mormons had named Moroni was well acquainted with LDS settler William Hamblin and once carried a piece of "glittering ore" to him for inspection. Hamblin was a two-year veteran of the California goldfields and quickly recognized the rock's potential.[3] Moroni, however, refused to tell where the ore came from, only indicating that "for years chosen members of his tribe had resorted to and used it as a paint." Moroni's dying father had warned him never to disclose the ore's location to white men, "lest they should come and drive the Indians from their hunting grounds to secure the riches thus exposed." Eventually, however, Hamblin wore down Moroni's resolve through continued friendship and the gift of a new rifle. Moroni led Hamblin, along with William Pulsipher, another Mormon settler, to the rugged side of a mountain where "the glittering ore cropped out above the ground."[4]

The three men dug down several feet to expose a well-defined vein and then laid claim to the spot. This very act set in motion a series of events that would drastically change the meaning of the dirt they dug. Within three years news of silver in southwestern Utah burst onto the national stage and mustered enough power in Congress to move the Utah/Nevada border one degree of longitude farther east.

For the time being, Hamblin and Pulsipher were satisfied with their find and returned home to tell others. Before long, local Mormon leaders expressed interest in the claim. In January 1864 Bishop Edward Bunker, head of the Mormon settlement at Santa Clara, organized an expedition to the site. Bunker's group surveyed and outlined a square claim but did not organize a mining district or post traditional notice, thereby leaving the ore open for the taking. Bunker reported his expedition to Brigham Young and sought the LDS leader's advice on the matter.[5]

Young responded favorably. He not only liked the idea of claiming the ore for Zion but also wanted the surrounding valleys as good grazing ground for Mormon cattle. He put it this way: "I think, all things considered, that it will be best for the brethren to branch out and occupy for their stock the valleys you mention [Clover Valley and Meadow Valley] and also claim, survey and stake off as soon as possible, those veins or ore that br. Hamblin is aware of, . . . all that are sticking out, or likely to be easily found and profitably worked."[6] Young also furnished Erastus Snow, an LDS apostle and leader of the Saints' colonizing efforts in southwestern Utah, with copies of "the rules and regulations observed by miners in locating and working mining claims" and instructed him to file on the site.[7]

However, almost before Snow had time to act, word of the strike spread beyond Mormon circles and put in motion a contest for silver that quickly grew to include General Patrick Edward Connor.[8] Upon hearing news of mineral deposits in southern Utah, Connor supplied non-Mormon prospector Stephen Sherwood with a copy of mining laws so that he could establish a proper mining district if he discovered ore on his planned trip south. After arriving in southern Utah in March 1864, Sherwood and his old mining friend, Jacob N. Vandermark, along with Mormons Thomas Box, Sr., Thomas Box, Jr., and Peter Shirts, gained Hamblin's confidence, and he led them to the silver. On March 18 they held a miners' meeting and organized the Meadow Valley Mining District. They adopted with a few alterations the set of mining laws borrowed from Connor. The group then parted ways; Box, Sherwood, Vandermark, and Shirts traveled to Salt Lake City for supplies while Hamblin returned home to Clover Valley.[9]

Despite the establishment of a proper mining district, the campaign over who would control the Paiute silver intensified over the next three months. Erastus Snow sent settlers to Meadow Valley 12 miles south of the new mines to hold the pasture and farmland there. The first of these Mormons arrived on May 5, and soon others followed, founding the town of Panaca, named for the Paiute word for "silver."[10] In the meantime, in Salt Lake City, Connor quickly involved his troops. On April 30 he issued orders to one company of soldiers to "scour the country" surrounding the "newly discovered silver mines in Washington County, . . . for the protection of miners and exploration of the resources of the country." Connor not only wanted his men to "afford protection to miners from Mormons and Indians" but also commanded, "You will thoroughly explore and prospect the country over which you travel, and if successful in finding placer diggings, you will at once report the fact to these headquarters."[11]

In addition to two companies of Connor's men, Sherwood, Vandermark, and an expanded group of mining buddies were also on their way back to the mines. They had spent the previous month gathering supplies, recording the district rules in a proper ledger, and noising news of their find around Salt Lake City. Close to the same time, Erastus Snow left St. George with a large contingent, intent upon solidifying Mormon claims to the area. All of these groups converged in late May and early June 1864 upon the once-isolated land that less than a year before had held meaning only for Southern Paiutes.[12]

Snow's entourage arrived first. Convinced that, by virtue of abandonment, the Sherwood group's claims were invalid, Snow organized a new mining district based upon the rules Young had given him. He staked new claims and placed Mormons in charge.[13] On the return trip to St. George, Snow passed several other parties in the area, including Sherwood, Vandermark, and company, as well as Captain Hempstead and his troops from Fort Douglas. Despite such competition roaming the countryside, Snow seemed satisfied that "our brethren . . . had laid claim to what they supposed, the principal leads for some miles."[14]

When Hempstead arrived at the mines, he found that Snow had indeed "gone in with a vim." Hempstead and his men discovered the whole mountain "covered and spotted with stakes." "So frequently did the St. George President's name appear stuck in the stakes," Hempstead reported, "that it looked as though there had been a recent *snow* storm on that mountain." Hempstead concluded that "the whole country has gone wild over the silver mines."[15] Upon further investigation, however, he determined that the Mormon attempt to establish a competing mining district was illegal. Hempstead convinced the Mormons that this was true, and they filed their claims with Sherwood, the legal recorder of the district. Hempstead remarked, "All is harmony, peace and good will, and Saint and Gentile are at work in happy accord, on Mount Panacker."[16]

Paiute Silver

While Gentiles and Mormons had at least temporarily worked out their differences, neither group had bothered to address the Paiutes' longer-standing claim to the area.

It was the Paiutes, in June and July, who attempted to reassert their control and in the process placed themselves at the center of the contest. As Sherwood recalled, "The Indians troubled us incessantly: stole and ran off our stock . . . they drew weapons on us . . . [and] threatened to kill us if we did not leave." Thomas Box, Jr., who claimed to know something of the Paiute language, said that he "understood the Indians, and they said they meant war." The hostility escalated to the point that in July 1864 the miners moved to the new town at Panaca, where they huddled together with the Mormons for mutual defense. Sherwood remembered that the miners were forced to guard "against an Indian attack" and alleged that for two years "the Indians were too hostile for us to do anything."[17] In response the miners held a meeting to amend the district laws so they could maintain their claims without actually working them until the Indian danger dissipated.[18]

In addition, the miners appealed for government intervention against the Paiutes and then eventually quit working the mines. It would be nearly six years before miners returned to the site and founded Pioche. In the meantime, Thomas C. W. Sale, an Indian subagent sent to investigate reported problems with the Southern Paiutes, himself generated excitement about another site farther west that gained national attention and led to the shift in the Utah/Nevada border.

Sale arrived at his new post in Washington County by the middle of November 1864 and spent considerable time "talking" with various Paiute bands at Panaca, Clover Valley, St. George, and Santa Clara. Sale, however, clearly had more on his mind than just the welfare of the Southern Paiutes. The region's potential mineral wealth soon enticed him into prospecting. By March 1865 Sale had planned a trip throughout the region to meet with Southern Paiutes in southwestern Utah, northwestern Arizona, and southeastern Nevada. He took with him a group of restless miners then staying at Panaca due to the Indian troubles. Included in the group were three of Stephen Sherwood's friends: Samuel S. Shutt, William McCluskey, and David Sanderson. John H. Ely, future partner in the Raymond and Ely Mine at what later became Pioche, also went along, as did a local Mormon, Ira Hatch, employed as Indian interpreter.[19]

As the group visited local Indians, they showed specimens of ore. One "old Indian" apparently told them that "he knew where there was plenty more of the same kind" and led them to what would become the Pahranagat silver strikes.[20] The men did not hesitate to organize a mining district, with Indian agent Sale elected as recorder, and then smothered the region with claims.[21] Perhaps Sale did as much as anyone to spread word of the find, which was no doubt aided by his belief that he had found the far-famed "Silver Mountain" for which so many had searched in vain.

In a letter to Utah's superintendent of Indian affairs, Orsemus H. Irish, Sale repeated the legend associated with the mountain and its ill-fated discoverers, the 1849 Lewis Manly party. The Manly group, headed for the California goldfields, took the southern route from Salt Lake City through Utah Territory along the Old Spanish Trail and then west to California. Manly, despite objections from his guide, opted for a supposed shortcut, leaving the Spanish Trail in southern Utah and heading west through the treacherous terrain of the southern Nevada desert. Unknowingly he led his group into some of America's driest and most barren expanses, including Death Valley—which the party named. Despite horrific hardships, Manly's group did reach California, losing only one life. According to legend, one member of the company, while passing through the Nevada desert, discovered "exceedingly rich mines of silver" and carried a specimen on to California, which proved nearly pure. Since that time, Sale went on to explain, many expeditions had searched for the source of the silver, but none had proven successful—that is, until Sale and his party came along. "'Silver Mountain' is found at last," Sale boasted; or, if not that, at least "something worthy of attention is discovered."[22]

News of the find spread quickly, swirling a whirlwind in motion that would forever redefine Utah's borders, in an effort to ensure that power to control "Silver Mountain" fell into the right hands. Early reports in national mining journals called the mines "richer than any yet discovered" and predicted that in less than a year "this district will roll out silver bricks in quantities sufficient to shame the Comstock in her palmiest days." One miner called the district "the richest I have ever seen," and another wrote that "there are no more promising mines to be found in Nevada."[23]

The early assumption was that the Pahranagat District *was* in Nevada, but no one really knew. In late 1865 two miners reported that "the Pahranagat District is believed to be in Washington county, Utah." Another miner a few months later wrote that the "exact locality" of the mines was still in question, but he believed that they were in the "extreme southeast portion of the State of Nevada."[24] Even as late as May 1866, at the same time that Congress settled the question with a boundary shift, Dr. O. H. Congar visited the region and made a series of scientific observations. He concluded that the center of the district was "a little more than thirty-one miles within the boundaries of Utah Territory by longitude, and also, about the same by latitude from the Arizona line."[25] Congar's measurements notwithstanding, Hiko, the district's easternmost town, actually fell 10 miles *within* the then borders of Nevada.[26]

With Pahranagat's promising discoveries of silver at stake, Nevada's politicians and the U.S. Congress could not leave things to chance. In May 1866 one of Nevada's senators, William M. Stewart, introduced a bill that instigated the boundary shift. It moved the border one degree farther east, from the 38th to the 37th meridian of longitude west from Washington. In the debate that ensued on the floor of the House, one congressman argued, "I hope we will by all means give Nevada a slice"; it is, after all, "well governed and is now yielding a very large revenue to the Government." Delos Ashley, the congressman from Nevada, added, "The reason why we want this territory for Nevada is that our people from Nevada have discovered mines in that degree of latitude, and we are occupying the country now." "The people of Nevada are a mining people," Ashley insisted, "while the people of Utah are an agricultural people." Regardless, he only knew of "but one Mormon living in that degree of latitude," anyway, and "the Mormons have always been averse to mining." The bill passed over the objections of Utah's territorial representative, William H. Hooper, not a voting member of Congress.[27]

Ironically the Pahranagat district never lived up to even the tamest of claims. By 1868 Mining Commissioner Rossiter W. Raymond reported that investors had poured nearly one million dollars into the district while extracting only about $20,000 worth of bullion.[28] While Pahranagat was a terrible disappointment economically, it did prove powerful enough to move a border. In addition, the spillover from the excitement that it generated drew renewed attention to the nearly forgotten ore at Meadow Valley. By 1869 a few miners from Pahranagat had shifted their focus there. They reorganized that district and renamed it the Ely Mining District after John H. Ely.[29] Within a year the site caught the interest of several California capitalists, including François Louis Alfred Pioche, who invested in the mines and was celebrated with the budding town named in his honor. By 1872 the mines at Pioche had reached an annual production peak of nearly $5.5 million, and the camp was in full swing.[30]

Even though, by the time the mines at Pioche had begun to yield their riches they had become a part of Nevada, southern Utah residents still benefited. Mormons from throughout the region freighted and traded at the mining camp, all the while denouncing the evil that they believed it embodied. In 1872 Bishop George H. Crosby of Hebron, a ranching outpost in the northwestern corner of Washington County near the Utah/Nevada border, reported that everything was "prosperous" at his town and emphasized the financial benefit that proximity to Pioche represented: the co-operative store paid a yearly dividend of 30 percent, residents enjoyed "a ready cash market for all their produce," and peddlers received "sufficiently higher" prices at Pioche "to make it profitable to the freighter."[31]

Hebronite Orson Huntsman concurred: "I made several trips to Pioche with lumber, in company with Father Terry and others from our place," he wrote. "Pioche proved to be a great camp, . . . [and] made a good market for lumber and other products or produce, also a great amount of labor." Over the next several years, Huntsman continued his forays to Pioche. He generally returned home well pleased with his cash payments, especially after one trip where he sold "one little horse" for $50 in gold. Despite the economic benefit, Huntsman still described Pioche as "a very wicked city," or, "at least," he explained, "there is some very wicked men and women in and around Pioche."[32]

Pioche eventually developed a more direct connection to southern Utah, especially after prospectors uncovered marketable ore at Silver Reef. Much of the early activity at the reef took place simultaneously with the events at Meadow Valley, Pahranagat, and Pioche. Silver Reef's development, however, lagged behind the other districts. By 1876, when William Tecumseh Barbee began publicizing his strikes at Silver Reef, Pioche had already started to decline. Many Pioche miners, merchants, and mill owners simply moved east to the new bonanza town—buildings and all—in what some called the "Pioche stampede."[33]

Before that move took place, however, prospectors had to discover the unlikely existence of silver in sandstone and convince others of its reality. Common wisdom dictated that silver did not exist in sandstone, at least not in any marketable quantity. Silver Reef was therefore somewhat of an oddity that required overcoming long-standing geological ideas before its riches could be extracted. That was nearly a decade-long process that involved a variety of colorful characters and spawned a smattering of folk legends.

Finding Silver in Sandstone

As chronicled in chapter 4, one account involves a Pioche assayer nicknamed Metalliferous Murphy. Another popular legend tells it differently. Apparently an unnamed weary traveler sought refuge one particularly cold winter evening in the home of a resident of Leeds, Utah, a small farming village a few miles south of what later became Silver Reef. As the traveler warmed himself by the side of the home-owner's sandstone fireplace, he noticed drops of white metal weeping from the stone. He collected enough of the substance to have it tested, and, when it proved to be silver, he laid claim to a nearby sandstone ledge and the subsequent fortune that it yielded.[35]

These accounts notwithstanding, mining records from Washington County suggest quite a different story. John Kemple, a long time prospector trailing a string of horses from his recent mining forays into Montana, deserves credit for the first discovery of silver in sandstone in 1866. Kemple took a room with Orson B. Adams, a settler at the small Mormon hamlet of Harrisburg. Kemple had brought with him his own assay equipment and spent the winter prospecting around southern Utah. He shortly moved on to the White Pine District in Nevada but returned to the Harrisburg area by 1868. He eventually found silver in *float* material (eroded rock from a lode, usually found at the base of a ledge or in a stream), which he assayed at more than $17,000 to the ton. He was initially not able to locate its source, however. Kemple did send a sample of sandstone rock to an assayer, H. H. Smith, then living at Shaunty in Beaver County, who refused to test it, exclaiming, "Kemple must be crazy to ask me to assay a sandrock."[36]

By 1870 Kemple had discovered ore on "a prominent white ledge sandstone west of Harrisburg and northward." Local settlers grew increasingly interested in Kemple's activities, and in February 1871, under his lead, several men organized the Union Mining District. The group elected Samuel Hamilton of Harrisburg as the recorder for the district, and over the following year, he noted 16 claims. Mormons from Harrisburg and Leeds figured prominently amongst the claimants (William Robb, Orson B. Adams, Elijah Fuller, William Leany, Brigham Y. McMullen, and E. W. Ellsworth) as did a few noteworthy leaders from St. George, including E. G. Woolley, Richard I. Bently, James Andrews, and Apostle Erastus Snow. A few settlers from Toquerville and Washington also tried their luck at mining in the new district.[37]

Despite this surge in activity, the would-be miners never developed their claims. It is possible that their only intention was to keep the ore from falling into Gentile hands. "The Saints have said these mines would be opened when the Lord's own time should come," the *Salt Lake Tribune* asserted in 1876. "The right time is now," it announced. "Christ was too slow," so "Barbee brought the time with him."[38] Brigham Young had long exhorted his southern Utah flock to avoid mining. He especially preached against Pioche and on one occasion "deprecated the desire of many to go to the mines and mingle with the wicked,—learn to swear—drink—gamble, lie and practice every other wickedness of the world."[39] It was not so much the act of mining that Young disliked but the individualism, instability, and social stratification that the search for wealth tended to breed. As long as mining operations existed for the benefit of Zion and could be managed on the cooperative plan, Young approved. In 1870 Erastus Snow explained the distinction this way: "If the mines must be worked, it is better for the saints to work them than for others to do it, but we have all the time prayed that the Lord would shut up the mines. It is better for us to live in peace and good order, and to raise wheat corn, potatoes and fruit, than to suffer the evils of a mining life."[40] Perhaps this notion best explains the inactivity among Mormon claimants—Snow included—in the Union Mining District.

Whatever the reason, the claims lay dormant until Kemple returned yet again to Harrisburg in 1874 and organized a new district, calling it the Harrisburg Mining District. This time interested parties gathered on 22 June at the Harrisburg schoolhouse and elected Orson Adams president; W. J. Early, secretary; and John Kemple, recorder. The new district encompassed 144 square miles with the schoolhouse at its center. Over the next three months, eager prospectors filed 23 claims, all of which were staked on the Pride of the West Ledge (later known as the White Reef). The claimants were again largely comprised of local settlers, although this time two groups of Mormon women from Harrisburg also filed strikes with the recorder. Kemple, however, was the only miner to work his location. He sank a 50-foot shaft and attempted to develop his mine. Meanwhile, the Mormons largely did nothing to advance their claims and eventually abandoned them. In fact, apart from Kemple, only Walter E. Dodge maintained his strike and later patented it.[41]

After the organization of the Harrisburg District it took another nine months for outsiders to pay any significant attention to the location. John S. Ferris and Elijah Thomas provided the impetus, and William Tecumseh Barbee spread the word. Ferris, much later in life, recalled those events: In 1874 LDS leaders sent him to quarry rock for the St. George Temple, and as a result, he spent the winter cutting stone in southern Utah. In March 1875 he took a week's leave of absence from his labors to go prospecting. Before doing so, he decided to visit Elijah Thomas, then living at Leeds. Thomas was a Mormon Battalion veteran who had been at Sutter's Mill in California when gold was discovered there. "We took a ramble up onto Quail Creek, north of Harrisburg," Ferris remembered, "and in returning back we went up on the great white reef. In climbing up the reef we came to a crevice of copper stain about mid way of the reef. We dug on it and opened up about a foot vein of ore of that class." Ferris took a rock sample with him, which eventually made its way to Pioche for testing. When it proved "high in silver," it generated excitement at Pioche, and Ferris eventually took prospectors from there to the site.[42]

Word spread quickly among miners at Pioche. Before long, news of the silver that Mormon prospectors had long kept secret spiraled in a variety of directions and brought an influx of outsiders to southern Utah. Joseph M. Coschina and his brother, Victor, two miners from Pioche, staked claim in April 1875, locating the Bonanza, Emily Jane, and Maggie Mines. Another Pioche resident, Robert Shaky, staked the Stormont, while several more men from Pioche followed suit.[43]

Rumors of silver in sandstone also circulated in Salt Lake City and piqued the interest of David, Joseph, Samuel, and Matthew Walker. The Walker brothers, as they were known, were prominent merchants, bankers, and mining capitalists who had earlier invested in the Emma Mine at Alta, Utah, as well as other ventures in the Ophir fields and at Butte, Montana.[44] The brothers soon decided to investigate the stories then swirling around Salt Lake City about silver strikes in southern Utah. In June 1875 they sent three men south to serve as their agents and inspect conditions there. William Tecumseh Barbee led the expedition while fellow miner Thomas McNalley accompanied him, as did Ed Maynard, an assayer. It was Barbee, however, who was destined to discover a rich vein of silver and instigate a mining boom at the reef.[45]

Barbee was an experienced miner by the time he arrived in southern Utah. He had been born in Kentucky but had spent considerable time in the West prospecting. He was influential in the discovery of silver in the Ophir District near Tooele, Utah, and later sold his Silveropolis Mine there for a reported $27,000. Barbee and his companions arrived in the Harrisburg Mining District in June 1875 and shortly began prospecting. By August Barbee had filed claim on a mine for himself and then recorded a location for Joseph R. Walker as well. In September he and McNalley excitedly returned to Salt Lake City to report to the Walker brothers about what they deemed to be a promising district. The Walkers, however, were not convinced and opted for cautious involvement, perhaps only by Joseph. Undaunted, Barbee returned to southern Utah and by November had struck a rich silver-bearing vein.[46]

It was one of those rare moments of good fortune for Barbee, thanks in part to the two sons of widow Nancy Jane McCleve of Leeds. The two young men had spent a fall day in 1875 gathering wood along Quail Creek and on the Buckeye Reef for their mother's winter supply. As they returned to Leeds, their well-loaded wagon scraped along soft rock for several feet at one fortuitous spot. Barbee happened to be the next person traveling the same route and immediately recognized the exposed "waxy brown mineral" as *horn silver* (a term used to describe an outcrop of ore that dust and wind have polished to the dull luster of a cow horn). Barbee filed on the site, naming it the Tecumseh claim, and shortly located several more lucrative strikes.

He also began promoting what he deemed to be the entire region's enormous potential through a series of reports to the *Salt Lake Tribune*. Barbee thereby touched off a rush to Silver Reef. In December 1875 he described his activities: "About one month ago I discovered the Tecumseh mine, situated on a hill by the same name. The vein is small, but rich, and easily worked," he explained. "On the northeast end of Tecumseh Hill is Silver Flat, chuck full of little chloride veins. Still further east we have Silver Reef and Silver Butte Hills, also full of chlorides by the acre." So rich were these strikes, Barbee claimed, that a man "can dig up the ore as easily as the farmer digs his potatoes."[47]

Barbee and other prospectors sent early shipments of mined ores to Salt Lake City for refining. His rock assayed at from $440 to $850 of silver to the ton, wonderfully rich in comparison to the usual $20 per ton typical of the district.[48] For his early shipments to the Utah capital city, Barbee earned a reported $17,000. By July 1876, however, he investigated Pioche as a closer and less expensive destination for his raw ore. He soon realized a $23,000 return from several shipments to Pioche, a fact that no doubt fueled interest among residents in the new Utah mines. Twenty-nine claims were recorded in the Harrisburg District by the close of 1875, many of them made by men from Pioche.[49]

These early strikes produced considerable wealth and proved easily workable without significant capital investment. They came to be called "poor man's mines" and prompted the *Salt Lake Tribune* to report in the spring of 1876 that "more crude wealth has been developed in a shorter period of time, considering the number of men engaged than any other mining district on the continent. Nine out of ten men who have assayed their fortunes are now in possession of good properties."[50] Barbee no doubt concurred and expressed a similar sentiment when he claimed that "the success of the mines of this district does not depend upon a rush of miners and adventurers that usually flock to new mining camps, and make hurrah times for a brief season; nor does it depend upon San Francisco, Salt Lake, or Eastern capital." Rather, Barbee bragged, "our mines are our capital; our banks are sand banks; we draw on them at will, and our drafts are never dishonored. We have no board of directors to consult, nor sinecure officers to pay." As proof of this, Barbee claimed that it was "a matter of perfect indifference" to the reef miners "whether any immigration come into our district or not."[51]

Even if Barbee's assertions were true initially, the easily worked surface strikes quickly played out, and significant capital investment was required to reach the deeper ore. In addition, Barbee's "perfect indifference" in regard to immigration proved rather imperfect because, by the end of 1876, he had had a town site surveyed at the base of Tecumseh Hill, which he named Bonanza City. Barbee's planned community included an assay office, a boardinghouse, a general store, a blacksmith shop, and a racetrack. Clearly he intended to profit not only from his mines but from real-estate development as well. Unfortunately for Barbee, however, outside capitalists were already investing in the district, and they managed to steer Silver Reef's development in a different direction.[52]

Pioche became the center of much of the early excitement about Silver Reef. Many miners and merchants from that town moved more than 100 miles southeast to the new strikes. The timing was right for a move anyway. By the mid 1870s, both the Raymond and Ely and Meadow Valley mining companies at Pioche had hit water in their shafts and fallen on hard times.[53] When news of Barbee's strikes circulated, it therefore touched off what some people described as a stampede. It proved a rather slow one at first, but it increased in volume and speed as 1876 wore on and grew to include even the relocation of buildings and mills from Pioche to Silver Reef.[54]

It took more than news of Barbee's success to startle the stampede into motion, however. Many Piochers kept a watchful eye on developments at the reef but waited for capitalists to make a move before they were willing to cast their lots with the new camp. Outsiders remained leery of the district's true potential, not believing that silver could exist in sandstone in sustainable quantities. Two Pioche merchants, Hyman Jacobs and Louis Sultan, however, were influential in changing these attitudes. The two businessmen received an ore sample from a Mr. Shepherd of Leeds, which they promptly forwarded to Charles Hoffman, then superintendent at the Almaden Quicksilver mines outside San Jose, California. The sample caught Hoffman's attention, so much so that he traveled to Pioche and brought with him a metallurgist, Professor Janney. These two men obtained their own sample from the Harrisburg District and, after testing and retesting it, became convinced of the region's potential. Hot on the heels of Janney and Hoffman's report, in October 1876, a group of San Francisco capitalists, including A. Borland, L. L. Robinson, W. L. Oliver, George D. Roberts and S. F. Gashwiler, organized the Leeds Mining Company. The new business promptly purchased a few different claims from local owners and appointed Hoffman as superintendent of operations.[55]

Jacobs and Sultan benefited indirectly from this new venture. The Leeds Mining and Milling Company soon had a town site of its own surveyed less than a mile above Barbee's Bonanza City. The company appointed Jacobs as its agent to sell town lots at the new location. He named the place Silver Reef and then demonstrated a significant level of commitment to the undertaking. He promptly boxed the merchandise from his Pioche store, dismantled the store itself, and moved the entire enterprise to Silver Reef. Jacobs's faith in the reef's potential, combined with the perceived security that outside investors seemed to promise, generated widespread excitement at Pioche, and the stampede began in earnest. By the spring of 1878, one former Pioche resident described Silver Reef as "flourishing" and estimated that more than 1,000 inhabitants lived there. He believed that "fully eight hundred of them are from Pioche and vicinity. Many of the old Piochers established here are doing well," he continued, "among whom are John Cassidy, Jacobs & Sultan and Peter Harrison." He went on to describe Harrison's large and commodious building, which served as a grocery store, furniture store, and "first-class lodging house." Other former residents of Pioche had moved saloons, hardware stores, and other businesses to Silver Reef and were all benefiting from the boom.[56]

The reef's early roar quickly attracted additional outside investors, who purchased strikes from prospectors eager to sell. Barbee sold some of his claims but held onto others and worked them with the aid of capital from the Walker brothers under the name of Barbee and Walker. He developed his most lucrative mine on the White Reef, and by March of 1878, he had even constructed the Barbee and Walker mill, a five-stamp facility within 50 feet of his mine. In June 1879, however, a fire destroyed the mill, and although he spent the time and money to rebuild, New York investors eventually bought him out. By June 1881 Barbee had given up his dreams for Bonanza City and left Silver Reef altogether. He moved on to other mining ventures in Humboldt County, Nevada, although he kept in touch with his old friends at the reef through occasional reports to the local newspaper.[57]

Under the direction of Milton S. Latham, New York businessmen continued to work the Barbee and Walker mines and mill. They launched their new venture after issuing 100,000 shares of stock at a par value of $10. Investors also appointed Richard L. Odgen resident director of the mines and rejoiced as the value of company stock rose to $41.60 per share within six months. By 1882, however, the high-grade ore at the Barbee and Walker mines was nearly depleted, and before the end of the year, company assets were sold at a marshal's sale.[58]

Leeds Mining and Milling proved to be another large operation at Silver Reef, among the top four companies there. To save time and money transporting its ore for processing, Leeds investors purchased the Old Maggie mill in a town south of Pioche. They then had the mill dismantled and moved to Silver Reef, where the company rebuilt it less than a mile from its mine. There the ten-stamp mill crushed sandstone ore and extracted silver in marketable quantities, an operation that enjoyed early success. However, as the company sank its shaft to around the 300-foot level, it hit a thick zone of "lower-grade copper-bearing silver ore," which only yielded $16 of silver per ton. The mill also suffered from breakdowns caused by the difficulty in refining the poor-quality ore.

By 1881 the bust part of the mining cycle was already haunting Leeds Mining and Milling. That year it issued an assessment of 25 cents per share in an effort to repair its mill and infuse new life into the company. However, additional problems plagued the mill, so by June 1882, the company was bankrupt. Its assets passed into the hands of the county sheriff. It had produced almost $800,000 worth of silver in its initial 20 months of operation but could not sustain such levels of profitability on low-grade ore.[59]

The Christy Mining and Milling Company enjoyed much more success. San Francisco investors, under the leadership of William H. Graves, organized this venture after purchasing Barbee's Tecumseh claim as well as the Buckeye, California, and Maggie Mines. The Stormy King and Silver Flat strikes adjacent to the Tecumseh also produced ore for Christy Mining. Directors selected Captain Henry Lubbock, an astute businessman and former superintendent of the Floral Springs Water Company at Pioche, to head the development of its mines at Silver Reef. Under Lubbock's skilled guidance, the Christy mines enjoyed a wonderfully productive boom and managed to outlast the reef's other companies. The Tecumseh alone produced $500 of silver per ton of rock in its surface ore and continued to yield rich quantities at lower levels as well. The company soon opened a five-stamp mill near its mines, situated on a five-acre mill site. The mill processed around 40 tons of ore per day, which on average yielded a $2,000 silver brick every other day. At its peak the company employed about 20 mill hands and 40 miners; in total those men extracted an estimated $2.5 million worth of silver. Production slowed after 1882, but Christy Mining continued operations until 1889, when its silver-bearing ore finally played out.[60]

The Stormont Mining Company found similar success at Silver Reef, although it did not last as long as Christy. New York financiers held the purse strings to this company under the leadership of J. R. Bothwell and W. C. Clark. Charles S. Hinchman of Philadelphia was the largest investor in the company and served as financial representative to the board of trustees. The Stormont controlled the Buckeye, Last Chance, and Stormont Mines on Buckeye Reef as well as the Thompson-McNalley strikes on White Reef. It also erected a "large and efficient mill" on the Virgin River. While the mill's river location made water power possible, the six-mile distance from company mines proved costly. Simply hauling the ore to the mill added $2.20 per ton in expenses and prompted managers to search for alternatives. They investigated the possibility of erecting a tramway that would have bisected present-day Interstate 15, passed over Purgatory Hill, and then descended to the river below. Nothing came of the plan, and the Stormont continued paying teamsters to haul its ore.[61]

A more pressing concern, especially for stockholders, was company management under Clark and Bothwell. When financial problems hit the company in 1881 and 1882, some stockholders, as well as business analysts, suggested that profiteering and mismanagement were to blame. In May 1882 Charles Hinchman, representing the board of trustees, and Schuyler Van Rensselaer, company secretary, traveled to Silver Reef to investigate such allegations firsthand. The two reportedly resolved the difficulties, and the Stormont successfully continued producing silver. By 1886, however, work began to slow, hampered by a flood on the Virgin that washed out the dam and temporarily closed the mill. As ore quality deteriorated at lower depths, especially in the Savage and Buckeye shafts, managers instituted vigorous underground explorations in the hopes of striking new silver-bearing veins. While these efforts were extensive, they were also costly and yielded little return. As a result, in August 1887 Stormont Mining announced its closing.[62]

Even though outside capital funded the major mines at Silver Reef, it did not call all the shots. The Kinner Mine on the Buckeye Reef is an important example. Colonel Enos Andrew Wall owned and operated the mine, albeit somewhat disastrously. Wall was born in North Carolina and raised in Indiana. In 1860 he traveled west, where he began his mining career at Pike's Peak, Colorado. He later established ties to Utah when he began a freighting business between Salt Lake City and the Montana goldfields. He no doubt heard of silver discoveries in the southern corner of the territory and arrived at Silver Reef in 1876. He discovered the Kinner strike and was apparently aided in its development by his father-in-law, a California mining man. By 1879 he was furnishing the Leeds mill about 15 tons of ore per day, and some advisors recommended that Wall build a mill of his own. However, he soon hit water in his mine and accumulated significant debt installing a pump and hoisting machinery. He employed about 30 men to keep his operation running, which also added labor costs to his financial stress. Wall owed at least some of his debt to Christy Mining, and the company eventually brought a suit against him in an effort to collect. In July 1881 Washington County Sheriff August Hardy auctioned Wall's mine to pay delinquent interest on a $728 note.

In addition, Wall's men were frustrated at his inability to pay them; many demanded back wages and soon incarcerated Wall at the Harrison House to prevent him from skipping town. However, a friend of Wall's, remembered only as Shauntesy, planned a daring escape. As predetermined, Shauntesy pulled up in front of the Harrison House in a buckboard hitched to a span of fast horses. Wall ran from the hotel, jumped on the wagon, and grabbed the reins while Shauntesy leveled a repeating Henry rifle at the bewildered crowd. Some versions of the story suggest that Wall called out that he would return and pay his men everything that he owed them. He went on to become very wealthy from his involvement in the Utah Copper Company and the development of the Bingham Copper Mine. Legend has it that he eventually made good on his promise to repay his men at Silver Reef.[63]

Besides Wall's Kinner Mine, a variety of other small operators worked among the big companies, although the details of their activities are largely lost to history. The Thompson and McNally claims were on the White Reef and had around 15 men working them in 1879. Two mines were also developed on the East Reef: the Duffin and Vanderbilt. These were largely worked by independent miners, who were granted leases to extract whatever they might find. As many as 30 men tested their luck at the Duffin Mine; some at least supported themselves for a time before moving elsewhere. In general, however, the bigger companies controlled both mining and milling at Silver Reef, and the men who worked the mines did so for wages, not any significant share of the wealth.[64]

Immigration and Town Development

Merchants no doubt made good on the town's boom years as well. By 1877 Main Street and the surrounding commercial district sported nine saloons, three restaurants, four assay offices, two barbershops, five merchandise and clothing stores, a furniture store, printing office, cabinet shop, tobacco store, and a variety of other businesses, including three Chinese laundries. The most notable among the new establishments was the impressive stone structure that John Rice constructed to house the Wells Fargo and Company Bank and express office. It was touted as one of the finest structures of its kind in southern Utah and still stands at Silver Reef, a solid reminder of the city's heyday.[65]

Not all was prosperous for the merchants, however. The heart of the business district experienced a terrible setback in May 1879 when fire tore through town. It started under the plank walkway outside Harry Wiest's barbershop. Despite the diligent efforts of most townspeople who formed bucket brigades, the fire did an estimated quarter-million dollars in damage. Flames destroyed the stately Harrison House, Jacobs and Sultan's store, and a variety of other businesses before firefighters could contain it. Within two months the Barbee and Walker mill and a significant portion of Silver Reef's Chinatown also burned. Arson was suspected in the latter blaze. Townspeople rallied, nonetheless, and appointed Captain Lubbock of Christy Mining to chair a cleanup and reconstruction committee. Both the town and mill were rebuilt and even improved in the process. Over the ensuing years, other fires plagued Silver Reef, but none to the extent of the 1879 blaze.[66]

By 1880, when the census taker passed through town, 1,046 people listed Silver Reef as their home. The demographics of this group were typical of a mining town but stood in stark contrast to the family-dominated Mormon agricultural settlements that surrounded Silver Reef. Of the total population, 459 were single men, a full 45 percent. As historians Douglas Alder and Karl Brooks note, "There were more single men in Silver Reef than there were in the rest of the county combined."[67] In addition, only 28 percent of the population was female, a stark contrast to women at the nearby Mormon community of St. George, who comprised 52 percent of the population. Besides being overwhelmingly male, Silver Reef was an adult community. Almost 75 percent of its population was 20 or older, compared with 45 percent in St. George.[68]

Of those who listed their occupation as miners, more than half were foreign born. Forty percent of those came from Ireland, while another 32 percent listed England as their place of birth. Germany, Scotland, Wales, Sweden, Peru, Poland, Italy, Mexico, Australia, Denmark, Canada, and Norway all had at least one native son searching for wealth at Silver Reef in 1880, as did a smattering of other countries. Of the native-born miners, the largest contingent—20 percent—hailed from New York, with Illinois, Kentucky, Ohio, Pennsylvania, and Utah making strong showings. While miner was the single-most-prevalent occupation, a plethora of other professions were well represented. Fifteen saloon keepers, 27 merchants, 17 blacksmiths, three bakers, seven butchers, two brewers, and a charcoal burner are on the list of jobs people attributed to themselves at Silver Reef.[69]

A large contingent of native-born Chinese endured racial prejudice to make the mining camp their home. The census record lists 50 foreign-born Chinese at Silver Reef in 1880, none of them working as miners. Rather, they owned laundries, grocery stores, and restaurants as well as operated slaughterhouses and meat markets. At various times the Chinese were derided for selling alcohol to the Indians, living in opium dens, and generating a stink with their pigpens. They were even accused of refusing to help fight the 1879 fire that destroyed the business district, a belief that suggests a possible motive for a blaze that later burned the Chinatown section of Silver Reef.

In addition, one Chinaman was targeted for his interracial relationship with a white woman. "China Joe," as the *Silver Reef Miner* called him, lost his home to fire and was forced into the street with his lover early one Sunday morning. "No one knows how the blaze originated," the newspaper reported, "but the Chinaman thinks somebody applied the torch to smoke out him and his girl. Anyhow, his loss is $2,000 by the fire, and he is now sloshing around town the maddest Chinaman this side of Hong Kong." Other Chinese residents had their houses stoned while the town press openly rejoiced at the passage of the Chinese Exclusion Act in 1882, a law that banned Chinese immigration into the United States for 20 years.[70]

Silver Reef's large Irish population, almost one-quarter of the town in 1880, fared much better. Because of this large contingent of Irish Catholics, the Catholic Church took an active role at Silver Reef. Father Lawrence Scanlan, an Irishman himself, had presided over many of the Catholic miners previously at Pioche. By the time of Silver Reef's boom, Scanlan had been appointed missionary rector over the Salt Lake diocese, a vast geographic area that included Silver Reef. Scanlan visited Silver Reef in 1877 and the following year sent Father Dennis Kiely there to preside. Scanlan later returned to Silver Reef and helped direct a fund-raising effort to build a church. He proved successful, and soon the new frame building was dedicated as St. John's Church. It housed its first services on Easter Sunday in 1879.

Scanlan also oversaw the establishment of a hospital at Silver Reef and secured five Holy Cross sisters to staff it. These sisters then expanded their outreach to include a day school, which they named St. Mary's. Besides the education that it offered, the school also put on programs and dramas which provided entertainment and uplifting social activities for youth and adults alike. By 1883, as Silver Reef started to decline, the hospital lacked enough miners to support it, and the sisters were recalled to Salt Lake City.[71]

Perhaps the most notable event to come from the Catholic Church's presence in southwestern Utah was an interdenominational mass held at the Mormon tabernacle at St. George. During his visits to Silver Reef, Father Scanlan struck up a friendship with John Macfarlane, a Mormon surveyor often employed there. In the course of this friendship, Macfarlane became aware of Scanlan's desire to hold a High Mass in southern Utah. Scanlan, however, lacked a choir and a large facility to accommodate the service. Macfarlane spoke with Erastus Snow about the matter. The LDS apostle offered the St. George Tabernacle, and Scanlan gladly accepted.[72]

As for the music, Macfarlane solved that problem, too. Since moving to St. George in 1868, he had formed a choir from local talent which had quickly gained a reputation for its fine singing. Songs for the High Mass, however, would have to be sung in Latin. Undaunted, the choir set to work memorizing the text of the songs and learning the music. Its members practiced every night for six weeks until they could sing in the ancient language with confidence.[73]

On the third Sunday in May 1879, the St. George Tabernacle filled to capacity with Catholic faithful and a throng of curious Mormon onlookers. Father Scanlan conducted the service, centering his sermon around the topic, "True adorers of God adore Him in spirit and in truth." He reportedly began his homily, "I think you are wrong, and you think I am wrong, but this should not prevent us from treating each other with due consideration and respect." Macfarlane's choir performed with its usual precision, and following the service, Scanlan complimented the group for singing the Latin as beautifully as he had ever heard it. The occasion proved an uplifting one for all involved.[74] More recently, on 16 May 2004, in a tribute to Father Scanlan, the Southern Utah Heritage Choir relived the 1879 event when it performed, in Latin, Peters's celebrated Mass in D. Elder M. Russell Ballard of the Quorum of the Twelve Apostles of the Church of Jesus Christ of Latter-day Saints and Monsignor J. Terrence Fitzgerald, vicar general of the Catholic Diocese of Salt Lake City, both spoke at the event and honored Scanlan for his efforts to establish positive relationships that stretched across religious boundaries.[75]

The High Mass notwithstanding, Mormon/Gentile relations at Silver Reef were less than friendly. The *Silver Reef Miner* occasionally poked fun at Mormon leaders, calling them "revelations sharps," and spoke contemptuously of the Saints' aversion to mining. Barbee even predicted that with an influx of miners to southern Utah, Silver Reef could provide the means of outvoting the Mormons and taking control of Washington County politics. At the very least, it offered a shout of "hurrah for Leeds and her hardy miners who braved the Prophet's bowie-knife, and laid the foundations for a prosperous Gentile community in Southern Zion." "We hope they will use Brigham's new temple for a smelter," it prodded.[76] Silver Reef's voting population, nonetheless, never reached numbers high enough to challenge Mormon control of politics in the county.[77]

Mormons did not view Silver Reef any more kindly. Brigham Young had long preached against Pioche and the evil he believed that it represented.[78] By 1877 he had expanded his definition of evil to include Silver Reef. That year, in his remarks at the dedication of the St. George Temple, he urged the saints to "go to work and let these holes in the ground alone, and let the Gentiles alone, who would destroy us if they had the power." "You will go to hell, lots of you, unless you repent," he announced.[79]

Such warnings notwithstanding, southern Utah Saints were quick to recognize the abundant economic opportunity available at Silver Reef. Joseph E. Johnson opened a drugstore and enjoyed a successful business. Other Mormons took advantage of the ready cash market for agricultural goods that the mining camp offered. Saints from throughout southern Utah peddled their produce at the mines and earned higher profits than elsewhere in the region. Other Mormons supplied rock salt, coal, hay, meat, and wine to the miners. Supplying lumber for the mines and mills also proved lucrative, and for a time a Mormon wood-haulers' camp existed on the edge of town. Indeed, as Alder and Brooks contend, "the economic impact of Silver Reef and silver mining on the county was fortuitous."[80]

The Mormons did not fare as well spiritually as economically at Silver Reef. They made an attempt at proselytizing, but the results were dismal, and the effort soon ended. Local church authorities sent missionaries from St. George, Leeds, and

Toquerville to Silver Reef on a weekly basis to preach. Occasionally small crowds turned out to hear the Mormon speeches. Eventually, however, Bishop George H. Crosby of Leeds asked permission of his superiors at St. George to discontinue the practice. At several meetings, "not one soul came," and the missionaries who had been sent felt that it amounted to "labor thrown away." St. George authorities concurred, and the Saints relegated themselves to interactions of an economic nature.[81]

Even those economic exchanges changed over time, and before the reef's ultimate demise, Mormons entered the workforce as miners. That story is largely bound up in the bigger economic context within which Silver Reef mines operated. Miners at the reef enjoyed wages of four dollars per day, equal to those at the more lucrative Comstock mines in Nevada and above the wages offered elsewhere in Utah. By 1881 a showdown occurred between labor and management that ended in reduced wages at the reef and a significant loss of power for the town's short-lived labor union.

Union Organization

It was February 1880 when miners at Silver Reef first organized themselves into a union. At the initial meeting, interested miners elected P. H. Shea temporary president and C. C. Reynolds secretary. The miners who wished to join the new organization signed their names, 110 in total, and agreed to adopt as applicable the constitution and bylaws of the Virginia Miner's Union of Virginia City, Nevada. Later the same month, members elected Matthew O'Loughlin permanent president. Under his leadership, the union began regular meetings and held frequent socials. Just one year later, however, the nature of those meetings changed drastically in response to several companies' attempts to slash wages.[82]

This effort apparently originated with Colonel Allen, superintendent of Stormont Mining. He became convinced that, given the deteriorating quality of ore being extracted from the Stormont mines, his company could no longer afford to pay four dollars per day. He met with Richard Ogden, superintendent at Barbee and Walker, and Captain Lubbock at Christy Mining and suggested that the three Silver Reef powerhouses unanimously cut wages by 50 cents per day. Ogden agreed, but Lubbock refused.[83]

According to plan, then, when the miners of the Stormont and Barbee and Walker companies showed up for work on 1 February 1881, they were given an ultimatum: either work for $3.50 per day or do not work at all. The miners walked away. After a one o'clock meeting, union members took action. They marched on Stormont's Buckeye Mine and took possession of it. Later that afternoon, faced with demands from Sheriff Hardy, the miners relinquished control of the mine. They did not, however, back down on the issue of wages. They began a strike that dragged on for three months and threatened the long-term viability of Silver Reef.[84]

In an effort to solidify control of the reef's labor, the union issued a circular requesting all miners to join before 15 February. As a result, union membership swelled to around 300. Union leaders began a round of talks with management teams from both companies aimed at compromise. However, by the end of February, the newspaper reported that negotiations had "broken off" and the situation had deteriorated into "a freeze out game."[85] The Miners' Union drafted a letter to the *Salt Lake Tribune* explaining its perspective: "Miners's trade takes longer to master than others. It takes more strength and endurance. He ends up broken when he should be in the flush of manhood. Danger is ever present," the union argued. Under such conditions the "miners feel about the same way to the superintendents as a mule to his driver."

The union went on to question management's handling of the wage issue. It wondered why the companies did not approach the union with the matter, rather than waiting for the men to show up at work "and say you can't go down unless you will go for $3.50 a day." The companies "could have at least told us the night before," the union complained; they "treat us like mules, this is the way they have ever dealt with us."[86] The union, however, found little sympathy, at least at the *Tribune*. It reminded the miners that while Comstock workers earned $4.00 per day, conditions at the Nevada camp were a "little hell" in comparison to the favorable atmosphere at Silver Reef.[87]

For the miners, the success of their strike largely depended upon Christy Mining as a holdout in refusing to drop wages. When rumors circulated in late February that Christy also planned to cut its daily rate, the union panicked. Allen, Stormont's superintendent and instigator of the whole affair, became its target. A committee of 12 men from the union paid Allen a hostile and unexpected visit. They ordered him at gunpoint to leave town. When Allen requested three hours to gather his things, they allowed him three minutes and then escorted him out of Silver Reef.[88] It was an act of desperation on the part of the union which eventually led to its own loss of power.

The strike was proving difficult for the miners and the companies. The Stormont closed completely while Barbee and Walker remained open, along with the other companies not involved in the dispute. Still, the strike affected the mood at Silver Reef and drastically slowed economic activity. One report put it this way: "A great many men are leaving town, the aspect of which is rather blue just at present."[89] Another predicted that "unless something unexpected should 'turn up,'" it would be "the dullest Summer ever experienced here since Silver Reef first flung its banner to the breeze as a bullion producing camp."[90]

Things only got worse before they got better. Allen did not take his dismissal from town lightly. He traveled directly to Beaver, Utah, and filed a complaint with the federal district court. Before long, a grand jury had handed down indictments against the miners who had forced Allen from town. On 16 March a posse, largely comprised of Mormons from St. George, headed by U.S. Deputy Marshal Arthur Pratt, invaded Silver Reef and rounded up 36 men, including many of the leaders of the Miner's Union. Pratt charged the arrested men with riot, conspiracy, and false imprisonment and placed them under guard at the town jail. The jail, however, proved too small, which forced law officials to house overflow prisoners at another building, where, according to one account, a line was drawn around it and "those under arrest were told that they would be shot if they crossed the line." The following day the suspects were taken to Beaver, where they eventually stood trial. The court convicted a total of 13 men, three of whom paid fines and were released; the remaining 10 served time in the penitentiary. O'Loughlin, union president, received a $75 fine and 20 days in jail; the court fined another union memer, Joe Carr, $50; and the remaining 11 were fined $100 each.[91]

In the wake of these activities, the Miners' Union, its power already drastically diminished, met in early April to decide upon a course of action. After considerable deliberation, members voted to accept the $3.50 wage. Management had won. Hot on the heels of this decision, Christy Mining reduced its hourly rate to match the other companies. Rehiring at the mines also curtailed the union's strength because some companies replaced striking Irish workers with Cornish ones. Mormons from surrounding towns also took advantage of the turnover in the workforce to begin jobs at the mines.[92]

While some local historians have suggested that the strike represented the beginning of the end for Silver Reef, silver extraction continued for more than 20 years. More than the strike, it was the simple bust of the mining cycle that brought an end to the reef. Ore quality continued to deteriorate over the course of the 1880s, and no new discoveries replaced the veins that were playing out. Each of Silver Reef's major companies had closed by 1890, but a local company, Wooley, Lund, and Judd, continued operations until 1898. The St. George businessmen for whom the company was named leased and operated the Stormont and Christy mill and mines, although at a significantly reduced pace. In 1898 the Brundage Company of Cleveland, Ohio, purchased Wooley, Lund, and Judd's interest at Silver Reef and made plans to revive the camp. It did conduct business there past the turn of the century; however, declining silver prices forced it to reconsider further investment. Independent miners and leasers continued to work sporadically at Silver Reef, refining tailings and looking for new strikes. The last recorded mill run took place in 1908.[93]

Mining's Impact on Southern Utah

It is difficult to estimate Silver Reef's impact upon southern Utah. Between 1875 and 1908, miners extracted an estimated $7.9 million worth of silver from the district. However, the camp's influence reaches well beyond that. The influx of hard cash into a cash-poor economy was vital to the stability of the surrounding agricultural settlements. Silver Reef provided a ready market for Mormon wine, fruit, beef, and grain and offered labor opportunities otherwise not available for teamsters, sawmill operators, carpenters, lumbermen, merchants, and miners alike.[94]

Silver Reef was by far the most successful extractive enterprise in Washington County. Prospectors scoured the countryside looking for similar riches but largely left disappointed. The Apex Mine west of St. George employed southern Utah men starting in 1890. Wooley, Lund, and Judd eventually controlled the mine and even built a smelter at St. George, but its output was minimal in comparison to the reef.[95]

Oil was the only other extractive industry to impact the region significantly. It did so for a short time, just after the turn of the twentieth century.

In 1907 Walter Spencer was working as foreman of the Rescue Consolidated Mine in Tonopah, Nevada, when he learned of peculiar burning rocks from a southern Utah farmer. After considerable questioning, Spencer ascertained the rocks' location, near the small town of Virgin City in southwest Utah. Soon oil men and drilling equipment invaded the sleepy little Mormon village and for a time startled it awake.[96]

After ordering geological research in the area, Spencer and other interested parties sent Elwood I. Hastings, an experienced eastern driller, to Virgin. Hastings recommended drilling, and soon Tonopah businessmen joined other investors to organize an oil company. They chose a spot almost two miles northeast of Virgin City to sink their first well. They struck oil at 480 feet and installed a pump, and the well began producing about 15 barrels a day.

Oil men continued to drill wells in the area, eventually sinking nearly 15 different holes, most of which were failures, producing little or no oil or gas. Three wells did produce—by some estimates as high as 36 barrels a day—but the financial panic of 1907 limited capital for further exploration, and the boom proved ephemeral.

In 1918 investors started pumping the three productive wells again and even built a small local refinery. A fourth well was soon added, and in September 1920 production from the Virgin dome totaled 20 barrels a day. The refinery handled about 800 gallons of dark brown crude per eight-hour shift, and the gasoline and kerosene it produced found a ready local market.

From 1920 to 1932, various companies dreamed of striking it rich in the area and continued drilling in Washington County. In 1924 the Gustaveson Oil Company erected a large refinery to handle the crude oil from the six wells it had in operation near Virgin that were producing nearly 300 barrels per day. Even as late as 1929, a variety of investors held high hopes for oil production in Washington County, but output never quite met expectations, and with the onset of the Great Depression, drilling slowed dramatically. Following World War II, investors revived interest in Utah's oil and gas possibilities, but yield at the Virgin field remained limited.

In the end southwestern Utah's extractive industry proved a powerful—albeit fleeting—force in the development of the region. The Pioche and Pahranagat strikes, for example, mustered enough power to move the Utah/Nevada border. Pioche went on to become an important regional economic center that attracted peddlers, freighters, and laborers from throughout southern Utah. As Pioche started declining, Silver Reef sprang to life and filled a similar role.

Mormons were heavily involved in early discovery of both Pioche and Silver Reef silver. In both cases, however, it took outsiders to actually work the claims and absentee owners to develop them. The Mormons still benefited, especially from the overflow of work and cash that mining brought into their midst. As a result, southwestern Utah's extractive industries were a boon not only to Utah's mining economy, but to its agricultural one as well.

12

Alta, the Cottonwoods, and American Fork

Laurence P. James
James E. Fell, Jr.

Alta and Snowbird! By the twenty-first century, their remarkable snow, challenging terrain, and spectacular scenery had made them world-famous ski areas. But their origin had nothing to do with skiing. Alta and Snowbird sprang to life because of the minerals industry, an endeavor little known and poorly understood in the history of Utah—and that is especially true of mining in Little Cottonwood Canyon, Big Cottonwood Canyon, and nearby American Fork Canyon. More than any other area, except Bingham Canyon across the valley, they brought mining in Utah to life, sustained it for a time, and left behind an important scientific legacy even as silver gave way to snow.[1]

Phase I: Discovery and Development

The first era of mining in the Wasatch stemmed from the Mormon colonization of Utah in the 1840s and 1850s. During that time the need for wood in the Salt Lake Valley drew timber cutters high into both Big Cottonwood and Little Cottonwood Canyons, located in the Wasatch Mountains south and west of Salt Lake City. The result was a series of seasonal logging communities and tiny sawmill camps. The individuals involved probably had little interest in mining, although they found at least one mineral prospect. Instead, they explored the canyons, named many features, and laid out the trails the miners would soon follow.

The discovery of minerals in the Cottonwoods resulted directly from the work of General Patrick E. Connor, the U.S. military commander in Utah during the 1860s.[2] Mining in the Cottonwoods developed within the context of Connor's explorations. In the fall of 1863 and the months that followed, miners and prospectors, including Connor and Mrs. Robert K. Reid, wife of the surgeon at Fort Douglas, became interested in discovering gold and silver. The soldier/prospectors, and even the

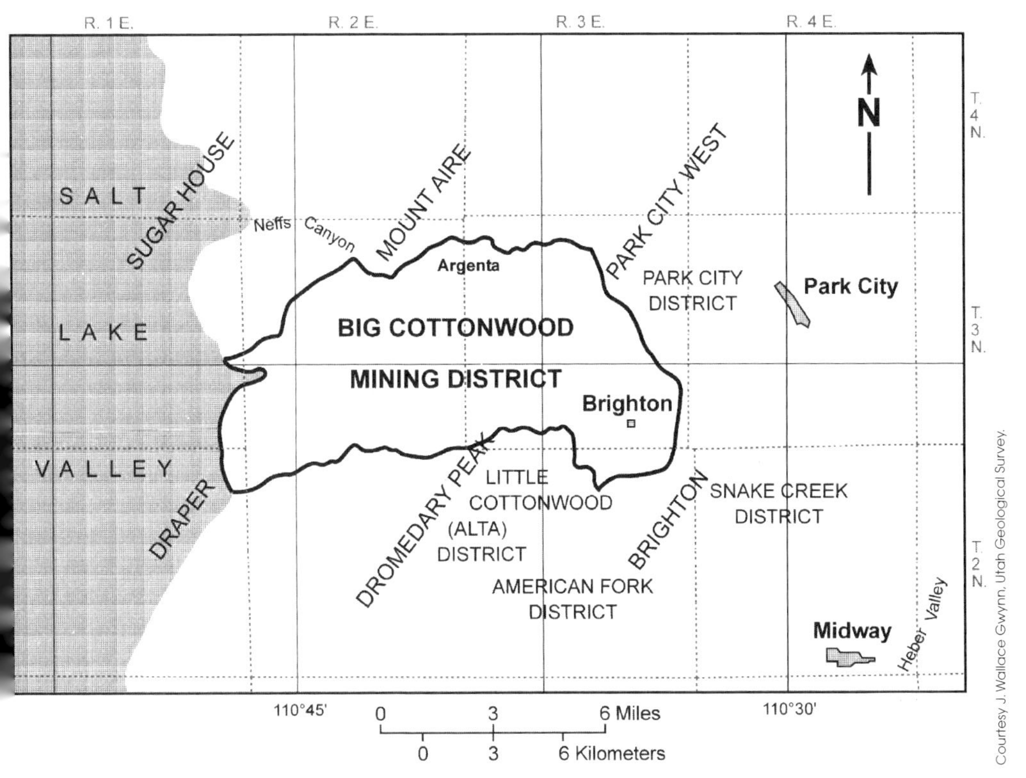

The Cottonwood and American Fork Mining Districts map shows the central part of the Wasatch Mountains and adjacent valleys, extending from Salt Lake City to Park City. The town of Alta (center of the map) is not shown. Diagonal labels are names of the U.S.Geological Survey topographic and geologic maps of the region.

Mormon Bishop Archibald Gardner, began creating a series of mining districts based on precedents set in California. These included the Mountain Lake Mining District, organized in November 1863.

The next year, at Connor's urging and with government supplies, some of his troops arrived in the Cottonwoods, where in July they found silver-lead ore. They quickly devised their own regulations on acquiring mining rights on federal land, and they staked a number of claims in both Little Cottonwood and Big Cottonwood Canyons. The news of silver spread rapidly, for everyone realized that these first discoveries might be the harbinger of a second Comstock lode, the great silver strike in Nevada then making sensational headlines all over the world.[3]

More prospectors arrived as the Civil War ended in 1865. They included one group led by Silas Braine (or Brain), who may have come to Utah in the 1850s, and another party led by an H. Poole, both working for Dr. O. H. Congar, who headed the New York and Utah Mining and Prospecting Company, which hoped to acquire rich mines. Both Braine and Congar filed claims in the district in August

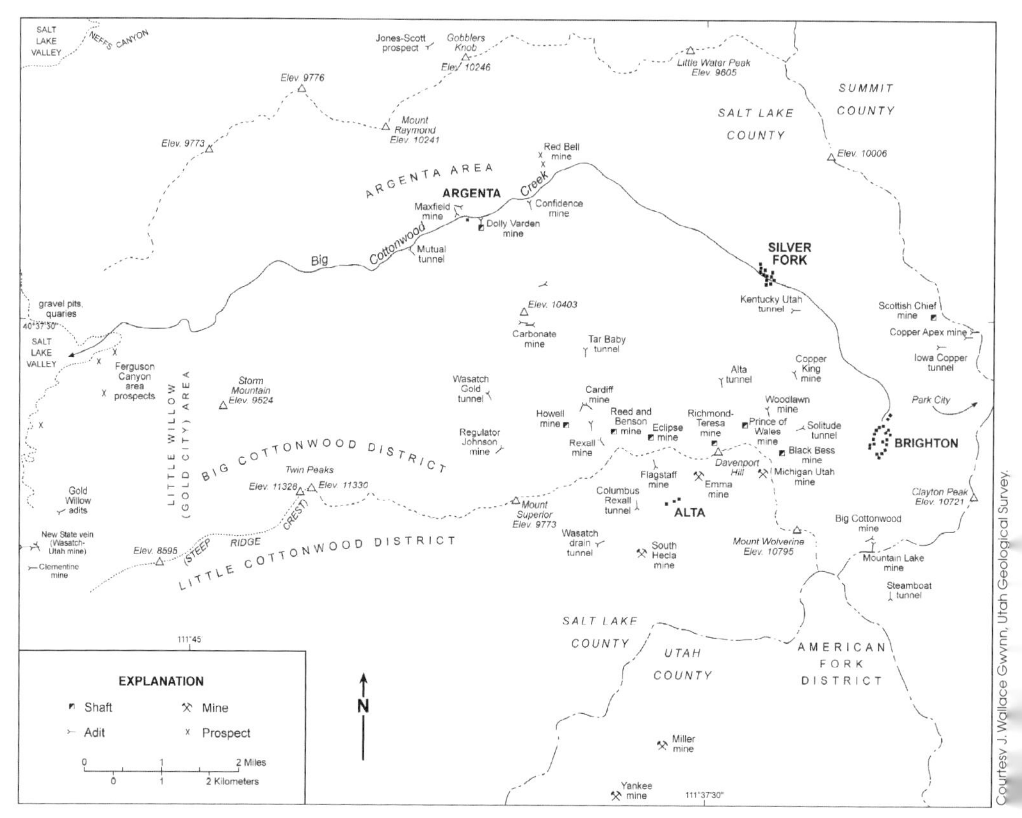

Courtesy J. Wallace Gwynn, Utah Geological Survey.

Map showing the Big and Little Cottonwood and American Fork Mining Districts and some of the mines and places in the text. The Park City Mining District lies to the east of Brighton.

1865. Encouraged by his own assays of the silver-lead ores, Congar apparently went to Philadelphia that fall. There he met James P. Bruner and other new investors, who provided the capital to open what became the North Star Mine. He also hired a metallurgist to build a smelter and reduce the ore to silver-lead bullion.

Operations began in 1866 at Central City, located on a flat high in Little Cottonwood Canyon, but despite the ballyhoo, neither the mine nor the smelter had much success. Although they failed by 1867, the Congar-Bruner operations left one legacy: they were the first real silver-lead mine and smelter in the Wasatch. But with these failures, along with conflicts over claims located by Connor's soldiers, the first era of mining in the Wasatch came to an end.[4]

Phase II: Alta and the Emma

Although the Cottonwoods now slipped into the backwater of life in Utah, they were not forgotten. Among those who remembered were James F. Woodman, Robert B. Chisholm, and Captain James F. Day. All three, it seems, had sprung from the lead-mining country of Illinois, then gone west to seek their fortunes in gold and silver in California and Nevada. Day became a superintendent at the Yellow Jacket Mine on Nevada's Comstock lode. Where the three met remains conjectural. What is important is that they formed a strong friendship that served them well in Utah.[5]

All three men made their way to Little Cottonwood in the first rush spawned by Patrick Connor. Day got involved with Congar's operations, took up some claims with him, and saw some of the rich ore that Braine had shipped to Congar's office for assay. Chisholm and Woodman probably saw Braine's ore as well, and given their background in the Illinois lead country, all three must have noted that the ore was embedded in limestone. It did not resemble the silver-gold veins of western Nevada.

But no matter how hard they worked in the Wasatch, Woodman, Chisholm, and Day failed to make a significant discovery. When the boom faded, they headed for Pahranagat in southern Nevada, but when their luck ran out there, they cast their thoughts again on Little Cottonwood. As experienced lead miners, they understood that Connor's boom had gone bust in part because of smelting problems. They also understood that the new transcontinental railroad, now approaching completion, would partly solve that problem by allowing ore to be shipped to distant smelters.[6]

In August of 1868, Day "furnished a little money to buy supplies for a ten days' trip," he recalled later, and the three returned to Little Cottonwood.[7] They found the canyon largely deserted of miners, but that deterred them not at all. They pushed on beyond decrepit Central City and finally made camp beneath the slopes of a steep hillside, eventually known as Emma Hill. Then they clambered up through the dense brush, where they found about 180 tons of galena-rich boulders protruding from soil and glacial till, simply lying on the surface of the ground. They loaded sacks and shipped what silver-lead ore they could, then found a seam that penetrated the hard rock nearby. With the 10-days trip about up and winter in the offing, the three decided that Woodman and Chisholm would remain to sink a prospect shaft into this promising site while Day furnished capital and supplies from time to time. The work now began on what became known as Woodman's shaft on the Monitor lode, but Day could not or would not furnish the money needed to keep digging, and that brought a new person into the equation, one James E. Lyon.[8]

Lyon was an investor, promotor, and bon vivant, to be sure. Others suggested that he was a schemer, even a con man. He lived mostly in New York, where he had earned a questionable reputation promoting mines in Colorado. But with mining there in collapse, Lyon was looking for his next chance, and he decided on a visit to Utah, perhaps because of the impending completion of the transcontinental railroad.

In the fall of 1868, just about the time that Woodman and Chisholm began sinking a shaft on the Monitor lode, Lyon set out for the Wasatch. He arrived in October, when, he recalled later, he first "became acquainted" with Woodman and

Chisholm. What happened next is controversial. What probably occurred is that Lyon furnished both capital and labor and worked with Woodman and Chisholm to develop the Monitor in return for a third of whatever they found. At the 40-foot level, they struck a pocket of high-grade silver-lead ore and took out 100 tons with a windlass. But if the ore was encouraging, the vein was not. The pay rock pinched out. Nonetheless, when Lyon departed in late October, he left Woodman a contract to sink the shaft another 50 feet. And while all this was going on, Lyon was apparently sinking (or paying someone to sink) his own shaft elsewhere on the Monitor lode.[9]

With Lyon gone, Woodman and Chisholm kept working as long as they could, but they could not stay in Little Cottonwood indefinitely. The moneys from Day and Lyon proved inadequate for the work. So in November Woodman and Chisholm made a new plan. Woodman decided to stay in Little Cottonwood as long as he could, guarding the site against claim jumpers. Chisholm would return home to Elgin, Illinois, to sell his farm to raise capital, and they apparently decided to freeze Lyon out of the property altogether.

In the spring of 1869, the Wasatch leapt to life again. As the snows melted, many people headed up the canyon, drawn by reports of new metal at old claims and the promised conclusion of the transcontinental railroad. Among the miners were Woodman and Day to resume work on their claim. They were soon joined by William W. Chisholm, who had come in place of his father, Robert, who was delayed by selling his land in Illinois. They set a crew to work on Woodman's shaft, and as they did, Chishom received letters from his sister, who implored him to return to his "dear Emma." The name must have struck a chord with these lonely miners far from home. When Robert Chisholm finally sold his farm and rejoined his colleagues, he found that the Monitor lode had evolved into the Emma Mine and the hillside into Emma Hill. And many new prospectors had joined the search for mineral wealth.

All that summer the Chisholms worked hand in hand with Woodman and other miners to sink their shaft deeper. Finally, in late summer, about 93 feet below ground, they blasted into a massive deposit of lead-carbonate ore. Assays revealed that it had more than 130 ounces of silver per ton, richer than the average grade on the Comstock lode. The miners sent for Day, who was incredulous. The Emma Mine seemed likely to make Little Cottonwood the second coming of the Comstock lode. The boom was on.[10]

During the next two years, the prospectors and miners who flooded up the canyon turned Emma Hill and the flat below it into an industrial city. Woodcutters stripped the hillsides to provide wood for homes and timber for mines. Business blocks rose to supply the wherewithal to support miners. Funicular railroads and then tramways brought miners and supplies up to the new mines and the ores down for shipment. Some prospect holes turned into productive mines. There were the McKay and Revolution, the North Star, the Montezuma, along with the Davenport and the Toledo. More important was the Flagstaff, which rivaled the Emma, and then there was the Emma itself, the most famous mine in Utah.[11]

As Day, Woodman, and the Chisholms continued work, they needed ever more capital to develop the mine. And other investors were right at hand, notably the four Walker brothers, English immigrants and now former Mormons, who had developed a mercantile, banking, and freighting empire in Salt Lake City. They followed mining developments closely, particularly because they did quite a lot of shipping in and out of Utah's mining areas. The firm of Walker Brothers "rawhided" some of the first ores down Little Cottonwood Canyon and shipped them to San Francisco for reduction. They also invested in the Prince of Wales Mine just over the divide in Big Cottonwood. They now approached the owners of the Emma, who still needed capital. As a result, Walker Brothers invested a reported $30,000 in the enterprise in exchange for a quarter interest, organized the Emma Silver Mining Company of Utah to take control of the property, and supplanted the discoverers with a young mine manager. He was Marcus Daly, Irish born, self-educated, and experienced on the Comstock lode. Daly systematized operations, found ever-richer ores (he shipped 4,200 tons in the second half of 1870), and enabled the new Emma Company to pay substantial dividends. As mining progressed, however, he warned the Walker brothers that he believed the deposit was pinching out. They decided that it was best to sell, and Daly soon moved on to other mines owned by Walker Brothers and eventually to his famous career in Butte, Montana.[12]

By the time they decided to sell, however, the Walker brothers had a new partner in the Emma Company. He was Warren Hussey, a former Colorado banker forced out by the collapse of the mining industry there. He now made his home in Salt Lake City, where he had taken over the Miner's National Bank, which he and others converted into the First National Bank of Utah. He bought Woodman's share of the Emma Company and became an important player in its destiny.

Meanwhile, the shipments from the Emma brought it to the attention of Lyon back in New York. As production mounted, he fumed that Woodman, Chisholm, and Day had frozen him out and went to court, represented by Senator William M. Stewart of Nevada, one of the most famous mining attorneys in America. Litigation over title clouded all aspects of development.[13]

By the spring of 1871, the success of the Emma Company and the desire of Walker Brothers to sell brought them all to the attention of one Erwin Davis of San Francisco, who acquired an option. He went to New York, where he interested two attorneys-turned-speculators in the property. They were Trenor W. Park, a banker from Bennington, Vermont, and General H. Henry Baxter, a Wall Street operator and one-time president of the New York Central Railroad, one of the nation's most famous lines. Intrigued by the speculative possibilities, Park and Baxter purchased a half interest for $370,000, then organized the Emma Silver Mining Company of New York, which held their half interest plus that of Hussey and the others. Then, after protracted negotiations, they settled (or thought they had settled) with Lyon to gain clear title.

The timing of these investors was both lucky and perfect. Rossiter W. Raymond, the new U.S. Commissioner of Mining Statistics, hailed the Emma as "one of the most remarkable deposits of argentiferous ore ever opened." Based on shipments received in England, the prestigious London *Mining Journal* called it "par excellence, the mine of Utah," even the "Potosi mine" of the Wasatch Mountains. Publicity like this was the stuff of dreams.[14]

In September 1871 Park arrived in England to sell the Emma. Senator Stewart arrived as well, ostensibly to protect Lyon but clearly to sell Lyon's shares and those that he, Stewart, had acquired. In London Park and Stewart met one Cyrus M. Fisher, a New York attorney promoting rail and mining securities and a man with close connections with the U.S. diplomatic legation. The three agreed to form a new enterprise to be established in Britain. To capitalize on the free publicity, they wanted a further expert opinion from an ostensibly unbiased source. Therefore, Baxter hired Benjamin Silliman, a professor of chemistry at Yale University in New Haven, Connecticut, and told him to name his own terms for an investigation. Yale was highly respected; so, too, was the Silliman name. And in October 1871, after a quick inspection, Silliman pronounced the Emma "one of the great mines of the world."[15]

Whether Silliman was naive, or inexperienced, or conned, or bought has long remained a controversial matter—he was paid the extraordinary sum of $25,000 for his efforts, something approaching a half-million dollars in today's currency. Whatever the truth, he must have understood that his report was supposed to be favorable. And it was telegraphed to London from Corinne, Utah, for more than $3,000 to be sure that it would be on hand for the stock offering. This was perhaps the most expensive telegram ever sent from Utah Territory. The promoters now had the imprimatur of American science and a prestigious university.

But here the promotion hit a snag. Fisher's firm was too small to do the job, and as a result, promoting the Emma fell to one Albert Grant, born Albert Gottheimer, a controversial promoter and member of Parliament. In early November 1871, Park and Stewart concluded secret agreements with "Baron" Grant, who for a fee and founder's shares now launched the Emma Silver Mining Company, Ltd., which took over all the property of the Emma Silver Mining Company of New York. Grant paid members of Parliament substantial sums to use their names on the prospectus. Park used his connections and financial legerdemain to land General Robert C. Schenck, U.S. minister to Britain and an authority on draw poker. The prospectus released on November 9, with Silliman's report attached, glittered with the names of the high and the mighty.

Yet the press reaction was mixed, and some commentators deplored Schenck's role as a conflict of interest, even in an age little concerned about that. Word of that conflict made it all the way to the United States and the highest levels in government when Secretary of State Hamilton Fish discussed it at a cabinet meeting at the end of the month. What in 1868 had been a promising mineral prospect now had a gleam that reached all the way to the White House.[16]

Schenck had, in fact, overreached himself. After discussing the matter with President Ulysses S. Grant, Fish asked Schenck to either resign his post or withdraw from the company. Schenck chose to resign from the company's board, but this did not become public knowledge until January, which gave him more than a month to sell his investment quietly at high prices. By then the U.S. Senate had adopted a resolution calling for an investigation of diplomatic personnel in foreign corporations. The Emma's impact was indeed far reaching.[17]

Schenck's problems notwithstanding, Park and Stewart moved ahead with their plans. With the stock selling well at high prices, they paid off Lyon (who arrived in London) for $250,000 and then gave him another $50,000 to keep quiet. They also bought out the 10 remaining shareholders of the Emma Company of New York, apparently for far less than the price of the shares in London. Then Park sold his vendor's shares, with Grant buying some to sustain the high price.

Park, Stewart, and Grant had begun selling none too soon, for bad news began appearing about the Emma scarcely a month after the public offering. As early as December 1871, some journal editors called for caution about the mine and the company, but they were drowned out in the euphoria. Nonetheless, various London speculators, including a disgruntled Lyon, began selling short and hunting for bad news to "bear" the Emma stock.

The news they wanted came in June 1872, when Lyon released letters from William Dalton, a mine manager. Dalton reported that cave-ins had closed most of the mine, 200 feet of water had flooded the workings, mining could not resume for another six months, and worst of all, the most recent shipments consisted of ore from the nearby Illinois, whose owners had sent armed men underground to seize the workings. The Emma Company's directors denounced the news, but the price of shares fell sharply that June.

Pressure on the company built up over the summer. Park finally had to admit that a cave-in had ended production. Park, Stewart, and others battled in the Utah courts to exclude the Illinois company from mining ore they claimed belonged to the Emma. Next they bought the Illinois Tunnel Company, whose claims they had once decried as "perfectly baseless." In December the mining journals published the new foreman's report that "the mine had been gutted and there was no ore in sight."[18] What buoyed the stock in the face of this bad news was the continual payment of monthly dividends—13 in all—but that same December the company announced that it would switch to quarterly dividends because of the usual "winter obstruction" in Little Cottonwood. Then in January 1873 came worse news. There would be no February dividend, owing to litigation, falling production, and the need for retimbering.

At the company's second annual meeting came the recriminations. Chairman Anderson had no explanation for high corporate expenses in London and about $250,000 owed the Illinois Tunnel Company and Trenor Park. Anderson also had to admit that the final dividend had been paid with money borrowed from Park. Further investigation revealed that the mine had produced very little new ore since the British had taken control in November 1871. Most of the dividends had come from ores already in transit, loans from Park, and cash on hand from the purchase itself.

Worse, Park was now threatening to seize the mine for nonpayment on his loans! (He did have it attached in September 1874.) By mid-1873, as the financial panic engulfed the United States, the press excoriated the whole Emma Mine promotion as a hoax.

The reverberations of the Emma crash echoed all through the 1870s. It discredited mines in Utah, particularly in Little Cottonwood, and caused needed British investment to dry up everywhere. Schenck's activities helped discredit Grant's administration, and Schenck was fired (although not until 1876 for political considerations). Finally, the Emma imbroglio produced a full-scale congressional investigation designed to embarrass Grant and the Republicans on the eve of the presidential election of 1876. The final result was litigation which went on for years. In the end Park took over the company for nonpayment of debt and, with the discredited Schenck, launched the American Emma Company. But the mine was a shadow of its old self. And Park himself remained hounded by litigation until the 1890s.[19]

Meanwhile, as the Emma, Flagstaff, and other mines were creating an international sensation, the Walkers saw other possibilities at Alta—real-estate promotion. As prospectors, miners, storekeepers, and others poured into the district, Alta had grown quickly beneath what had become known as Flagstaff Mountain and Emma Hill. By the early 1870s, the town may have had a summer population of as many as 8,000 people, although that figure seems high. Overall they comprised a cross section of American society, along with its beliefs, values, and practices. The town itself lay on a slight slope covered with glacial gravels that was public land. No gold was found in the gravels. Given that there was no visible mineral outcrop and no suggestion of any mineralization below ground, almost any prudent miner knew that the federal-government surveyor would reject any mineral-patent application. But this fact deterred none of the individuals rushing to the new boomtown. Everywhere on Alta flat along Little Cottonwood Creek, there were hotels, taverns, and buildings of every sort rising to meet the demand of mining. And all of these structures were occupying public land.[20]

The Walker brothers had probably wanted to acquire land in Little Cottonwood when the boom originally unfolded in the late 1860s, and the person who paved the way was one Robert Nagler, who lived in a log house in Central City. Nagler had prospected in the canyon and recorded claims as early as July 1869, and he may have worked as a silent front man for the Walkers part of this time. What is important is that in 1871 he decided to apply for government title to the land that comprised the town of Alta. Just at that time, however, he was approached by three other individuals—Joseph R. Walker, Henry W. Lawrence, and William S. Godbe—all of whom had mining interests in Little Cottonwood and wanted to make a formal town site of this location. Nagler agreed, and the group had the site surveyed and platted.

Reports soon surfaced in the *Salt Lake Herald* that a Salt Lake company had acquired the Alta and Central City town sites via the preemption laws and would sell lots for $300 each. This caused a furor in Alta, but nothing happened, and the commotion died out, although Nagler did sell some lots as early as March 1871 as "attorney in fact" for one Amanda Brown (meaning that he had power of attorney for her).

But the report apparently revealed as much as it obscured. What probably happened is that the outcry scared off Lawrence and Godbe but not Walker and his brothers, who now turned to a different strategy. In 1854, nearly 20 years before the Alta boom, Congress had begun to extinguish the Indian title to land in Minnesota. In doing so, it authorized the federal government to create various paper instruments with monetary value, among them what was indelicately known as "Sioux half-breed scrip." What this meant was that, when "halfbreeds or mixed bloods of the Dacotah or Sioux Nation" had their lands surveyed, those individuals received scrip which allowed them to exchange their lands in Minnesota for 40, 80, or 160 acres of non-mineral lands anywhere else in the public domain.[21]

As the Emma, Flagstaff, and the other mines made the town lands of Alta ever more valuable, the Walker brothers took full advantage of the scrip situation, probably using the nearly penniless Naylor as their front man. He quietly located and recorded 160 acres of land in Little Cottonwood Canyon—the exact quarter section covering most of the Alta and Central City town sites. Then he secured control of scrip that entitled Amanda Brown, a Sioux Indian living in Sibley County, Minnesota, to acquire 160 acres of land from the public domain. Nagler arranged to apply for the land in her name, and on 3 January 1873, she received title to the land that encompassed the Alta and Central City town sites. Nagler acted as her attorney in fact in this action. Then at the end of January, Amanda Brown sold her property, again with Nagler as her attorney in fact, to Samuel J. Lees, who, of course, was the manager of the Walker Brothers store in Salt Lake City. She received $2,500, a considerable sum of money in that day.

The news of this quiet, skillful operation did not become public until the summer of 1873, when it appeared in the inaugural edition of the *Cottonwood Observer*, Alta's first newspaper. And with the news came reports that the Walker brothers intended to charge high prices for property that many people thought was theirs. There were charges of fraud, the Walkers were denounced, and the federal government was assailed. But passions cooled when federal officials declared the unpopular transaction free of fraud and J. R. Walker himself visited Alta to assure townspeople that the firm would settle up with home and business owners for reasonable prices. That was apparently what happened, and the matter died down.

The Walker brothers may have thought they now controlled the Alta town site, but they were wrong. Minnesotans clouded the title. In October 1873, when Amanda Brown married John W. Heines, she and her husband sold the land to others, who in turn sold part of it again. Finally, to secure clear title for the Walker brothers, in late 1873 and early 1874, Lees had to travel all the way to Minnesota and buy out all of the claimants. Ultimately, the Walkers transferred ownership of much of Alta to the Alta Mining and Development Company, an enterprise that they controlled. In the long run, Amanda Brown and the Walkers' bold strike proved more important than the town's angry residents realized. A century later this clear title helped make possible the unencumbered creation of Utah's first ski resort amid a tangle of patented mining claims and untouchable federal land.

Meanwhile, in the late 1860s, the boom at Alta spilled into other areas, notably American Fork Canyon, which lay south of town across what was called the Bullion Divide. To get there, miners could walk up Collins or Peruvian Gulches or take a longer route through Albion Basin east of Alta, then hike over a hillside next to a peak called Devil's Castle. All the routes were hard. They crossed rugged passes that possessed spectacular scenery but were dangerous in bad weather. These access problems, however, scarcely deterred the people who opened the first mines in this remote section of the Wasatch.

Among the earliest prospectors venturing south to American Fork were Jacob and William Miller. Near the summit of a peak near the canyon's headwaters, a place soon known as Miller Hill, they found an outcrop of rich lead-carbonate ore bearing some gold. This discovery produced a rush and development. The most important mine was the Miller, which soon passed into the hands of the Miller Mining and Smelting Company of New York.[22]

Ore also came from other places, including Mary Ellen Gulch, which branched to the northwest from American Fork Canyon. Here in the 1870s, prospectors located the Live Yankee and Bellorphan claims. Though narrow, the veins there produced silver, lead, and gold, but their output during the 1870s was small. Their time lay in the future.

To serve the mines and miners of American Fork, the rough town of Forest City arose below a steep gorge at the far south end of the district. It was very isolated, a long, arduous wagon haul from Utah Valley, but it was a beginning. As mining expanded, the investors in the Miller Company built the Sultana smelter to reduce ores in the district and then the American Fork Railroad, a narrow-gauge line designed to connect with the Utah Southern Railroad coming south through the Salt Lake Valley. The expensive new smelter, however, was short lived. It operated only a few years until it became possible to ship ores to the growing smelting industry in the Salt Lake Valley.[23]

Phase III: Disaster and Borrasca

Despite the sensation created by the Emma and other mines above Alta, this second era of mining in the Cottonwood region also proved to be short. Winter remained a challenge, transportation problems stopped production of lower-grade ore, and nature seemed to conspire against the industry. The spectacular collapse of the Emma Company discredited mining everywhere. And other, more fundamental circumstances assailed the industry. There were simply too many problems for the limited tonnage of high-grade ores mined near the surface to overcome, and those problems became ever more paramount as the 1870s unfolded.[24]

Natural disasters created one impasse. During the 1870s snowslides thundered down the steep hillsides of the mining country, especially at Alta. They routinely demolished buildings, ripped out tramways, and killed everyone in their path. They also shut down the rickety mule railroad built to serve Little Cottonwood. Fires broke out all too often in homes and business blocks at Alta and other camps, raced unchecked through the dry, wood structures, and destroyed innumerable buildings. The Alta fire of August 1878 was the worst, and it seemed to signal the end.[25]

A more serious problem was the steady fall in the price of silver. During the late 1860s and early 1870s, the market price held above the coinage rate of $1.25 an ounce, and the industry presented so little silver to the mint that Congress demonetized the metal. As silver prices slid during the Panic of 1873 and after, owing largely to increasing production, this policy became known as the "Crime of 1873" and produced the first political demands for the free and unlimited coinage of silver at a ratio of 16 to one with gold, which would have boosted the price back above $1.25 per ounce. Silver was remonetized in the Bland-Allison Act of 1878, but the mint did not make enough coins to stem the price slide. So the decline in the value of silver became a problem for miners in the Cottonwoods but not the all-encompassing problem that it turned into later in the century.[26]

There were even more fundamental problems causing the early collapse of mining in the Cottonwoods and American Fork. For one thing, the narrow, high-grade ore bodies proved to be discontinuous, which caused management consternation and corporate failure. Rich-appearing ore discoveries in this highly faulted geological setting tended to disappear when miners extended the underground workings. In other words, the ore pinched out, and no one knew where to tunnel to find pay rock again, or if there even was any. For another, the miners had problems with ore reduction. All of the silver-lead ores of Alta had to be smelted to recover the valuable metals, but smelting was in its infancy in the 1870s. The plants were small, the technology was uncertain, and the costs were high. While they had some success in working the oxidized ores mined near the surface, the Utah smelters of the 1870s could not handle the different kinds of sulfur-rich ores found below the zone of weathering. When the weathered, enriched silver ores ran out, the deeper, poorer sulfide ores below had no market, despite their silver and lead content.

And so one by one the mines in the Cottonwoods and American Fork shut down. In November 1878, with no more ore in sight, the once-rich Flagstaff discharged its last miners and all but closed its doors—the largest of the early mines to fold. A few lessees remained at the rock face mining what ore they could, but the once-great mine was now a shadow of its former self. Most of the remaining miners left the district.

It was the same over in American Fork. The town of Forest City largely disappeared. The once-big Miller mine, its ores exhausted, closed down. The Sultana smelter quickly followed. And that meant that there was virtually no business for the narrow-gauge railroad. It shut down, and its tracks vanished. New entrepreneurs turned the right-of-way into a toll road for seasonal prospectors, but the up-canyon rates were so high they discouraged work. Many would-be miners denounced the new road as nothing more than "an institution of highway robbery."[27]

Despite the collapse, this epoch of mining still prompted the first regional studies of the geology, ore deposits, and mining and smelting techniques of Utah. They began in the 1870s with the 40th parallel survey led by Clarence King of the new United States Geological Survey. Coupled with that were the annual reports of Rossiter W. Raymond, the U.S. Commissioner of Mining Statistics. Their work ultimately led in the 1880s to the work of Dwight B. Huntley, a mining engineer, whom King sent to Utah to study conditions in the Cottonwoods and other mining districts in the territory. Various authors also published articles on the Wasatch Mountains in the *Transactions of the American Institute of Mining Engineers*, the *Engineering and Mining Journal* (both influenced by Raymond), and other publications. If the first two eras of mining in the Wasatch had largely collapsed, the scientific studies of the region had only just begun. In the long run, they would contribute significantly to the development of the mineral sciences.

The mines of the two Cottonwoods and American Fork now slid into an era of *borrasca* for more than 20 years. A pre-1900 colloquial miner's term, *borrasca* means "storm" in Spanish, and it was a stormy time for mining in the region. The price of silver continued to slump throughout the 1880s and early 1890s, then plunged 25 percent from 80 to 60 cents an ounce in the silver crash of June 1893. Once-rich mines sat idle, their saleable mineral exhausted. Capital became hard to find as great new districts like Leadville, Colorado, and Butte, Montana, attracted vast new investments. The collapse of the Emma haunted the Cottonwoods and drove investors away. It was an age of *borrasca*—a time when dark clouds of failure enveloped the district.

Despite the downturn, there were some efforts at mining. The Emma, which seemed to have nine lives, led the way. Despite the spectacular collapse of the London Emma Company, which tarnished reputations on both sides of the Atlantic, the fame of the mine and the memory of the rich bonanza ores of the early 1870s would not go away. First a new English company, then a Scottish company tried to find the faulted ore body. They reorganized the corporate structure, raised what capital they could, and rebuilt the surface plant after multiple avalanches had carried it off. They drove thousands of feet of workings into limestone—all of it barren. By the 1890s the Emma's investors had finally had enough. They chose to seek gold in Australia, while litigation continued on endlessly in London, draining company coffers even lower.[28]

Other individuals refused to give up on Alta—men like John S. Johnson, known in the canyons as "Regulator" Johnson. With his string of mules, he was characteristic of this in-between era of mining in the Cottonwoods. A big, strong immigrant from Sweden, he had joined the Alta boom in 1868, and his Regulator claim high in Grizzly Gulch northeast of Alta paid very well for a few years. It was so rich, he bragged, that it would eventually regulate the price of silver, a claim that earned him his nickname. He was wrong, but the name stuck. But like most of those in the district, by the mid-to-late 1870s, he found his hopes in *borrasca*.

Johnson was not discouraged, however. He kept prospecting the snow-filled gulches in Big Cottonwood. People joked in a mock-Swedish accent about how he had planned to regulate the price of silver; he probably reacted negatively and became something of a recluse, but he kept working on his hopes and dreams, like many others. But unlike many others, he had some success: he found a rich pocket of gold/silver ore in the Silver King Extension claim in Mineral Fork. He had the ore sorted by hand, stuffed into cloth sacks, and then tied to mules for the laborious five-mile journey down to Big Cottonwood Canyon. From there he sent 168 tons by wagon to the Germania smelter in Salt Lake City. He made something when most who remained in the Cottonwoods made nothing. Though he later lost one eye and many fingers in a dynamite blast, the crippled giant with a gold watch cut a large figure as he continued mining and prospecting.[29]

People like Regulator Johnson kept the region alive in hard times. Near Alta the German-born miner Fritz ("Old Baldy") Rettich extended the Frederick tunnel deep under older workings searching for ore beneath the Frederick and Crown Prince claims but had little success. James Monk prospected in Big Cottonwood from his home, the sole surviving building at Argenta town site. And Edward Hines became known as "the hermit of American Fork" as, almost alone, he continued prospecting from his home, a shack at the abandoned Sultana smelter. Other prospectors and mine leasers packed in seasonally, but they had little more success than these individuals who refused to give up and leave. For nearly 20 years, the Cottonwoods and American Fork languished in this twilight world of *borrasca*.[30]

Phase IV: The Revival

Beginning in the 1890s, important economic developments preceded a fourth era of mining in the Cottonwoods. As the Intermountain West began to assume greater importance in American commerce, and as metals prices recovered from the financial recession, there were important new mineral discoveries at nearby Park City and Tintic. After the Mormons formally renounced polygamy, Utah finally achieved statehood in 1896. And when some of the territorial appointees departed, they also left their idle mining properties in the Cottonwoods and American Fork—a fact not lost on wealthy investors in Salt Lake City and Provo. New technology, notably the power drill, made it cheaper and more efficient to drill holes to plant dynamite. New steam- and then gasoline-powered hoists appeared in shafts in the Wasatch. Also important was the silver crash of 1893. One key result of the debacle, both in Utah and throughout the West, was a renewed search for gold, whose price was fixed by law at $20.67 an ounce.

A key prospecting technique in the new era was tunneling. Tunneling for exploration had been impossibly slow and expensive in the days of hand steel, but by the late nineteenth century, the newly evolving drills, powered by steam or electrically driven air compressors, enabled operators to drive adits below old workings to search for ore. Various entrepreneurs drove such adits into Emma Hill and many sites in Big Cottonwood, hoping to strike rich ore at depth beneath old mines. The results were mixed. Sometimes they hit ore, but mostly they didn't. But rich finds were made in mines like the Maxfield at the old town site of Argenta in northwestern Big Cottonwood Canyon. This success paid $118,000 in dividends in the early 1890s, buoyed the hopes of many people, and inspired more tunnel ventures.

The fourth era of mining in the Cottonwoods was now about to begin. The hard times that followed the Panic of 1893 had come to an end. A new prosperity emerged in Utah, and there was a local market for mining stocks. For the new state's mines, the sophisticated metallurgical industry in the Salt Lake Valley created greater opportunities for working more-complex, lower-grade ores to recover base metals like lead, copper, and even zinc.

As the twentieth century approached, the mines of the western Wasatch began to receive new attention from Utah entrepreneurs. Some, like the Goshen Valley rancher Jesse Knight and Salt Lake Mayor Ezra Thompson, were newly wealthy Mormons who had grown up in Utah. They had studied the mining industry first as freighters and suppliers before getting involved as entrepreneurs. Others, like the well-known miner Enos Wall and Utah Senator Thomas Kearns, were non-Mormons who had come from other states and gotten involved in mining in the Wasatch. Before the 1890s, capital had come from out of state, and Mormons and non-Mormons rarely worked together to develop mines. When they did collaborate, it was mostly to negotiate property deals. But by 1900 something of a political and religious truce had evolved, and these once-antagonistic groups began working more closely together in every phase of mining. The great investment potential and the wealth of the mines in places like Park City, Bingham Canyon, and Tintic attracted both groups. And new workers arrived from all over the world to make the ethnic mosaic more complex than ever.[31]

The new century also brought new smelters to the Salt Lake Valley. The American Smelting and Refining Company (AS&R), formed in 1899, acquired several small smelters, then began building one of the world's largest and most technologically advanced lead-smelting plants at Murray, south of Salt Lake City, once the Guggenheim family took control of the enterprise. The Guggenheims worked with large investors from New York and Cripple Creek, Colorado, to develop the great open-pit copper mine at Bingham Canyon. And investment capital poured into more new smelters in Midvale, northwest Murray, Garfield, and Tooele. Given the new capital, lower costs, and technological expertise available, the depressed camps in the Cottonwoods and American Fork began to receive new attention, both from entrepreneurs and the smelting companies themselves. Ever seeking ores, these enterprises built expert staffs of geologists and engineers to consult on mining projects and even take stock in promising prospects.

As the twentieth century dawned, all of this financial energy had important consequences in the Cottonwoods. George Tyng, a veteran Texas entrepreneur, made a rich strike at the old Miller Mine in American Fork. In 1901 Tony and Al Jacobson, young Mormon miners from Tintic, raised enough money in Salt Lake City to acquire the old Yankee Blade claim on the steep hillside just north of Alta. Here they began developing a prominent, though low-grade, lead-zinc sulfide deposit in what was known as the Braine fissure, named for now forgotten Silas Braine of the equally forgotten New York and Utah Mining and Prospecting Company of the 1860s.[32]

Encouraged by their early work, the Jacobsons approached the well-known Salt Lake banker, W. S. McCornick. He was interested, and so now with the financial backing of W. S. McCornick and Company, the Jacobsons quietly purchased more old claims near the Yankee Blade and folded them into a new enterprise, the Columbus Consolidated Mining Company. Like the firm's namesake, they were bent on discovering a new world—a mineral empire under the mountain northwest of the old town of Alta. Moving quickly, they extended two tunnels, the Columbus and the Howland, deep into the hillside along the Braine fissure. Less than a year later and more than 1,000 feet beneath the ground, they broke into several sizable ore bodies. Excitement mounted, though it was tempered by the realization that the ore was low grade, too low, in fact, for shipping and smelting.

Given the low-grade quality, the only alternative the Jacobsons had was to mill the ore into a concentrate at Little Cottonwood before shipping it down to one of the smelters near Salt Lake City. Concentration was nothing new in the West, although methods were improving. The problem was that there were no mills at Alta. So, operating on a limited budget, the Jacobsons bought used equipment from the Silver King Mine in Park City, then built their own concentrating plant on a flat just west of Alta. This plant crushed the ore, then separated the valuable material from the worthless using jigs and shaking tables. The resulting concentrates were sold either to AS&R's new smelter in Murray or the new United States smelter then building in Midvale.

But building a concentrator solved only one problem the Jacobsons faced. For mining and milling, they needed more power to run the machinery, the mine pumps, and the air compressors. Even the decrepit town of Alta would pay for electric lighting. To fill the need, the Jacobsons and the other key investors in the Columbus incorporated the Wasatch Power Company and built a hydroelectric plant some four and a half miles below Alta. To run the turbines, they laid pipe and built penstocks up Little Cottonwood Creek and into a nearby gulch. The water ran down to two turbine wheels housed beneath a granite powerhouse above Tanner's Flat. The mine and mill got the power they needed, and electric lights burned for the first time at Alta in June 1904.[33]

Mining and milling now began in earnest. The Jacobsons brought in electric locomotives to take miners into the workings and bring out the ore and waste rock. They also instituted the eight-hour day, a goal of labor in that era, and so worked three shifts of miners around the clock. Theirs was one of the first companies in Utah to do this. No doubt they hoped to avert labor problems then disrupting work in many districts. They were veteran miners themselves. Their largely Mormon crews trammed ore to the mill, and the mill men shipped silver-lead and zinc concentrates to the smelters. By 1905 the Columbus Consolidated was a large operation, the first to work at Alta in decades. The town grew again and briefly even boasted a newspaper, the *Alta Independent*.[34]

Other entrepreneurs followed quickly on the heels of the Jacobsons. One new enterprise was the Continental Alta Mining Company, which focused on Grizzly Gulch, northeast of Alta. There it reopened some complex lead/copper/silver deposits which eventually evolved into the Michigan-Utah Mine, another good early twentieth-century producer. This development led to the building of a four-and-a-half-mile aerial tramway, one of the longest in the world, a line that carried people, the U.S. mail, explosives, and other mining supplies up the canyon, and ore and an occasional coffin down during the mine's heyday.

Still another entrepreneur was Jesse Knight. A stolid Mormon rancher who had struck it rich by finding new mines at Tintic, he was perhaps the richest man in Utah by the early twentieth century. He bought up old, but now promising, mines in American Fork and the Cottonwoods and set his own crews to work driving long, deep tunnels. He also built hydroelectric plants. His schemes were risky, but he was philosophical about mining in the Wasatch. "If I don't find ore," he said, "I still will hire my bretheren and make water for farmers." The Knight Investment Company even acquired the old Emma Mine and sent a few miners to repair the workings, but even "Uncle Jesse," as Knight was known to many, concluded that the venture was too risky.[35]

And then there was the rugged challenge of Big Cottonwood. It was here that a young carpenter and promoter, the Welsh-born Fred Price, focused his energies. With several veteran prospectors, Price acquired control of a group of claims along a strongly mineralized fissure or fracture zone. The group then raised enough money to begin sinking a small shaft to explore their Mountain Cheaf claim at the head of what later became known as Cardiff Fork. Problems assailed their efforts. An avalanche destroyed several buildings and killed the mine horses; the workers underground barely escaped with their lives. The near catastrophe forced Price and his company to abandon the shaft in favor of a tunnel in a safer location as the only feasible way to explore the Mountain Cheaf. To raise the money, in 1906 the group organized the Cardiff Mining and Milling Company, named for Price's birthplace in Wales, then floated shares on the Salt Lake Stock and Mining Exchange.

Although the Panic of 1907 and the recession that followed slowed work, the new tunnel exposed enough ore to arouse public interest. As a result, new investors, including Ezra Thompson, former mayor of Salt Lake City, bought into the company and financed an even-deeper tunnel. But this project ran into a host of problems, the most serious being no ore. Management levied assessments on the shareholders, forcing out the small players. Years passed as the company appeared to throw good money after bad, and the stock price languished in despair on the Salt Lake Mining and Stock Exchange.[36]

Then fortunes shifted suddenly. In 1914 the onset of World War I boosted the prices of lead, zinc, and copper, and then in October the lower tunnel struck a massive ore body rich in silver, copper, lead, and zinc. Shaped like a "flattened cigar," said one geologist, the Cardiff proved to be the largest source of ore ever found in the Cottonwood and American Fork Districts. It was the stuff that dreams were made of but so few achieved. As operations progressed, the owners took advantage of new technology: recently developed off-road trucks and high-voltage power lines, both of which enhanced efficiency. The Cardiff proved immensely profitable, particularly during its early years of production during World War I. It paid more than $1.2 million in dividends by the mid-1920s.[37]

Like the Columbus across the divide in Little Cottonwood, the Cardiff stimulated more work in Big Cottonwood. One individual who cast his lot there was Morris R. Evans, a venturer from the pioneer era of the 1870s. Three years after the Cardiff burst into its great ore body, the redoubtable Evans opened the rich gold/copper deposits of the Columbus-Rexall mine nearby. Nearly under the divide itself, it was worked from a long tunnel extending north from the Columbus Mine at Alta. Announcement of this high-grade discovery of gold and copper buoyed the stock market and stimulated interest in the Wasatch as nothing had before.

These discoveries in particular and some lesser ones as well, coupled with record-high silver prices, led to a level of investment interest unprecedented in the Wasatch. Prospectors pushed in to search everywhere possible. Ore production surpassed the famous years of the 1870s, and the Wasatch contributed a large share to Utah's all-time record silver production of 75 million ounces in 1915.

But these new mines also had problems, and one of them was water, which had to be pumped out of the workings, usually at great cost. From the late nineteenth century forward, investors came to like deep tunnels that would do three things: explore for minerals at depth, permit mining from below, and dewater old shafts. As electric power, compressed air drills, and ventilation fans became available, plans for these deep tunnels permeated the Wasatch. Several Park City companies began driving south toward the Cottonwoods.

One promoter who proposed a deep bore beneath Alta was Frederick Valentine Bodfish. He was well known for managing high-altitude mining projects at Cripple Creek and in Colorado's San Juan Mountains. Once in Utah he took up land in Big Cottonwood, then organized the Alta Tunnel and Transportation Company to fund his grand scheme. He brought the needed equipment and power lines to Silver Fork in Big Cottonwood and began driving south toward Alta, beneath the old Prince of Wales Mine. But like many of these major projects in many districts, the Alta Tunnel encountered mainly *borrasca*. It found only small ore bodies despite 25 years of work funded through the Salt Lake Stock and Mining Exchange.[38]

An even more ambitious work was the Wasatch Drain Tunnel below Alta. The Jacobson brothers initiated work on their Snow Bird Mine, located on a patented claim in Little Cottonwood, to drain the Columbus. From here, decades later, the tunnel was extended to the Cardiff workings over in Big Cottonwood. The builders also intended to drive it beneath the Emma and other mines, but those plans never materialized.

The deepest of the tunnel projects originated in Big Cottonwood. One was the Mutual Metal Mines adit, driven from the mouth of Mineral Fork. A second was the Golden Porphyry at Little Willow Creek on the Wasatch Front, a project whose name tried to capitalize on the now-famous copper-porphyry deposits elsewhere like references to the Comstock lode in Alta's first era. Some things never changed.

Some other strong advocates of Alta mining also achieved success in this era. One was George H. Watson. The son of a family of mine operators, for decades he had raised enough money from stock sales on the Salt Lake Stock and Mining Exchange to explore at depth beneath the mountains south of Alta. Coming from the copper mines of northern Michigan, where subzero months challenged the Arctic, he found the long winters in the Wasatch more liveable than most local people did, and he succeeded in selling stock in both Utah and Michigan. The mine names he chose reflected both mining areas: the South Hecla and the Alta Quincy were two, the Hecla and the Quincy being famous Michigan copper mines. Eventually his Alta United Mines Company consolidated most of the old patented claims surrounding the town.

Just like the long-gone days of the 1870s, the new boom also attracted the unsavory, the shady, and the corrupt. In the early twentieth century, a time of almost negligible securities regulation, fortunes could be made through stock manipulation or simple fraud. One such operator was George Graham Rice. Best known for his book *My Adventures with Your Money*, written while he was in jail in Atlanta for securities fraud, he teamed with former Treasury Department official William Barrett Ridgeley to acquire the old Emma Mine and reinvent it as a brilliant investment. His capable mining staff found the faulted ore body and shipped $1 million worth of ore from it, but the endeavor didn't work financially. Rice eventually fled the country and reportedly disappeared in the Philippine Islands during World War II.[39]

Another stock player was Alois Phil Swoboda, a New York and Chicago bodybuilder turned stock promoter. During the 1920s he raised huge sums of money through stock promotions for his cult, the Swobodans. They took flyers on silver mines in the Wasatch, as well as in Mexico, but nothing amounted to anything, and the investors lost most, if not all, of their money, except Swoboda, of course.[40]

This era of base-metal mining also required far better transportation. The famous, but rickety, tramway built by the Continental Alta Mining Company was only one technological answer. Prior to World War I, the Salt Lake and Alta Railroad laid new standard-gauge track up the old 1870s grade from Midvale to the mouth of Little Cottonwood Canyon. From there the Little Cottonwood Transportation Company put down new narrow-gauge rails over the old grade of the Wasatch and Jordan Valley mule line. This enterprise later purchased geared Shay locomotives to pull cars up the very steep grade and ease the loaded ore cars carefully down. Funding part of this effort was Minor Cooper Keith, founder of the United Fruit Company. The narrow-gauge rails finally reached George Watson's South Hecla Mine at Alta in 1917.

Later, a spur reached the new lower terminal of the Michigan-Utah aerial tramway near Peruvian Gulch.[41]

From the outset, however, the railroad up Little Cottonwood proved precarious, plagued by steep grades and poor engineering. Starved for capital and working with seasonal, intermittent ore producers, its equipment frequently broke down, notably the brakes on the steep runs, which led to some tragic accidents. Avalanches and snowslides played havoc during the winter. And some of its secondhand equipment arrived already worn out. To carry passengers, food, and explosives to Alta, local entrepreneurs introduced gasoline-powered railbuses. These made travel to Alta a lot easier, although not a lot safer.

By the late 1910s and early 1920s, Alta had indeed changed from its decrepit character of the late nineteenth and early twentieth century. The Columbus and other mines had turned it back into a small, but active, mining-district hub, comprised of boardinghouses, plus a few tents and homes, overshadowed by mine dumps and compressor buildings. Its miners and general store flaunted the law as the new age of Prohibition arrived and enhanced the economy. Illicit shipments of alcohol arrived quietly by rail, and like many mining towns in distant places, Alta served as a distribution point well beyond the easy reach of the law.

There was another, almost-imperceptible change as well. Despite the rough, unpaved roads, the automobile made it possible for city dwellers to acquire claims and visit their prospects to conduct the annual assessment work then required by law. These men and women were mostly hopeful mine finders, dreamers, and recreational miners, often funded by unlisted stock companies. They could escape the hot city and valley for weekends of drilling, blasting, and mucking in their miniature tunnels in the Cottonwoods—and grab a drink, go shooting, and enjoy an isolated cabin. Perhaps their backers helped them invest in an old an air compressor and drill. Mainly they did the annual labor required to hold onto their claims in hopes that this arduous tunneling and mining might someday lead to the big bonanza. Although they rarely struck anything of any consequence, they were the harbingers of change in the new automobile age—the age of second homes and recreation in the Cottonwoods. But that would have to wait. When the heady times of the 1920s gave way to the Great Depression of the 1930s, the economic collapse curtailed all mining activities dramatically.

This epoch of mining also witnessed resurrected work to the south in American Fork Canyon. As gold attracted more attention because of the collapse of silver, the long-inactive Live Yankee and Bellorphan claims in Mary Ellen Gulch drew renewed interest. In the early twentieth century, they passed into the hands of Liberty Holden, the principal owner of the *Cleveland Plain Dealer*. An avid investor in Utah mines, he worked the properties for a time. Then they were handed on to Carl Bernard Ferlin, a Swedish-trained German engineer who drove a tunnel into the hillside. When Ferlin died in 1927, control of the mines moved on to a Murray, Utah, physician with no mining experience.

Meanwhile, commencing about 1913 as the twentieth-century industrial epoch unfolded, some growing mining and smelting companies with access to investment capital and new technology began to send experts into the field to look at potential mineral prospects. Two such roving experts arrived in Mary Ellen Gulch, where they examined the Live Yankee and Bellorphan. One was Henry C. Carlisle of the Tonopah Mining Company of Nevada. The other was Allan Hay Means of the American Smelting and Refining Company (AS&R), now a huge corporation with international operations. Based upon their global experience, both men came to believe that the geological features of the Live Yankee and the Bellorphan suggested the veins might overlie large *manto* ore deposits in limestone not far below. Similar to the Cardiff find and the underground ores of Bingham and Park City, manto ore deposits were high-grade "plums" easily concealed beneath barren limestone. The convictions of these two company men, who could now almost access this remote area by automobile, led to a personal and corporate competition. Alan Means and AS&R outbid the Tonopah firm, and the smelting company took control.

In the late 1920s, AS&R began development, confident of rich ores to feed its nearby Garfield and Murray smelters. To solve the acute transportation problems of American Fork, it hired the Riblet Tramway Company to build an all-steel aerial cableway from the Live Yankee portal, over the deep gorges of the Wasatch, to Deer Creek Flat lower in the canyon, where trucks could then haul the ore to the railhead. Although the tramway proved a technological marvel, the sudden economic collapse caused by the Great Depression frustrated AS&R's plans just after mining began. With losses mounting everywhere, the enterprise leased the mine and tramway to mining engineer Earshel Newman and two other former key employees involved in the project. By cutting back operations, reducing overhead, and taking advantage of the new price of gold (nearly doubled to $35 an ounce by federal law in 1935), Newman and company finally got the mine into profitable operation by the mid-1930s. With access provided by the tramway, the Live Yankee and several smaller mines shipped ore and allowed the exploration of other claims in this remote district until about 1938.[42]

And like the new road up the old railroad grade in Little Cottonwood, the work of Newman and his colleagues suggested the fundamental changes to come. To get to the mine, men and sometimes their wives or girlfriends could easily ride up the cableway in the empty 500-pound-capacity ore buckets—a trip not for the faint of heart because the line crossed above Major Evans Gulch, a 2,000-foot drop. And Newman soon found that the best way to visit his fiancée was to ski from the mine up to the head of Albion Basin, then schuss down to Alta to pick up a ride down Little Cottonwood Canyon. It was the sign of things to come.

Meanwhile, the opening of these new mines brought a renewed interest in geological and other scientific studies of the region. U.S. Geological Survey teams mapped the area, and their detailed studies of the mines, along with work by others, brought the first widespread recognition of what geologists call *thrust faulting*. This geological concept enhanced understanding of the evolution of the area, along with the development of petroleum reserves in the Intermountain West. But the vast metal resources developed in the Bingham District across the Salt Lake Valley, partly identified by old Alta hands like Colonel Enos Wall, have vastly overshadowed any mining achievements in the Cottonwoods.

Phase V: Petering Out

The fifth epoch of mining in the Cottonwoods began in the mid-1920s. And it began with optimism. The infrastructure needed for successful mining was all in place. Railroads and trucks provided the means to carry ore and concentrates to mills and smelters—and four enormous smelters stood ready in the valleys to buy almost any metallic mixture from the Cottonwoods and every other place where miners could produce ore. And in the Cottonwoods, expectations ran high that miners could find ores to feed this network and launch an age of renewed, perhaps unprecedented, prosperity.

But the 1920s crushed these hopes. Even with rising metal prices for part of that time, the production from the innumerable small mines in the Cottonwoods fell far short of expectations, so short that it failed to support even the railroads serving Alta. New ore bodies were not found. Several spectacular and fatal locomotive and passenger-bus accidents on the impossibly steep grades destroyed confidence in the future of the narrow gauge. The trains stopped, and rail service to Alta came to an end. Snowslides again ripped out the aerial tramway. While larger mining centers in Utah prospered, the Alta-American Fork region floundered throughout the 1920s. And given that capital investment was proportional to capital availability, not metal prices, the Great Depression of the 1930s devastated most of the activity that remained.

The region did provide scrap metal. The rails to Alta disappeared as flux in the smelting furnaces to which they had once carried ore. Scrap iron, readily stolen day and night from idle mines and mills, became a valuable commodity; in the late 1930s, junk dealers hauled pieces of old equipment down the canyons and shipped them to Japan to feed a growing war machine.

Consolidation had long gone on in the district, and the Depression might have provided the opportunity for more. But for the most part, it did not. There seemed to be little future for the Cottonwoods, and additional consolidation only led to higher tax bills for the owners, most of whom couldn't pay anyway. Salt Lake County foreclosed on some properties.

Hope lingered after the Depression ended. Huge corporations, such as Kennecott, Utah Copper, and the Anaconda Copper Company, tested new ideas by core drilling for undiscovered ore deposits. So, too, did the U.S. Bureau of Mines as government tried to help post-World War II industry. Leasers did some work here and there, notably from the Wasatch Drain Tunnel, the well-equipped, deep-access portal on the Snow Bird claim. They had little success, although the Cardiff Mine made significant production returns from its deepest levels via the tunnel during the 1950s and 1960s, one exception to the generally depressed conditions in the Cottonwoods. And here and there a few independent miners replaced rotting mine timbers and scratched out what ore they could in hope of a better day.

And so after World War II, mining in the Cottonwoods and American Fork just petered out. The demand for uranium for nuclear bombs produced a new mining and stock-promotion boom in southeastern Utah. This craze boosted some old Alta stocks, but none of these companies found anything of consequence. Some successful uranium miners like the famous Charles A. Steen (see chapter seven) explored and extended deep workings in the Cottonwoods, but they found no new silver bonanzas. An army of prospectors toting Geiger counters ventured into the old workings, but they soon found that there was no sign of uranium in the mining camps of the Wasatch. The old districts had a few industrial metals such as tungsten, molybdenum, and bismuth, possibly in minable grade and quantity, but uranium was not one of them.

Finally, beginning in 1949, Utah's silver-lead industry began a precipitous decline. The custom smelters at Murray and Midvale shut down first; then the Tooele smelter, the last holdout, locked its doors in 1972. So, too, did the Midvale custom flotation mill, which had been the hope of the surviving small mines in the Wasatch and elsewhere. Out-of-state smelters would rarely take crude ore, historically the mainstay of Alta miners.

The high silver and gold prices of the early 1980s sparked some new interest around Alta, but it was short lived. One group built a mill at Orem, Utah, planning to recover a plethora of metals from ore shipped from the old South Hecla Mine, but this failed project probably produced more lawsuits than tons of metal. By the mid-1980s, after another plunge in silver, mining in the Cottonwoods and American Fork had ended once again. There were no ore shipments for the next quarter century into the early 2000s.

Remaining on the landscape, however, were the ruins and waste rock piles. Buildings collapsed, machinery rusted, and tramways fell down. Small mine dumps dotted countless hillsides. Artists and tourists enjoyed these relics. Environmental legislation in the late-twentieth century prompted regulatory agencies and their attorneys to sue luckless owners whose properties were the alleged source of any pollution. That also deterred prospecting and exploration, although government investigations revealed that mining in the Cottonwoods had created few significant environmental problems. The small ore bodies, coupled with the relative lack of acid-generating pyrite and, most importantly, the reactive limestone host rocks, provided natural pollution control. The old miners had unintentionally done a good job of blending their operations into the rugged natural setting. Old adits became water reservoirs for newer communities built atop old mill sites, and the larger mine dumps served as waste rock for road construction and parking lots for the Cottonwoods' exciting new industry.[43]

Phase VI: From Mines to Skis

As mining faded away in the Wasatch, another industry gradually took its place—skiing. Back in the nineteenth century, a few prospectors and miners, particularly Scandinavians, had traveled here and there on skis, or what were sometimes known as Norwegian snowshoes. Some were imported; some were made in the United States. They became a feature of winter life throughout the Mountain West. Most skis were seven to ten feet long, up to six inches wide, and nearly an inch thick. They could be heavy, easy to break, and hard to maneuver with the single balance pole the skier carried, but they provided a means to travel and transport supplies through deep snow. They also created a new sport. Some skiers enjoyed cruising down short runs or racing their friends. Daredevils took to jumping over trains in railroad cuts.

As time passed, fewer and fewer people used Norwegian snowshoes for travel, but the downhillers and daredevils gave rise to an increasingly popular twentieth-century pastime. By the 1930s downhill skiing and ski jumping had emerged as competitive sports, and recreational skiing had grown because of the automobile, new roads, and advances in equipment. Old western mining towns, particularly those with long winters and open slopes, became a focus of the emerging industry. Aspen in Colorado was a pioneer. Alta and Big Cottonwood in Utah would not be far behind.

Two developments emerged out of mining that made the Cottonwoods one of the meccas of world skiing. One was the accumulation of patented claims, which are fee land. The other was the new use of old tramway technology.

The most prominent person accumulating mining claims at Alta was George Watson. He had come to Utah with his mother in 1902, first working as a "swamper" in a bar, the person who swept up after it closed. In Salt Lake City, he learned about mining and stock brokerage. Then in 1907 he became incorporator and an officer of the Alta and Hecla Mining Company. He also got involved with the South Columbus Mining Company during the Jacobson brothers' era, and when that enterprise fell on hard times, he merged it with others, including the Alta and Hecla, to form the South Hecla Mining Company. As he continued acquiring Alta companies, young Robert F. Marvin became his chief engineer and right-hand man. "He would attend the stockholders' meetings of troubled small Alta companies in the 1920s, give an inspiring speech, and win enough votes to take over management," Marvin recalled years later. Then he would merge the companies into ones that he controlled. Watson thus accumulated a host of mines and claims until 1923, when he consolidated his companies into the appropriately named Alta Merger Mines Company, a forerunner of the Alta United Mines company. This enterprise controlled most of the land south and substantial amounts of property both north and east of Alta. Watson even tried to reopen the deep workings of the famous old Emma, which had flooded years before.[44]

Watson gained control of an ever-larger number of claims. He endured allegations that he was nothing more than a crook. The magic silver-boom era was gone, and he managed to raise only enough money during the late 1920s and early 1930s to do minimal underground work. As a result, he became more the promoter and acquirer than the serious miner as he talked up his favorite mountain town as "Romantic Alta." For Watson the hope of new ore bodies and wealth always seemed to lie in the offing. He did gain control of the Little Cottonwood narrow-gauge railroad in 1928 and tried to introduce a big gasoline-powered bus for tourists: the Alta Scenic Railway! Later, it was Watson who sold the road for iron-smelter flux.[45]

Watson told all who would listen about his two visions. First, he said he wanted to build a home for "indigent and crippled old prospectors," the objective of his Great American Prospectors Association. It certainly sounded altruistic in an age when there were few pensions and no social security. His other vision, no doubt more important, was to hold onto the mineral rights in the vast land empire he had assembled. When the Great Depression ended, he expected a new heyday for mining in the Cottonwoods. But that vision came a cropper. The hard times of the 1930s forced his hand. He couldn't pay the taxes on the vast merged property holdings running from American Fork to Alta.[47]

Meanwhile, recreational skiing began to emerge in the United States. Though small, and hardly an industry, it spread among various groups of enthusiasts, some in Salt Lake City. Small numbers of people began driving the 20 or so miles to Alta to enjoy the open terrain and powder snow. Some were skiers trained in Europe, but most were not. One was S. Joseph Quinney, a Salt Lake City attorney. A ski enthusiast, in the late 1930s, he became head of the new Salt Lake Winter Sports Association, and he got to know Watson, who had begun to talk of using his lands for skiing some years before.[47]

Everything now coalesced to provide a new direction for Alta and the Cottonwoods. By the late 1930s, Watson's financial resources were depleted. Quinney and his colleagues wanted to develop a ski area on part of the Watson property. The Forest Service, the Works Progress Association, various Utah state agencies, and other groups were seeking to stimulate business and create jobs in hard times. Negotiations ensued, and in 1938 a three-way deal among Watson, the Forest Service, and Quinney's group emerged. In return for tax forgiveness, Watson deeded most of his lands to the Forest Service but retained the mineral rights for future development. The Salt Lake Winter Sports Association got the right to develop a ski area on Watson and Forest Service lands. This also meant that the Alta town site, privatized nearly 75 years before by Walker Brothers, using the scrip of Amanda Brown, was a valuable island of developable property looking out onto Forest Service lands.

Private and public forces now combined to serve the burgeoning ski area. The Sports Association's William O'Connor, an executive with AS&R and president of the Michigan-Utah mine at Alta, oversaw construction of one of the world's first ski lifts; parts of it looked suspiciously like the old Michigan-Utah tramway. This primitive Collins chairlift, the second in the United States, had problems; it was uncomfortable and had questionable safety features (although it had no swaying ore buckets), but it was the start of things to come. In future years Colorado mine-tramway and European resort experts helped build a newer generation of bigger, faster, and safer lifts.

Other ski areas emerged as well. Two small early developments began at Brighton in Big Cottonwood Canyon almost simultaneously with the development of Alta. Solitude Resort, developed by Utah uranium money, came next. And finally, much later, there was Snowbird back in Little Cottonwood, which evolved to serve the steep slopes high above the Wasatch Drain Tunnel, portaled on the old Snow Bird claim. Eventually its massive, Swiss-built tramcars to lift skiers surpassed anything the hard-rock miners had ever envisioned to move ore. And as the years passed, the international fame of Alta and Snowbird largely obscured the powerful legacy of silver, lead, and copper mining in the Cottonwood-American Fork region.[48]

Conclusion

For more than a century, three historical circumstances drove mining in the Cottonwoods and American Fork. First, the relatively small, but sometimes very rich, ore bodies like those in the Emma spurred prospectors, miners, and promoters to drive tunnels, sink shafts, and search for mineral riches ever deeper in the Earth. They made possible the raising of investment capital. Second, the region's close proximity to the largest ore-dressing and smelting complex in the world provided a ready market and competing ore buyers, along with occasional investment capital and technical expertise. And third, the rugged mountainsides, faulted terrain, and long winters combined to prevent any easy determination of the real potential of the deposits. These facts combined to create hope, and that hope kept mining, leasing, and stock promotion alive.

Mining histories have tended to focus on the particular or the spectacular, from the gung-ho rushes to the red-light districts. And the Wasatch had its share of these: the Brigham Young–Pat Connor fight, the larger Mormon–non-Mormon struggle, and the Emma scandal, to name only a few. But fundamentally, development in the Cottonwoods reflected the general contours of the western industry. Mining emerged in the search for gold and silver spawned first by California gold and second by Comstock silver. The Wasatch had both metals, but the first booms foundered on complex ores, inadequate reduction technology, and bad transportation. Base metals became ever more important to American industry and so to mining in the Wasatch. Tunneling, ore reduction, and transportation technologies advanced, each slashing production costs. Price cycles (notably the general slump in silver), the shifting industrial and wartime demand for base metals, and overall economic trends had their impact on rising or falling production. Overall these factors combined to make the mining industry in the Wasatch a cyclic enterprise for more than a century. Mining did not end when the rushes ended, or even when the nineteenth century ended, as most people tend to believe. Instead, it endured for many more decades, well into the twentieth century, until a time when entrepreneurs could no longer raise money to fund development.

Despite its John Wayne image of cowboys, critters, and criminals—the American creation myth—the American West was in fact a place of scientific and technological advancement, something little understood and rarely acknowledged in histories of the region. From the late nineteenth century forward, the steady publication of scientific and technological studies based on research in the Cottonwood-American Fork area contributed to the larger knowledge and understanding which made the American minerals industry the foremost in the world. That research and the publications that resulted are the most significant, enduring legacy of mining in the Wasatch Mountains and the metallurgical industry in the Salt Lake and Tooele Valleys.

Courtesy *Deseret Morning News.*

An abandoned mine car stands in front of an equally forlorn Wells Fargo building in Silver Reef in 1982.

Courtesy *Deseret Morning News.*

The southern Utah desert, not a booming mining town, is all that can be seen through the sandstone arch of an abandoned building in Silver Reef in 1976.

Courtesy Utah State Historical Society.

Main Street of Silver Reef included the Elk Horn Salon on the left and James N. Louder's store on the right.

Courtesy Utah State Historical Society.

The Savage shaft in Silver Reef was part of the Stormont Mine.

Courtesy Utah State Historical Society.

Margaret Grambo, owner and operator of the Cosmopolitan Restaurant, in Silver Reef, maintained she served the best hash in Utah.

The remains of steam hoisting equipment and a boiler at Baby McKee inclined shaft in Cardiff Fork in Big Cottonwood Canyon reveal changes in mining technology. Built by Atlas Engineering Works in Indianapolis, this equipment was installed in the 1890s or early 1900s. The nearby Monte Cristo prospecting shaft (not shown), sunk during the 1917 silver boom, featured an electrically powered hoist. Photo by L. P. James (1974).

Ruins of the Wasatch Power Company plant stand below Tanners Flat in Little Cottonwood Canyon. Installed in 1904 by Columbus Consolidated Mining Company, two water-powered turbines generated electricity for the air compressors, mills, and miners' dwellings at Alta. Photo by L. P. James (1963).

Remnants of a Davis horse whim (animal-powered hoist) at the Cardiff Mine in Big Cottonwood Canyon were used to explore the Mountain Cheaf claim. The shaft followed a vein that connected, at lower depths, to the large Cardiff ore body. Photo by L. P. James (1974).

Courtesy C. H. Malmborg (1976).

Prospector or recreational miner? Charles H. Malmborg leans into an air-powered, "drifter" percussion drill, completing a blast hole in hard limestone in Big Cottonwood Canyon. The drill, mounted on a homemade support leg, is in his Lost Emma tunnel west of Silver Fork. Malmborg, a baker and the son of a longtime area mine superintendent, drove more than 1,100 feet of workings here beginning in 1951. Federal authorities later evicted him from his unpatented claim.

Courtesy *Deseret Morning News*.

The lower terminal of the Michigan–Utah aerial tramway, the predecessor to ski lifts, carried ore from several Alta mines to storage bins at Tanners Flat. Ore wagons and sleighs then hauled the ore to railroads in the Salt Lake Valley. In this 1915 photograph, the idle Continental Alta gravity concentration mill, erected along with the tramway about 1905, extends beneath the terminal building. The tramway was rebuilt in the 1920s, and this terminal was dismantled. The rebuilt tram extended only between Grizzly Gulch and a new terminal on the present site of Snowbird resort, where ore was loaded onto cars of the Little Cottonwood narrow-gauge steam railroad. In the 1930s machinery from the aerial tramway was incorporated into the Collins chair lift at Alta, the second ski lift in the country.

Courtesy Park City Historical Society and Museum.

At almost five hundred tons, the Cornish pump was one of the largest and most powerful steam-operated pumps ever built. It pumped almost four million gallons of water a day out of the Ontario Mine, lifting it from one thousand feet underground to the six-hundred-foot level, where it flowed out the Ontario Number-One Drain Tunnel.

Courtesy Park City Historical Society and Museum.

Built in 1877 at the mouth of Ontario Canyon, this mill loomed over the south end of Park City for most of a century. A late-night explosion brought down one of the brick smokestacks in 1949. The remnants of the mill were demolished in the early 1970s. Today a runaway truck lane occupies the site.

Courtesy Park City Historical Society and Museum, Pop Jenks collection.

Out of this machine shop came parts used in the Judge Mine and mill. The shop helped Park City make the transition from mining to skiing. In the mid-1940s it made parts for lifts at Snow Park (now part of Deer Valley). In the early 1960s, it made components for J-bars and equipment to pack snow and cut moguls at Treasure Mountains (now Park City Mountain Resort).

Courtesy Park City Historical Society and Museum.

R. C. Chambers, here in his office in January 1900, served as superintendent at the Ontario Mine, and acquired interests in other Park City mines.

Courtesy Park City Historical Society and Museum.

Near the turn of the 20th century, a crew in slickers poses at the Ontario Mine. The sign reminds unmarried miners that state law requires them to live at the mine boarding house. Mine owners who wanted a work force near the mine obtained the legislation, which was repealed in 1901.

Courtesy Park City Historical Society and Museum.

Mine foremen used four-wheeled cycles, also called track bikes, to travel inside the mines. A fast-moving light alerted miners that the boss was on his way, prompting them to work harder. This 1920s photo was taken at the mouth of the Spiro Drain Tunnel.

Courtesy Park City Historical Society and Museum.

The word "tipple" doesn't do justice to this building, which towered over Park Avenue for 80 years. Buckets of concentrate moved via aerial tram from the Silver King mill to a hopper in the top of the building, then dropped into train cars below. The building was devoured by a spectacular fire in 1981.

Courtesy *Deseret Morning News* from a negative donated by Clark Wilson.

This photo by Hal Rumel shows miners preparing fuses for blasting in the Mayflower Mine.

Courtesy Park City Historical Society and Museum, Pop Jenks collection.

The Silver King Mine as it looked about 1935. The sampler on the far left, two bunkhouses on the ridge, and many other buildings are now gone. However, the remains of the large mill and flotation tank at lower right still pique the curiosity of skiers at Park City.

Courtesy L. Tom Perry Special Collections, Harold B. Lee Library, Brigham Young University.

Hard-rock miners assemble in Park City in 1893.

Courtesy L. P. James.

Emil J. Raddatz stands by the Bullion Coalition mining camp at Bauer, near Stockton, about 1910 or 1912. Raddatz worked up from mine captain to mine superintendent and was in charge at the Honorine, one of Patrick Connor's mines. Raddatz's great success came when, through early use of exploration geochemistry (panned concentrates), he found a good prospect in East Tintic and formed the Tintic Standard Mining Company. Years of sinking shafts and raising money led to the discovery of what geologist Waldemar Lindgren called "the richest silver mine in the world."

Courtesy *Deseret Morning News.*

The Eagle and Blue Bell surface plant near Eureka in 1977. Photo by P. Mogensen.

Courtesy *Deseret Morning News*.

The head frame of the Bullion Beck Mine at Eureka is seen from the south in August 1975. Photo by P. Mogensen.

Courtesy Utah State Historical Society.

The San Francisco Mining District's distinctive beehive kilns were 20 to 24 feet in diameter and 19 to 22 feet high, with walls a foot or more thick.

Courtesy Uintah County Library Regional History Center.

The Cowboy vein of gilsonite at Dragon. Gilsonite veins range from a fraction of an inch to large ones like this that is 22 feet wide.

Courtesy Uintah County Library Regional History Center.

Hauling gilsonite from Bonanza about 1900. Note the tandem wagons, which were common practice at the time. Gilsonite was hauled from the Uinta Basin to railheads at Wellington where it was loaded on trains to Grand Junction, Colorado.

Note the black faces of this gilsonite mining crew at Bonanza, Utah, about 1906. Gilsonite was picked from the veins and packed into sacks that were hauled to the railroad by wagon. While it breaks into workable pieces easily, the fractured gilsonite creates small, needle-sharp slivers that torment the miner's body.

When mined, gilsonite fractures into small pieces, and the resulting dust is highly explosive. There were several fatal explosions in the Uinta Basin, but the worst, shown here, occurred on 4 November 1953 and claimed the lives of eight miners.

Courtesy Uintah County Library Regional History Center.

The Black Diamond Mine at Fort Duchesne, Utah, had a wire track with buckets to haul the gilsonite from the mine to the loading dump, where it was sacked and hauled by wagon to Wellsville.

Courtesy Uintah County Library Regional History Center.

The Dyer Mine, located 26 miles above Vernal on Anderson Creek, was the most significant gold and silver mine in the Uinta Basin. It operated from 1889 through 1941, but the best deposits of ore were depleted by 1904.

13

Park City

Hal Compton
David Hampshire

"Perhaps the most striking feature of the Park City mining district is that it does not look like a great mining district," geologist M. B. Kildale wrote in 1956. "Despite the fact that several large waste dumps are scattered throughout the twenty square mile area which includes the mining district, it is difficult, even today, when looking down from the higher peaks over the lower beautiful woods and canyons to believe this area is honey-combed beneath the surface by many miles of underground workings."[1]

Today, with three major ski areas, several golf courses, and hundreds of luxury homes, Park City looks even less like a great mining district. Nevertheless, between the early 1870s and the early 1980s, about 300 area mines produced ores valued at more than $500 million at the time. The mountains surrounding Park City are laced with an estimated 1,000 miles of underground workings.

Present-day Park City is located near the junction of the Wasatch and Uinta Mountains. This conjunction, created centuries ago by plate tectonics, resulted in the formation of a massive bulge called an anticline. Igneous activity thousands of feet below caused a superheated solution to carry concentrated minerals and deposit them in veins, fissures, and limestone beds within the anticline.[2] Early prospectors discovered these ancient deposits in the form of gold, silver, copper, lead, and zinc.

The first settlers in this area were not miners but Mormon pioneers. They came under the leadership of Heber C. Kimball, Jedediah M. Grant, and Samuel C. Snyder and in 1853 established a settlement near the present Silver Springs subdivision in Snyderville.[3] Those early settlers operated sawmills, gristmills, and quarries and raised crops and farm animals to ship to the Salt Lake Valley.

Early Prospectors

Lacking any governmental agencies or a newspaper to document it, early Park City mining history is fragmentary and obscure. According to geologist J. M. Boutwell, a

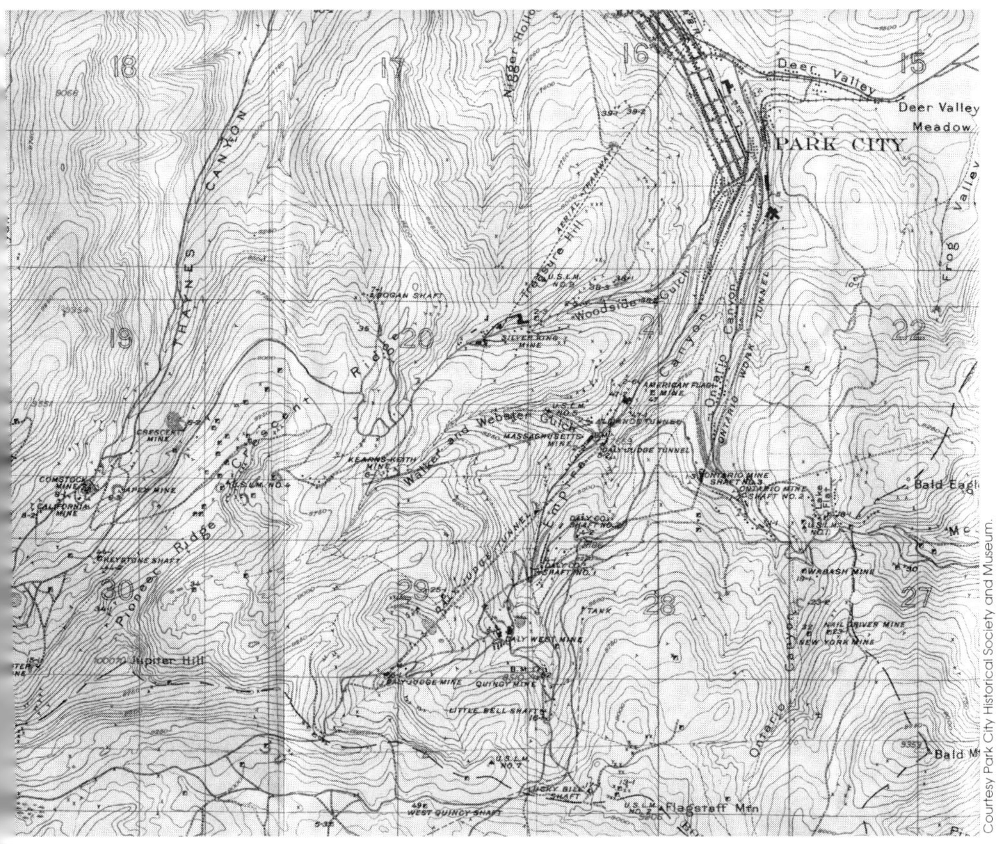

Courtesy Park City Historical Society and Museum.

his map, labeled "Topographic Map of Park City District, Utah," originally appeared in J. M. Boutwell's ook, *Geology and Ore Deposits of the Park City District, Utah,* which was published in 1912. The Park tah and several other major mines on the east side of the mining district are not shown. They were eveloped after the book was published.

group of prospectors, including Ephraim Hanks, Rufus Walker, and Louis Simmons, camped near Snyderville in 1868 while they roamed the nearby mountains prospecting for precious metals, against the wishes of Mormon prophet Brigham Young.[4] They reportedly discovered the Young America, the Yellow Jacket, the Pinyon, and the Green Monster claims on Pinyon or Pioneer Ridge and the Walker-Webster claim in the gulch below. This area is now part of Park City Mountain Resort.

At about the same time, another very different group of men was playing an important part in the discovery effort. These were U.S. Army soldiers stationed at Camp Douglas in Salt Lake City under the command of Colonel Patrick Edward Connor. Three of the soldiers traveled over the high mountain pass from Big Cottonwood Canyon east of Salt Lake City to a place called Bonanza Flat, just south of Park City. Encouraged by Connor, they began prospecting for gold and silver and soon discovered a rich deposit of silver and lead. They named their claim the Flagstaff.[5]

The following year the soldiers sold the claim to James Kennedy for $5,000, and in 1871 it gained the distinction of being the first mine in Park City to ship ore. Several years later Kennedy sold the Flagstaff to Edward P. Ferry for $50,000.

The early mines in Park City were combined with those in Big and Little Cottonwood Canyons to become part of the Mountain Lake Mining District. However, there were so many new discoveries that three new mining districts were formed: the Uinta, containing the Park City mines, in 1869, and the Blue Ledge and the Snake Creek, both in Wasatch County, in 1870. The name Park City Mining District was later used to describe the Uinta and parts of the Blue Ledge and Snake Creek Districts.[6]

Word of the new discoveries near Snyderville spread throughout the West. The transcontinental railroad, which was completed in northern Utah in 1869, helped bring a large workforce to the mines. Men from many different countries converged on the area. The 1900 U.S. Census for Park City lists 854 foreign-born residents among a total population of 3,759. Summit County, which includes Park City, lists 2,125 foreign-born residents including 859 from England, 276 from Ireland, 167 from Sweden, 154 from Scotland, 104 from Denmark, 75 from Norway and 71 from China.[7]

Although many Chinese lived and worked outside of town in the boardinghouses near the mines and mills, they were not allowed to work in the mines themselves. Park City's Chinatown was located east of Main Street on what was then called Grant Street.[8] There the Chinese lived in their small shacks, when they weren't operating the laundries, the hotels, and the boardinghouses. In the summer they grew vegetables in small gardens and peddled them around town.

Some of the town residents didn't feel comfortable walking through Chinatown on their way to and from Rossie Hill, east of Main Street. In 1886 a long wooden bridge was built high above Chinatown, connecting Main Street with Rossie Hill. It became known as the China Bridge, a landmark in Park City for many years. After the first bridge was destroyed by fire, a second replaced it; it stood until it was demolished in 1954.

Many of the new mining claims were in the mountains south of Park City's present location. The McHenry, Lucky Bill, Parley's Park, Lady of the Lake, and Flagstaff were among them. To live close to their work, many miners settled on the shores of a small lake surrounded by mountains at the head of McHenry's Canyon.[9] At one time Lake Flat was home to more than 700 people. However, weather at this 8,000-foot elevation was severe, and the little settlement was eventually abandoned. The lake was accidentally emptied when a drain tunnel was constructed below it.

The year 1872 was significant in Park City history for several reasons, not the least being the naming of the town. George and Rhoda Snyder had been living in nearby Wanship when they learned of the discoveries in the mountains south of them.[10] Realizing the potential for financial gain, they moved to the mining area and homesteaded 160 acres. Using a horse-drawn sledge, they cleared the trees and brush on a flat next to a beautiful little stream that Mormon explorer Parley P. Pratt had christened Silver Creek. The Snyders built the first house at the bottom of what became Main Street. Legend has it that, during the Fourth of July celebration of 1872, George and Rhoda raised a homemade flag and declared that the new town would be known as Parley's Park City.

The development of local mining was relatively slow until 1872, when one of Park City's richest mines was discovered in Ontario Gulch. Although four prospectors were involved in the discovery, Rector Steen is generally recognized as the principal one. Soon after, George Hearst, a successful mining man from the goldfields of California and the silver mines of Nevada, arrived, looking for a rich claim to purchase. Hearst also represented two San Francisco friends and partners, James Ben Ali Haggin and Lloyd Tevis. On Bonanza Flat, Hearst met a friend, Marcus Daly, who suggested that he investigate a new discovery in McHenry Canyon at the base of Bald Mountain.

Hearst rode his horse down the canyon to the mine but immediately saw that the claim had large amounts of underground water, which meant high development costs. Daly then recommended the new discovery in Ontario Gulch. Hearst found the four prospectors working the claim and received permission to borrow a sample and have it tested in town. He found that it contained very high grades of silver and lead and made the prospectors an offer. He was told that the claim was already spoken for, but that if it became available, he would have the first chance to buy it. Hearst arranged to have someone, perhaps Marcus Daly, keep an eye on it and returned to San Francisco. Not long after, the claim did become available, and Hearst purchased it for $27,000 and named it the Ontario Mine.

Hearst hired a friend from California, Robert C. Chambers, as mine superintendent. Chambers represented the Hearst interests with great efficiency and success until his sudden death in 1901. During the time the California group owned the Ontario Mine, it produced $14 million of tax-free dividends.[11] The Ontario Mining Company not only produced great wealth for its owners but also contributed significantly to Park City's economy for many years, employing hundreds of miners.

The Michigan Bunch and the Immigrant Entrepreneurs

In 1873 a group of entrepreneurs from Grand Haven, Michigan, arrived in Park City to be close to their mining investments and develop new ones. Edward P. Ferry, David McLaughlin, James Mason, and Frederick Nims soon became known as the Michigan Bunch.[12] They were later joined by Ferry's brother, Colonel William M. Ferry. Over the years they would have a great influence on mining and the development of the town.

Soon after his arrival, Edward Ferry purchased the Flagstaff Mine for $50,000; it became the first holding of the Marsac Silver Mining Company. The name Marsac came from a Grand Haven family friend, Sophie de Marsac.[13] Ferry then constructed the Marsac mill (also known as the Daly mill) just east of Main Street to process the ore. It was a local landmark until it was torn down in 1904.

Members of the Michigan Bunch soon learned that very few people occupying land in town held legal title to it—they were in fact squatters—and saw an opportunity to make a great deal of money, perhaps more than in mining. They surveyed and platted the land, prepared legal descriptions, took the documents to the territorial offices in Salt Lake City, and filed for ownership. After paying a nominal fee, they became owners of most of Park City.[14]

Each member of the Michigan Bunch was allocated a number of lots in the town site to market and sell to the public. They gave the people living on their newly acquired land an ultimatum: purchase the land or get off. Many were outraged, but they had no other choice. In addition to bringing a little order to the growth, the Michigan Bunch made some other positive contributions, such as surveying and planning the side streets, constructing a waterworks, and building boardwalks on Main Street. Having made their fortunes, Nims and Mason returned to Grand Haven. McLaughlin, a trained attorney, married a local woman and became very active in business and mining. However, he died suddenly in 1901 at the age of 46.

Colonel William Ferry was one of the original owners of the Quincy Mine, later merged into the famous Daly-West Mine. He built a fine mansion at the mouth of Thaynes Canyon that was later moved to Monitor Drive in Park City. It is still known by many as the Ferry Mansion. He was a generous philanthropist, donating the land for Westminster College in Salt Lake City and contributing to its financial support. Ferry died in 1905 at the age of 81. His remains were shipped to Grand Haven for burial in the family cemetery.[15]

In addition to the Flagstaff Mine, Edward Ferry owned an interest in the Woodside Mine, the Anchor Mine, and the American Flag Mine, and he consolidated all of the mines on Pioneer Ridge to form the Crescent Mine Company. He built a fine home in Salt Lake City near his friend Thomas Kearns. He died in 1917 at the age of 80.[16] Edward Ferry's son, William Mont Ferry, was managing director of the Silver King Mine from 1905 to 1919, a member of the Salt Lake City Council, president of the Utah State Senate, and mayor of Salt Lake City. He died in 1938 at the age of 67.[17]

Meanwhile, R. C. Chambers was developing the Ontario Mine and removing many tons of silver-lead ore. Some of it was so rich it could be shipped directly to Liverpool, England, for smelting and still produce a profit. Chambers later leased the Marsac mill and the nearby McHenry mill to process the Ontario ore. In 1877 the Ontario completed its own mill south of town at the mouth of Ontario Canyon.[18]

As his miners went deeper into the earth, following rich veins of silver, they encountered more and more underground water. Some of the Park City mines began using steam-operated water pumps. Chambers and Hearst knew it would take a very special pump to dewater the Ontario Mine, so they ordered a large pump modeled after one developed to use in the wet mines of Cornwall, England. In 1881 they contacted an old friend and water-pump expert from the mines of Virginia City, David Keith, to prepare the shaft and supervise the installation.[19]

Weighing almost 500 tons, the Cornish was one of the largest and most powerful pumping engines ever built. The flywheel alone was 30 feet in diameter and weighed 56 tons. It pumped almost four million gallons of water a day from 1,000 feet underground to the 600-foot level, where it flowed through the newly completed Ontario Number-One Drain Tunnel from the Ontario Shaft Number Three to a point just below the Ontario mill. Water from the tunnel supplied the Marsac mill and generated power.[20] To fuel the boilers for the pumps and provide timbers for the mines, most of the surrounding mountains were soon stripped of their trees. Timber was even shipped from Kamas, Heber, and the Strawberry Valley. The treeless hillsides were prone to avalanches, and a number of Park City residents lost their lives over the years. When wood became scarce, the pump boilers converted to coal.

After supervising the installation of the pump, David Keith left the Ontario to make his fortune on his own. Born in Nova Scotia on 27 May 1847, Keith had gained a lasting interest in mining from working in local mines while growing up. When still a boy, Keith heard about the gold discoveries in California. After a long, grueling journey, he arrived there to find the streets were not paved with gold. Meanwhile, great silver discoveries were occurring in nearby Nevada. Keith went to work as a common miner in the Comstock Mine. He worked his way up to pump man and finally foreman. He was working in the Gold Hill District in Virginia City when he received his call from R. C. Chambers.[21]

Not long after arriving in Park City, Keith met Thomas Kearns, a son of Irish immigrants who had originally settled in Ontario, Canada. Kearns had gotten his early mining experience in the Homestake gold mine in South Dakota and mines around Tombstone, Arizona. Said to have arrived in Park City in 1883 with only 10 cents in his pocket, Kearns was offered a job by Keith as a mucker at the Ontario Mine, removing the debris after the explosives had done their work.[22]

After working at the Ontario for several years, Kearns began exploring the southern end of Treasure Hill west of Park City and in 1889 discoved a vein of silver in Woodside Gulch near the Mayflower Mine. To develop the prospect, he formed a partnership with David Keith that would last as long as they lived. They eventually built and lived in mansions close to one another on South Temple Street (then known as Brigham Street) in Salt Lake City. Keith died in April 1918 and Kearns six months later. Keith is buried in Mt. Olivet Cemetery and Kearns in the adjoining Mt. Calvary Cemetery.[23]

In 1881 John J. Daly was a miner at the Ontario Mine, having gained his experience as a prospector and miner in Montana and Nevada. On his days off, he roamed the canyons and mountains around Park City; he developed a theory that the rich Ontario ore body extended west into Empire Canyon and gradually acquired 24 claims. It was later rumored that Daly, to purchase the claims, received money from R. C. Chambers.[24]

Daly left the Ontario Mine and began work on what became known as the Central Tunnel. He soon struck the rich ore fissure he knew he would find. He hired miners to sink a shaft into the ore body and erected buildings containing the very best hoisting and mining machinery. In a short time, the Daly Mine was producing 800 tons of high-grade ore a month for processing in the Marsac mill. Daly soon quietly deeded a half interest in the mine to R. C. Chambers.

Daly acquired another 40 claims in Empire Canyon and organized the Daly-West Mining Company, which eventually became one of Park City's three richest, thanks in part to the great, bedded ore deposits which were discovered at the adjoining Quincy and Silver King Mines. Chambers became president of the new company, and Daly was vice president.[25]

In 1886 Daly was awarded a contract to construct a 6,000-foot tunnel to drain a large amount of underground water in Edward Ferry's Anchor Mine, just above Daly Mine in Empire Canyon. It was completed three years later; the flume, three feet wide and two feet deep, carried a steady flow from the mine. Known today as the Judge Tunnel, it supplies Park City's culinary water system. The most respected and well liked of all Park City mining leaders, Daly announced in 1907 that he was retiring from active mining because of ill health. The family soon sold his fine mansion on Brigham Street in Salt Lake City and moved to Los Angeles, where Daly died in 1927. Many in Park City mourned his death.[26]

Irish immigrant John Judge was known as a hands-on leader who enjoyed nothing more than working underground with his men. Judge was a foreman in the Daly Mine when he was awarded a contract to construct a drain tunnel to the Sampson Mine in Walker-Webster Gulch, just west of Empire Canyon, in 1889. The 4,576-foot tunnel was completed in 1890 and began to drain the Sampson as well as the Rebellion and other nearby claims. Known as the Alliance Tunnel, it provided water to Park City for a few years until a money dispute caused the city to change to Daly's tunnel.

Judge joined John Daly in acquiring the Anchor Mine (later named the Daly-Judge Mine). He later founded the Judge Mining and Smelting Company. The Judge mill was built in lower Empire Canyon to process ore from the Daly-Judge and Daly-West Mines. However, Judge paid the price for his years underground, dying of miner's consumption (silicosis) in 1892 at the age of 48. After Judge's death, his wife, Mary, also built a beautiful mansion on Brigham Street in Salt Lake City. She also helped fund construction of the Cathedral of the Madeleine, Judge Memorial Home for Miners (now Judge Memorial High School), and the Judge Building in Salt Lake City. Mary died in 1909 and was buried beside her husband in Mt. Olivet Cemetery. Rather than see the mansion turned into a boardinghouse, the Judge family had it demolished.[27]

By 1888 the Cornish pump at the Ontario Mine had outlived its usefulness. The ore had been mined to the 1,000-foot level, and the company wanted to follow the ore body deeper. The Cornish pump was incapable of operating any deeper, so work was started on another drain tunnel at the 1,500-foot level: the Ontario Number-Two Drain Tunnel, also known as the Keetley Drain Tunnel after John Keetley, who supervised its construction. In 1894, after six years of construction, the tunnel was completed to Ontario Shaft Number Two, and 13,000 gallons of water a minute was flowing out of the mine.[28] The tunnel was so straight that a miner could stand at the shaft end and see daylight three miles away. The tunnel was later extended to Ontario Shaft Number Three and the Daly-West Mine. In 1904 the Cornish pump was cut up for scrap. The tunnel, with its portal on the edge of the Jordanelle Reservoir, still functions today. A large filtration plant processes the water prior to its journey down the Provo River to towns and cities along the Wasatch Front.

In the early days, ore was hauled by horse-drawn wagons from the mines to the mills at a cost of $1.50 per ton, and much of the ore consisted of waste rock.[29] One of the main ore-hauling routes was down unpaved Main Street, where deep ruts developed, especially in the winter and spring. The mines also used horses underground; they remained there as long as eight years, pulling ore cars and performing other heavy labor.

The Crescent Mine Company developed a different means of hauling ore. In 1883 the company contracted to construct a narrow-gauge railroad, the Crescent Tramway, from the mine in Thaynes Canyon to the Crescent mill in town.[30] A small Shay steam engine, christened the "Maud Withey," hauled 60 tons of ore a day from the mine to the mill. However, because of deep snow, it could operate only in the summer and ceased operation in 1898.

Conventional trains also came to Park City: the Union Pacific and Utah Eastern in 1880 from the transcontinental line at Echo (the Union Pacific soon acquired and closed the smaller railroad) and the Utah Central from Salt Lake City in 1890 (acquired by the Denver and Rio Grande Western in 1897).

The mine owners finally realized that if they separated the good from the worthless near the mines they would save even more in transportation costs. They started building mills in the mountains near the mines and shut down the ones in town. The Ontario, the Marsac, the McHenry, and the Union concentrator on Daly Avenue all closed. Residents of Park City were not sorry to see the end of the mill smokestacks that had once filled the air with highly corrosive acidic smoke.[31] Most of the mills also dumped their tailings into Silver Creek, which flowed through town and northeast through the Snyderville Basin. The stream, polluted with mill and human wastes, became popularly known as Poison Creek.

In 1892 Thomas Kearns and David Keith joined together to organize one of Park City's greatest mining companies, the Silver King.[32] They acquired and merged five mines in Woodside Gulch: the Mayflower, the Woodside, the Northland, the Silver King, and the Tenderfoot. They next constructed a main shaft and connected the five mines on the 700-foot level. The two friends installed the finest in modern technology and equipment. Officers of the Silver King were David Keith, president; Thomas Kearns, vice president; and John Judge, W. V. Rice, W. H. Dodge, and Albion Emery, directors. Judge died soon after, and his wife, Mary, sold his stock to others, including James Ivers. Albion Emery, who had reached his lofty position with the Silver King through the timely purchase of a large number of shares of stock in the Mayflower

Mine, died in 1894 and was replaced on the board of directors by his wife, Susanna Bransford Emery, later known as Utah's "silver queen." She eventually had a succession of husbands, including Colonel Edwin Holmes, Dr. Radovan Delitch, and Prince Nicholas Engalitcheff. She died in 1942 at the age of 83, virtually penniless.[33]

Near the Silver King shaft house, Keith and Kearns built a state-of-the-art concentrator to process the ore. In 1901 they added an aerial tramway to carry the concentrate from the mill to the lower terminal near the tracks of the Denver and Rio Grande Western. Here the concentrate was loaded into mine cars and transported to the smelter in Salt Lake City. The tramway reduced the cost of transporting ore and concentrate from $1.50 to $.22 a ton.[34]

In addition to the problems caused by rapid growth, many of Park City's early ups and downs reflected the fluctuation in the national price of metals. An increase in the supply of silver, thanks largely to the production of western mines, coincided with a movement in Washington to adopt the gold standard for backing currency instead of a combination of gold and silver. Silver prices dipped sharply in 1893 and by 1897 had dropped so low that several mines, including the Ontario, suspended operations.

Disasters and Town Development

On 19 June 1898, an entirely different problem almost brought an end to Park City. A fire erupted in the kitchen of the American Hotel on upper Main Street at 4:00 AM, when most residents were in bed. The fire, driven by canyon winds out of the south, quickly flared out of control, raced down Main Street, and spread to Park Avenue and Rossie Hill. At sunrise it ran out of fuel at Heber Avenue at the bottom of Main. The losses were staggering: 200 homes and businesses destroyed, including the city hall and the new opera house. Seventy-five percent of the town was destroyed at a loss of a million dollars. Although 500 people were homeless, there were no deaths from the inferno. With the town still staggering from the impact of depressed silver prices, many said Park City would never rise from the ashes.[35]

"Aside from the great loss involved in yesterday's fire, it will have a depressing effect upon those who suffered losses and all others interested in the town directly," the *Salt Lake Herald* wrote in an editorial on June 20. "With business so slow as it is now, there will be a disinclination to build again."[36] In spite of the newspaper's gloomy assessment, Park City merchants wasted no time in going back to work. By 2 July 35 new buildings were under construction. Within 18 months, the city was almost completely rebuilt.[37]

On 15 July 1902, the worst disaster in Park City mining history occurred at the Daly-West Mine. It was 11:20 PM, and the night shift was hard at work in the mine. Suddenly, workers on the surface heard what sounded like a muffled explosion and felt a sudden tremor. A station tender, who was brought up on the hoist from the 900-foot level, reported that there had been a tremendous explosion below where he had been working. An investigation revealed that the miner operating the powder magazine on the 1,200-foot level had dropped either a candle or the coals from his pipe into the magazine containing powder, blasting caps, and fuses.

The explosion and poison gas killed 25 miners in the Daly-West Mine. Nine others died when poison gas drifted through a connecting tunnel into the adjoining Ontario. Morticians from Salt Lake City traveled to Park City to assist the local one. Funeral processions down Main Street were a common sight. The mine company paid for burials and headstones and gave $2,000 to each family of a victim and $500 for each child. As a result of the disaster, legislation was passed outlawing the storage of explosives underground.[38]

In the early days of mining, holes for explosives were drilled by hand, either by one miner using a drill and a four-pound hammer called a single jack, or by two miners, with one holding the drill and the other using an eight-pound hammer called a double jack. In 1890 mechanical drills operated by compressed air were introduced in the Park City mines. Air drills greatly improved productivity but were deadly for the miners. The drills, which became known as widow makers, created clouds of rock dust that the miners breathed. The sharp particles cut the membranes of the lungs, creating what the miners called "the con" or miners' consumption (silicosis), for which there was no cure. Many miners made the long hard trip down Parley's Canyon to Salt Lake City for medical treatment. Later, water hoses were attached to the drills to suppress the rock dust.[39] Miners also risked death from cave-ins and poison gas. If they lived to be 40, it was considered old age.

Until the passage of a state law in 1896 limiting their workday to 8 hours underground, miners worked 10 hours a day, six days a week, with only two holidays a year: the Fourth of July and Christmas. Unskilled muckers were paid an average of $2.75 a day, the miners $3.00, and the timber men a little more.[40] Payday was once a month; in between paydays, miners lived on credit from the local merchants. Their rare time off was often spent in one of Main Street's many saloons. Each nationality had its favorite bar, where the men could gather and sing the songs of their native countries, tell jokes and stories, gamble, and drink.

Like most mining towns, Park City had its red-light district, which was located on upper Heber Avenue on the road to Deer Valley. In the early 1900s, several houses were operated by Rachel Urban, called Mother Urban by most of her regular customers. She was good to her girls, providing them with food, clothing, shelter, medical treatment, and a small salary. In return she required that they stay out of the saloons on Main Street and not date men outside of working hours. Rachel was kind, generous, and one of the first in town to support a worthy cause. She died in 1933 and is buried in the city cemetery.[41]

Until 1901 unmarried miners working for large companies were required by Utah territorial law to live in the company bunkhouses in the mountains near the mines. The legislation was passed at the request of mine owners, who wanted a workforce near the mine in case of emergencies. Miners were charged one dollar a day for room and board, a third of their wages. Married miners could live wherever they wanted, and most chose to live in town. Not long after Utah became a state in 1896, the law was repealed, prompting the construction of many new boardinghouses in town.

Affordable medical care and burial insurance posed a problem for the unmarried mine workers in town. Some community leaders knew that fraternal organizations could provide both of the services along with a more wholesome social life. Early in 1878 efforts were begun to organize the Park City Order of Freemasons. It prospered and was granted a charter in 1880 to become Uinta Lodge Number 7. Other fraternal organizations followed, including the Independent Order of Odd Fellows, the Knights of Pythias, the Ancient Order of United Workmen, the Woodmen of the World, the Ancient Order of Hibernians, the Benevolent and Protective Order of Elks, and the Loyal Order of Moose. In 1885 several of the organizations joined together to acquire five acres of land and establish the Glenwood Cemetery. The fraternal organizations were allocated plots, which they sold to members for $10 or $27 each. About 900 people are buried in the privately owned Glenwood Cemetery.[42]

Union Organization

Labor historians say the influx of European workers during the nineteenth century helped introduce socialist ideologies to semiskilled and unskilled laborers in America. Among the first labor organizations to have an impact on Park City was the Knights of Labor, founded in 1869, which reached a national membership of more than 700,000 in the mid-1880s. Park City sent three delegates to the Knights of Labor convention in Salt Lake City in August 1886. In May 1887 the Loyal League of Park City, which was associated with the Knights of Labor, tried unsuccessfully to pressure mill and mine owners into discriminating against Mormon and Chinese workers.[43] However, it wasn't until the formation of a local of the Western Federation of Miners in 1895 that Park City miners began to have a unified voice. In June 1896, in a show of solidarity, about 450 Park City miners held a Miner's Day parade.[44]

In 1903 miners, mine-union officers, city officials, and concerned citizens met to discuss the need for a hospital in Park City to serve the miners. Finally, they reached a consensus, and most of the miners signed subscriptions for one dollar a month, deducted from paychecks, to help pay for construction and later their treatment. Land for the hospital was generously donated by Eliza Nelson, while citizens, businessmen, and mine owners added contributions.[45]

Located on what was called Nelson Hill, the hospital opened in 1904 and served the miners and the public until 1971, when it became a skier flophouse. It was moved to the center of the city park in 1979 and renovated as the Park City Public Library in 1982. Today it is the Park City Community Center. The Shadow Ridge Hotel now occupies the land where the hospital once stood.

Although much of the early mining activity took place in Ontario Canyon, Empire Canyon, Woodside Gulch, and Walker-Webster Gulch, Thaynes Canyon, although more remote, also saw a great amount of activity. Among the significant mines were the Jupiter, the California, the Comstock, the Crescent, and the Keystone. Two mills, the California-Comstock and the Keystone, operated for many years near the mines.

An unnamed gulch between Thaynes Canyon and Treasure Hollow contained the Silver King Consolidated Mine, known in Park City as the King Con, developed by Solon Spiro and Salt Lake City financier Samuel Newhouse.[46] The two men acquired the Bogen Mine claims and then built a large complex, consisting of a shaft house, three boardinghouses, and a machine shop.

Deer Valley, also called Frog Valley, was the location of a number of mines and claims, including the Park City Consolidated and the Queen Esther. Upper Ontario Canyon had the Wabash, the Naildriver, and the New York Mines. At the east end of the mining district, Glenco Canyon contained the Mayflower, the Park Galena, the Glen Allen, and the Star of Utah. McHenry Canyon, at the base of Bald Mountain, had the Wasatch, Hawkeye, McHenry, Parley's Park, and Lady of the Lake Mines. About 300 mines produced ore worth $500 million in the Park City Mining District.

In 1907 the Silver King Mine Company acquired the Odin property, the Keith and Kearns property, the Pinion and Crescent claims, the Creole Mine, and the Alliance Mine and formed the Silver King Coalition Mining Company. Its officers were David Keith, president; Thomas Kearns, vice president; and W. S. McCornick, James Ivers, and William Ferry, directors. Mike Dailey was mine superintendent.[47]

Imperfect milling methods failed to extract all of the silver and lead from the ore before the tailings were dumped into the streams or carried by slurry pipes to the empty land north and east of town. Beginning about 1910, several mills such as the Beggs, the Broadwater, and the Miller-Beggs were constructed to reprocess the tailings. The Big Four mill was built along the Union Pacific spur line near Atkinson, at the head of Silver Creek Canyon.[48]

It was also discovered that zinc, a by-product of silver mining, could be sold for a profit if processed properly. Believing that it might have some value in the future, John Daly had asked his miners to save the zinc, once considered worthless. Daly's vision was rewarded when a method of processing zinc was perfected for use in paint, batteries, and medicine. The Grasselli zinc mill was built on a railroad spur north of town in 1909 to process zinc ore from the Daly-Judge Mine, which was producing 500 tons of the metal a month. In 1916–17 the Judge electrolytic mill was built in Deer Valley to process zinc.

In October 1915 the Silver King Coalition Mine was hit by the first strike in its 30-year history when 400 employees walked out in a dispute over the mine's plan to force them to join its workingman's compensation association. Among other concerns, the strikers believed this move would undermine their union's support of the local Miners Hospital and absolve the mining company from any liability for on-the-job injuries. However, the protest was short lived. After 15 days a majority of the strikers voted to return to work, apparently without having won any major concessions from the company.[49]

By 1917 the United States had become involved in the Great War in Europe, which stripped the mines of many of its skilled workers and drove prices higher, both for the mines and the miners. Labor shortages and high costs collaborated to suppress the output of area mines in 1918 and set the stage for a period of serious labor unrest shortly after the war.

Early in 1919, in an effort to control costs, the mines cut wages by 75 cents a day, a significant amount when the average wage was about five dollars. Sam Raddon, the editor of the *Park Record*, saw trouble coming. "War prices for food stuffs and pre-war wages is a combination that causes unrest and much dissatisfaction," he wrote.[50]

In May 1919 miners and mill workers walked off the job in Park City, forcing the first complete shutdown of the local mines in 50 years.[51] The *Salt Lake Tribune* reported that the strike involved 800 or 900 men. Among other demands, they wanted a six-hour workday and a pay raise from $4.50 to $5.50 per day. "Every mine and mill in Park City District is idle, and one property, the famous old Ontario, that since its dewatering more than four years ago, has produced many thousands of pounds of ore every week, and given employment to scores of men, has been put out of commission perhaps for all time, because of the strikers edict to call out the pumpmen and the consequent flooding of the expensive lower workings of the property."[52]

The mining companies refused to budge, blaming the strike on outside agitators—in particular, the Industrial Workers of the World (IWW), an offspring of the Western Federation of Miners, which one author described as the "wrecking crew" of the labor movement.[53] IWW members, known as Wobblies, were said to sympathize with communist causes and advocate the overthrow of capitalism. "No doubt exists in the minds of outside observers but that the strike was brought about by I.W.W. agitators and their kinsmen, the Bolshevick, a rotten gang whose presence should not be tolerated for a moment in any American gang or city," said the *Salt Lake Mining Review*. Many local residents, including Sam Raddon, shared that dislike of "those flag-hating anarchists."[54]

The mine owners infiltrated union meetings and carefully monitored the debate between IWW members and less radical miners, many of whom were married and could ill afford to stay out of work for long.[55] After about six weeks, the strike collapsed. In late June the miners voted about two to one to return to work, and the mines gradually resumed operations.[56] However, the strike had forced many miners to look for work elsewhere, aggravating the local labor shortage. In July, in an effort to attract qualified miners and mill workers, the Park City mines raised wages by 75 cents a day (ironically, the amount they had earlier cut).[57]

Shipping and Water Problems

The water problems that plagued the mines in Ontario and Empire Canyons also frustrated the miners on the west side of the Park City Mining District, including the employees of Silver King Consolidated. In the spring of 1916, Solon Spiro, the company's president and general manager, launched construction of an ambitious tunnel designed to drain the major portion of the company's property and, it was hoped, provide access to new ore bodies. At the same time, the company started converting the old Grasselli mill into a new concentrating plant and building a 10,000-foot-long aerial tramway to move ore from the mine to the mill. By October the tramway was running, and the tunnel had penetrated 700 feet into the mountain. "The company [tramway] saves 75 per cent of the original cost of shipping by wagon, and more than 85 per cent on the cost of upfreight, besides the inestimable value of being able to keep its transportation system open both winter and summer," said the *Salt Lake Mining Review* in January 1917.[58]

Although the tramway was a success, Spiro's drain tunnel was a disappointment. By January 1922 it had reached a depth of more than 15,000 feet without passing through substantial ore deposits. It did drain the company's largely undeveloped properties, but it also drained the pockets of Spiro and the company's shareholders.[59]

In the summer of 1917, the *Park Record* announced the incorporation of a new company, the Park Utah Mining Company. One of its goals was to develop property on the east side of the Park City Mining District. The company acquired several hundred acres of existing claims and worked out agreements to gain access to some of them using tunnels owned by the Ontario and Daly mining companies. The Park Utah immediately began work inside the Ontario Mine's Keetley Drain Tunnel. The managing director of the new company was George W. Lambourne, who was already the general manager of Judge Mining and Smelting.[60]

In 1918 Lambourne added another title, general manager of the Daly-West Mining Company, when a group of dissident stockholders installed a new board of directors and slate of officers, all of whom had ties to Judge Mining and Smelting. Although he shunned the limelight and left his name on no public buildings, Lambourne became almost as familiar in Park City mining circles as David Keith, Thomas Kearns, and John Judge at the turn of the century.

A Salt Lake City native, Lambourne had left his first job with Zion's Cooperative Mercantile Institute (ZCMI) to take a job as secretary of Daly-West Mining in Park City. After the retirement of John J. Daly, Lambourne became manager of the Daly-West. "A retiring, unassuming man, Mr. Lambourne was virtually unknown in public life, and yet in the past four decades he held a commanding position in the western mining industry," the *Salt Lake Tribune* said in a front-page story at the time of his death in August 1935. "He and his associates pioneered the eastern Park City mining district and from small initial holdings developed the Park Utah Consolidated Mines company, which has paid more than $40,000,000 in dividends."[61]

In January 1921 the Silver King Coalition's concentrator was destroyed by a fire believed to have started in the boiler room. The imposing wooden structure on Treasure Hill west of town had been in almost continuous operation since the turn of the century. With no mill to process the lower-grade ores, the company wasted no time in planning a replacement. By April the board of directors had approved plans for a new $200,000 steel structure using part of the foundation of the old mill. By December the new plant was almost complete.[62]

In January 1922 the *Salt Lake Mining Review* revealed the first details of an impending "gigantic merger of big mines" in the Park City Mining District. According to the writer, L. E. Camomile (a former editor of the *Park Record*), the merger would involve the Park Utah, Daly, Little Bell, Daly-West, and Judge Mines.[63]

As it turned out, the merger took several years to consummate. The first step was the formation of the Park City Mining and Smelting Company in February 1922 with George Lambourne as president and general manager. Through an exchange of shares, the holdings of Daly-West Mining were conveyed to the new company, and a letter was sent to the stockholders of Judge Mining and Smelting inviting them to join the new enterprise. "The benefits to be derived by the stockholders of the Judge Mining & Smelting Company and by the stockholders of the other companies concerned, in the judgment of your directors, would be numerous and should result in a greatly enhanced value to their holdings," said the letter.[64] By the end of 1922, the Judge properties were also operating under the umbrella of Park City Mining and Smelting.

Meanwhile, there were glowing reports of a "sensational bonanza" at the Park Utah, which was continuing to develop its holdings through the Keetley Drain Tunnel and had built a camp at its mouth.[65] The mine's prospects looked so promising that in 1923 the Union Pacific Railroad built a six-mile spur to the mouth of the tunnel.

Thanks to the emergence of the Park Utah and renewed development at the Silver King Coalition, the Ontario, and the Park City Mining and Smelting properties, production from the Park City Mining District doubled between 1921 and 1922. "Today the old camp is rolling out a larger tonnage of ore than at any previous time in its history," the *Salt Lake Mining Review* reported on 15 January 1923.[66]

The process of consolidating the ownership of Park City's major mines continued in 1923 when George Lambourne negotiated the purchase of the venerable Ontario, whose drain tunnel he had been using to reach the properties owned by the Park Utah. "Through the purchase of the Ontario the last gap in the chain of mines extending from the portal of the Ontario drain tunnel on the east, to Brighton, in Big Cottonwood canyon on the west—a distance of about ten miles—has been closed and Mr. Lambourne's dream of years has become a reality," said the *Salt Lake Mining Review* on 30 December 1923.[67]

In the summer of 1925, shareholders approved the merger of Park City Mining and Smelting and Park Utah Mining. The new company, the Park Utah Consolidated Mines Company, controlled about 4,300 contiguous acres of mining claims. Lambourne's largest remaining competitor, the Silver King Coalition, made a move of its own in May 1924, acquiring its struggling neighbor, the Silver King Consolidated.

The 1920s were a time of prosperity for the major Park City mines. By 1925 the Park Utah was the largest silver-producing company in Utah and in 1926 was Utah's largest producer of silver, gold, and zinc and the second largest of lead. At that time Utah was leading the country in silver production and was second in producing lead.[68]

In 1928 a group of investors acquired about 600 acres of largely undeveloped mining claims at the head of Deer Valley, less than a mile east of Park City, and formed the Park City Consolidated Mines Company. Mining began in October, and the following March the *Park Record* reported that the mine had made one of the richest strikes in recent history. Investors scrambled to get a piece of the action, sending the price of a share of stock from $1.55 to $3.25 in a single day. In the summer of 1930, the Union Pacific and Denver and Rio Grande Western Railroads joined forces to build a one-mile spur to the new mine. In August more than 1,000 people celebrated the completion of the spur, including Utah Governor George H. Dern, who also happened to be the president of the mining company.[69]

The Great Depression

However, by this time there were signs that the giddy prosperity of the 1920s was losing its momentum. The price of silver, which stood at about 65 cents an ounce in 1925, had slipped to less than 53 cents by 1929. Then on 8 January 1930, it plummeted to 44 cents an ounce. By the end of January 1930, the Park Utah, citing an oversupply of silver, lead, and zinc, had reduced its workforce by about 60 percent in its mines and was operating its mill only one shift a day.[70]

As the Depression swept the country and demand for metals continued to drop, the news got worse. In February 1931 the price of silver dropped below 26 cents an ounce, the lowest level on record. Zinc prices also reached historic lows, and lead prices were the lowest in 18 years. In May 1931 underground metal mines statewide cut wages by 25 cents a day.[71] In response to the depressed prices, the Park Utah scaled back production even further and in the spring of 1932 suspended all operations except basic maintenance. The Park City Consolidated closed in June 1931, less than a year after the completion of the railroad spur, reopened in the spring of 1932, then closed again at the end of the year after silver prices dropped to 24¼ cents an ounce.[72]

By January 1933 only one mine, the Silver King Coalition, was reporting ore shipments from the Park City Mining District. Unlike their local counterparts, the mine management had decided it could stay in business by temporarily abandoning the production of lower-grade ore and reducing expenses wherever possible. Through such efforts, the mine managed to keep about 500 men steadily employed.

"But to do so, they had to ask the people to take a pretty heavy reduction in wages," recalled James Ivers, who was then working in the Silver King Coalition company store. "And, as they [mine managers] explained it, 'If we close down, you've got no choice for a job but to move on. If we can stay open, even though the wages are minimal, you at least have a choice. You can move on, or you can stay and take the minimal wages.'" Ivers said the decision to sell the high-grade ore at depressed prices ultimately cost the company millions of dollars when prices started to recover a few years later. "They mined out an awful lot of good ore during that period of time. . . . But that ore was taken out expressly to keep the mine vital and to keep the people there that needed jobs."[73]

In spite of the Silver King's best efforts, the closing of the other local mines sent the town reeling. The *Park Record*'s delinquent tax list, published in December 1932, was, according to the newspaper, "the longest, by far, ever published in Summit county." In June 1933 the newspaper announced the downgrading of the local post office under the headline, "Park City Slips to Third Class." In November 1933 244 unemployed Park City workers went on the payroll of the new federal Civil Works Administration.[74]

By this time the federal government had also begun taking steps to support the price of silver. And in December 1933, President Franklin Roosevelt announced that the government would buy newly mined silver at a price of 64½ cents an ounce. That news sent silver-mining stocks soaring, prompted the Park City Consolidated to resume operations, and induced the Silver King Coalition to increase wages twice in 1934, in January and again in September.[75] By October 1934 silver was selling for more than 53 cents an ounce, its highest level since 1929. However, the Park Utah properties stood pat. "The policy of 'no production' during the period of low metal prices has been continued throughout the year," the company told its shareholders at the end of 1934. "Your management is convinced of the wisdom of this policy and of the ultimate profit to the stockholders of conserving their ore resources until consumption by industry effects a recovery in the market price of the base metals produced by your company."[76]

Even another increase in federal supports in the spring of 1935, which drove silver prices up to 81 cents an ounce, wasn't enough to change the minds of the Park Utah's management. "Are they waiting for it to reach $1.29 before starting up, or, have the officers sold us old stockholders out in the interest of newer mines in which they are interested?" one shareholder wondered in a letter to the *Park Record*.[77] The explanation, according to Charles Moore, the owner of another Park City mine, was simple: "The public, generally, thinks that 70 per cent to 80 per cent of all mining profits are made from the production of silver and gold; and as a matter of fact, 70 per cent or 80 per cent of all the mining profits are made from the mining of zinc, lead and copper."[78]

Typically the deeper the mines went, the lower the grade of silver they found, and the more they relied on other metals to make a profit. The 1934 annual report for the Silver King Coalition shows that lead and zinc together contributed more than 55 percent of total receipts from ore sales, and gold another 3 percent, while silver totaled about 41.5 percent. By 1942 the combined value of lead and zinc had climbed to 67 percent of total receipts, while silver had slipped below 30 percent. And by 1947 lead and zinc together accounted for 74 percent of total receipts, while silver was below 23 percent.[79] After World War II, the local mines paid much more attention to the price of lead and zinc than silver.

By the end of 1935, higher demand for lead and zinc, combined with higher silver prices, were enough to induce Park Utah management to resume production. "Events of the past four years have proved the wisdom of those in charge of the company's affairs in their policy of not wasting its capital resources by producing and marketing its products during a period when the metals could not be sold except at a great sacrifice," the company told its shareholders in March 1936.[80]

In the spring of 1936, the Silver King Coalition announced plans to sink a new shaft east of the old California-Comstock shaft in Thaynes Canyon. At the 1,800-foot level, it would connect with the western end of the Spiro Tunnel. The shaft reached the Spiro Tunnel in May 1939, helping ventilate some of the old workings and providing access to new ore bodies.

As the mines began to recover, the miners, now being represented by the International Union of Mine, Mill and Smelter Workers, began to agitate for better working conditions and a greater share of the profits. Union representatives asked for a 50-cent-per-day increase in pay and rigid enforcement of the "collar-to-collar" law that defined an eight-hour shift as beginning when a miner reported to work at the top of the shaft or the mouth of the tunnel. When they couldn't reach an agreement with mine management, they called a statewide strike. At midnight on 9 October 1936, about 2,500 mine and smelter workers from Park City, Bingham, Eureka, Tooele, and Lark walked off the job. In Park City management prepared for a long strike.[81]

Workers in Eureka and Tooele returned to work about a month later after management offered them an increase of 25 cents a day. But others, including those from Park City's major mines, stood firm. In early December there were charges that a professional strikebreaker was recruiting nonunion workers in the nearby farming communities of Kamas and Heber. And when Park City mine operators announced that they would start accepting applications from new workers, more than 300 union members and sympathizers paraded through the business district carrying signs declaring, "50 cents a day or nothing," "We don't want scabs," "Let's stand together," and other slogans.[82]

Watching these developments with more than casual interest was Ephraim Adamson. During the last major labor disruption in 1919, Adamson had been identified by management moles as a strong union man. But this time he had a different role: he was sheriff of Summit County. When a confrontation seemed inevitable, Adamson met with Utah Governor Henry H. Blood to discuss law-enforcement strategy. The showdown came about 2:00 PM on Saturday, 12 December, when an automobile caravan approached Park City with about 125 nonunion men from Heber City. Sheriff Adamson and his deputies met the caravan and checked for weapons before allowing the men to proceed up Park Avenue toward the bottom of Main Street. Waiting at the corner of Main Street and Heber Avenue were an estimated 400 union men. Unable to pass, the caravan stopped. Among the onlookers was a reporter for the *Salt Lake Tribune*.

> Immediately the mass of pickets closed in from both sides. Nonunion men were snatched from the machines and the two first cars were overturned by the pickets. Other caravan members alighted from the cars, to be met with the flying fists of the union men, and to suffer bruised and bleeding heads and faces.
>
> The battle raged in the street for 10 or 15 minutes, then apparently ceased of its own volition. By this time, Sheriff Adamson had arrived with his deputies, but attempts to stop the struggle were futile. The pitched battle renewed and women, lined along the outskirts, shouted encouragement to the fighting union men, imprecations at the Wasatch county contingent.
>
> Minor encounters were noted along the fringes of the pitched battle, which centered around the caravan. After 40 minutes of fighting, the volunteer workers were beaten and staggered down the hill.[83]

According to news reports, about 40 men were injured in the fighting.

Fearing another showdown the following day, Adamson deputized about 100 local men and gave them nightsticks and orders to maintain peace under any circumstances. However, there were no other attempts to defy the union pickets.

Meanwhile, union negotiators and mine operators met with Governor Blood in an attempt to break the impasse. And on Tuesday, 15 December, a new proposal was presented to union members. It included the same 25-cent-a-day raise that they had turned down in November and a promise that the mines would consider basing the future wage scale on the price of metals or some other recognized index. And it also offered to give preference in reemployment to the men who had been working at the time the strike began. The miners voted to accept the proposal, and work resumed the following day.[84]

However, relations between miners and management were never quite the same, according to James Ivers. One of the casualties of the dispute was the company store where he used to work. "When the union came along, though, it was a *company* store, so they insisted it be closed down," Ivers said. "The Silver King also had a program at that time to better the living conditions, and they were just getting ready to implement a housing project on the lower end of Park City. And the union said, 'You can't have that. That's company housing. We won't go for it.'"[85] In March 1937 union miners voted to approve a sliding scale where wages were tied to the price of silver, lead, and copper. By the winter of 1937–38, clouds were once again looming on the horizon in the form of low lead and zinc prices. By March 1938 prices had dropped so low that the Park Utah had to restrict production and run its mines only one shift per day.[86]

In April the news got worse. The Silver King Coalition, which, with the exception of the strike, had run continuously through the depths of the Depression, announced that it was closing for an indefinite period. A week later the Park Utah followed suit. The two closures put almost 1,000 men out of work.[87] Of Park City's three major mines, only the Park City Consolidated kept producing ore. The closures lasted about a year. The two mines reopened in the spring of 1939, bolstered by higher prices for lead and zinc and new federal price guarantees for silver.

By the late 1930s, another company, the New Park Mining Company, had begun making ore shipments from the east side of the Park City Mining District. Organized in May 1932, the New Park was a consolidation of several old companies, including the Park Galena Mining Company, the Star of Utah Mining Company and the Mayflower Mines Corporation. To develop new ore bodies, the New Park began extending the Mayflower tunnel toward the Park Galena. Like some of the other mines on the east side of the district, the Mayflower tunnel started to uncover ore with an unusually high gold content. "The gold content has risen until now its value is greater than that of the silver," the New Park told its shareholders in the spring of 1940. In the summer of 1941, the Union Pacific Railroad extended service, building a two-mile spur from the line already serving the Park Utah's Ontario Tunnel to the north. By the early 1940s, production from the New Park rivaled that of the Park Utah and the Silver King Coalition. By the end of World War II, the New Park was regularly shipping more ore than all the other major local mines combined.[88]

World War II created an unusual dilemma for Park City mines. While the war opened up new markets for their metals, the War Production Administration and other government agencies set the price they could command. War industries and the armed services also took away many experienced miners, creating a serious labor shortage. "At one period employment was down to 55 percent of maximum employment," the Silver King Coalition told its shareholders in March 1943. In the fall of 1943, soldiers were assigned to work at the Silver King Coalition, the Park Utah, and the New Park. Nevertheless, by the end of 1944, Utah lead production had dropped to 65 percent of prewar levels.[89]

World War II and the Late Twentieth Century

While some segments of the Utah economy flourished in the years following World War II, metal mining was not one of them. According to one report, the labor shortage was more acute at the end of 1946 than at any time during the war. Conditions in the mines and mining camps apparently were driving away would-be miners. It took another year before the Park City mines began to report an adequate supply of workers.[90]

In April 1948 five Utah mining companies, including all of Park City's major mines, announced they were terminating contracts with the International Union of Mine, Mill and Smelter Workers because the union had not filed affidavits stating its officers were not members of the Communist Party. Two months later the 700-member Park City Local Number 99 voted to withdraw from the union and negotiate its own contracts. According to the *Park Record,* both owners and miners remarked on the "unusual harmony" during subsequent negotiations for a new contract. In April 1949 the Park City local voted to join the Progressive Metalworkers' Council.[91]

If, with the benefit of hindsight, you picked a year when Park City began its steady slide from prosperous mining camp to potential ghost town, you might well choose 1949. Certainly the town struggled in the 1930s, but so did the rest of the country. However, in 1949 when the rest of the country was feeling the euphoria of a postwar economic boom, Park City started to slip behind.

The first hint came in early March, when prices for lead and zinc started to drop in national markets. With lead and zinc ores now generating 75 to 80 percent of the income at the Silver King Coalition and Park Utah, it was an ominous sign. By the third week of June, both mines had started to curtail production, putting about 160 men out of work. By the end of the month, the Silver King Coalition, Park Utah, and New Park had suspended operations entirely, bringing total layoffs at area mines to 875. Many of these men would never collect another paycheck from Park City mines.

Mine spokesmen blamed the layoffs on the disparity between wages and metal prices. "Employees, because of declining values for lead and zinc, were asked to take a $2.50 a day basic wage cut," the *Salt Lake Tribune* reported. "They declined to accept the proposal, contending living costs would not permit pay cuts."[92]

By September metal prices had recovered enough to convince management at the New Park to resume production, but the Silver King and Park Utah retained only a skeleton crew for maintenance work. It was another year before both the Silver King and Park Utah were producing ore again, and even then with only a fraction of their former workforce. The Silver King, which had employed about 400 people in 1949, hired back fewer than 200.[93]

Production at the Park Utah and Silver King was interrupted in the summer of 1951 by labor troubles. And in the summer of 1952, a wage dispute and low metal prices combined to close both mines yet again. Mining officials blamed the low prices on unfair competition from foreign imports. Though local residents had no way of knowing it, after August 1952 the Silver King aerial tramway would never deliver another bucket of ore to the turn-of-the-century tipple on Park Avenue.

That December, according to the Employment Security Office, employment in Park City reached its lowest ebb in eight years. "However, dropping employment was not reflected in increasing unemployment due to the fact that many people were leaving the area as fast as jobs were terminated," said a story in the *Summit County Bee*.[94]

The Park Utah and Silver King were still closed in May 1953 when shareholders of the two companies approved a proposal to merge them into a new corporation, United Park City Mines Company. Combining the two properties gave United Park 8,700 acres of contiguous mining claims. "United Park City Mines Co. also has another asset, the good will and the buoyant hopes of the former employees of both Park Utah and Silver King Coalition," said the *Salt Lake Tribune*.[95]

Unfortunately, neither the merger nor the buoyant hopes of former employees could overcome the impact of depressed metal prices. As 1953 became 1954, the properties of United Park City Mines stayed closed. Only the New Park managed to keep producing ore.

In March 1954, with Park City showing no signs of righting itself, the Summit County Commission appealed for help in an open letter to Utah's congressional delegation. The letter painted a grim picture of life in the old mining camp. Since the 1941–42 school year, it said, the enrollment in the local schools had dropped from 1,052 to 449. More than 200 properties had been sold to the county for taxes in 1953, and another 48 were due to be sold in 1954. "Many of these properties are homes on which, because of the closing of the mines and no income, the owners have been unable to pay taxes. Many of these homes are vacant and rapidly going to ruin. Others are being lived in but the occupants pay no rent." The commissioners blamed the situation on the national government's willingness to "permit the influx of low cost foreign metals practically duty free." They called for a revamping of the tariff structure to protect domestic industries "rather than to enrich foreign producers."[96]

That refrain would be repeated many times in the 1950s and early 1960s. Every president from Harry Truman to Lyndon Johnson was asked to support the lead-zinc industry, either by imposing tariffs on imports, supporting domestic production, or both. But nothing they did could satisfy their critics or turn the domestic industry around.

United Park City Mines resumed production on a limited basis in September 1954. About 200 men were put to work at the Keetley unit, producing ore through the mouth of the Ontario Number-Two Drain Tunnel. This meant that the two largest surviving mining companies in the Park City District, United Park and New Park, were shipping most of their production from the east side of the district in Wasatch County. It also meant that the mine portals were just as accessible to workers from the nearby towns of Kamas and Heber City as those from Park City. And, in fact, the mines were attracting a growing proportion of their workers from nearby communities in Summit and Wasatch Counties. According to a 1956 newspaper story, about two-thirds of the New Park's 290-man workforce lived in Wasatch County and the Kamas Valley.[97]

After suffering several consecutive years of losses in the late 1950s, officials of United Park City Mines began to look for creative ways to pump money into their mining operation. The new bonanza, they decided, was recreation. What if they could turn their spectacular mountain property into a year-round playground for skiers, golfers, and horseback riders? Encouraged by the results of a feasibility study, they applied for a low-interest loan from the federal Area Redevelopment Administration. In August 1962 they were notified that their loan application had been approved. Within days men with chain saws and bulldozers were at work on the mountains west of town, clearing trails and lift lines for the new recreation area. On 21 December 1963, Park City Mayor William P. Sullivan cut a ribbon officially opening Treasure Mountains Resort. "Park City Dead? Not Yet, Podner," said a headline in the *Salt Lake Tribune* as the resort prepared to open.[98]

Meanwhile, the management of the New Park had signed a contract leasing its Mayflower Mine to the Hecla Mining Company of Wallace, Idaho. In the spring of 1962, Hecla began construction of a semiautomated $650,000 mill to process ore from the mine.[99] Thanks to the discovery of new ore deposits that yielded substantial quantities of gold and silver, Hecla managed to keep the Mayflower running for another decade, employing as many as 200 people. However, the ore eventually dwindled, and Hecla terminated its lease at the end of 1972. The mine's surface rights were sold as a potential recreational development in the spring of 1973.[100]

On the other side of the mountain, United Park City Mines was having problems with cash flow at its new recreation area. As the 1960s came to a close, it still hadn't made money. "Salt Lake [skiers] took to Park City's ski resort great," said Jim Ivers, who was named president of United Park City Mines in 1965. "So on Saturdays the hill would be quite crowded. Sundays, it would drop off to about half that. And the other five days there was nobody up here."[101]

To bring midweek skiers to the mountain, Ivers said, Park City needed to become a destination resort. Unfortunately, the mining company was unable to develop lodging on its own. Under the terms of the 1962 financing package used to create the resort, 75 percent of the proceeds of any property sales had to go toward paying off the federal loan and two large investors.[102] So the mining company started looking for someone to buy the resort, someone who could develop the mountain and build overnight accommodations. That "someone" turned out to be Royal Street Land Company of New Orleans.

In a meeting in April 1970, the mining company's board of directors granted Royal Street an option to purchase the recreational facilities for about $6 million. Another significant piece of news came out of that same meeting: The board agreed to lease all its mining properties and facilities to a partnership composed of two large shareholders, Anaconda Company and American Smelting and Refining Company (ASARCO), which together owned about 31 percent of the mining company's stock. Shareholders approved both transactions at the company's annual meeting in July.[103]

During the next five years, the partnership, under the name Park City Ventures, spent almost $20 million on the project, upgrading the underground workings and building a new 750-ton-a-day mill at the site of the Ontario Number-Three Shaft. After the opening of the mill in the spring of 1975, the mine increased its workforce to about 350 people.

But the prosperity was short lived. On Friday, 13 January 1978, the mine announced it was suspending operations. Bill Norem, general manager of Park City Ventures, told the *Park Record* that severe problems with groundwater and soft rock had caused production costs to skyrocket. For the first time in more than a century, the Park City Mining District was completely shut down. Hardest hit by the closure were the communities of Kamas and Heber City, home to most of the 350 workers. Also reeling were the partners in Park City Ventures, who had lost more than $26 million on the project since 1970.[104]

A Canadian company, Noranda Mines Limited, took over the lease from Park City Ventures in August 1979. However, Noranda had no more success than its predecessor. In May 1981 it stopped producing ore and laid off 155 workers because of "insufficient production levels and inadequate metal prices." Two months later the abandoned Silver King Coalition tipple on Park Avenue burned in a spectacular fire. Exploration and development at the Ontario continued until April 1982, when Noranda terminated its lease with United Park City Mines.[105] After a spectacular run of more than 110 years, mining had apparently run its course in the Park City Mining District.

Park City Today

In December 1995 the Ontario Mine opened to tourists, with interpretive displays on the surface and guided tours underground. Although the exhibits attracted almost 500,000 visitors in four years, company officials announced their closure in November 1999, citing poor attendance during the winter months.[106]

Today a skeleton crew still performs routine maintenance in the mines, particularly in the Judge and Spiro Drain Tunnels that provide drinking water for the residents of Park City. In town there are few clues to the industry that was its lifeblood for so many years. Skiing, not mining, now drives Park City's economy. United Park City Mines is now working to cover or remove many waste dumps and tailing piles so the land can be developed for luxury homes. But venture into the mountains west of town, and you can still find reminders of the industry that brought so much wealth: the timber skeleton of the California Comstock, the tattered hulk of the Silver King mill, the rusting steel towers that once carried ore to the tipple on Park Avenue. For that matter, just walk down Main Street in Salt Lake City and look for the names of Park City's mining magnates on the stately old buildings.

14

Tintic Mining District

Philip F. Notarianni

The Tintic Mining District lies on the western and eastern slopes of the central portion of the East Tintic Mountains, which includes portions of Juab and Utah Counties. They are aligned with the Oquirrh Mountains to the north and merge on the south with the Canyon Range and the Gilson Mountains. The East Tintic range is bordered on the west by the Tintic and Rush Valleys, and on the east by Dog Valley, Goshen Valley, and Cedar Valley.[1]

In the Tintic area, gold, silver, lead, and copper were the primary minerals. Zonal descriptions of mineralization best facilitate an understanding of geographical location. The Centennial-Eureka Ore Run includes mines in the town of Eureka proper, namely, the Centennial-Eureka (Blue Rock), Eureka Hill, Bullion Beck, and Gemini. The Mammoth-Chief Ore Run embodies the Victoria, Eagle and Blue Bell, and Chief Consolidated Mines, all in Eureka's immediate vicinity. The Plutus, Godiva, and Iron Blossom Ore Runs are all in East Tintic and round out the five groups.[2]

Early Development in the District

Native Utes, under the leadership of Chief Tintic, for whom the area is named, roamed this region. White intrusions in the 1850s led to armed resistance by the Native inhabitants, primarily resisting the use of their land by cattlemen.[3] In the 1860s a group of Mormon cowboys found a piece of *float*, silver ore brought to the surface, in the area. Illuminated by beams from the sun, the outcrop led to naming the claim "the Sunbeam." Shortly after this discovery, those involved formally organized the Tintic Mining District on 13 December 1869.[4] An influx of prospectors and miners followed the Sunbeam discovery. In the wake of this activity, new properties were located, and work commenced.

Tintic's development progressed like other metal-mining regions of the West. Once ore had been discovered, prospectors funneled in to work the surface operations. When these early outcrops were depleted of high-grade ore, individual effort gave

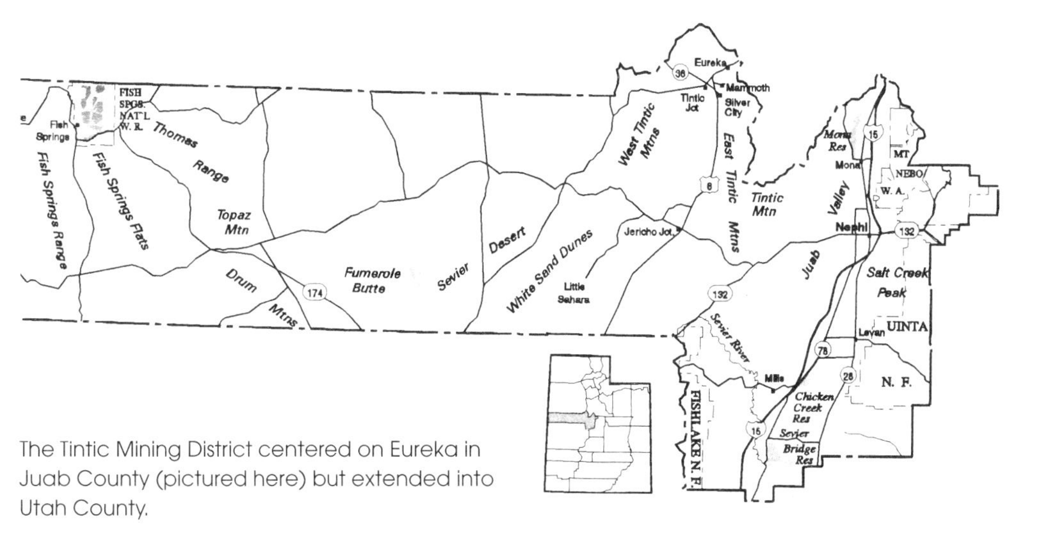

The Tintic Mining District centered on Eureka in Juab County (pictured here) but extended into Utah County.

way to corporate interests which had the large outlays of capital necessary to mine the depths.[5] Thus, in the early 1870s, Eureka's main mines were staked, and by 1899 the Eureka Hill, Bullion Beck and Champion, Centennial-Eureka, and Gemini became known as the "Big Four" and ensured the area's growth.[6]

Studies of metal-mining areas in the West describe growth as progressing from camp, to settlement, to town.[7] Key factors in these transitions were sustained growth, permanence, and the creation of a regional or local economic center. The settlement process in Tintic followed a specific sequence. First, the area of igneous rock in the southern portion of Tintic, near the towns of Diamond and Silver City, southwest of Eureka and near the Sunbeam, was staked. Second, miners flowed into the area north of that, Mammoth Basin, also southwest of Eureka. Third, farther to the north, separated from Mammoth Basin by a high limestone ridge, lay Eureka Gulch, the site in 1870 of tents pitched by teamsters who hauled ore from the newly staked Eureka Hill Mine.[8] Areas such as the city of Eureka entered the camp phase in 1870 and, as the Big Four progressed, transportation and smelting provided the major impetus for fostering the second stage of development: settlement.

However, due to poor transportation facilities, the Tintic mines did not develop rapidly. Rich ores were shipped to San Francisco, California; Reno, Nevada; Baltimore, Maryland; and Swansea, Wales.[9] Later, most of the Tintic ores were sent to Argo and Pueblo, Colorado, and the smelters in the Salt Lake valley.[10] Lower-grade ores were treated at early area mills and smelters, but with indifferent success. Ores mined from the immediate ground surface were handled somewhat better in amalgamation plants. Amalgamation was the process where gold and silver were extracted from pulverized ores by producing an amalgam, a union with mercury, from which the mercury was then expelled. Later, treatment of ores by roasting or chloridizing proved more successful.[11]

The Impact of the Railroads

Building railroads in the Tintic District increased both the extent of mining and productivity. Railroad development had a great impact upon the output of Tintic since the district's key product was first-class shipping ore. Production increased as railroad facilities improved. Of course, production also depended upon mineral prices and the general economy. The first attempt at railroad construction into the district was a joint venture by a group of Mormon and Gentile businessmen who chartered the Lehi and Tintic Railroad Company on 28 October 1872. Proposed to be routed west about 50 miles into Tintic Valley from a junction with the Utah Southern Railroad at Lehi, the venture ultimately failed in the financially disastrous years of 1872–73.[12] In 1878 the Utah Southern extended its operation into Ironton, approximately five miles southwest of Eureka. The result was a marked increase in output for 1879, nearly doubling the previous year's production.[13]

On 30 May 1881, the Union Pacific Railroad incorporated the Salt Lake and Western Railway Company and among its goals was the construction of a line into the Tintic Valley. The route as planned would follow the old Pony Express Trail through Fairfield and Camp Floyd, south to Vernon, then on to Ironton, and farther south before turning west to Cherry Creek, and finally on to Nevada. An amendment to the papers of incorporation designated a branch line to be built from Ironton to Eureka and Silver City. Work commenced in 1881, and the main line to the town of Tintic, south of the main district, was finished in 1882; the branch line from Ironton to Silver City was completed by the end of July 1882.[14]

Branch lines were extended into Mammoth shortly after the Silver City spur and reached Eureka in 1889. In 1890 the Salt Lake and Western consolidated with the Utah Nevada, Utah and Northern, and Echo and Park City lines under the general management of the Oregon Short Line, a subsidiary operation of Union Pacific. In July 1903 the San Pedro, Los Angeles & Salt Lake Railroad, known later as the Los Angeles and Salt Lake, assumed control of the Oregon Short Line branch from Salt Lake to Eureka as part of its through line to Los Angeles.[15]

In 1891 the Denver and Rio Grande Western Railroad (D&RGW) entered the Tintic area with a branch line from Springville through Santaquin and Goshen to Eureka and on around to Mammoth and Silver City. The Tintic Range Railway Company, which built the line, was the D&RGW's construction subsidiary. An agreement signed in 1892 allowed the D&RGW to operate the line. The entrance of the railroad was heralded as one of the most important events of the year since it gave Tintic the advantage of two competing lines.[16]

Development of Major Companies with Mills and Smelters

Even early in the development of Tintic, many kinds of ores were mined, including principally the carbonate and sulfide of copper and siliceous gold and silver ores containing small amounts of copper. These types of ore produced numerous complications in their sale and treatment, however.

The first stage of mill and smelter development occurred prior to the 1890s. A small mill and smelter were erected almost simultaneously at Homansville, approximately one mile east of Eureka. The smelter was constructed by the Utah Smelting and Milling Company and was called the Clarkson. Operations began on 17 June 1871 with a reported 172 tons of silver-lead bullion produced in 60 days. The ores smelted in the Homansville furnaces came from the Scotia Mine, the Swansea, near Silver City, and the Eureka Hill. The plant was closed and moved in 1872. A smelter was erected at Diamond in 1871, and two furnaces were built at Goshen, east of Eureka, in the fall of 1874.[17] Likewise, smelting operations began at Mammoth, but, typical of many such operations, they were ephemeral in nature.[18]

Milling occurred simultaneously with the smelting efforts. Again, a mill was built at Homansville for treating Eureka Hill ore, utilizing the amalgamation process. This plant was equipped with a crusher and 12 revolving stamps. It was ultimately moved to a site eight miles south of the Mammoth Mine.[19] A second mill erected at Homansville was the Wyoming, built in 1873 to process ore from the old Wyoming Mine immediately south of Eureka. Other mills included the Tintic Company plant near Diamond; the Shoebridge; the Ely mill; Mammoth-Copperopolis; and one near the Crismon Mammoth Mine.[20]

The years from 1869 to 1890 constitute the formative period of the Tintic District. Mining in the 1890s and later sought to strike below the oxidation zones. Above this zone, oxides were encountered, but below lay the sulfides. The result was the opening of new workings in old ground.[21] Prior to the 1890s and especially in the 1880s, the mining operations and camps in Tintic developed in a steady manner with the number of mines growing rapidly.[22] Camps began to acquire some of the features that characterized their later existence. By 1879 the town of Diamond could boast a shoemaker, blacksmith, butcher, physician, surveyor, assayer, and even a dressmaker.[23] As of 1880 the business establishments of Eureka included William Hatfield, general merchandise; Williams and Cusick, general merchandise, saloons, and billiards; and W. W. Mathews, saloon and billiards.[24]

In the 1880s litigation between the Bullion Beck and Champion and the Eureka Hill interests over property rights temporarily caused a snag in the area's development. During 1886 mining activity was practically at a standstill; however, both companies made valuable use of this pause in operation to remodel and do additional and valuable development work in their mines. At the Bullion Beck, the hoisting works were remodeled, and new machinery was installed, as well as the method of sinking and prospecting by drifts and winzes. Likewise, the Eureka Hill installed an entirely new hoisting engine and gallows (head frame), along with other machinery that allowed the company to mine as deeply as needed.[25] Finally, in June 1888 the two enterprises arrived at a compromise, and the *Salt Lake Tribune* heralded the event for injecting new life into Eureka, one sign of which was the increase in the number of residences and business establishments.[26]

Mining to the depths sparked a rebirth for Tintic in the 1890s. Mills were erected, and production increased, yet the vicissitudes of the mining industry, subject as it was to sudden economic downturns, caused the district to suffer agony as well as joy. Tintic's ability to rebound from adversity indeed forced it to come of age. Following the compromise between the Bullion Beck and Eureka Hill Mines, Tintic appeared again to be on the road to prosperity. In 1889 reports maintained that Eureka had experienced a large increase in population and the addition of numerous houses and several business establishments. Stimulus for this activity was attributed to a new railroad spur from near Mammoth Hollow to Eureka. The railroad gave "new life to mining about Eureka."[27] In 1889 Eureka's key mining enterprises also acquired water. The Bullion Beck and Centennial-Eureka mining companies purchased the lower springs at Homansville. Water lines and pumps were installed to transport this needed commodity to the mines.[28]

The Development of Towns

Ushering in a new decade, Tintic in 1890 made "greater progress" than ever before with a "rush to the district." That progress was evident in the marked improvements in Eureka. The number of houses rose above 300. Among the more substantial improvements at Eureka during 1890 were a two-story, stone Independent Order of Odd Fellows (IOOF) Hall, erected at a cost of $9,000; a two-story brick block built by McCrystal and Company for $10,000; the Beauman building additions to the Tomkins Eureka Hotel and the Hatfield House; improvements at the Meyer's Hotel (Keystone) and the Anna Marks store; erection of the Pat Shea boardinghouse; and the establishment of a COD store and the George Arthur Rice and Company Bank.[29]

Precipitating this upswing in the district were the successful operations of various mines, including the Bullion Beck, Godiva, Snowflake, and Iron Blossom. By 1892 Tintic was feeling prosperous with activity burgeoning at an incredible pace. Eureka continued to build, fast becoming the district's center. A total of 150 dwellings were added in 1891. They were "mostly of small size, but of much better and neater form than of former years."[30] Juab County erected its courthouse in Eureka in 1891 at a cost of $5,000. The Catholic Church built a school for $3,000, and benevolent societies flourished. In addition, the D&RGW's entrance into the town signaled progress.[31] New hoists and improved surface plants were erected in 1892 at the Centennial-Eureka and Gemini Mines.[32]

Mills and smelters sprung up around the landscape, but due to the fluctuations in types of metal mined, mills and smelters opened and closed as specific grades of ores were located and treatment methods worked out. Just as significant in explaining the sporadic nature of mineral processing were the battles waged by competing smelters and railroads over smelting and shipping rates. William Bredemeyer, a mining engineer and U.S. surveyor, called attention early to the "exorbitant charges made by the Tintic mills on the custom ores brought to them by the mines. Opposition here is sadly needed; it must and will come."[33] The *Salt Lake Mining Review* published numerous articles dealing with the "smelter trust," the need for better smelting facilities and the erection of an independent smelter, and problems with the railroads.[34] In a September 1900 article, the publication, obviously sympathetic to the mine operators, stated that smelting charges often totaled more than 50 percent of the ore value. Citing the Centennial-Eureka, the mining tabloid asserted that smelting charges for 20,000 tons of ore equaled approximately 40 percent of the value of the ore.[35]

Jesse Knight, the Mormon mining entrepreneur who plunged into the Tintic scene with the Humbug Mine in the mid-1890s, exemplified the tenacious owner bent upon breaking trusts by creating his own. The Knight smelter, narrow-gauge railroad (built in east Tintic to battle exorbitant rates during the 1907–8 years), power plant, Tintic Drain Tunnel, Knight-Dern mill, and Knight dry farm all represented his efforts at self-sufficiency in mining operations.[36] "Uncle Jesse," as he was affectionately called by most, had his town, Knightsville, about one mile east of Eureka, as well as coal interests near Helper in Spring Canyon. Knight discovered through basic desires or corporate need that if no one would supply certain services at the right price, he would just have to supply them himself.

Ethnic Populations and Organizations

Smelter and mill building affected more than the landscape and economics; owners had to bring in massive numbers of unskilled workers. In Tintic's case, these were southern and eastern Europeans whose presence outraged the predominantly northern European Eureka population. The influx of Austrians (i.e., Serbs and Croats) and Greeks intensified nativistic sentiments in the area.[37]

The timing factor proved most significant in peopling Tintic. Irish, Cornish, Welsh, English, and Scandinavian immigrants pressed into the cedar-covered hills after the initial discoveries. Irish railroad workers left the finished transcontinental line in 1869 and entered the mining field. The Chinese who had labored on the railroads also came to Tintic to run laundries and cook. The Chinese Exclusion Act of 1882 stopped the Chinese influx into the United States, so Chinese workers and merchants represented the early immigration period in Tintic but a minor part of the Chinese experience in Utah.[38] In 1892 Sing Chung, proprietor of Chinese and Japanese Goods, located on Eureka's Main Street, advertised, "Chinese Cooks and Laborers Furnished at Short Notice," thus functioning as a labor agent.[39]

Cornish, Welsh, German, Finn, and primarily Irish workers were the first to seek riches in Tintic. Many of these mining pioneers were veterans of mining or railroading ventures elsewhere in the West, Midwest, or East and had migrated to the newly opened district. Thus, the available immigrant workforce tended to be comprised of these northern Europeans and their American-born descendants. By the late 1890s, a shift in immigration sources occurred. Southern and eastern Europeans began leaving their countries in huge numbers in search of the American dream. The "distant magnet" pulled a variety of peasant peoples to the United States, which, in a period of rapid industrialization, needed unskilled labor.[40]

Subsequent census statistics for Juab County further illustrate the scarcity of southern and eastern European people in Tintic. Since the district attracted most of the unskilled laborers in the county from these groups, figures for the entire county are accurate measurements of alterations in ethnic populations. In 1910, of a total population in Juab County of 10,702, Eureka accounted for 3,829. The total foreign nationality count, both for native whites born of mixed parentage and foreign-born whites, was 4,064. Of this number, Austrians accounted for 72, Greeks, 22, and Italians, 98. By 1920 the foreign-born white population in Juab numbered 1,180, with 17 Austrians, 12 Yugoslavs, 24 Greeks, and 32 Italians. In 1930 those foreign born in the county totaled only 588, with 2,323 of foreign and mixed parentage. In this latter category were 17 Austrians, 11 Yugoslavs, 11 Greeks, and 19 Italians. The number of Mexicans was in the fifties during the 1910s and 1920s, but they resided for the most part in Dividend, on the Utah County side of the Tintic District.[41]

The breakdown shows a unified population in Eureka based upon national origins favoring Anglo peoples. Combined with the intensification of community identification demonstrated during the 1893 strike and after, this homogeneity helped create social boundaries within Eureka. Concern over even the meager number of Chinese blossomed in 1903, when a smallpox epidemic forced the cancellation of many local events but caused the *Eureka Reporter* to quip, "Why don't the local board of health stop the Chinks from celebrating Chinese New Year?"[42]

By 1912 the local scribe was crying out against "Austrians," primarily Serbs, Croats, and Slovenes. In asking, "Is Eureka to Be a Bohunk Town?" the paper stated: "Their presence in a mining town, a smelter town, or any other town for that matter, adds nothing to its dignity, its wealth or its importance."[43] Prior to that blast, the *Reporter*, in praising the local labor union, made a point to note that "four-fifths of its members are married men, citizens of character and responsibility, well read and intelligent men and sturdy pillars of the community. The majority of them are American born."[44]

Population increases and the rise of camps created demands for various institutions and social services. Father Lawrence Scanlan, whose mission embraced all of Utah, visited Tintic in 1873, bringing Catholicism to Eureka. By 1884 Eureka's Irish residents had asked for a resident priest, and in 1885 Reverend Denis Keiley erected St. Joseph's School, operated by the Sisters of the Holy Cross.[45] Members of the Church of Jesus Christ of Latter-day Saints gradually settled in Eureka, organizing a temporary branch in November 1883, with a permanent one following in 1884 at the Bullion Beck Mine. In 1902 the ward erected a Gothic-style frame church on upper Main Street in Eureka.[46] Methodism entered Eureka when Dr. Thomas C. Iliff visited and preached in the town in June 1890. He immediately secured $700 in subscriptions toward a church, followed by an appropriation of $525 from the Mission Conference of 1890. Reverend W. A. Hunt was appointed pastor and was succeeded by Dr. J. D. Gillilan, who finished building the church in the summer and fall of 1891. A publication on Methodism exhorted, "The church moves up or down with the price of silver."[47] Other denominations in Eureka included the Lutherans, Baptists, and Episcopalians.

Activities of fraternal and benevolent orders, organizations, and societies have always played a central role in the social life of the Tintic Mining District, especially Eureka. In the 1890s among the groups in existence were the Godiva Lodge Number 8, Knights of Pythias; Tintic Lodge Number 9, Free and Accepted Masons; Keystone Encampment Number 8, IOOF; Eureka Lodge Number 12, IOOF; Court Eureka Number 8503, Ancient Order of Foresters of America; Oquirrh Lodge Number 19, Ancient Order of United Workmen; and Columbia Lodge Number 2, Daughters of Rebekah.[48]

Development of Labor Unions

Prosperity appeared eternal in the years 1891 through 1893, but it proved fleeting as the economic downturns of 1893, coupled with labor strife, brought Tintic to its knees. The national Panic of 1893 and the ensuing depression brought about the repeal of the Sherman Silver Purchase Act of 1890, which had required the Federal Treasury to purchase four and one-half million ounces of silver per month. Accompanying this action, the silver market became highly unstable, and the price fell. In Tintic this coincided with difficulties over the company boardinghouse and store. The result was a labor strike destined to become the most severe in the district's history.

Declining lead and silver prices and an apparent victory of labor by the Coeur d'Alene Mine Owners Association spurred western mine owners to launch an attack on wages and unions in the winter of 1892–93.[49] The Bullion Beck of Eureka initiated the action in Tintic.

The mine closed in early February, and 200 men were thrown out of work. The company offered to pay $2.50 per day until silver reached 95 cents an ounce, then after one month the $3.00 a day wage would be restored. The Eureka Miners Union, established in the late 1880s,[50] countered by proposing a $2.75 per-day wage and $3.00 restored when silver was 90 cents an ounce. A compromise failed to materialize, and the strike followed with both sides adamantly holding their respective positions.[51]

Bullion Beck officials, predominantly Mormons, and the union proved to be the principals in the affair. The company was headed by Moses Thatcher, president; John Beck, vice president; William B. Preston, treasurer; with A. E. Hyde and George Q. Cannon completing the board of directors. Hyde was general manager and W. J. Beattie, secretary. The Eureka Miners Union was represented most intently by John Duggan, its secretary.[52] Late in February the Eureka miners, represented by the union, attempted to break the impasse by proposing a wage scale based on a fixed weight of silver with the average price for a month being a factor in determining wages per day—in effect, a sliding scale tied to silver prices. Hyde viewed this proposal as a signal that the union was about to give up the struggle and declined the offer, planning to utilize strikebreakers.[53]

The controversy over the company boardinghouse and store became entwined with the wage issue. In an open letter to the *Salt Lake Tribune*, the union maintained that near the end of March 1891, board at the house run by Hyde and W. H. Smith increased from $26 per month to $1 per day for food of the "worst kind." Grievances were presented and seemingly heeded by the company; however, Hyde and Smith were able to have the store bills of Bullion Beck employees collected through the company's office. The union further alleged that Hyde had discharged a number of men, none of whom appeared on the store's ledgers.[54] Thus, they were not shopping at the company store.

On 7 March the Bullion Beck started operating with about 40 men, mostly scab labor. Union men met the train at the mine and verbally attempted to dissuade those brought in from going to work. Apparently they were successful. Three days later a group of women marched to the Beck Mine in an unparalleled move to parade their discontent about the opening of the mine. John Duggan reported the event and stated that it reminded him "of an old-time Welsh wedding procession." About 40 ladies participated under the direction of Miss Annie Kelly, and in the words of Duggan, "as they reached the mine the car dumper was warmly saluted by having his ears pulled and told what he was thought of. . . . Up to lunch marched the earnest, exuberant band, and down again they came to once more serenade the grub-filled 'scabs' [replacement workers who were given a free lunch]."[55] While such unruly action was abhorred by management and "frowned upon" by the union, it nevertheless typified public support for the miners' position.

Tensions were mounting on both sides. In mid-March deputy United States marshals and special marshals were on duty in Tintic at the request of the Beck people. Cries about the pro-mine owner partiality of these marshals were heard. The situation was labeled "most serious," but nothing materialized. Dr. Charles W. Dark, acting mayor, asserted that city authorities could handle the matter. This pointed to a significant aspect of the affair. The Beck management sought to offset or neutralize local support for the miners by summoning federal marshals. Scrutiny of the Eureka criminal justice dockets for 1893 indicates that there was no unusually high arrest rate. For a mining town experiencing a strike, the number of arrests for violence was surprisingly low.[56]

A second point of contention was the role of the Mormon Church in escalating the tension. Reports alleging church interference on behalf of the Beck Mine were rampant. A Marshal Benton, in command of the federal group, supposedly sent a letter to "Brother Hyde" recommending one Hugh Roark for employment. In March Wilford Woodruff, LDS Church president, replied to a letter from the union protesting against Mormon bishops acting as employment agents. The letter stated that he had no knowledge of the matter; furthermore, if a call was indeed given to aid the needy, any action would be completely justified. On the other hand, Woodruff asserted, if a bishop merely acted as an agent, the action would be improper. Duggan replied, imploring the president to use his influence to stop church "meddling" in the labor problem.[57]

A church policy toward unions had developed during the 1880s with the entrance of the Knights of Labor, to which the Eureka miners then belonged. The union became a competitor with the church on several key levels. These were defined in the following way:

> The Knights of Labor, like The [*sic*] Church of Jesus Christ of Latter-day Saints, promoted its own cooperative movement, and it required prospective workers to pass through a secret initiation ritual. To Mormon leaders the cooperative ideas and secrecy of the Knights threatened to compromise the loyalty of their own members. In addition, church authorities faulted the violent methods used by the union to eliminate the Chinese and the radical pleas of some union representatives urging the overthrow of the capitalist system.[58]

Farmers from valley towns were entering the Beck Mine, adding credibility to charges of church intervention. Newspaper reports, primarily from the *Salt Lake Tribune*, told of the recruitment of farmers. One account charged that the bishop of Payson had instructions from church authorities to supply men for the mines.[59] Dennis Sullivan, Beck Mine foreman, was quoted as saying, "Farmers are farmers, and miners are miners, but I decline to say anything about it."[60] Dr. F. E. Bostwick, Eureka physician, in assessing the strike asserted that "it is not 'Manager Hyde' so much, now, but it is the whole Mormon church power that is behind this matter of the reduction of wages and its results."[61] Bostwick's charges provoked an instant reply from the church organ, the *Deseret News*. It blasted, "The hand of the Church is not now manipulating and has not at any time manipulated the affairs of the Bullion-Beck or any other mine."[62] The physician countered by emphasizing that the church owned the Beck Mine and statehood "would now be detrimental to Gentile labor and capital in Utah."[63] Whatever accusations were bantered about, the fact remained that the Beck Mine was operating primarily with farm labor from towns in the valley. Evidence pointed to Mormon influence, which was also consistent with early church policy.

Friction between strikers and scabs began to intensify. Fears of a general melee on St. Patrick's Day led to increased activity by the marshals, but no trouble occurred. However, on 31 March Hiram Hyde, the brother of A. E. Hyde and an employee of the Beck Mine, exchanged shots with A. Collins and Bat Sullivan, striking miners, in the lower part of Eureka. Injuries resulted but no deaths. Finally, on 5 June the houses of two nonunion miners were blown up by dynamite, just southwest of the Beck workings. Again no deaths resulted. Such affairs brought stern denunciations by Hyde and other mine owners, who sought protection from Governor Thomas of the Utah Territory, who declined the plea for assistance.[64] A grand jury was impaneled to investigate the strike situation, but not the dynamiting. Riot charges were issued for other incidents such as verbal attacks on working miners. Both men and women, predominantly Irish, were indicted, and all pled not guilty.[65]

The strike was defeated, but its significance lay in providing a catalyst for community identification. During the first week of April, a citizens' committee from Eureka, consisting of Reverend G. W. Comer, pastor of the Methodist Episcopal Church, Ben D. Luce, and George Hanson, began to gather public opinion about the strike controversy. Forty-six respondents were questioned, and their almost total pro-union stand illustrated that the miners enjoyed the support of professionals, merchants, and artisans.[66]

Most of the responses focused on the issues of wages, company board and room, and the peaceable conduct of the union. Local merchants and professionals recognized their reliance on the patronage of the workers. It was to the advantage of the entrepreneur for miners to be free of the company store. Even though the dynamiting incident occurred after the opinion poll, the peaceful orchestration of the strike by the union and its conservative approach set a standard for the expected behavior of the miners.[67]

Additionally, the Mormon issue emerged as important. After the strike, the Beck Mine management and business approach changed, becoming more compatible with community interests.[68] Mormons began entering the mines, but the chasm that might have been expected between Mormons and non-Mormons did not materialize. Ethnic homogeneity, identified community interests, and the presence of the well-regarded Mormon entrepreneur Jesse Knight explain this phenomenon. Knight had established a town inhabited primarily by Mormons only one mile east of Eureka; it was reputed to be the only saloon-free, prostitute-free, privately owned mining town in the country.

The Beck management change signaled that the company recognized existing problems involving officials, tactics, and interests. Melvin Dubofsky, in an article on western working-class radicalism, observed that originally "workers and local businessmen were not split into hostile camps." However, Dubofsky thought the modern corporation had driven a wedge between the "workers and their non-working class allies."[69] In fact, the Beck Mine had tried to do this but failed. The 1893 strike, contrary to Dubofsky's assertion, drew Eureka's working class together and allied it with the varied interests of the community.

The strike succeeded in eliciting expression of a common concern about community support and endorsement by business. Some merchants had climbed the ladder from labor to business still bearing the stamp of miner and were able to relate to low wages and poor working conditions. Abrupt class distinctions that would have been exacerbated by the presence of diverse cultural groups considered "foreign" and possibly produced class conflict did not exist in Eureka. The cultural uniformity proved of signal importance.[70]

The Bullion Beck made other changes during 1894 as a result of the strike. The mine erected a mill on the side hill north of the hoist, on the opposite side of the gulch. The plant rested on four excavated, largely stone steps and covered an area 110 feet long by 125 feet wide; the tower rose 105 feet above the lower section. As in other cases, an increased water supply accompanied the construction.[71]

Growth at the End of the Nineteenth Century

Renewed mining activity in the mid-1890s brought Tintic to the forefront of Utah's mining districts in production by 1899. The *Salt Lake Mining Review* branded Tintic as "among the leading mining sections of the intermountain region."[72] The *Eureka Reporter* asserted, "This big district, one of the greatest in area in the state—is carving its way toward becoming one of the richest and largest producers of the entire country."[73]

The *Salt Lake Tribune* announced that faith in the district had been greatly revived. A primary reason was the renewed interest in milling, which proved highly significant because it made possible the treatment of second-class or low-grade ores. Upon successful completion of the Mammoth mill in 1893, similar undertakings commenced at the Eureka Hill and Bullion Beck Mines. During 1894 the Eureka Hill, practically idle all year long, built a 100- to 120-ton-capacity mill with plans and machinery furnished by Frazer and Chalmers Company. An increased water supply also aided this endeavor.[74]

By 1897 Tintic's economy had begun to recover, exports and prices were rising, and the flow of gold continued to increase, spurring investment. Heated battles over the free and unlimited coinage of silver waned in the postelection year since William McKinley, running on a "gold platform," had won the presidential campaign in 1896 and eventually made gold the sole standard of currency. Tintic enlarged its gold output and reaped the benefits of more investments.

By 1895–96 the mill construction had sparked recovery from the strike in 1893, fire, and flood in 1896, and depression. In 1895 in Eureka, for example, the Bullion Beck erected a two-story, $15,000 brick store, known as the B & B, on lower Main Street. W. H. Wood added a $25,000 business block, while the Eureka Hotel built a brick addition. A new $14,000 schoolhouse appeared in 1896. W. D. Myers remodeled the Keystone Hotel at a cost of $1,500, and housing increased by approximately 100 residences that same year. The George Arthur Rice banking house failed in 1897 but was replaced by McCornick and Company of Salt Lake City.[75] Conditions were such that the *Salt Lake Mining Review* wrote that Eureka only 10 years earlier had been one "'straggling winding and narrow street,' however in 1899, it is a little metropolis of several thousand [about 3,500] inhabitants, and its enterprising citizens point with pride to its many fine business blocks, its tasty and comfortable residences, its churches, schools and newspapers."[76]

The year 1899 indeed became a banner year for Tintic. In that year *Salt Lake Mining Review* noted that Tintic was attracting the attention of capitalists and investors.

Production—the value of ore sold or treated in the district in one year—reached $5,228,575, the highest in its history to that point. From 1870 to 1899, Tintic had produced approximately $35 million in ore value.[77] So impressive was the district's performance in light of the state of the economy that an 1899 mining directory stated, "The present high price of lead and copper, together with the boom in gold mining and the many new mines opened to production in the district, have easily rendered Tintic one of the foremost districts in America. . . . it has a very large area yet undeveloped and offers to the miner, capitalist or speculator a magnificent field for investment and today few spots of the earth are more promising to old or young in the mining field than Tintic."[78]

Among mining districts, Tintic had reached maturity; it had survived the economic crisis of the 1890s, labor strife, and physical crises. More importantly, Tintic was an area abundantly endowed with minerals, and it possessed many mine owners who sought to guard its potential wealth by sound development of their mines. Such conscientious development ensured the longevity of the district and its success in the twentieth century.

Eureka Becomes a Full-Fledged City

Local newspapers and editors are in many ways the lifeblood—or at least the pulse—of a community. Eureka claimed the *Eureka Chief*, first published in 1889 by Charles C. Higgins; E. H. Rathbone's *Tintic Miner*; and the *Eureka Reporter*, edited by C. E. Huish, and C. E. Rife. The *Reporter* eventually became one of the top newspapers in its class in the United States.[79]

Eureka did contain the ingredients of a "little metropolis." Incorporation as a city meant establishment of a local government with ordinances and enforcing bodies. A mayor, recorder, treasurer, council, and marshal directed Eureka's affairs. Interestingly enough, the marshal's monetary compensation was fixed at $1,200 per year and "one Dollar ($1.00) for each arrest where the costs are paid by the defendant." Not surprisingly, scrutiny of the criminal justice docket ledgers reveals an abundance of arrests, most of them not for serious crimes. For example, Sophie Rice (alias Molly Brown, Jane Doe, etc.), the local madam, and her "ladies of the night" were arrested monthly and fined for "maintaining a house of prostitution" or "advertising her vocation as a prostitute." The fines were promptly paid, and Sophie and her charges were set free. The marshal had his dollar per arrest, the city coffers were financially ahead, and the madam's business and customers were only temporarily inconvenienced.[80]

Business activity reflected the general prosperity of the decade. By 1903–4 Eureka alone contained some 90 business listings in the *Utah State Gazetteer and Business Directory*. The growth of commercial activity was of course commensurate with the district's revitalization beginning in 1899. By 1903–4 the McCornick bank was firmly established in Eureka, making the city the financial hub for the entire Tintic District, explaining why Eureka assumed the role of Tintic's business center.[81]

Social organizations developed in an atmosphere that proved conducive, namely, that of a mining town fairly isolated from other areas yet prosperous enough for its inhabitants to seek camaraderie with others who shared their interests, religion, or socioeconomic level. Mormons maintained the Mutual Improvement Association and the Relief Society. Catholics joined the Knights of Robert Emmet, Society of the Blessed Virgin Mary, Society of St. Patrick's Church, and Knights of Columbus. Methodists belonged to the Masons, Order of the Eastern Star, and Daughters of Rebekah.[82] Other social groups active during the decade included the Ladies Aid Society, Eureka Home Dramatic Company, Blue Rock Club, Eagles, and the "Socialists Club of Eureka."[83]

Significant in Tintic's growth were schools and education. A 1909 school census enumerated pupils in Eureka as follows: 541 boys, 513 girls, 15 female teachers, and 3 male teachers. These figures represented an increase of 133 pupils from the preceding year.[84] Figures included both the public schools and St. Joseph's Catholic school, the only private school in Eureka.

Total production in 1900 had reached more than $7 million in value, dropping to only $3,700,000 in 1902, but then increasing again by 1903. Output by 1906 was valued at more than $8 million, largely due to the increase of lead ore. The Panic of 1907 caused another downturn, with production plummeting to about $5,300,000. Prices of metals were low and smelting rates, high. By 1909 mining had recovered with production near $8,250,000, and the district enjoyed a record output of lead. Production dropped in 1910–11 to near $7 million, but by 1912 it had jumped to $9,800,000.[85] The economic seesaw would continue.

The chief factor in this vacillation was, of course, the fluctuation in mining. The entrance in 1909 of the Chief Consolidated Mining Company into Tintic proved important. Walter Fitch, Sr., had entered Tintic earlier, purchasing shares of the Little Chief Mining Company. On 21 January 1909, the Chief Consolidated was incorporated, and by March the operation was launched. Fitch erected a home near the Chief Consolidated surface plant, southeast of Eureka, later known as Fitchville. In July he organized the Eureka City Mining Company, with the explicit purpose of prospecting under the Eureka town site. The owner of an acre of ground was to receive 1,000 shares of stock for mineral rights to the property. Records at the Juab County Recorder's Office indicate that mineral rights to most properties were purchased in 1909. By July 1910 the Chief Consolidated and Eureka City Mining Companies were merging.[86] Fitch's ventures had proven successful.

The Twentieth Century

The second decade of the twentieth century was also characterized by periods of economic fluctuation. Such a pattern placed Tintic in a situation similar to the rest of the country. War and its attendant labor shortages affected the area as it did other parts of the United States. In addition antiforeign sentiment penetrated the Tintic Mountains. Production, in value of ore, reached an all-time high in 1912, soaring to nearly $10 million. At this point in time, Tintic became a zinc producer for the first time with mixed carbonate and silicate ores shipped from area mines. Fluctuations in ore values, usually ending on the down side, occurred until 1916, when a value of over $9,000,000 was regained. [87]

Labor activity also proved significant during the decade. Accompanying the prosperity of 1912 were demands by labor for better wages, but for the most part, labor-management relationships were characterized by a general lack of conflict. Higher wage demands flared again in 1917, but the problem was solved, and Tintic has been relatively free from labor strife for most of its history.

Mining reached new heights during this second decade of the century. In value of production, 1918 attained a peak of $11,183,506; 1919 declined to $10 million; and 1920 achieved an all-time high of nearly $12 million.[88] New operations were launched, mills were erected, and surface plants were improved. The rich ore body of the Tintic Standard Mine was discovered in 1916, sending many patient Eureka stockholders to the banks rich. Dividend, just east of Eureka, blossomed as the company town of the Tintic Standard.[89]

Tragedy, storms, and disease marred various years during this period. In September 1914, 12 men were trapped when the Oklahoma slope caved in at the Centennial-Eureka mine. Eleven eventually died in the worst accident in Tintic history.[90] A winter storm crippled mining operations in 1917, and for the first time in the history of the railroads in the district, a storm brought them to a standstill. Snowslides inflicted considerable damage, and trains were literally buried just east of Eureka.[90] Nor was the region spared by the worldwide influenza epidemic, which spread like wildfire through Tintic's towns in 1918.[92]

In the third decade of the twentieth century, production remained at an encouraging level. Figures for production in Tintic during the 1920s were as follows:

1921	$9,801,712
1922	11,911,501
1923	14,015,916
1924	13,043,031
1925	16,187,583
1926	15,011,520
1927	11,188,934
1928	9,828,823
1929	10,545,407
1930	6,875,688

By 1933, however, production had plunged to $1,881,637.63.[93]

Commercial activity boomed in the early and mid twenties. Politics also maintained a position of importance, with the Socialist Party of Eureka an especially strong local. The middecade optimism permeated all facets of life as exemplified in a *Eureka Reporter* article that discounted the notion of Eureka's demise. In best booster fashion, the newspaper pointed out that merchants were remodeling, and Eureka's residents continued to have faith "in its phoenix-like come-backs."[94]

Despite that faith, the decade ended with the Great Crash that soon turned into a Great Depression, and the mining towns of the West could not be spared. With the onslaught of bad times, the character of Tintic began to change. Employment lagged, payrolls declined, production practically ceased, and most commercial enterprises suffered greatly from 1929 to 1934. Minerals existed, but there was no market; and a decline in mining operations affected taxes, power consumption, and even the circulation of books from the Eureka Public Library. Employment dropped with workers leaving the area and moving to valley towns.[95] Thus, this impact on taxes, power, and migration made it clear that Tintic's influence transcended its immediate borders.

By the beginning of the 1940s, ore shipments from the Tintic District were continuing to decrease, and the drop in mining affected the local economy. When the Safeway grocery store closed, the *Eureka Reporter* noted, "This indicates a trend which has been apparent for years, and it also indicated the fact that . . . this town is going down by degrees."[96] In July of 1940, the city of Eureka bought a half interest in six wells from Chief Consolidated Mining to provide water for the city. In that same month, members of the International Union of Mine, Mill, and Smelter Workers voted to strike but agreed to wait for further negotiations. Five weeks later an agreement guaranteed "recognition of the union as exclusive arguing agent for a period of two years; a wage increase, to be retroactive to 1 June, although the amount was not divulged; provisions for vacations with pay; and grievance procedures designed to prevent strikes and lockouts."[97]

In February 1941 the 70-year-old Mammoth Mine closed, and in April the Blue Rock Mine also shut down. In June, however, the Tintic Standard Mine "opened a new ore body running high in lead and silver, being one of the most important discoveries in the property in recent years."[98]

World War II pulled many of the young men in the area into the armed services, creating a labor shortage.[99] At the same time, the United States Defense Department appropriated $190 million to build the Geneva Steel plant in Utah County to help refine ore into more usable metals.[100] Nonetheless, "the war changed lives and economies."[101]

While mining continued, new businesses developed, and the economy of the area diversified, Tintic remembered its roots. The Tintic Historical Society was founded in 1973, and in 1979 the Tintic Mining District was added to the National Register of Historic Places as a multiple resource area. The creation of the Tintic Mining Museum, a history of the Tintic Mining District published in 1982,[102] and preservation of several significant homes and buildings won Tintic the 1983 Albert B. Corey Award from the American Association of State and Local History, the highest honor in the nation for a local historical society. Some mining operations continue in the Tintic area today, often dealing in minerals unimaginable to those cowboys who first saw silver in the sun.

Mining Trends to the New Century

During the last half of the twentieth century, mining in Tinitic followed the patterns of upswings and downturns reflected in the national economy and price of metals. Work continued into the 1950s and 1960s, slowing in the 1970s. Chief Consolidated Mining held onto its properties in the district. Kennecott Utah Copper worked the Bergin and Trixie Mines in the Dividend area but by the 1980s had moved out of active mining. The Sunshine Mining Company even minted some silver coins in the early 1980s from the Dividend end of Tintic.

At the same time, and extending into the early 1990s, an Australian-based company, Centurion Mines, actively pursued reprocessing the expansive mine dumps of the district. Earlier, the North Lily Mining Company had established a leeching facility at the site of the old Tintic smelter in Silver City. Thus, mining the dumps brought Tintic's history full circle. Centurion also emphasized mapping the district and collecting more of the historical record to document specific areas of mining activity and the grades of ore mined.

The boom Tintic residents were expecting never did materialize. World markets and the price of metals ensured mining would not reach the prominence of earlier years. In the 1990s the United States Environmental Protection Agency declared Eureka a superfund site. High levels of lead in soil samples led to a complete face-lift. Potentially contaminated soil has been removed, replaced, and covered, creating new landscape features for the town.

A once bustling center of the Tintic Mining District, Eureka clings to life with tenacious citizens, some 650 in number, who find the lifestyle there appealing and essential. On 29 September 2004, the Tintic Mining Museum celebrated its 30th anniversary of documenting, collecting, and exhibiting Tintic's valuable collection of mining documents and artifacts. As EPA bulldozers move rocks and alter the face of the district, the Tintic Historical Society continues to ensure that those changes will be seen in the context of Tintic's valuable and unforgettable history.

15

San Francisco Mining District

Martha Sonntag Bradley-Evans

Introduction

About 15 miles due west of Milford, the San Francisco mountain range forms a swath down the western half of central Utah. After its organization on 12 August 1871, the San Francisco Mining District ran down both legs of this massive range. Grabbing the attention of even the most casual viewer, and dominating the silhouette against the eastern sky, is Frisco Peak. At 9,660 feet high, it is one of Beaver County's most distinctive visual features. Toward the southern end of the range, Frisco, once a rousing and colorful mining town of the nineteenth century, prospered because of the riches of silver, gold, lead, and zinc extracted from its mines. The district became an important source of silver and lead during the mid-1870s.

In May 1872 the Congress of the United States passed a general mining law called, *An Act to Promote the Development of the Mining Resources of the United States*. The significance of this act was its provision that mineral deposits in lands "belonging to the United States are free and open to exploration and purchase by citizens of the United States, according to provisions detailed in the law, and also according to local customs and to the rules established by miners in various districts."[1] Moreover, the law validated the mining districts established to govern mining activity in various local areas of the United States. And the law gave mining districts the authority to govern the method of locating and recording claims, which provided important jurisdiction over local matters as long as they didn't conflict with the laws of the United States.[2] During the last three decades of the nineteenth century, at least 90 mining districts were organized in Utah Territory, and many of these filed bylaws with the General Land Office in Salt Lake City.

Considering the San Francisco range as a totality, from north to south it is roughly 20 miles long and 4 miles wide. Moving out of Millard County to the north, where it connects with the Cricket Mountains, it stretches into the center of Beaver County.

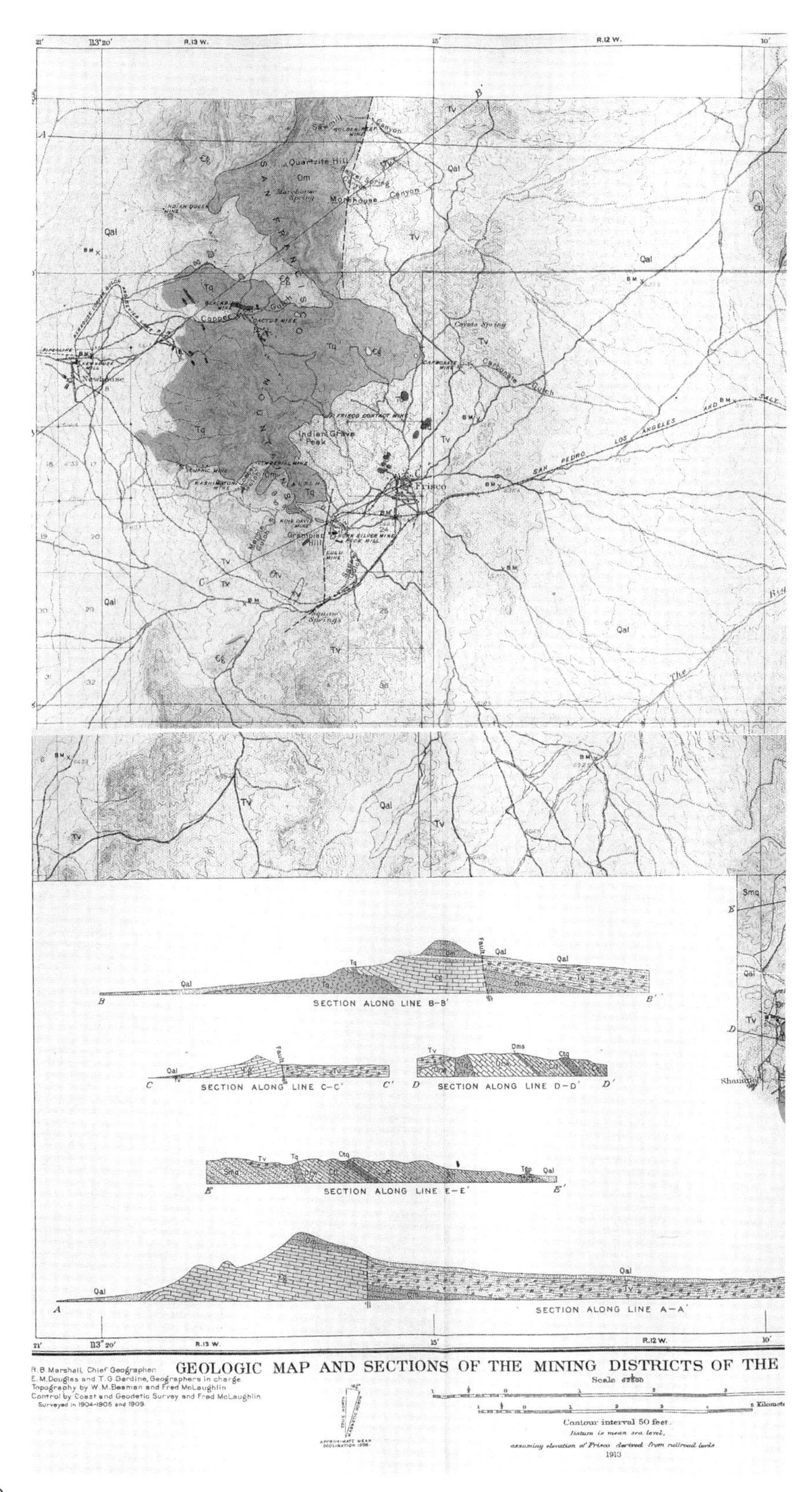

113°20'
R.13 W.
15'
R.12 W.
10'
Quartzite Hill
Morehouse Canyon
Copper Gulch
Newhouse
Indian Grave Peak
Frisco
Grampian Hill
Squaw Springs
Coyote Spring
Carbonate Gulch
SAN FRANCISCO MOUNTAINS
LOS ANGELES
SAN PEDRO
SECTION ALONG LINE B-B'
SECTION ALONG LINE C-C'
SECTION ALONG LINE D-D'
SECTION ALONG LINE E-E'
SECTION ALONG LINE A-A'
GEOLOGIC MAP AND SECTIONS OF THE MINING DISTRICTS OF THE
R.B.Marshall, Chief Geographer.
E.M.Douglas and T.G.Gerdine, Geographers in charge.
Topography by W.M.Beaman and Fred McLaughlin
Control by Coast and Geodetic Survey and Fred McLaughlin.
Surveyed in 1904-1905 and 1909.
Contour interval 50 feet.
Datum is mean sea level,
assuming elevation of Frisco derived from railroad levels
1913

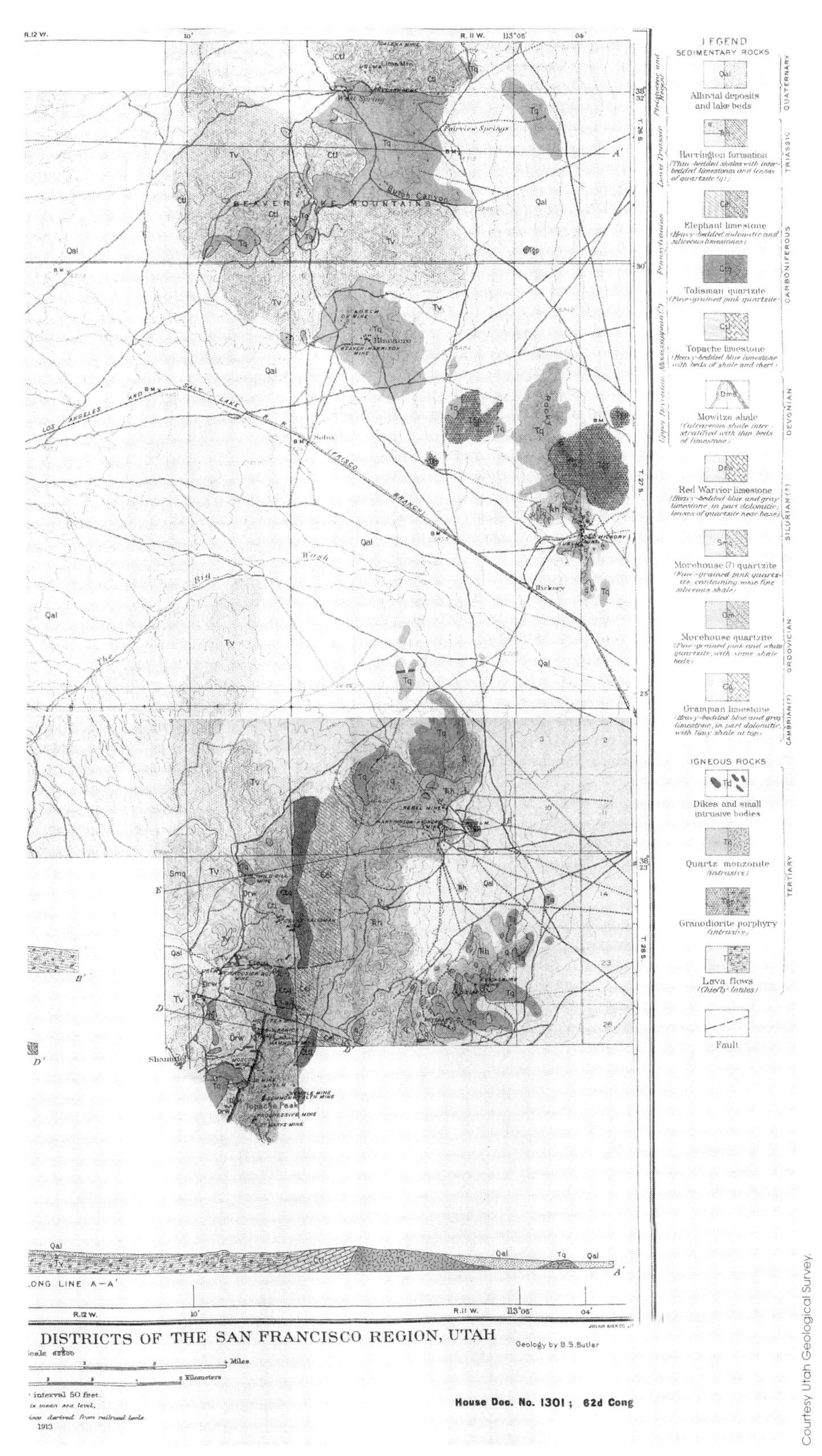

Courtesy Utah Geological Survey.

The range is primarily Precambrian rock, except for its southern end that includes Ordovician sedimentary stone and Tertiary intrusive rock. The western slope is Cambrian. Frisco Peak itself consists mainly of Precambrian metamorphic and quartzite rock. The peak offers twisting, weathered bristlecone pines and one of the best views of the Great Basin in Utah, stretching to the east and west as far as one can see.

The San Francisco Mountains are a playground for antiquarians and scavengers of memorabilia from the past. Nineteenth-century remnants from the mining heyday still remain at various points along the sides of the range—mines, kilns, and narrow-gauge railroad tracks. Evidence of the powerful lure of the promise of riches beneath the earth's surface, foundations of houses raised by prospectors or miners trace footsteps of their efforts to survey the landscape of the range. The Indian Queen and Imperial Mines lie to the west, and to the east are found the King David, Golden Reef, and Horn Silver Mines. Springs that originate in the San Francisco Mountains carve slices down the slopes—the Pitchfork, the Horse, Crystal, Morehouse, and Tub Springs, as well as small, seasonal creeks that run down into the valley below through Morehouse, Sawmill, and other canyons.

Rich mineral veins ran through the San Francisco Mountains, and discoveries in the region made Beaver one of the most important mining centers in the western United States. Despite Brigham Young's admonition to avoid mining and the enticements of quick riches,[3] Mormon farmers were tempted by mining for the same reasons as outsiders—the promise of quick riches rendered other arguments insignificant. The wide range of valuable and desirable natural resources found extensively through this mountain range included gold, silver, lead, copper, zinc, bismuth, and sulfur, as well as marble, granite, sandstone, potash, and other materials used for building or various commercial ventures.

Beaver County had been settled two decades before mineral deposits were first located in the San Francisco Mountains. Small towns had 20 years of farms and businesses already in place. Nevertheless, not far from the sites where minerals were located, new towns sprang up to support local miners, and older towns rushed to service the population of miners that moved through the region. For many these towns were only temporary homes. But over time they became more stratified. Their principal focus was mining and the incidental industries related to extracting, smelting, and producing various minerals rather than attaining a community ideal or meeting a common political objective. When the mines were exhausted or veins dried up, the towns were abandoned. These towns were as fleeting as the dream of wealth itself. Some boomtowns and ghost towns vanished within a single generation. Others, such as Milford, which benefited from the mining economy, rode the various waves of mining activity and came out on the other side still thriving because their agricultural identity remained intact.

The *Beaver County Blade* and statewide publications such as the *Daily Union Vedette* published the news of local mining activity—strikes, busts, and profits were noted like the rising stars of the stock market. More than a local commentary, these weeklies worked like boosters, drawing a picture of mineral wealth to tease investors into the territory and tempting them with the chance to strike it rich. And their strategy worked. It is clear that Beaver's mining activities were sensational enough to put the county on the map and attract the attention of prospectors throughout the West eager to make their fortune. Whether dream or reality, the story of mining in the San Francisco Mountains was like a magnet. Entrepreneurs from Salt Lake City and throughout the nation invested their money in mines in the San Francisco District, betting on the richness of mineral veins hidden beneath the surface of the mountain slopes. The odds were irresistible.

The San Francisco District lies in the center of Beaver County about 225 miles south/southwest of Salt Lake City and 98 miles northeast of Pioche, Nevada. Although the district was first created in 1871, it was not until five years later that the most significant discovery was made, a year after the Horn Silver Mine had been established and the settlement of Frisco completed.

According to historian Miriam B. Murphy,

> The story of the Horn Silver Mine, one of the great producers in Utah and American mining history, reads like pulp fiction: Two prospectors casually discover a rich ore body, a bankrupt financier promotes the venture, the boomtown of Frisco becomes one of the wildest mining camps in the West with a murder or two every evening, a tough lawman who shoots on sight begins to clean up the town, after producing millions the huge mine collapses, and Frisco becomes another ghost town.[4]

Discovery of Silver and Lead

The discovery of silver and lead ores in Beaver County was the result of a particularly propitious accident. Not long after, Frisco was settled in 1875 at the southern tip of the San Francisco Mountains. Samuel Hawkes and James Ryan first prospected what would become the Horn Silver Mine while they were working the Grampian Mine, a source of galena ore. Each day as they walked to the mine, they passed by a huge outcrop of limestone which they eventually tested. After finding that it was a solid ore body, they staked a claim to it. Discovering that this was a good source of anglesite ore, they developed the mine to the 500-foot level. After that favorable strike, they sank a 25-foot shaft through seemingly solid ore. Fearful that the mine would fail to produce, they sold the claim to A. G. Campbell, Mathew Cullen, Dennis Ryan, and A. Byram on 17 February 1876 for $25,000. They thought they had struck a good deal, but as it turned out, they had not. In fact, what the new owners called the Horn Silver Mine proved to be enormously profitable, a rich vein which produced silver valued at $100 per ton.

Banker Jay Cooke and other Salt Lake moneyed interests bought the Horn Silver Mine in 1879 for $5 million and induced the Utah Southern Railroad to build a line to Milford and Frisco to transport ore to distant markets. That same year the

United States Annual Mining Review and Stock Ledger called the Horn Mine "unquestionably the richest silver mine in the world now being worked."[5] The mine was a good investment; in fact, it revitalized Milford and much of the surrounding area. Frisco became a boomtown almost overnight.

The new owners built a smelting works before they sold any ore or developed the mine any further. In time they sold the mine to investors in the New York Mining Company and the Salt Lake Company for $5 million. The mine's new proprietors dug deeper, extending the mine to the 800-foot level. It produced a healthy profit until 1884, when the mine caved in.

William A. Hooper, a Utah businessman, recorded the mine's significant production figures in 1879:

> The quantity of ore extracted up to February 1, 1879, is given as 22,712 tons. During February about 90 tons daily were raised, making a [monthly] total of 25,000 tons of ore. The extraction of this 90 tons was barely enough to keep the mine in good shape and prevent the breasts of ore from crowding too much upon the timbers. The present expense of mining is low. We have as the cost of taking out 90 tons daily:

Labor	$144.00
Timbering	$73.00
Superintendent	$6.00
Supplies and expenses	$50.00
Total	$273.00
Cost per ton	$3.05[6]

Ryan and Hawkes's first mine, the Grampian, was near the Horn Silver Mine. Gold near the surface and trailing down 80 feet varied from one to eight feet of ore. Originally a vein of ocherous ore assayed at about $50 of silver.[7]

Several large ledges of pyrite and copyrite ore resulted in very high-grade silver in claims called the Comet, Cactus, and Copper Chief in the northern part of the San Francisco District. The South Utah Mines and Smelter Company owned the Cactus Mine, located in Copper Gulch about two and a half miles northeast of Newhouse. The main body of ore originated about 6,450 feet above sea level or 200 feet above Newhouse. Prospectors first identified the Cactus Mine in 1870, so it was one of the earliest in the district. Despite a series of ambitious efforts to reap its wealth, a series of investors and companies failed to capitalize on its potential. Even a small smelting plant, which was built in 1892, proved to be a failure and produced only a small amount of ore.

Because of the difficulty in transporting the raw ore from the site to distant markets, smelting became an important incidental and related industry to the district's mines. Facilities such as the Williams smelter, built after the Horn Silver Mine was discovered, used innovative construction methods in an attempt to improve safety, be more environmentally responsible, and trap fumes and particles before they ran through the chimney into the air above. Experiments in Frisco centered on smelting techniques, but because of scarce water and difficulty getting sufficient charcoal, success was limited.

Town and Railroad Development

Revenues from the San Francisco District fed small towns in the area with a ready supply of customers and residents. Milford, for one, was an agricultural town settled by families who had moved out of Beaver under the leadership of Arvin Stoddard and his family in 1880, although crops had been planted in the area as early as 1859. Familiar with what it took to start a town, Stoddard, himself a surveyor, platted the town site.[8] Many of the earliest inhabitants were farmers, but from the first, Milford was also important as a supply station that served freighters and soon fed off the mines. In an area of the territory dependent on agriculture rather than commerce, the influx of business customers proved a boon to local enterprise. Trade boomed from the first, and by 1890 Milford had a significant commercial Main Street, which was lined with the typical service establishments of a town colored by association with the mines—saloons, mercantiles, hotels, and a variety of businesses like blacksmith shops and livery stables. What had been a typical Mormon town laid out along the grid structure of the City of Zion became a lively mining town, diversified in its services and people. Alton Smith provided one account that there were six saloons in Milford at the turn of the century: the Atkins Bar, Milford Saloon, Crescent Bar, Oxford Saloon, Long Tom Martin Saloon, and East Side Saloon. There miners found home-cooked meals, easy liquor, and the entertainments offered by a free and easy, male-dominated society—gambling, dancing, and eventually, vaudeville.[9]

A Scotch Canadian company called the Harrington-Hickory Consolidated Mining Company constructed the Milford stamp mill (also known as the A. G. Campbell mill) for $45,000 in 1873. An important processor of ore from the Hickory Mine between 1873 and 1874, it produced from $9,000 to $12,000 in bar bullion although it was only in business for five months.

Milford became a railroad terminus for the Utah Southern Railroad; after the railroad came to town, Milford also became a major loading place for southern Utah cattle. The sweeping arid landscape of central Utah was difficult for farming but ideal for stock raising. Throughout the 1880s and 1890s, grazing lands in this part of Beaver County were exploited fully. Milford scored a major coup in being selected as a railroad terminus; that made success as a community a sure thing. Inevitably, an endless series of hotly contended races by outside companies searching for suitable locations for their rail lines continued.

Beaver County's mining industries had drawn the Utah Southern to the area. Less than a decade after it first arrived, in fact, by January 1879 the company announced that it would build an extension line to Frisco, home of the Horn Silver Mine, and mine owners pledged to pay a quarter of the construction costs. Before the Utah Southern extension reached Frisco, the roadbed was graded outside of 10 miles of Milford early in 1880.

Equally as hotly contested as the location of rail lines was the competition over the location of the county seat. Yet another way to draw local money or attention to a town was by placing county government there. Frisco's citizens attempted to play this card and petitioned the county court in the late 1870s for a local jail. Beaver selectman James Low met with representatives from Frisco to discuss the possibility of locating both the county seat and a county jail there, as well as to determine what local businessmen could contribute to the building process. The construction of a line through Milford by the Union Pacific with its interests in the Utah Central Railroad also figured prominently in the discussion. Much of this campaigning and politicking was successful. As a result, by the fall of 1879, rail lines had been laid as far south as Deseret in Millard County. The county seat was established in Beaver City, however.

Milford's first settlers laid out homesteads like those in hundreds of other Mormon rural villages. Named for the crossing of the Beaver River by freighters coming and going to mines and mills to the west, Milford was an unpolished frontier town with a marked contrast between those associated with the mines and people who had businesses, farms, or families to raise nearby. The people who arrived with the Union Pacific, which ran north and south through the center of town, reflected the town's importance in transporting goods throughout the region. Also important for shipping coal to Los Angeles for export to Pacific Rim countries, Milford was a place whose identity was diverse and tied to a variety of different industries.

Electrical power was extended from Beaver to Milford to Frisco in 1880. In fact, Frisco was the largest town in Beaver County at that time.[10] Instead of locating the railroad station in the original plat, city builders drained swampland east of the original survey and laid out a second series of streets on a diagonal to the original grid. Frisco drew miners, merchants, and investors by the hundreds. Other mines were soon discovered in the area—the Carbonate, Rattler, Golden Reef, and Gampion. Each of these very successful mines had its own smelter, and five beehive-shaped charcoal ovens were built to serve the smelters, using local wood like cedar, dwarf pine, mountain mahogany, and sagebrush.

The Carbonate Mine rivaled the Horn Silver for success and was located one and three-quarter miles northwest of Frisco at an elevation of 6,750 feet; it consisted of 11 claims and fractions. First discovered in 1878, and sold the next year for $10,000, the mine produced concentrates with an average assay of 43.63 percent of lead and 94.09 ounces of silver to the ton. The tailings yielded significant profit as well because they were also rich with minerals.

Unlike Milford, from the first Frisco was an unabashed mining town complete with rapidly built structures, a largely male and transient population, and a dependence on a single industry—mining. One historian described Frisco as "as wild and tumultuous a town as any in the Great Basin, . . . and the wildest camp in Utah. Twenty one saloons had so many killings the undertakers wagon made daily rounds."[11] Marshal Pearson of Pioche, Nevada, came to Frisco to clean things up; he had heard that Frisco was so wild that he promised to shoot on sight anyone breaking the law.[12] Frisco had 23 saloons, false-front stores, boardinghouses, and restaurants within only a few years, plenty enough to make happy a population of miners that reached as high as 6,500 between 1880–85. The local newspaper, the *Frisco Times*, tracked successes and failures in the mines and detailed the activities of an increasingly diverse population and the immensely entertaining and colorful nature of local politics. According to Phil Notarianni, an 1879–80 directory listed 33 businesses and different services, but by 1900 the number had dwindled to only 14. Population totals and businesses fluctuated wildly and reflected the relative value of the wealth extracted from the mines.[13]

Fred Hewitt's letters to his wife during his journey to Frisco and his work in a mine there provide a rare and rich contemporary commentary. A mining engineer from California, Hewitt traveled first by railroad and then by stage to Frisco. He began working for the Champion Silver Mining Company but eventually switched to the Horn Silver Mining Company. Hewitt painted a vivid picture of life in a nineteenth-century mining camp in a series of letters, including one describing a strike staged by miners at the Horn Silver Mine when wages were reduced. Dated 1 February 1880, his first letter describes the railroad trip to the end of construction in Juab County and the stagecoach ride on to Frisco:

> The stage left at 1 PM. I wrapped my feet up as well as I could and put the shall [*sic*] around my shoulders and after riding about half an hour found my feet so comfortably warm that I congratulated myself on the nice arrangements. Jerkey stopped, driver looks in and sees my fellow passenger on front seat. I was on the back. "You front." So my companion with some grumbly takes the back seat. Of course this somewhat disarranges our things. Then the driver gets in to handle the bag of corn and walks on our legs in so doing. The corn laid on the seat and the driver back in his place we start again, somewhat colder than before. We are evidently going to have a cold night. The breath freezes to our whiskers. Pretty soon the bag of corn comes off the seat on to our legs and it takes considerable exertion to get it back again. It keeps coming off until some time in the night the driver takes it away. Not very long after the advent of the corn, another stoppage, "gentlemen you will have to get out and help me over the railroad crossing." So the pins come out of my blanket the wraps are laid aside and out we get. The drivers tries to cross the track which is here in an unfinished . . .
>
> Nothing particular happened only the usual stage bumping. First it bucks you up a foot or so, then a quick jerk sideways another from the other direction, then a twisting jerk that seems to go all around you, then a few

minutes of ordinary jolting and then we are bucked up again and the side jerks and twists are repeated. We became colder and colder until it seemed as if I could not bear it much longer. The stage stops we think we will get out at station and warm, but no, no stoppage here only taking on another passenger. This at half past two in the morning. On we go again until about 4 o'clock we hear the drivers hallo to the next station. We are to have breakfast here but the people are not up, and so we stamp about in the snow for some time until the door is opened. The table is set, there is a bed in the room from which the man and his wife have just risen she is buttoning or hooking something as we enter, a child is in the bed. A good wood fire is soon going in the stove but it takes us a long time to thaw out. After about an hour breakfast is ready such as it is. I could hardly eat any, but the warmth and the coffee was worth the 50 cents it cost. After that we did not get so cold and did not stop again until we arrived at Frisco about 9:30.

February 7, 1880

After reaching the [Belcher] mine somewhat behind time, as the mining time is half an hour ahead of town time, I found that I was expected. I was taken into a back room of the office. A flannel shirt was handed to me and a pair of flannel pants fastened with a string round the waist. The whole outfit like a Coney Island bathing suit. I was requested to strip, when I inquired how much I was told everything. After dressing as requested, they made a parcel of my money and took care of it, they gave me socks and opened a long box filled with shoes from which I fitted myself. Then I went out to the shaft house the costume being very airy for winter, and was put in charge of the pump man. Visitors to mines do not generally go down the pump shaft but go in the large cage down the regular hoisting compartment. My object being more to examine the engineering arrangements I went down the pump shaft. There was a little cage if it may be so called, a little shelf or bracket guided only on the back on which we stood. It was about 16" wide one way, just large enough for the two of us to stand on with feet close together and standing up straight. The word was given and we went 900 ft on that, slowly and stopping at different stations to examine pumps, balance bobs etc. At the 900 ft station we got off and walked down an incline of 31 degrees to the 1600 ft station. Walking down stairs, it was all steps for 700 ft is pretty fatiguing work [*sic*]. It was very warm down this incline, the perspiration pouring off of me. A regular Turkish bath arrangement. At this point we drank ice water and rested going to a point where cold air came through for this purpose. Then we took a large car and were lowered down to the 2400 ft station that being the lowest part of the pumps. The mine is 3000 ft deep but we did not go to the bottom. The water that came up from the pumps was too hot to be able to bear the hand in, and in some of the drifts it felt like being in an oven.

> February 11, 1880
>
> About the mine is a collection of shanties mostly belonging to the Co. I think. One or two of [*sic*] you could almost call houses. The Co's boarding house and the shanty over the mine shaft. I have had a 9 x 12 shanty allotted to me. It is furnished with a bedstead of pine wood made by the carpenter, a very rough table and sundry boxes, one of which I use for a seat. Like all such shanties this one is full of cracks. There is a fire place at one end made of rocks and mud and so far I have been supplied with fire wood. I had to buy blankets at the store—also bed tick which I filled with hay at Co's stables, also wash basin and tin cup. I moved over here Monday morning, and started a fire in the afternoon, and kept it up all evening. Last night I had to move my bed to get out of the snow. Mud and gravel also blows in.
>
> July 20, 1880
>
> The new superintendent of H. S. Mg. Co. A Mr. Hill made an attempt to cut down the miners wages, resulting in a strike yesterday. The men last evening went in force and compelled them to stop the smelters and declared that no other work should go on and Mr. Hill backed down and sent word for them to go to work at the old wages. The miners are getting slightly higher wages than in some other places $3.50 per day, but it ruins the health of every man that works in this mine, and I think that a man that once gets "leaded" will never be the same again.[14]

Hewitt's letters more importantly create an image of the backdrop of human life. Wages, the quality of housing, and working conditions mattered greatly to individual workers.

Smelting and Beehive Kilns

Far more curious and structurally unique than the mines or smelters were 36 beehive charcoal kilns that fueled the furnaces of the San Francisco District. The kilns cost between $500 to $1,000 to build according to one author. Michigan engineer J. C. Cameron designed them in 1868. The *Utah Mining Gazette* described Cameron's kilns as shaped like a "parabolic dome, with a base of twenty to twenty-four feet in diameter and altitude of nineteen to twenty-two feet," at a cost of about $700.[15] Within 6 to 18 miles of Frisco, six to eight groups of charcoal kilns were separately managed and operated in conjunction with the mining activities. Distinctive because of their unusual cone shape, the kilns were constructed of granite float, which was extracted from nearby areas, and cemented together with lime mortar. These kilns were nearly as high as they were wide and had walls from 12 to 14 inches thick at the top. One entered the kiln through two massive iron doors. The door at ground level was four by six feet, and the side door, which was located two-thirds of the way to the apex, was three feet by four feet in diameter. The kilns had three rows of vent holes located near the ground. Piñon pine from nearby mountains was the primary wood burned in the kilns. It was cut for $1.25 per cord, brought to the site on wagons or sledges, and valued at $1.50 to $2.50 per cord.

Originally workers produced the fuel for smelting in pits. However, the Frisco cone-shaped ovens were much more efficient at producing a higher-grade of charcoal choppers. Single men included 14 of the group who lived in boardinghouses.

The heyday of peak productivity for the kilns was between 1879 and 1884. Operators fired them from the center of the bottom, which drew the flame to the top through a small space above the door. The vent holes allowed operators to regulate the fire that was typically maintained over a period of three to seven days; then the charcoal cooled for three to six days and was shipped to smelters in racks at a cost of from three to five and a half cents per bushel.[16]

W. S. Godbe and Benjamin Y. Hampton, as superintendent, managed the Frisco Smelting Company and its five distinctive beehive charcoal kilns. M. Atkins was agent for the firm.[17] When the Frisco Mining and Smelting Company reorganized, it published a capital stock of $2 million in 80,000 shares. The property included the smelting plant at Frisco, the Carbonate Mines, the Cave Mine in the Bradshaw District, and an iron-flux mine in the Rocky District.[18] The *Tenth Census of the United States* indicates that in 1880 the Frisco smelter was a "complete one," which consisted of a Blake rock breaker, a Number-5 Baker blower, two horizontal boilers, one 40-horsepower horizontal engine, numerous pumps, a shaft furnace and flue-dust chamber, a reverberatory flue-dust slagging furnace, and five charcoal kilns.[19]

The *Tenth Census* produces a picture of the work environment as well as the workers themselves. It says that there were 21 coal burners, seven stonemasons, one brick mason, two wood contractors, and five wood contractors.

Coke replaced charcoal as the cheaper and more efficient fuel after the railroad came through the county. This shift is easy to recognize because the *Utah Gazetteer* simply stopped mentioning Frisco Mining and Smelting.

Prosperity and Depressions

These mining communities routinely experienced dramatic shifts in fortune. When in 1885 the Horn Silver Mine caved in, both Frisco and Milford suffered. Miners fled the area, and the mills and charcoal kilns shut down and lay dormant periodically, although 10 years later the mines still had produced a total of $54 million. Regardless of the fact that a new shaft was drilled farther down to 900 feet at the Horn Silver, only a few brave families stuck it out in Frisco. Eventually vacant buildings and a proverbial ghost town were all that remained although the mines continued to operate well into the twentieth century. When a fire broke out at the Horn Silver Mine on 5 April 1894, John Franklin Tolton wrote in his diary, "This is a hard blow to Southern Utah and Beaver County in particular, and will be keenly felt as it was a means of circulating a great deal of money in this and other communities."[20]

Prior to that the first railroad engine had stormed into Milford on Saturday, 15 May 1880, it had signaled future prosperity and business. This was without a doubt one of the town's most important days. Governor Eli Murray and other dignitaries traveled to Milford to celebrate and stayed the night in the Stoddard Hotel. Not long after the coming of the railroad, new businesses, hotels, stores, feed yards, and other facilities sprang up. One of the first, the Consolidated Implement Company, supplied materials used by freighters and miners throughout southern Utah. Located near the west side of the tracks on the road south to Beaver, it was proof positive of the importance of the mines to Milford's growth.

As soon as the railroad came to Milford, discussion began about continuing the line even farther south, although that eventually faded away. But clearly, the railroad had a positive impact on Milford, greatly enhancing its importance and key position in the San Francisco District. Because of the bullion that came from Nevada smelters and the continuing healthy mining in the San Francisco District as well as the Star District, even more new supply businesses sprang up which featured tools, wagons, buggies, and other products servicing the freighting industry.

The Horn Silver Mine continued to produce well after the owners installed a new shaft in 1886. In 1911, however, the Salt Lake Company sold its interest in the enterprise to its partner, the New York Mining Company, which sent W. H. Hendrickson to manage the mine and serve as superintendent until 1943, when the Metal Producers Company of Los Angeles assumed control under three men: George W. Clemson, general manager; James H. Wren, superintendent; and W. H. Hendrickson, who remained as mining engineer.[21]

The *Weekly Press* described the Hub's finds in the Star District, which were also surprisingly rich. The article concluded, "Another year will see Beaver County one of the most active mining Sections of Utah, and a goodly tonnage of copper, silver, lead and gold is promised. With men like Samuel Newhouse and the Knights, 'blazing the trail' the results have been assured for sometime past."[22]

Despite the dip during the depression of 1893 that caused minor changes in mineral values, as Beaver County entered the twentieth century, mining continued to be an important part of the local economy. Mining entrepreneur Samuel Newhouse purchased the Cactus Mine in 1900 and absorbed it into the Newhouse Mines and Smelting Corporation. It operated until 1910, when it was reorganized as the South Utah Mines and Smelter Company. The mine began producing in 1905 and continued until 1909, generating 19,419,319 pounds of copper, 7,510 ounces of gold, and 176,365 ounces of silver. The Beaver River Power Company furnished electrical equipment and power, although originally operations at the mine were powered by steam.

Two decades into the twentieth century, after World War I, new mines developed at a slower rate than during the boom times of the late nineteenth century. In the 1920s finds in the vicinity of the Horn Silver Mine indicated that rich veins stretched in several directions. The owners of the King David and Frisco silver-lead mines pumped in fresh revenue to sink shafts and drive drifts.

By the 1920s the Horn Silver Mine consisted of the original claim, 1,440 by 600 feet; two five-acre smelter sites in Frisco; a complete, three-stack smelting plant; a refining works in Chicago, Illinois; iron-flux mines near Frisco; charcoal kilns; a 40-mile telegraph line to Beaver; two large stores in Frisco; and other less-important property.

The King David and Frisco Mines were also prospering. According to the *Richfield Reaper*,

> At the King David, six mineralized veins, striking toward the Horn Silver, have been penetrated by a long crosscut to the north on the 750-foot level of the main working shaft. The most promising leads are being developed through a raise. Further west on the same zone, within less than 300 feet of the surface, the Frisco Silver-Lead has opened shoots of high-grade silver-lead ore from which many shipments have been made to the smelters.[23]

The shear zone led directly into the Horn Silver Mine; it was 4,200 feet long and 500 feet wide and contained many veins and deposits of pay ore in its fractures. A discovery originally called the Beaver Carbonate, located on a fault north and south of the Horn Silver Mine, became the Quadmetals by 1930.

Regardless of changes in market values or less output from the mines, mining companies continued to prospect for new deposits. Joe and Bernett Swindlehurst opened the Gold Basin three miles above the old Rob Roy Mine in the Indian Creek area, where they struck a large body of gold-bearing quartz at a depth of 12 feet.[24] That same week R. J. Finley of Los Angeles began assessing a large body of galena-lead ore, which also had the potential to produce good silver and copper in the West Mountains. Finley, who had been raised in Beaver, had located this site 13 years earlier, about four miles west of the Fortuna Mine. Reportedly, this was one of the largest bodies of ore ever discovered in this section—the outcrop was nearly 1,000 feet long and ran several hundred feet deep. Assays along the vein averaged $40 per ton in lead in addition to containing silver and copper.[25]

By 1931 the Fortuna Mine, first developed in 1914, was producing significant amounts of lead, which renewed interest in mining in the Indian Creek area. "The success at Fortuna this year is giving an impetus to mining on Indian Creek, the west mountains and other mineralized zones, with the result that new discoveries and rock assays are being reported daily," observed the *Beaver Press*.[26] Mine manager John Bestelmeyer stated that the Fortuna area "shows undeniable promise. Both in the igneous rocks or in the sedimentaries remarkable mineralization can be found. Surface indications are splendid, but not enough work has been done at depth."[27] The gold values ran from $.60 to $108 a ton in gold. One boulder yielded $854 a ton in gold. The company stored ore in a 100-ton bin on its property.[28] Thomas and Fay Harris, who were working a claim just west of the Utah Gold Mining Company's property at Fortuna, were so excited about their claim that they built a log cabin at the site and worked continuously through the winter.[29]

In addition to what was happening at the Fortuna Mine, work began during the same decade at the Oak Leaf. A tunnel which produced good quartz that assayed for $20 dollars in gold reached 308 feet and broke into Buckskin limestone with an eight-foot vein of $35 to $40 in gold ore. The manager of the site boasted that nearby miners had also opened a three-foot vein of manganese ore, six feet under the ground, that ran more than $25 in gold to the ton and promised even better results as the vein widened.[30]

Generally, there was enthusiasm in the early 1930s about the potential still lying in the county's mineral-rich mountains. According to the *Beaver Press*, "There is an apparent optimistic trend in the mining situation over the country that is being noticed in Beaver County as well as in other sections. That the price of silver is due to come back to a point where mining of that commodity will again be profitable in not a far distant future is confidently felt by the mining fraternity."[31]

Plans were under way in 1931 to reopen sulfur mines 22 miles north of Beaver.[32] R. E. Ellingsworth, Jim Hemby, Don Workman, and Bert Nichols, all from Milford, installed a gasoline hoist at Big Project in the Bradshaw District, seven miles southeast of Milford. Earlier in 1931, they had found a two-and-a-half-foot ledge of lead-silver-zinc ore which they determined was at least 2,500 feet long and several feet deep.[33] Also in 1931 a new prospect was developed by the Horn Silver Company in an area called the Buckhorn shaft.[34] Sulphurdale's mines opened in June 1932, and 15 families moved back into homes nearby.[35] The principal product manufactured by the company was sulfur dust, used extensively in California for coating melons, lettuce, and other vegetables in fields to prevent mold.

An editorial in the *Beaver Press* on 21 July 1933 expressed the common sentiment: mining was helping to turn the tide of bad times and promised future profits.

> With metal prices again at a profitable level and moving higher, the west is preparing for a genuine old-fashioned mining revival. Talk of reopening old properties is rife and the prospector and promoter is beginning to venture forth again after several years of inactivity in silver, lead, zinc and copper mining. . . . In this transition a mine has been turned from a liability into a profitable venture once more. As yet, however, the margin of profit is small and producers feel that it will be better reopening their mines and placing production again ahead of consumption. By late summer and early fall this condition should be classified and a number of producers will undoubtedly see their way clear to reopen properties and thousands of men will be returned to their normal occupations. Reopening of the mines will be followed by the reopening of the smelters. The railroads will again be moving long trains of ore cars, supplies, etc., and the farmer will begin to find a market for more of his products. The start toward all this has been made, now the conclusion is up to the industries themselves. Men must be put to work now to perpetuate the improvement. This is no time for timidity. If

> consumption is to improve, it must be made to improve [by] placing men back to work.[36]

The mining industry continued to inspire entrepreneurs with ideas for new ventures. Businesses feeding off the revived mining activities also opened in the county. D. W. Jeffs, John M. Bestelmeyer, and John M. Broomcamp, among others, established a rod mill for handling ore in Beaver in September 1932. They created the mill to handle ore from Utah Gold Mining out of Fortuna but milled for other mines as well.[37] The Forrester balanced-rod mill could grind, elevate, and classify ore for immediate amalgamation, flotation, or concentration and did so with four to seven horsepower per 50-ton unit.[38] The first shipment of ore from Utah Gold Mining arrived at the mill in November 1932. It consisted of 42 tons of ore, which yielded $38 per ton.[39]

Several mines were sold outright. Edward Schoo sold 18 mining claims, known as the Prosper group, in 1935 for $250,000 to Harry Murtha, a mining engineer from South Africa.[40] The owners of the Sheep Rock Mining and Milling Company leased their property to E. Bissell of Beaver and Charles A. Sihler of Glendale, California, who anticipated commencing work within the next month.[41] By the next fall, they were shipping high-grade gold and silver ore to smelters in northern Utah for processing.[42]

In April 1935 mining activities stepped up at the Horn Silver Mine as well as the King David.[43] The next month, in May 1935, news of a rich new strike in the Horn Silver Mine became public. Allegedly, the vein was eight feet wide and contained gold, silver, and lead.[44] By 1935 one *Beaver Press* headline described the area as "teeming with mining activity."[45] The heaviest-producing mines were the Lincoln, Moscow, Carbonate, Rob Roy, Shamrock, Beaver Copper, Old Hickory, Montreal, and Horn Silver, which by 1935 had produced more than $50 million. The rising price of silver in part explained the renewed activity—but success in terms of new sites, high yields, and general optimism about future efforts proved contagious and spread through the district.[46]

In 1935 a group of Chicago investors leased the Quadmetals Mine in the San Francisco District after it had been closed for several years. They planned to bring in new equipment, dewater the mine, and commence work as soon as possible. [47]

The *Beaver Press* described the San Francisco District in 1937 as "alive with mining activity." That same year discovery of an extension of the $50 million vein system of the Horn Silver Mine drew the attention of national mining engineers, mine operators, and others who visited the district to examine the find.[48] One visitor, lead smelter owner E. R. Phelps, described the claim as convincingly rich in potential yields: "Conditions south of the Horn Silver are so nearly identical with those in the Horn Silver that I [anticipate] ore deposits for considerable distances."[49] The Bonanza Mining Company drove a tunnel to tap the vein at 110 feet. In 1929 the American Smelting and Refining Company had driven a 900-foot shaft for what was then called the Lulu. Work on the site had stopped with the stock market crash, and the mine had lain idle until 1937.[50] San Francisco Mines, Inc., also was chartered in 1938 to carry on mining activities near the Horn Silver Mine.[51]

In the 1930s Beaver County mining districts produced $489,155 worth of ore. Despite the devastation to the mining industry caused by the Depression, higher prices and revived production stimulated new activity. By the mid-1930s, and since the inception of mining activity in 1860, Beaver County had produced 453,422,708 pounds of lead and 23,354,296 ounces of silver. Copper, next in line with 53,946,296 pounds, and zinc, with 42,123,360 pounds, were both important sources of revenue. The Horn Silver Mine had its best year in a decade in 1939—producing 10,590 tons at a gross value of $128,000. This total included 1,444,000 pounds of lead, 77,330 ounces of silver, 1,470 ounces of gold, and 139,299 pounds of zinc.[52] By the 1940s the Horn Silver Mine had produced approximately 190,192 tons of lead, 17,104,544 ounces of silver, 33,000 ounces of gold, 9,177,853 pounds of copper, and 19,192 tons of zinc.[53]

The Late Twentieth Century

Rich deposits of scheelite or tungsten, having an estimated value of $10 million, were found at the Old Hickory Mine in 1940.[54] Clarence H. Hall, engineer of the U.S. Vanadium Corporation, headed the work, which included core drilling on the tungsten vein, drifting in the McGarry shaft to crosscut it, and working the surface to try and determine the width, extent, strike, and value of the vein on the leased property.[55] M. M. Ward and Edith Ward, who owned one-half interest in 24 claims known as the Scheelite group, operated a second tungsten claim in the West Mountains. Miner E. A. McCarry worked for them on the project and was sinking a double-compartment shaft to the 200-foot level. They also hired a number of engineers to study the site.[56] Tungsten was of particular benefit to the war effort, which increased interest in the thriving mining climate.[57] Used for filaments in electric-light globes and hardening and toughening steel, tungsten was also a vital element in modern industry.

In an event described by the *Beaver Press* as appearing like an "Arabian Knights Fable; Affects all Beaver County's Future," in 1943 Lew Lessing discovered a rich tungsten deposit in a tunnel dug more than 50 years earlier. The *Press* described the find in glowing language: "It remained for Lewis Lessing, a comparatively young prospector in his late 30's while prospecting the surface ground for indications of tungsten to enter the old abandoned workings with his 'Aladis' [fluorescent] lamp and discover a veritable enchanted chamber, shimmering and scintillating with a billion tungsten crystals."[58]

Beaver County tungsten mines were remarkably productive during the 1940s. On 3 March 1944, Strategic Metals Incorporated shipped three carloads of tungsten from the Granite Mining District to U.S. Vanadium in Salt Lake City.[59] Reportedly, this mining district, which was under intensive prospecting, showed "a tremendous granite-lime contact, geologically conducive to the existence of tungsten ores." Local miners believed at the time that future development would reveal even larger deposits of shipping and milling grades of this particular strategic metal.[60]

Outside investors attracted by the increased mining success in the West Mountain District were also important to the county. In 1942 a group from Chicago leased the Garnett property owned by E. A. McGarry, and James E. Robinson's claims—the Rattler—were leased to outside investors as well.[61] A group of investors from Pennsylvania joined with others from Utah to form the Penn-Utah Mining Company, purchase mining properties and leases, and begin operations in the Frisco District. At the time of their incorporation, they had 20 promising claims to investigate.[62] The Metal Producers Company acquired a lease on part of the Horn Silver Mine and cleaned out the King David shaft, repaired it to a depth of 800 feet, and extended a lateral from that level of the shaft 2,000 feet along the Horn Silver ore channel. There they discovered new ore and expected a 200-ton output daily.[63]

Despite the war, throughout the 1940s, considerable mining activity continued in the San Francisco Mining District. Seven major mines were under various levels of development—the Horn Silver, the Moscow, the Wah Wah, the O.K., the Harrington Hickory, the Gold Reef, and the Old Hickory. Five other mines were preparing to produce, and 90 local men were employed in mining activities for an average monthly payroll of $30,000. The average daily production was between 175 and 200 tons.[64] Milford became increasingly central to southern Utah's mining activities during these years.[65]

For instance, a new 400-ton processing plant was scheduled to begin operating in the fall of 1947 to process low-grade ores from the Horn Silver Mine. Huge quantities of low-grade copper, lead, and zinc ores adaptable for milling were to be charted and blocked out for processing at the new mill. For the Metal Producers Company headquartered in California, which had run the old Horn Silver Mine in western Beaver County since 1941, the facility represented an accommodation to the amount of ore produced locally.[66] But in 1947 a presidential veto of the metals-subsidy bill broke the mining momentum, and the Horn Silver Mine closed down altogether.[67]

Under the management of Jack Lowe, the mine reopened in September and began plans again to construct a mill to handle the low-grade ores.[68] The mill was nearing completion in February 1948, and it was generally believed that this $300,000 facility would benefit the entire district. A several-thousand-gallon reservoir and a 16-inch well would supply the mill that was powered by two large, diesel-electric motors and would employ 25 local men.[69] By July the mill was grinding out ore at the rate of 500 tons per day and producing concentrates which were then shipped to Salt Lake smelters.[70]

Most recently, the Western Utah Copper Company formed to consolidate the principal copper/gold deposits of southwestern Utah into a single entity and ensure that they were developed. Copper mining has always been important to Utah's economy. The company recognized the continuing potential of the Milford area of the

San Francisco Mining District as a producer of copper, at first supplemental to what was supplied by Bingham Canyon. Western Utah Copper secured ownership of mining rights in the copper-producing mines in the district as well as in the Beaver Lake Mining District and the Rocky Mining District, both also in Beaver County. [71]

The story of the San Francisco Mining District mirrors that of Beaver County generally. Changing national markets and economic conditions dramatically impacted local conditions. Cycles of boom or bust created outrageous fortunes and overwhelming disappointment that made or broke lives. This contrast is represented by the towns of Milford and Frisco—Milford persisted stubbornly through successive mining booms to become a town more diversified in its services and resources, but Frisco's exotic conical kilns are poignant reminders of a time long past and dreams forever dashed by the harsh realities of life.

16

Uinta Basin

John Barton

The basin called Uinta in the northeastern part of Utah is a huge depression surrounded by mountains. It is approximately 125 miles long and varies between 40 and 60 miles in width. This unique region, with a variety of notable geographic features, includes Uintah and Duchesne Counties and spills over into western Colorado.[1] The Uinta Mountains and basin are part of the larger physiographic region known as the Colorado Plateau Province. The Uinta Mountains are the dominant feature of the region, and they form the basin's north rim. The Uintas, the highest range in Utah, are rugged mountains that angle east to west, unlike most mountain ranges in the world, which run north to south.[2] The Wasatch Mountains form the western rim of the basin; the Tavaputs Plateau, commonly called the Bookcliff Mountains, lies to the south, and the Rocky Mountains of Colorado to the east. The basin was once the lakebed of ancient Lake Uinta 150 million years ago. As the lake drained during the Mesozoic time periods, it created swampland, making it the home of the dinosaurs for which the region is famous. Eons later, some 50 to 36 million years ago, the Rocky Mountain (sometimes called the Laramide) orogeny (upthrust) created the Uinta Mountains.

The same geological upthrusts and glacial action that created the Uinta Mountains and carved them into their present form left deeply buried rich veins of precious ore. Most of the mining and mineral wealth of the Uinta Basin takes the form of oil and natural gas, exotic hydrocarbons, and phosphate. Of the total jobs in the basin, combined mining accounts for approximately 15 percent, and the larger portion of these positions earn the highest wages in the region, making mining a significant contributor to the overall economy.[3]

Significant amounts of precious metals have never been found in the Uinta Basin, in spite of folktales that claim the richest gold deposits in North America, if not the world, are located on the south slope of the Uinta Mountains. Wealth untold, according to Uinta Basin folklore, lies in the Uinta Mountains. These tales grow with retelling from generation to generation, and if they were true, the amount of gold deep within the south slope of the Uintas would topple the world's economies and devaluate the price of gold. The frequently repeated and often printed stories, although they lack historical documentation, include accounts of the Spanish discovering gold and forcing the Indians to work mines in the Uinta Basin. There is little doubt that the Spanish explored the basin for gold. Spanish bridle bits, cannonballs, ancient diggings, rock smelters, rusted helmets, swords and breastplates, and tree and rock inscriptions all confirm their presence but do not clearly establish when they were there nor their successes or failures at finding gold.[4]

The truth of these stories will possibly never be fully determined, but many believe them. Hundreds of people go each year to Rock Creek and other places throughout the Uinta Mountains to search for gold and the lost mines, including the legendary Rhoades Mine. To date no significant gold discoveries have been made, and, given the absence of corroborative physical evidence of Spanish mining in the Uinta Mountains, accounts of the lost mines must remain the stuff of folklore (see chapter four).[5]

Metals Mined in the Uinta Basin

Throughout the Uinta Mountains and basin, individuals have done a great deal of prospecting for precious metals with little significant success. Deposits of commercial-grade ore are scarce in the region. There are trace amounts of many minerals, including gold, silver, copper, lead, iron, molybdenum, and uranium,[6] but presently their quantities are not sufficient to make them minable. Flake gold is visible in the Green River's oxbows and sandbars but, again, not in enough quantity to justify the time and energy necessary for placer and/or dredge mining. One of the many attempts to placer-mine the Green River occurred in 1908. The Uintah Placer Mining and Exploration Company opened an immense dredge and sold shares in the company to investors. The company planned on buying several additional dredges from its anticipated profits to garner further wealth from the Green River, but the expected profits failed to materialize. The dredge operated on the second bend below Split Mountain.[7] Later, in 1913, the Gold Placer Mining Company attempted a dredge operation several miles downstream near Horse Shoe Bend, but again costs exceeded the amount of gold recovered.[8]

Historically there have been only a few commercial mining operations in northeastern Utah for metals. The Dyer Mine, located about 26 miles north of Vernal at an elevation of 9,000 feet, was the most successful. In 1889 a smelter was constructed on Anderson Creek, and mining was undertaken for the next several years. The Dyer Mine produced an average of 50 percent copper, some lead, and trace amounts of silver—26 ounces per ton—and gold—0.255 ounces per ton.[8] Between 1887 and 1900, the mine produced more than $3 million in copper. From 1891 to 1917, the Dyer churned out 4,377 tons of ore that yielded $395,655 in copper, $63,497 in silver, and $18,857 in gold. By 1904 most of the pockets of high-yield ore had been exhausted, but intermittent mining continued until 1941.[9]

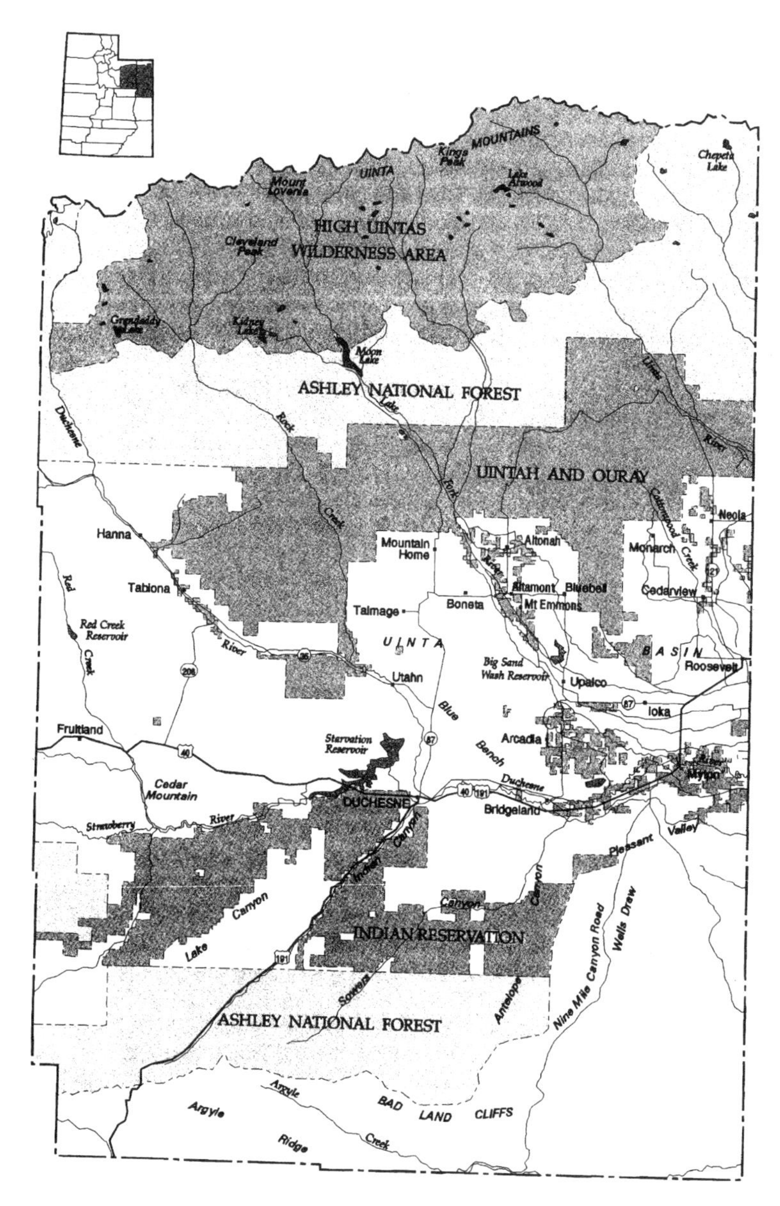

The Uinta Basin covers most of two Utah counties, Uintah (facing) and Duchesne (above), and extends into Wyoming and Colorado.

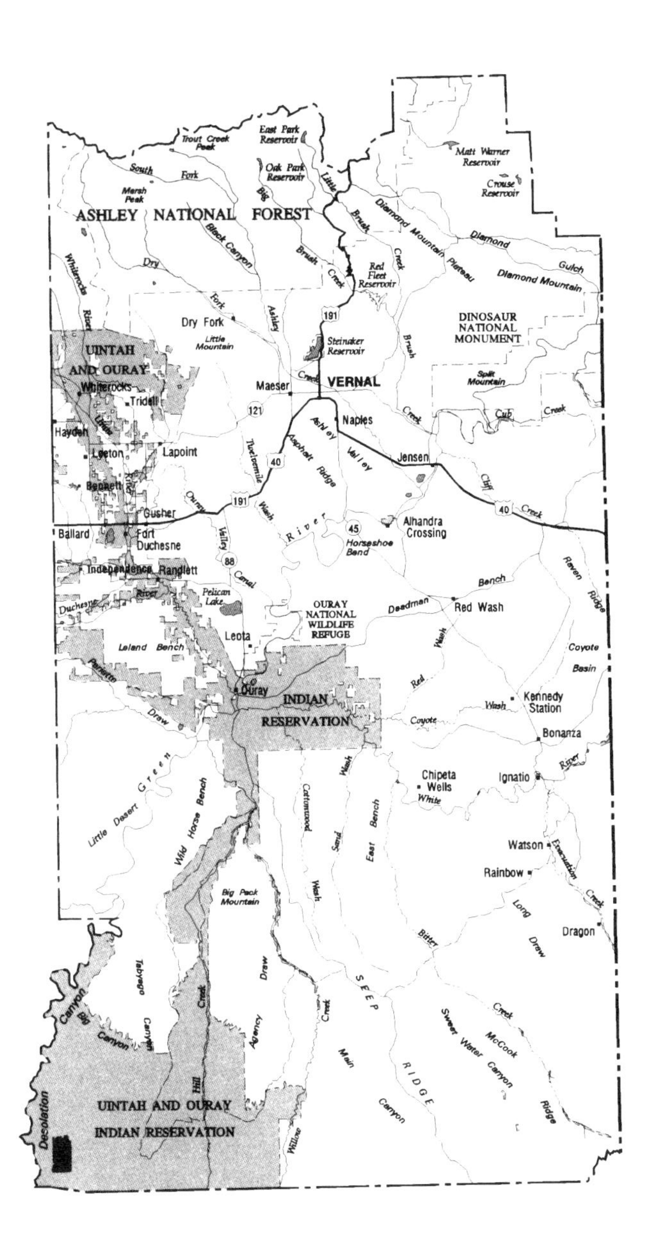

Trout Creek Peak
East Park Reservoir
Oak Park Reservoir
Matt Warner Reservoir
Crouse Reservoir
South Fork
Marsh Peak
ASHLEY NATIONAL FOREST
Black Canyon
Little Brush
Diamond Mountain Plateau
Diamond Gulch
Diamond Mountain
Whiterocks River
Dry
Brush Creek
Red Fleet Reservoir
Dry Fork
Little Mountain
Ashley
Steinaker Reservoir
DINOSAUR NATIONAL MONUMENT
UINTAH AND OURAY
Whiterocks
Tridell
Maeser
VERNAL
Split Mountain
Hayden
Naples
Cub Creek
Lapoint
Ashley Valley
Asphalt Ridge
Jensen
Leeton
Bennett
Gusher
Ballard
Fort Duchesne
Alhandra Crossing
Horseshoe Bend
Cliff Creek
Independence
Randlett
Pelican Lake
Duchesne River
OURAY NATIONAL WILDLIFE REFUGE
Deadman Bench
Red Wash
Raven Ridge
Leland Bench
Leota
Coyote Basin
Ouray
INDIAN RESERVATION
Kennedy Station
Bonanza
Chipeta Wells
White
Ignatio
Little Desert
Green
Wild Horse Bench
Cottonwood Wash
Sand Wash
East Bench
Watson
Rainbow
Big Pack Mountain
Long Draw
Dragon
Bitter Creek
SEEP RIDGE
Sweet Water Canyon
McCook Ridge
Desolation Canyon
Big Canyon
Tabyago Canyon
Hill Creek
Agency Draw
Main Canyon
UINTAH AND OURAY INDIAN RESERVATION
Willow Creek

The Silver King Mine is located only about a mile and a quarter from the Dyer. Its peak production year was 1947 with 34 tons of lead-zinc ore shipped. Prior to 1947 only 12 tons of hand-sorted ore had been sent. The ore assayed at 25 percent lead, 17 percent zinc, 1.5 percent copper, 2 ounces of silver, and 0.02 ounces of gold per ton.[10]

Additional claims have been worked, including the Commonwealth Mine, also on Anderson Creek. A placer claim, this 1908 operation boasted flumes, a mill, a smelter, mining pits, and bunkhouses. No records detailing the successes or failures of the Commonwealth survive, but had it been successful, more information would surely exist. Copper was discovered in 1896 in the Split Mountain area near the Utah/Colorado line. Only 5 tons of hand-sorted ore were shipped to Park City for smelting. It yielded 56 percent copper and 69 ounces of silver to the ton, but no further ore was produced. Iron deposits were located in the Brush Creek drainage. The Pope deposit contains high-grade red hematite, but the quantities are too limited to justify mining.[11] There are hundreds of additional claims, many of which were worked to a small degree, others not at all, throughout the basin but none worthy of further study.

Oil and Natural Gas

The Boom

As early as 1900, former straight-shooting sheriff of Uintah County John Pope drilled the first known well for natural gas or oil in the basin. Drilling to a depth of 1,000 feet showed no sign of gas or oil, and John Pope Number-One well was abandoned.[12] Additional wells were drilled on the west side of Asphalt Ridge in 1911, but again only dry holes resulted from the efforts. In 1917 the *Roosevelt Standard* reported that six different companies were planning on drilling for oil the next summer.[13] However, the excitement was short lived when the oil companies failed to make an appearance. During the next several years, the *Standard*'s headlines made wild claims about oil possibilities: "Mountains of Oil in the West," "Riches in Oil Shale," "The Uintah Basin Is the Greatest Underdeveloped Oil Field in the West."[14] But no drilling occurred. During the 1920s Earl Douglass, the famed paleontologist who discovered a cache of dinosaur bones in what is now Dinosaur National Monument, spent considerable time and effort touting the oil potential of the Uinta Basin. With Douglass's urging, a well was drilled east of Vernal in 1925 and made an excellent strike of natural gas: 10 million cubic feet. This started the development of the Ashley field, which was the first natural-gas producer in eastern Utah.[15]

The first oil well drilled in Duchesne County occurred in 1928, but there was no follow-up for 21 years. There was some limited oil drilling in Uintah County in the early 1930s, but as the Depression deepened, it ground to a halt. Again in the 1940s, with the outbreak of the war, it was hoped oil drilling would add jobs and augment the agricultural economy of the region. Between 1945 and 1947, the Standard Oil Company of California, Continental Oil, Gulf Oil, Carter Oil, and Union Oil all showed interest in eastern Utah, but a small company out of Salt Lake City under the leadership of J. L. Dougan brought in the first well. The Equity Company's strike unleashed a small oil boom in Uintah County. By 1949 there were 26 wells in production, and the Gusher and Roosevelt fields were opened that same year. In 1955 limited drilling had begun in the Bluebell field, but there was little follow-up for another 15 years.[16]

Then, in 1970, Miles Number One, the first new well in the Altamont/Bluebell oil field, was drilled just outside the town of Altamont. As large quantities of oil were found, the Uinta Basin became excited about the possibilities of the oil industry and the potential wealth it would bring to the area.[17] No other industry has a history of such wide swings between boom and bust as the oil business. Nationally and internationally, the oil industry has gone from scarcity to glut, and the prices of oil have reflected those extremes. The factors that brought about the development of the Uinta Basin's oil were not mere happenstance. The oil companies had known for decades that oil was available, but the nation's needs were not sufficient to justify drilling small fields such as those in the Uinta Basin.

Halfway around the world in October 1973, the Organization of Petroleum Exporting Countries (OPEC) declared an embargo on shipping crude oil to those countries supporting Israel in its conflict with Egypt. The Arab oil embargo triggered renewed drilling in the United States. Oil exploration and production in the basin and elsewhere in the West increased significantly when the nation and most of western Europe felt the sting of the embargo.

Speculation and development in the basin reached unprecedented proportions during the next few years. The Uinta Basin soon experienced a rapid increase in population, matched only by the opening of the Uintah Reservation to homesteading in 1905. The decade of the 1970s saw boom times in most oil-producing states: Texas, Alaska, Oklahoma, Colorado, Wyoming, Montana, and, to a lesser degree, Utah.[18] As the rest of the nation was reeling with unexpected jumps in the price of gas, coupled with lines of customers blocks long to purchase a rationed amount, the Uinta Basin had its first period of prosperity. Seemingly overnight, hundreds of rigs were drilling around the clock, each with a several-man crew, support crews, and services. Motels and restaurants could barely meet customer needs; often they were crowded to capacity and beyond. Traffic reached new levels as hundreds and then thousands of new residents and oil-related businesses used the roads and highways. When the oil boom started, there was not a single stoplight in Duchesne County, and Uintah County only had two.

For the first time in the basin's history, jobs were plentiful, and wages were good. Entry-level workers could make double their former teachers' salary by going to work on oil rigs. Support businesses for the oil industry flourished, including roust-about crews, heavy construction, work-over services, trucking companies, hot-oil trucks, fishing-tool services, oil-hauling companies, finishing rigs, wax cutting, pumping, pipelines and pump stations, and finally, a refinery in Roosevelt. The boom times brought sudden and new prosperity along with the growth. Real-estate prices soared. Local businesses—grocers, builders, auto dealers, clothing and furniture stores, movie theaters, and drugstores—all expanded and hired additional help. Many new, non-oil-related stores and businesses sprang up, such as mechanic shops, tire-sales stores, and additional gas stations and convenience stores.

As a result of the oil boom, the population of Duchesne County multiplied from 7,299 in 1970 to 12,537 in 1980, a 58 percent increase, and Uintah County experienced similar growth.[19] The sale of mobile homes did a booming business. Grocery stores and restaurants added new shifts, and many built additions to accommodate the growing needs of the county. More teachers were needed to meet the doubling of students in the local schools. The increased population brought new demands for professional services: doctors, lawyers, realtors, and insurance personnel.

The state's oil production, of which the combined Duchesne and Uintah Counties account for some 42 percent,[20] shows a marked increase in barrels (figured on a 42-gallon size) between 1948 and the high in 1985. In the last 35 years, the state's oil production peaked at 39.36 million barrels and had an estimated value of just over $1 billion. The greater Altamont-Bluebell Duchesne field reached 12.266 million barrels in 1978. The cumulative production total through 1978 for the field was 115.14 million barrels, making it the second largest oil field in the state.[21]

The Bust

As quickly as the oil boom struck, so did the bust. International oil prices and production significantly changed oil exploration in the county. Beginning in 1985, the price of crude oil fell from a record of near $40 a barrel just a few years earlier to $20 a barrel. Just as the international oil market brought about the conditions that resulted in the boom, so, too, did they contribute significantly to the bust. In 1985 several member nations of OPEC reshaped the oil market, their production causing a glut in the international market. The relatively high production costs of Uinta Basin oil severely reduced new drilling in the area.[22]

The total number of rigs drilling in the county plummeted as companies cut exploration back. Many workers lost their jobs, and a devastating ripple effect set in as oil and service companies, restaurants, retail stores of all types, banks, real-estate companies, city and county governments, schools, and social-service agencies were all impacted by the price drop.

Surcharges and fees imposed by the state and federal government, coupled with high drilling costs, made oil exploration and production less profitable in the basin when international prices dropped. The average cost per foot for drilling in Utah is $87.68, while the national average is $67.87.[23] Finally, most oil produced in the basin has a high wax content that necessitates heated pipelines to keep the oil flowing through the winter. This adds to the cost of production and makes the region's oil less desirable to produce.

High production costs of Uinta Basin oil led large international oil companies such as Shell, Gulf, and Chevron to sell their oil fields to smaller companies, including Linmar and Keystone. New drilling was almost unknown between 1986 and 1992, but in 1992, 44 new wells were drilled in Duchesne County, while in Uintah County 238 wells shut down that year.[24] The sale of basin oil fields and the rapid decline in the number of new oil wells drilled created a severe economic downturn in the region. Many oil-field service companies filed for bankruptcy, leaving local businesses financially impaired. The downturn of the county's oil industry resulted in a basinwide economic recession.[25]

House Bill 110 and the Revitalization of an Economy

The rapid decline in oil exploration and production left the basin economically challenged. After a political battle that many throughout the state thought futile, the gas and oil industry got a much-needed tax break for new and existing wells. As oil prices fell, it simply was not worth the oil companies' time and effort to keep local wells producing and pay the high taxes that the state demanded. Nor were the oil companies motivated to drill new wells. In 1990 Utah Representative David Adams, along with Representative Beverly Evans from Mt. Emmons (House District 54), Representative Dan Price from Vernal (House District 55), and State Senator Alarik Myrin from Altamont (Senate District 26), combined their legislative efforts to pass House Bill 110, providing a tax waiver and exemption for oil-production regions of the state, including Duchesne and Uintah Counties.[26] The tax waiver varies depending upon the international price of oil at the time but provides incentive to stimulate existing well production and did initiate some new drilling. The free fall of the region's economy halted with the slight but immediate difference House Bill 110 made. The 1990s and early 2000s have seen slight rises and dips in local oil production but have been for the most part stable.

Oil Shale and Tar Sands

In the Uinta Basin lies a wealth of yet untapped oil—shale oil and tar sands. Of all the world's oil reserves, a huge percentage is locked into sandstone and shale deposits. This oil, though as pure and refinable as any other, is far too costly to extract to be competitive on the open market at present prices. During the administration of George W. Bush, as conflict in the Middle East and the destruction from Hurricane Katrina helped push oil prices to all-time highs, research into the possibility of extracting oil from shale and tar sands increased greatly.[27] Commercial production of oil from bituminous sandstones or *tar sands* represents a considerable reserve for the future. *Oil shale*, similar to tar sands, is oil that has seeped into shale deposits. These "heavy oil" deposits have presented complex extraction problems for at least 17 companies that have conducted geological and engineering studies.

It is estimated that more than 2,000 square miles of basin lands have oil-shale deposits beneath their surface. A total of 21 impressive deposits of tar sands and oil-impregnated rock lie beneath the Uinta Basin, including the Asphalt Ridge deposit near Vernal, the P.R. Spring deposit near Hay Canyon in the Bookcliff Mountains, and the Sunnyside deposit. These three have been classified as "giants" and have an estimated reserve exceeding 500 million barrels of oil each. "Very-large" deposits of 100 to 500 million recoverable barrels include those at Raven Ridge and Dragon Wash. "Large" deposits indicate reserves with 10 to 100 million barrels and include Myton Bench, Whiterocks, Chipita Wells, Deep Creek, and Rim Rocks. "Medium" reserves range between half a million and 10 million barrels and include the Hill Creek and Yellowstone Lake Fork deposits.[28]

The oil-shale and tar-sand deposits will have a significant impact on the population and economy in the Uinta Basin when they are fully developed. Eastern Utah contains the highest grade of oil shale in the state, and it is estimated the Uinta Basin alone can produce some 100 billion barrels of oil once extractive processes have been perfected and their production costs become competitive with traditional drilling. Most of the present processes include some form of thermoextraction to separate the oil from the sand or shale. The energy costs of creating sufficient heat make them prohibitive at present. Various companies hold claims to these deposits and await a rise in the cost per barrel or new methods of extraction to lower production costs to the point where they can realize a profit.

Coal

The millions of years of swamp vegetation in the Uinta Basin created expansive coal deposits. Although there are significant coal-mining operations in the Uinta Basin, they are across the border in Colorado. Ironically, much of the coal mined there is burned in Utah fueling Deseret Generation and Transmission's power plant in Bonanza, about 35 miles south of Vernal.

Small deposits of coal exist in the northwest portion of Ashley Valley called Coal Mine Basin. At present there are no working mines there, but from the early days of settlement through the Depression, the mines operated on a small scale to supply coal for the local homes and businesses of Vernal.[29] The years from 1903 to 1905 saw the maximum productivity of Utah's Uinta Basin coal production reaching 10,000 to 13,000 tons. Production fell off gradually until 1948, when oil became cheaply available. By 1961 all coal mining in Ashley Valley had ceased.[30]

The first coal mine in the area was the Pack-Allen. Its main shaft was 900 feet with additional drifts of 1,400 feet. It employed some 18 men in the winter months and produced an annual average of 4,500 tons of coal during its 33 years of operation. Coal sold for $6.50 a ton, and miners were paid $2.50 per ton. All the coal produced was sold and used in Ashley Valley. The Government Mine, owned by the federal government, operated for 15 years and employed only three to four men annually. It supplied coal to the military post at Fort Duchesne.[31]

Carl Gardner, a century old and a longtime resident of Vernal, says that during the Depression, he was unable to find employment; so, taking matters into his own hands, he determined to support his family with what tools and skills he had. During December he arose early in the morning, about 4:00 AM, harnessed his team of horses and hitched them to his wagon in the minus-30-degree temperature, and drove them the six miles to Coal Mine Basin. He hand-loaded coal into the wagon and drove to Vernal's Main Street, where he went to every business offering to sell a day's worth of coal and promising to return the next day with more. He farmed what few crops would grow in the summer with the drought of the 1930s and continued to deliver coal every day to stores and businesses in the winter throughout the Depression.[32]

Gilsonite

Gilsonite, a rare hydrocarbon found in commercial quantities only in the Uinta Basin, is a unique solid petroleum substance and perhaps the area's most significant contribution to mining history.[33] It is shiny, hard, and black and resembles slick coal. Gilsonite is used to seal beer barrels and as a base for paints, inks, and perfumes. Additional uses are insulating pipelines, waterproofing and undercoating, paving, roofing, and other jobs utilizing crude asphalt; however, gilsonite's high quality makes it vastly more valuable than asphalt. It is also made into filler for rubber products, floor coverings, and tiles, among many other uses. Gilsonite appears in pure veins that vary from less than an inch to 17 feet wide, are 141 miles long, and total an estimated 30 million tons.[34] The larger veins are located some 20 miles both south and west of Vernal.

Several early travelers through the region found veins of gilsonite and were intrigued by them, but few had any notion of what to do with the mineral. Some tried burning it like coal, but the thick, dark smoke discouraged further fuel use. In 1885 Henry Gilson became interested in gilsonite and performed many home experiments to determine how to use it; to his wife's dismay, he ruined several of her pans and smoked up their home as he tried to melt it. His interest in the substance led to its being named after him.

The first boomtown for gilsonite mining was known as the Strip, later renamed Moffat. In 1888 a small section of land, about one mile east of newly constructed Fort Duchesne, was removed from Ute Reservation lands by an act of Congress. After the lobbying effort of brewer Adolphson Coors, who thought that gilsonite would be an excellent liner for his beer barrels, Congress agreed to remove the land if the Utes were willing. Plied with illegal liquor, the Utes were asked to sign an agreement allowing a 7,040-acre strip to be removed from the reservation to allow gilsonite mining by the Raven Mining Company. After the tribe looked the land over and found nothing of value, it agreed. The tract of land was about one mile wide at the north by three miles to the south and about four miles long; after it was removed from the Ute Reservation, gilsonite mining began.

A wild, lawless boomtown soon sprang up, starting as a tent city of miners with saloons and a red-light district; in a short time, frame buildings replaced the tents. The Strip was outside the boundaries of the reservation; therefore, the army at nearby Fort Duchesne had no jurisdiction there. And because it lay at the far west end of Uintah County 20 miles from Vernal, county law officers rarely visited. Outlaws such as Butch Cassidy and Elzy Lay frequented the place. Gambling and prostitution flourished, providing diversion for miners from their hard work and long hours. At its height the Strip possessed four houses of ill repute and five saloons. As wild as any western mining town, the Strip had at least 17 deaths from gunfights during its short existence.[35]

In 1899 more than 2,000 tons of gilsonite were shipped from the Strip at an average price of 50 cents per ton from the railhead in Wellington. The cost of production and hauling by wagon from the Uinta Basin over the Nine-Mile Road to Wellington averaged $21 per ton.[36] Hauling gilsonite from the Uinta Basin to Wellington was a flourishing business at the turn of the century. The gilsonite was placed in large canvas bags for shipping, and after loading their horse-drawn wagons with an average of 3,500 pounds, the men drove them from Vernal or Fort Duchesne to the bridge at Myton, south through Nine-Mile Canyon over Soldier Canyon, and then on to Wellington. The trip took an average of five days from Fort Duchesne or six days from Vernal. Freight costs were one dollar per hundred weight, and a good freighter made as much as $80 a week, although he had several overhead costs to pay from that figure.[37]

At the turn of the century, gilsonite miners earned an average of four dollars a day, which was a good wage for the time.[38] Most of the work was done with pick and shovel. Work in the gilsonite mines was hard, and conditions were appalling. When struck with a pick, gilsonite easily fractures into small workable pieces, and a goodly amount can be produced in a short time. But the fractured gilsonite creates small, needle-sharp slivers that torment every crease in the miner's body. To try and combat this problem, many miners rubbed themselves with lard. This of course was not popular with the women who washed their clothes. Mining ground to a halt in Moffat by 1904 when the veins were exhausted. In 1921 the town name was changed to Gusher because the citizens hoped nearby oil exploration would bring in a "gusher" to make the town rich and revitalize its economy.

As the gilsonite veins played out at Moffat, additional discoveries were made in the basin. In January 1905 Raven Mining was given 60 days to locate and file on 100 mining claims of gilsonite and asphalt on federal lands. Likewise, the Florence Mining Company was given a preferential right to locate 640 acres of mineral lands. These mining claims were a sore spot for the thousands of waiting settlers who planned on entering the Uinta Basin in August when the lands were opened for homesteading. *The Daily Sentinel* of Grand Junction, Colorado, denounced the affair as a "land steal and a farce."[39] Nevertheless, these claims were filed upon, and Raven Mining became the largest gilsonite-producing company in the world for the next several years.

In 1904 the Barber Asphalt Corporation started mining and shipping gilsonite from the town of Dragon, Utah. In 1912 the Uintah Railway was completed, linking the tiny town with the Denver and Rio Grande Western Railroad. Its primary purpose was to haul gilsonite ore from the Uinta Basin. The Uintah Railway was a narrow-gauge line that operated as the only railroad in Utah's Uinta Basin until 1939.[40] In 1912 the mining camp at Dragon moved to Rainbow, where mining continued until 1938, when the new camp of Bonanza was established. Most of the buildings from Rainbow were disassembled and reconstructed at Bonanza. The small town flourished until the late 1970s, when the last residents moved into Vernal and surrounding towns. The newly completed Bonanza highway made travel in car pools from Vernal easy, so miners could drive and have their families live in larger communities with greater shopping, educational, and cultural opportunities.

In the 1950s a slurry line was laid from the Cowboy Mine, 40 miles south of Vernal, to Gilsonite (near Fruita), Colorado, some 72 miles away. Cost of the line and refinery was $16 million. Ore was crushed to minus one-eighth inch, slurried into the six-inch pipeline in a concentrate of 60 percent water and 40 percent solids, and pumped at 350 gallons per minute.[41] By 1973 the slurry line was no longer used for gilsonite, and trucks carried the ore out of the basin.

Mining for gilsonite is still done in the Bonanza area by the American Gilsonite Company, Ziegler Chemicals, and Lexico Chemical and Mining Company. These are the only gilsonite-mining operations in the world. American Gilsonite is the largest. It began in 1942 when Chevron Oil Company purchased Barber Oil Company and formed American Gilsonite as a subsidiary. In 1991 Chevron divested itself of all non-oil-producing companies, and American Gilsonite became a private stock company. At present American Gilsonite ships 35,000 to 50,000 tons annually, and gilsonite sells for an average of $350 per ton. It is trucked to Salt Lake, where it is loaded on trains to be shipped anywhere in the world. American Gilsonite employs around 60 people in full-time positions. Many of the miners are the sons and grandsons of gilsonite miners.[42]

Mining gilsonite is fraught with danger and hardship. The hard, slick substance defies most mechanical means of mining. Use of spark-ignited explosives is dangerous due to the volatile nature of gilsonite. Heat from friction causes it to melt and gum up jackhammers. It dissolves in lubrication oils and fouls the drives or motors of the equipment. Rippers and chain saws have been tried with limited success. If the vein is large enough, front-end loaders simply scoop yards of fractured gilsonite up, but this is rare. Commercial mining operations now use hand-held compressed-air slushers to break the brittle ore from its veins after electronically detonated explosives have loosened and fractured it. The highly compressed airstream further fractures the ore into near-powder fragments of one-sixteenth inch, and then the ore is vacuumed by pneumatic lifts out of the mines.

Gilsonite, when fractured, becomes highly unstable. Explosions in gilsonite mines during the years 1894 to 1908 killed six miners. In 1945 14 shafts exploded in a Bonanza mine, igniting the largest gilsonite vein in the United States. The blast threw debris from the shaft into the town a half mile away as timber and rocks rained down upon homes and mine buildings. Miraculously no lives were lost. The worst mining disaster in the region occurred in the mines at Big Bonanza on 4 November 1953, when an explosion killed eight miners.[43]

Of the estimated 30 million tons of gilsonite deposits in the Uinta Basin, there are some 10 million still thought to be available.[44] All gilsonite deposits on federal lands are considered nonlocatable minerals, and permits for mining must be obtained. All the major deposits are on federal lands.

Elaterite and the Hope Mine

Perhaps the largest elaterite mine in the basin was the Hope Mine. Elaterite, another rare hydrocarbon similar to gilsonite, though pale in color, was prospected for and mined throughout the county in the early part of the twentieth century. The Hope Mine was just a mile south and slightly east of the Strawberry Pinnacles, about 15 miles west of Duchesne. It was the home and workplace of more than a dozen families and upward of 20 men. Complete with a pony-powered track line to get the ore out of the mine and down the mountain to the road, it was an engaging operation. The families lived in small log cabins, and there was a short-lived school to serve the children's educational needs.[45]

Phosphate

Some of the largest phosphate deposits in the world are found in the Park City formation some 15 miles north of Vernal. The deposits average 20 percent P_2O_5 at a thickness averaging 20 feet and covering more than 20 square miles on Brush Creek. There are an estimated 700 million tons of reserves of phosphatic rock in the Big Brush Creek deposit. Its principal use is in agricultural fertilizer. There are additional significant phosphate deposits on Ashley Creek and Diamond Mountain, but they are not mined at present.

J. H. Ratliff discovered the Brush Creek deposits in 1915. He and a partner, A. E. Humphrey, filed a claim, but little mining was done for the next several decades. In 1958 the San Francisco Chemical Company purchased the operation and shipped 15,000 tons of ore that year. By 1960 the company was trucking ore to Wyoming, but lack of a railroad in the area for large-scale shipments hindered production. By 1965 more than 180,000 tons had been produced, and the operation employed more than 200 workers.[46] In 1969 Stauffer Chemical Company bought the mine and planned a huge expansion to 300,000 tons annually. In 1980 Chevron Resources Company purchased the mining and ore-processing facility on Brush Creek and increased production to 450,000 to 750,000 tons per year. A fleet of 24 trucks hauled phosphate from the mine around the clock. In 1986 a slurry line was completed to pump the pulverized phosphate, mixed with water, over the Uinta Mountains to Rock Springs, Wyoming. In 1991 Chevron sold the operation to J. R. Simplot Company and Farmlands Industries, Inc. A subsidiary company, S. F. Phosphate, was formed and owns the mine at present. The company slurries nearly 300,000 tons a year and employs around 125 people.[47]

Conclusion

In spite of the unique, varied, and complex geological features of the Uinta Basin, the mining wealth and contributions to the economic well-being of the region come from oil and unusual minerals. Even without profitable metal mining, the mining industry is still a significant addition to the basin's economic well-being, however. Most mining jobs, whether they are at S. F. Phosphate, American Gilsonite, or in the oil industry, provide the highest wages in the area. Gilsonite and phosphate mining in the Uinta Basin is unique to Utah and western mining. The Uinta Basin is the only place in the world where gilsonite is mined commercially. Some of the largest phosphate deposits in the world are also located in the area. Oil and natural gas are a mainstay for the local economy also, and the Uinta Basin contains some of the largest shale-oil and tar-sands deposits in North America. Wealth in the form of precious metals on a significant scale has yet to be located.

Glossary of Geologic and Mining Terms

accretion: Growth of an inorganic body, like an island or a **continent,** by gradual addition of small land fragments or wedges of **sediment.**

abutment: The solid walls along the sides of a **mine** which support the **roof** or arch of **strata** spanning the roadway.

acid deposition or rain: A wet and dry mixture containing a higher than normal amount of nitric and sulfuric acids from both natural sources, such as **volcanoes** and decaying vegetation, and man-made emissions, primarily sulfur and nitrogen oxides resulting from **fossil-fuel** combustion.

acid mine drainage: Acidic runoff from mine **waste** dumps and **mill tailings ponds** containing **sulfide minerals;** also groundwater pumped to the surface from **mines.**

acid waters: Water, including runoff and rain, containing an excess hydrogen ion and/or sulfuric acid, usually due to weathering of iron **pyrites.**

active workings: Any place in a **mine** where **miners** are normally required to work or travel and which is ventilated and inspected regularly.

adit: A horizontal or nearly horizontal opening driven into the side of a mountain or hill providing access to a **mineral deposit;** horizontal adits are called **tunnels,** while vertical adits are **shafts.**

air split: The division of an air current into two or more parts.

airway or air course: Any passage through which air moves.

alluvial plain: Alluvium refers to **sediments** deposited by rivers. The sediment may be deposited in riverbeds, alluvial fans at the base of mountains, or floodplains.

alteration: Any physical or chemical change in a rock or **mineral,** subsequent to its formation, that is milder and more localized than metamorphism.

amalgam: The combination of **mercury (quicksilver)** with one or more **metals** after **milling.**

anemometer: Instrument for measuring air velocity; used to ensure **ventilation** in **mines.**

angle of draw: In a **coal mine,** this angle bisects the angle between the vertical and the **angle of repose.**

angle of inclination of the descending lithospheric slab: Angle in relation to the horizontal plane of the descending **lithosphere** at a **convergent plate margin.**

angle of repose: The maximum angle from the horizontal where a material will rest on a surface without sliding or rolling.

anode: Impure copper cast into a specific shape to be used in an electrolytic refinery for final purification.

anticline: An upward **fold** or arch of rock **strata**, often shaped like the crest of a wave.

apex: The edge or crest of an **ore vein** that is nearest the surface; basis of the apex law, which grants ownership of an entire vein to the individual or corporation that owns the claim where that vein comes closest to the ground surface.

aquifer: A water-bearing bed of porous rock, often **sandstone.**

arching: Fracture processes around a mine opening, or brick or stone forming the **roof** of a **mine.**

***arrastra*:** A Spanish device used for **milling** gold or **silver;** a circular pit faced with rocks where **ore** was spread and then crushed by heavy stones dragged across it.

assay: A chemical test performed on samples of **ores** or **minerals** to determine their contents.

auger: A rotary drill that uses a screw device to penetrate, break, and then transport the drilled material.

auriferous: Ground containing particles of gold or gold in combination with other **mineral**(s).

autogenous grinding: A secondary grinding system where material is rotated in a revolving chamber that uses no rods or balls.

auxiliary operations: All supportive activities that don't contribute directly to mining.

azimuth: A surveying term for the angle measured clockwise from an established line of reference.

back: That part of a **lode** nearest the ground surface in relation to any part of the **workings;** see also **roof.**

backfill: Waste material used to fill the void created by mining an **ore body.**

ball mill: A horizontal, rotating steel cylinder filled with steel balls into which crushed **ore** is fed; as the mill rotates, the balls grind the ore into fine particles.

barren: Rock or **vein** containing no **minerals** of value, or too little to be workable.

barricading: Enclosing part of a **mine** to prevent the inflow of gases from a fire or explosion.

basement: A complex series of **igneous** and **metamorphic rocks** beneath the oldest **sedimentary** ones.

base metal: Any nonprecious **metal** (e.g., copper, lead, zinc).

beam: A bar girder which supports a span of **roof** between two support props or walls.

beam building: The creation of a strong, inflexible **beam** by fastening together several weaker layers of rock.

bearing: A surveying term which designates direction; the acute horizontal angle between the meridian and the line.

bearing plate: A plate which distributes a load, for example, in the **roof** of a cavity.

bedding: The arrangement of **sedimentary rocks** in layers.

bedrock: The solid rock underlying the looser materials of the Earth's surface.

bench: A separation in a **coal seam** created by **slate** or formed by cutting the coal.

benching: Creating benches or steps to work **open-pit mines** or small quarries.

beneficiation: Treating mined material to make it more concentrated or richer.

berm: A pile of material large or dense enough to restrain a vehicle.

binder: A streak of impurity in a **coal seam.**

black damp: A mixture of carbon dioxide and nitrogen, or air depleted of oxygen.

black lung: See **pneumoconiosis.**

blacklist: List of suspected individuals or organizations, often refused business or employment, usually in relation to labor-management disputes.

blasting agent: Any material consisting of a fuel and an oxidizer.

blasting cap: A device that detonates explosives; it contains a charge of detonating compound which can be ignited by electric current or a spark from a fuse.

blasting circuit: An electric circuit that can fire **detonators** or spark a charge to light an igniter cord.

blast furnace: A furnace that combines solid fuel with air to **smelt ore.**

block caving: An inexpensive method of mining where large blocks of **ore** are **undercut,** causing it to break or cave in under its own weight.

blossom rock: Float ore that has decomposed from **outcrops.**

bonanza: (Spanish, "blue skies") The discovery of rich **ore.**

borehole: Any deep hole drilled into the ground to obtain geological information.

borrasca: (Spanish, "barren rock") A worthless **claim;** the opposite of ***bonanza.***

bottom: See **floor.**

brattice or brattice cloth: Fire-resistant fabric or plastic used to partition a mine passage and force air into the working area; also called *line brattice*, *line canvas*, or *line curtain*.

break line: The line along the **roof** of a **mine** that can possibly break; it usually follows the rear edges of **pillars.**

breakthrough: A **ventilation** passage cut through **pillars** or between rooms.

breasting: The process of removing **ore** from the face of a **drift** or **stope.**

breccia: A rock made of coarse, angular fragments, usually surrounded by a mass of fine-grained **minerals.**

brow: A low place in the **roof** of a **mine;** a place with insufficient headroom.

brushing: Digging up the **bottom** or taking down the **top** in a **tunnel** to provide more headroom.

Btu: British thermal unit; the amount of energy required to raise the temperature of one pound of water one degree Fahrenheit.

bug dust: The fine particles that result from boring or cutting the mine **face** with a drill or machine.

bulk mining: Any large-scale, mechanized method of mining involving many thousands of tons of **ore** being brought to the surface each day.

bullion: Gold or silver that has been **milled** and **smelted** to a state approaching purity, then molded into bars.

bump or burst: A violent dislocation of the **mine workings** attributed to severe stresses in the surrounding rock.

cage: An elevator compartment raised and lowered in a **shaft** to move people, equipment, or **ore.**

calcareous mudstone: A dark, clay rock containing calcium **carbonate.**

calorific value: The quantity of heat that is released from one pound of **coal** or oil measured in **Btus.**

caltrop: A spiked metal ball to impede horses or vehicles; sometimes used in labor disputes.

cannel coal: A noncaking block **coal** with a fine, even grain and a conchoidal **fracture** which has a high percentage of hydrogen, is easy to ignite, and burns with a long, yellow flame.

canopy: A protective covering on the cab of a mining machine.

cap: A **miner**'s safety helmet, or a highly sensitive, encapsulated explosive to detonate larger, but less sensitive, explosives.

car: A railway wagon, especially those adapted to carry **coal**, **ore**, and **waste** underground.

carbon monoxide: (CO) Gas that may be present in the afterdamp of a gas or **coal-dust** explosion or the gases given off by a mine fire; also called **white damp.**

car dump: The mechanism for unloading a **car.**

carbonate rocks: Limestone and **dolomite** composed of calcium carbonate or calcium and magnesium carbonate.

cast: In strip mining, the action of throwing the **overburden** from an area currently being mined to a previous one.

caving: See **undercut.**

***cazo* process:** An old Spanish method of treating **silver ore** by boiling it in a copper-bottom vat or kettle with salt and **mercury** to speed the process of amalgamation.

channel sample: A sample composed of pieces of a **vein** or **mineral deposit** that have been cut out of a small trench or channel, usually about 10 centimeters wide and 2 centimeters deep.

check curtain: Sheet of **brattice cloth** hung across an **airway** to control the passage of the air current.

chemical reduction: The process of removing oxygen by adding electrons to chemical compounds.

chili mill: A refinement of the ***arrastra*** that utilized stone wheels in place of rocks.

chlorination: Injecting chlorine gas into milled and roasted **ore** to remove gold from its **gangue.**

chock: Large hydraulic jacks to support the **roof** in **long-wall** and **short-wall** mining systems; see also **crib.**

chute: An opening, usually constructed of timber and equipped with a gate, through which **ore** is drawn from a **stope** into mine **cars.**

claim: The boundaries that encompass the ownership of a parcel of ground presumed to contain a **mineral vein** or **placer deposit.** See also **location, claim jumping.**

claim jumping: The illegal appropriation of another person's **claim.**

Clean Air Act: A federal law governing the nation's air quality originally passed in 1970 (and since amended) to address specific problems such as **acid deposition,** urban smog, hazardous air pollutants, and stratospheric ozone depletion.

clean-coal technologies: Technologies designed to burn **coal** more efficiently and enhance environmental protection.

cleat: The vertical cleavage of **coal seams;** usually the main set of joints along which coal breaks when mined.

cleavage: A tendency in some rocks and crystals to split in specific directions; see also **schistosity.**

closed shop: A business or company which hires only **union** members; the opposite of an **open shop.**

coal: A solid, brittle, stratified, combustible carbonaceous rock formed by decomposition of vegetation and classified by its degree of hardness, amount of moisture, and heat content; graded from hardest to softest: anthracite, bituminous, subbituminous, lignite.

coal dust: Particles of **coal** that can pass through a number-20 sieve.

coal gasification: The conversion of **coal** into a gaseous fuel.

coke: A hard, dry carbon substance produced by heating **coal** to a high temperature in the absence of air.

collar: Timbering or concrete around the **mouth** or **top** of a **shaft**.

colliery: A **coal mine.**

column flotation: A precombustion **coal**-cleaning technology where particles attach to air bubbles rising in a vertical column and are removed at the top of the column.

commercially mineable ore body: A **mineral deposit** that contains enough **ore reserves** to be mined with an economic advantage.

comminution: The breaking, crushing, or grinding of **coal**, **ore**, or rock.

competent rock: Rock capable of sustaining openings without any structural support except **pillars** and walls during mining.

complex ore: Ore containing a number of **minerals** of economic value. The term often implies that there are difficulties in liberating and separating the valuable **metals.**

concentrate: A fine, powdery product of **milling** containing a high percentage of valuable **metal.**

concentrator: A **mill** or plant where **minerals** are separated from **waste** rock.

cone crusher: A machine which crushes **ore** between a gyrating cone or crushing head and an inverted, truncated cone known as a *bowl.*

conglomerate: A **sedimentary rock** composed of rounded boulders, cobbles, and pebbles, often with some **sand.**

contact: A geological term used to describe the line or plane along which two different rock formations meet.

continent: A large landmass composed of **igneous** and **metamorphic rocks** with an average composition of granite covered with a thin veneer of **sedimentary rock.** The low average density of continental rocks causes them to rise above the deep ocean floor, while the margins of continents, the continental shelves, are often submerged beneath sea level.

continental glacier: An ice sheet, sometimes of great thickness, that covers a significant portion of a **continent;** the ice sheets covering Greenland and Antarctica are modern examples.

continuous miner: A machine that constantly extracts **coal** while it loads it.

contour: An imaginary line that connects all points on a surface having the same elevation.

convection cell: When applied to circulating water, it refers to heated, less-dense water rising and cooler, denser water sinking to replace it.

conventional mining: The first fully mechanized underground mining method which inserted explosives into a **coal seam,** blasted the seam, and used a loading machine to remove the coal onto a **conveyor** or shuttle **car**.

convergence: Convergent **lithospheric plates** are colliding, producing crustal shortening and thickening, volcanic activity, and earthquakes.

conveyor: An apparatus for moving material from one point to another in a continuous fashion; there are two kinds: a **belt conveyor** is a looped belt supported on a frame; a **chain conveyor** moves material along in solid pans or troughs by using scraper crossbars attached to powered chains.

coral-building organisms: Marine organisms that live on the sea floor and construct large edifices from their shells that are dominantly made of calcium **carbonate,** the main constituent of **limestone.**

core: The Earth's nucleus or kernal below a depth of about 2,900 kilometers, composed primarily of iron and nickel and responsible for the Earth's magnetic field.

core sample: A cylindrical sample, generally one to five inches in diameter, drilled out of an area to perform a geologic and chemical analysis of the **overburden** and **ore.**

Cornish pump: A device, based on the simple farm well pump, used in deep **mines** to remove water that seeps onto the **sump.**

country rock: The rock that encases a **vein,** usually rock with no **mineral** content; see **gangue.**

cover: The **overburden** of any **deposit.**

coyoting: A method of **placer mining** that utilizes a vertical **shaft** with radial **tunnels** bored out along the surface of the **bedrock.**

cradling: A variation of the **panning** method of washing gold where the cradle is rocked while a stream of gold-laden dirt and water funnels through, and the gold is caught in cleats on its bottom.

craton: The stable portion of a **continent,** usually in the interior, composed of **basement** and a thin veneer of **sedimentary rocks** and not involved in youthful mountain building.

creep: Very slow movement of slopes downhill; or, in **underground mining, pillars** sinking into a soft **floor** from the weight of a **roof.**

crevicing: The simplest, but usually least effective, method of mining, involving only prying bits of **mineral** loose from cracks and crannies in rock with a strong knife.

crib: A **roof** support of prop **timbers** or ties, laid in alternate cross layers, log-cabin style, which may be filled with debris; also called **chock** or *cog*.

crop coal: Coal at the **outcrop** of the **seam.**

crossbar: Horizontal **roof timber** supported by props.

crosscut: A horizontal opening driven from a **shaft** at (or near) right angles to explore a **vein** or other **ore body** or provide **ventilation** or communication within a **mine.**

cross entry: An **entry** running at an angle to the main **adit.**

crusher: An apparatus which breaks **ore** into progressively smaller pieces.

crust: The outermost layer of the Earth, which consists of the lowest-density rocks that contain the most silicon and aluminum in their average composition.

cut and fill: A method of **stoping** where **ore** is removed in slices, or lifts, and then the excavation is filled with rock or other **waste** material **(backfill)** before the next slice is extracted.

cutter or cutting machine: A machine, usually used in mining **coal,** that cuts a slot to allow room for expansion of the broken coal; the operator of the machine is also called a cutter.

cyanide process or cyanidation: A method of precipitating exposed gold or **silver** grains from crushed or ground **ore** by dissolving it in a weak cyanide solution; see also **leaching.**

cycle mining: A system of mining in more than one place at a time, such as moving from one **face** to another while the first is being shored up.

decline: An underground passageway connecting one or more **levels** in a **mine.**

delta: A generally triangular-shaped **deposit** of **sediment** at the mouth of a river.

deposit: A natural occurrence of a useful **mineral** or **ore.**

deposition: Material deposited from the atmosphere; may be influenced by **acid rain.**

depth: Vertical distance below the surface, or the distance down from the beginning of a **shaft** or hole.

detonator: A device to set off a small charge that will in turn ignite a larger explosion, such as **blasting caps,** exploders, electric detonators, and delay electric blasting caps.

development: Work carried out to prepare a **mineral deposit** or **ore body** for mining.

diamond drill: A rotary type of drill that cuts a core of rock into long cylindrical sections two centimeters or more in diameter.

diffuser fan: A fan mounted on a **continuous miner** to direct air delivery from the machine to the **face.**

diffusion: Blending of a gas and air or two or more gases.

dilution: Rock removed with the **ore,** subsequently lowering its **grade.**

dip: The angle at which a **vein, seam,** structure, or rock bed is inclined from the horizontal as measured at right angles to the **strike.**

disseminated ore: Ore carrying small particles of valuable **minerals** spread more or less uniformly through the **host rock.**

dolomite: A mineral composed of calcium, magnesium, and **carbonate;** also commonly the rock composed of dolomite.

dore: Unparted gold and **silver** poured into molds when molten to form buttons or bars.

double jacking: A process by which two (occasionally three) men hand-drill blasting holes; one holds the drill rod while another hammers it; the rod is turned after each blow.

dragline: A large excavation machine used in **surface mining** to remove **overburden.**

drainage: Removing surplus ground or surface water by pumps or gravity.

draw slate: A soft **slate, shale,** or rock located immediately above **coal seams** which falls when the coal is removed.

dredging: A variation of **panning** where a shallow barge crawls over a water-soaked surface believed to be rich in **ore,** washing a steady stream of gravel and depositing the **waste** behind in a winnows.

drift: A horizontal underground opening that follows the length of a **vein** or rock formation as opposed to a **crosscut** which crosses the formation.

drive: An underground passage used to **explore, develop,** or **work** an **ore body.**

dross: Scum that forms on the surface of molten **metals.**

due diligence: A financial and technical investigation before deciding whether a **mine** will be sufficiently productive to be worth the time, money, and equipment to extract the **ore.**

dummy: A bag filled with **sand** or clay for stemming a charged hole.

dump: Piles of **waste** rock from a **mine**, or the place where waste rock is deposited.

entry: An underground horizontal or near-horizontal passage for **haulage, ventilation,** or movement.

environmental impact study: A written report, compiled prior to a production decision, that examines the effects proposed mining activities will have on the natural surroundings.

epithermal deposit: A **mineral deposit** consisting of **veins** and replacement bodies, usually in **volcanic** or **sedimentary rocks,** containing **precious metals**, or, more rarely, **base metals.**

evaluation: Gaining knowledge of the size, shape, position, and value of **coal** or **ore.**

exploration: Work involved in searching for **ore,** usually by drilling or driving a **drift.**

face: The solid, usually unbroken surface at the end of a **drift, crosscut,** or **stope** where work is taking place.

face cleat: The principal cleavage plane or joint at right angles to the stratification of the **coal seam.**

face conveyor: Any **conveyor** parallel to a **working face** which delivers **coal** or **ore** into another conveyor or **car.**

factor of safety: The ratio of the ultimate breaking strength of the material to the force exerted against it.

Fair Labor Standards Act: Also called the Wages and Hours Law; passed in 1938 to eliminate "labor conditions detrimental to the maintenance of the minimum standards of living necessary for health, efficiency and well-being of the workers."

fault: A slip surface between two portions of the Earth's surface that have moved relative to each other.

fault zone: A **fault** which is not a single clean **fracture** but an area hundreds or thousands of feet wide with many interlacing small faults.

feed: Crude **ore** brought into a grinding **mill** or treatment plant.

feeder: A machine that puts **coal** or **ore** onto a **conveyor** belt evenly.

feldspar or felspar: Any of a group of crystalline, rock-forming **minerals** consisting of silicates of aluminum with potassium, sodium, or calcium.

fill: Any material that is put in place of the extracted **ore** to provide ground support.

final voids: The openings which remain after **ore** and **waste** rock have been extracted.

fines: The smallest particles of **minerals** or **coal** in a classification or sampling process.

fire damp: See **methane.**

fire setting: A Spanish mining method where the **vein** is heated, usually by lighting a fire against it, and then cold water is thrown against the rock, breaking it up by thermal stress.

fissure: A **fracture** in rock that is open to water flow in the subsurface.

fitcher: The moment when the surrounding rock grips a drill being pounded in to make a blasting hole.

fixed carbon: The part of carbon that remains when **coal** is heated in a closed vessel until all of the volatile matter burns off.

flat lying: Deposits and **coal seams** with a **dip** of five degrees or less.

float: Pieces of rock that have broken off and been moved from their original location by natural forces such as frost or glacial action.

float dust: Fine **coal-dust** particles, often carried in suspension by air currents, that can pass through a number-200 sieve.

floor: The **bottom** or underlying surface of an underground excavation.

flotation: A milling process separating **mineral** particles from **waste** rock by mixing ground **ore** with water, chemical reagents, and air; valuable mineral particles become attached to bubbles and float, while other, less-valuable particles sink.

flue-gas desulfurization: Commonly called **scrubbers,** a chemical or physical process that removes sulfur compounds formed during combustion.

flume: A trough, usually wooden, designed to carry water.

fly ash: Finely divided particles of ash suspended in gases resulting from fuel combustion.

fold: A bend in a layer of rock; see **anticline** and **syncline.**

fold mountain belt: Mountainous terrain consisting of folded **sedimentary** layers, **thrust faults**, and **igneous intrusions,** generally linear in outline and believed to form at **continental plate margins.**

footwall: The rock on the underside of a **vein** or **ore** structure; the block of rock below an inclined **fault** surface or **ore body;** see also **wall rocks.**

formation: An assemblage of rocks which have some characteristic in common, such as origin, age, or composition.

fossil fuel: Any naturally occurring fuel of an organic nature, such as **coal**, crude oil, or natural gas.

fracture: A break or split in the rock which allows **mineral**-bearing solutions to enter. A **cross fracture** is a minor break extending at more or less right angles to the direction of the principal ones.

free milling: Ores from which **precious metals** can be recovered by concentrating without resorting to pressure **leaching** or other chemical treatment.

fuse: A cordlike substance to ignite explosives.

galena: Naturally occurring lead **sulfide,** the most common **ore mineral** of lead, often associated with **silver.**

gallery: A horizontal or nearly horizontal underground passage, natural or artificial.

gangue: The base material in and around a **vein** that holds more valuable **minerals** in place. See **Country rock.**

gasification: A process by which **coal** is turned into low, medium, or high **Btu** gases.

geophysical survey: Indirect methods of investigating the subsurface geology using physics, including electric, gravimetric, magnetic, electromagnetic, seismic, and radiometric principles.

ghost town: Remnants of a community that grew up around a mining operation after the **ore** is gone or is no longer worth mining and the area has been deserted.
glacier: A mass of ice that has become so thick that it flows spontaneously.
global positioning satellites: A series of satellites orbiting the Earth widely used to locate geographic points precisely.
gob or goaf: The loose **waste** in a **mine** or in that part of the mine from which the **coal** has been removed and the space more or less filled up with **waste.**
grab sample: A sample from a rock **outcrop** that is **assayed** to determine if valuable elements are present; a grab sample is not intended to be representative of the **deposit,** and usually the best-looking material is selected.
grade: The average **assay** of a ton of **ore,** classified according to the worth of its **mineral** content.
grain: In petrology, that quality of the texture of a rock composed of distinct particles or crystals which depends upon their absolute size.
grizzly: Coarse screening or scalping device made of rails, bars, and beams that prevents oversized bulk material from entering a transfer system.
ground control: Measures taken to prevent **roof** falls or **bumps.**
gunite: Cement sprayed on the **roof** and sides of a mine passage.
gypsum: A naturally occurring, hydrated form of calcium sulfate used to make plaster of Paris.
halite: Rock salt.
hanging wall: The rock on the upper side of a **vein** or **ore deposit**; see also **wall rocks.**
hard rock: Ore that must be blasted as opposed to ore soft enough to be worked with hand tools.
haulage: The horizontal transporting of **ore, coal,** supplies, and **waste.**
head frame: A frame, usually wood or steel, at the top of a **shaft** that supports the **cage** for lowering or raising.
head grade: The average **grade** of **ore** fed into a **mill** or **concentrator.**
heading: A **vein** above a **drift,** or an interior **level** or **airway** driven in a **mine.** In **long-wall workings,** it is a narrow passage driven up from a gangway.
head section: That portion of the **conveyor** that discharges material.
head works: The buildings near or around the **adit** to a **mine,** such as the shaft house, hoist house, pump house, or offices.
heap leaching: A process which percolates a **cyanide** solution through crushed **ore** heaped on an impervious pad or base to dissolve out **minerals** or **metals**.
heaving: The rising of the **bottom** of a **mine** after the **coal** has been removed.
high grade: As a noun, rich **ore;** as a verb, selective mining of the best ore in a **deposit.**
high wall: The unexcavated face of exposed **overburden** and **coal** in a **surface mine,** or in a **face** or bank on the uphill side of a **contour mine.**
high-wall miner: A mining system consisting of a remotely controlled **continuous miner** which extracts **coal** and conveys it via augers or **conveyors** to the outside.

hoist: The apparatus that raises or lowers the **cage** in a deep **mine** to transport **miners** or material.

horizon: In geology, any given definite position or interval in the stratigraphic column or the scheme of stratigraphic classification; generally used in a relative sense.

horse: Waste rock within a **vein.**

horseback: A mass of material with a slippery surface in the **roof,** often resembling a horse's back.

host rock: The rock surrounding an **ore deposit.**

hydraulic mining: A form of **placer mining** that uses powerful jets of water to wash away hillsides where **alluvial deposits** of gold are found.

hydrocarbon: Chemical compounds containing carbon and hydrogen atoms in various combinations, found especially in **fossil fuels.**

hydrograph: A plot of a lake elevation as a function of time.

hydrometallurgy: The treatment of **ore** by wet processes (e.g., **leaching)** that turns a **metal** into solution so it can be recovered.

igneous intrusions: Rocks formed by cooling and crystallizing **magma** underground so that it intrudes into surrounding rocks.

igneous rocks: Rocks formed by cooling and crystalizing **magma** composed of melted silicate **minerals.**

immediate roof: The **roof strata** immediately above the **coal** bed.

impermeable: Rocks that block the flow of underground water.

inby: In the direction of the **working face.**

incline: Any **entry** to a **mine** that is not vertical **(shaft)** or horizontal **(adit).**

incompetent: Strata or rock of insufficient firmness and flexibility to transmit a thrust and lift a load by bending.

Industrial Workers of the World: A trade union organized in 1905 and noted, often feared, for its radical tactics; its members were known as **Wobblies.**

inferred resources: See **reserves.**

in situ: In the natural or original position.

intake: The passage through which fresh air is drawn or forced into a **mine.**

intrusive: A body of **igneous rock** formed by the consolidation of **magma** intruded into other rocks; a contrast to lavas, which are extruded upon the ground surface.

isopach, isopachous line, or isopachyte: A line on a map drawn through points of equal thickness of a designated unit.

jackleg: A percussion drill mounted on a telescopic leg which has an extension; the leg and machine are hinged so that the drill need not be in the same direction as the leg.

jig: A concentrating machine which separates very heavy particles from lighter **(waste)** ones by using the principle of hindered settling. A bed of heavy material (for example, lead or steel shot) inside a box is agitated up as a stream of **ore** particles in water percolates down. The heavy particles settle deeper into the bed, while the waste particles wash off the top. The **concentrate,** drawn from the bottom of the jig, can be upgraded further on a **table.**

joint: A divisional plane or surface that divides a rock and along which there has been no visible movement parallel to the plane or surface.

kettle bottom: A smooth, rounded piece of rock, cylindrical in shape, which may drop out of the **roof** of a **mine** without warning.

kerf: The **undercut** of a **coal face.**

lagging: Planks or small **timbers** placed between steel ribs along the **roof** of a **stope** or **drift** to prevent rocks from falling, rather than to support the main weight of the overlying rocks.

lamp: The electric light on a **miner's** cap, or the flame safety lamp in **coal mines** to detect **methane** gas concentrations and oxygen deficiency.

Laramide mountain-building event: Mountain building in western North America that took place between about 70 to 50 million years ago. Mountain building was thick skinned and involved ancient **basement** rocks.

layout: The design or pattern of the main roadways and **workings.**

leaching: Removing minerals from **gangue** by dissolving them in chemicals or percolating water. See **cyanide process** and **lixiviation.**

lead: (pronounced like "greed") The visible course of a **vein.**

ledge: A visible portion of rock that contains rich **ore.** See also **lode.**

lens: A body of **ore** that is thick in the middle and tapers toward the ends.

levels: The horizontal "stories" of a deep **mine;** it is customary to work mines from a **shaft,** establishing levels at regular intervals, generally about 50 meters or more apart.

lift: The amount of **coal** obtained from a **continuous miner** in one cycle.

limestone: A bedded, **sedimentary rock** composed mainly of calcium **carbonate** which often includes the shells of reef-building organisms.

liquefaction: The process of converting **coal** into a synthetic fuel similar to crude oil and/or refined products, such as gasoline.

lithology: The character of a rock described in terms of its structure, color, **mineral** composition, grain size, and arrangement of component parts.

lithosphere: The rigid outer layer of the Earth that includes **continental** and oceanic **crust** and the uppermost **mantle.**

lithospheric convergent margins: Margins of **plates** of **lithosphere** that are colliding.

lithospheric plates: Regions of **lithosphere** that share boundaries; these **plates** move independently of one another and may diverge or collide.

lithospheric slab: A portion of a **lithospheric plate** that, due to converging with adjacent **lithosphere,** is diving down into the **mantle.**

lixiviation: The process of removing **silver** from refractory **ores** by roasting and chlorinating, followed by **leaching** with water and then with sodium hydrosulfite and cuprous sodium hyposulfite to precipitate the final product; See also **cyanide process.**

load: To place explosives in a drill hole, or transfer broken material into a **haulage** device.
loading pocket: Transfer point at a **shaft** where bulk material is loaded by bin, hopper, and chute into a **skip.**
local: The branch of a **union** serving a specific **mine, smelter,** or factory.
location: The act of legally appropriating a parcel of land to establish **claim** to those **minerals** it contains.
lode: A **mineral deposit** in solid rock, usually a **vein** of larger than normal proportions; often used synonymously with vein.
long tom: A wooden sluice used in **placer mining,** fitted with cleats at the bottom.
long-wall mining: An underground **coal**-mining method that uses a steel plow (a rotation drum) which is pulled mechanically back and forth across a **face** of coal and allows the loosened coal to fall onto a **conveyor** for removal.
loose coal: Coal fragments larger in size than **dust**.
magma: Melted silicate **minerals**; usually consists of liquid rock, suspended crystals of silicate **minerals,** dissolved water, and other gases.
manhole: A safety hole in the side of a gangway, **tunnel,** or **slope** where a **miner** can be safe from passing locomotives or **cars;** also called a **refuge hole.**
mantle: The region of the interior of the Earth between the base of the oceanic or **continental crust** and the boundary of the outer **core.**
***manto*:** (Spanish, "cloak") A flat-lying **ore deposit.**
man trip: A carrier of mine personnel to and from the work area.
manway: An **entry** used exclusively for personnel, not machinery.
marine: Refers to **sediments** that accumulate in the ocean.
matte: A mixture of sulfur, iron, and copper from a reverberatory furnace.
meander: A broad bend in a river often characteristic of a floodplain.
Mendocino triple junction: The point off the coast of northern California where the following three tectonic **plates** intersect: the Gorda plate, which is subducting under North America to the east; the oceanic Pacific plate, adjacent to the Gorda plate along the Mendocino Fault; and the North America plate, adjacent to the Pacific plate along the San Andreas Fault.
mercury (quicksilver): A **metal** derived from cinnabar; liquid at normal temperatures, it is used in amalgamation of gold and **silver.**
meridian: A surveying term that establishes a line of reference.
metal: Any of a class of elements having a distinctive luster, malleability, ductility, thermal and electrical conductivity, and the capacity to form positive ions.
metamorphic rocks: Rocks that have been transformed from their original texture, structure, and **mineral** content by the application of heat, pressure, and mineralizing fluids such as water.
meteoritic water: Water that falls naturally on the surface of the Earth from precipitation in the atmosphere in the form of rain and snow.

methane: A potentially explosive gas produced by decaying vegetative matter; also called **fire damp.**

midocean ridge: Region of the oceanic **lithosphere** where two **lithospheric plates** diverge, producing a broad uplift with a central **fracture** zone and numerous cross fractures; new oceanic **crust** forms here.

mill: A processing plant that produces a **concentrate** of the valuable **minerals** or **metals** contained in an **ore.** The concentrate must then be treated in some other type of plant, such as a **smelter,** to recover the pure metal.

milling ore: Ore that contains sufficient valuable **minerals** to be treated by **milling.**

mine: An incursion into the Earth's **crust** to extract **ore, coal,** or other natural materials.

miner: One whose occupation is extracting **ore, coal,** precious substances, or other natural materials from the Earth's **crust;** includes auxiliary workers such as office staff or mechanics.

mineral: A naturally occurring inorganic compound in the Earth's **crust** with a distinctive set of physical properties and definite chemical composition.

mineralized material or deposit: A mineralized body which has been determined by sampling to possess a sufficient tonnage and an average **grade** of **metal**(s) to mine economically.

mineral resource: A **deposit** or concentration of natural, solid, inorganic, or fossilized organic substance in such quantity and at such **grade** or quality that its extraction at a profit is currently or potentially possible.

mineral rights: Legal ownership of the **minerals** contained in a specific piece of land; may or may not be included in general property ownership; see also **severance.**

Mine Safety and Health Administration (MSHA): The federal agency which regulates factors that can endanger **miners.**

mining engineer: A trained engineer with knowledge of the science, economics, and arts of mineral location, extraction, concentration, and sale, and of the administrative and financial problems connected with profitable mining.

mouth: See **adit.**

muck: Ore or rock that has been broken by blasting; its removal is called **mucking.**

mud cap: A highly explosive charge fired in contact with the surface of a rock after being covered with mud or **sand,** without any borehole; also termed *adobe, dobie,* and *sandblast.*

native metal: A **metal** occurring in nature in its pure form, uncombined with other elements.

net-profit interest: A portion of the profit remaining after all charges, including taxes, have been deducted.

net smelter return: A share of the net revenues generated from the sale of **metal** produced by a **mine.**

neutralized: Consumption of the excess hydrogen ion in **acid water.**

nonmarine sediments: Generally refers to **sediments** that accumulate upon a **continent** without interaction with the ocean in rivers or lakes, or sediments deposited by the wind.

normal fault: A steeply inclined **fault** where the upper or hanging wall block has slipped down with respect to the lower block.

oceanic lithosphere: The **lithosphere** that generally underlies the ocean basins and consists of oceanic **crust** and underlying **mantle.**

open cut, open cast, or open pit: A **mine** where **waste** or **overburden** is removed to extract the **minerals** that lie near the surface.

open shop: A company or business that hires both **union** and nonunion employees; the opposite of a **closed shop.**

ore: Minerals that contain valuable chemical elements such as **silver,** gold, copper, lead, or zinc.

ore body: A natural concentration of valuable material that can be extracted and sold at a profit.

ore pass: Vertical or inclined passage for the downward transfer of **ore** connecting a **level** with the **hoist shaft** or a lower level.

ore shoot: The portion, or length, of a **vein** or other structure that carries sufficient valuable **minerals** to be extracted profitably.

outby or outbye: Toward the **mine** entrance; nearer to the **shaft** than to the **working face.**

outcrop: An exposed **vein** at the surface of the ground or rock.

overburden: Layers of soil and rock covering a **coal seam** or **ore body;** overburden is removed prior to **surface mining** and replaced after the coal or ore is taken from the seam.

overcast or undercast: Enclosed **airway** which permits one air current to pass over or under another without interruption.

oxidation: A chemical reaction caused by exposure to oxygen that results in a change in the chemical composition of a **mineral,** which is then referred to as oxide **ore.**

oxidation states: Chemical elements may occur in different combinations with oxygen or a different number of electrons.

padrone: A contractor who provides laborers for a business or industry, usually for a fee paid by each worker.

panning: Washing for gold in **placer mining** with a shallow metal object much like a bowl.

participating interest: A company's interest in a **mine,** which entitles it to a percentage of profits equal to the percentage of the capital it invested for the project.

parting: A small **joint** in **coal** or rock; a layer of rock in a coal **seam**; or a side track or turnout in a **haulage** road.

pass: A mine opening through which **ore** is moved from a higher to a lower **level.**

passive: A **lithospheric plate margin** such as a **continental** margin where **crust** is neither created nor consumed.

patent: The ultimate stage of holding a **mineral claim** in the United States; after it no more assessment work is necessary because all **mineral rights** have been earned.

patented mining claim: A parcel of land originally located on federal land whose title has been conveyed from the government to a private party pursuant to the requirements of the General Mining Law.

pay rock: Rock containing sufficient **ore** to make mining profitable.

peat: Partially decayed plant matter found in swamps and bogs; one of the earliest stages of **coal** formation.

percentage extraction: The proportion of a **coal seam** or **ore vein** removed from a **mine.**

permeable: Rocks that easily transmit the flow of water.

permissible: Equipment and explosives which meet **MSHA** safety standards.

permit: A document issued by a regulatory agency that gives approval for mining operations.

phosphate: Chemical combination of phosphorous with oxygen to form **minerals.**

piggyback: A bridge **conveyor** to carry **coal** or **ore.**

pillar: A block of solid **coal, ore,** or other rock left in place to structurally support the **shaft,** walls, or **roof** of a **mine;** see **room and pillar mining.**

pinch: A compression of the walls of a **vein** or the **roof** and **floor** of a **coal seam** to squeeze out the desired material.

pinning: See **roof bolt.**

pitch: The inclination or rise of a **seam.**

placer mining: Extracting **ore** from **alluvial deposits** by washing streambed gravels in moving water to take the lighter gravel away, leaving the heavier gold at the bottom. It utilizes one or more of the following methods: **crevicing, panning, rocking, sluicing, hydraulicing,** or **dredging.**

plan: A map showing features such as **mine workings** or geological structures on a horizontal plane.

plate: A section of **lithosphere** floating on underlying **mantle** that moves independently of adjacent plates.

plateau: A region of high elevation often bordered on one or more sides by steep cliffs.

plate margins: The outer edges of **plates.**

playa: A flat, dried-up area, especially a desert basin, like much of Utah.

pneumoconiosis or black lung: A chronic disease of the lungs arising from breathing **coal dust.**

porphyry: Any **igneous** rock where relatively large crystals, called *phenocrysts*, are set in a fine-grained ground.

portal: The structure surrounding the immediate entrance to a **mine.**

potash: Salts of potassium such as potassium chloride

Precambrian shield: The oldest, most stable regions of the Earth's **crust;** the largest is the Canadian Shield.

precious metals: Those **metals** deemed of high value, such as gold or **silver.**

primer or booster: A package or cartridge of explosive which does not contain a **detonator** but simply transmits detonation to other, usually larger, explosives.

prospect: A mining property, the value of which has not been determined by **exploration.**

proximate analysis: A quick, possibly imprecise, physical or nonchemical test of the constitution of **coal** or **ore.**

pulp: Slime composed of finely crushed **ore** and water.

pyrite: (FeS_2 or iron disulfide) A hard, heavy, shiny, yellow mineral; also called *iron pyrites, fool's gold, sulfur balls.*

quartz: An **ore** composed of gold or **silver** in a **gangue** of crystallized silicon dioxide; sometimes applied to any hard gold or silver ore.

quartzite: A **sandstone** that has been tightly cemented with **quartz;** often the result of metamorphism.

quicksilver: See **mercury.**

radioactive age dating: Determining the age of a rock by measuring the quantity of an original radioactive element that has decayed to daughter elements.

radioactivity: Spontaneous disintegration of certain elements and isotopes with the emission of nucleons or electromagnetic radiation.

raise: A vertical or inclined **underground working** that has been excavated from the bottom up.

rake: The trend of an **ore body** along the direction of its **strike.**

ramp: A secondary or tertiary inclined opening connecting **levels** in a **mine.**

reagent: A chemical used in **assaying** or **flotation** to produce a specific reaction.

reclamation: The restoration of a site after mining or **exploration** has been completed.

recovery: The percentage of valuable **metal** in the **ore** that is recovered by metallurgical treatment.

red dog: A nonvolatile combustion product of the **oxidation** of **coal** or coal refuse.

reducing agents: Chemical substances that remove oxygen or add electrons to compounds.

refuge hole: See **manhole.**

regulator: A wall or door to control the volume of air in an **air split.**

replacement ore: Ore formed by a process during which certain **minerals** have passed into solution and been carried away, while valuable minerals from the solution have been deposited to replace those removed.

reserves: Measured amounts of **minerals** calculated to occur in a property; may be computed partly from inspection, sampling, analyses, measurements, and reasonable geologic projections to determine the viability of mining; also called **inferred resources.**

resource: The calculated amount of material in a **mineral deposit,** based on the results of limited drilling, or the percentage of **ore** relative to the surrounding rock.

respirable dust: Dust particles small enough to be inhaled, usually five microns or less.

retreat mining: A system of removing or extracting **pillars** of **coal** or rock, previously left for support, from the back of the **mine** toward the **adit** after the **ore** has been removed and the mine is about to close; see also **room and pillar mining;** also called **robbing.**

return: The air or **ventilation** that has passed through all the **working faces** of a split.

reverberatory furnace: A **smelter** that melts **concentrates,** then draws off **slag** and taps molten, **metal**-bearing **matte** for further processing.

reversal in the drainages: Change in direction of flowing water in a river often due to uplift of new mountain ranges.

rib: The side of a **pillar,** or the wall of an **entry,** or the solid **coal** or rock on the side of any underground passage.

rider: A thin **seam** of **coal** overlying a thicker one.

riffle: A series of cleats in the bottom of a rocker or **sluice** or **table** to catch ore passing through, often using **mercury** to aid in the process.

rift: A system of **faults** resulting from divergence or extension.

ripper: A **coal**-extraction machine that works by tearing the coal from the **face.**

river mining: Damming and/or diverting a stream to expose **ore deposits,** usually gold, in its bed.

robbing: See **retreat mining.**

rock bolting: Supporting openings in rock by anchoring steel bolts in holes drilled especially for this purpose.

rocking: See **placer mining.**

rock mechanics: The study of the mechanical properties of rocks, which includes stress conditions around mine openings and the ability of rocks and underground structures to withstand these pressures.

rod mill: A **mill** which grinds **ore** by rotating it in a horizontal steel chamber that includes steel rods.

roll: A high place in the **bottom** or a low place in the **top** of a mine passage, or a local thickening of **roof** or **floor strata,** causing thinning of a **coal seam.**

roll protection: A framework, safety canopy, or similar protection for the operator in case equipment overturns.

roof, back, or top: The overhead surface of a **mine**'s working area.

roof bolt: A long steel bolt driven into the **roof** of underground excavations to provide support; also called **pinning.**

room-and-pillar mining: A method of mining flat-lying **ore deposits** where approximately half of the **coal** or rock is left in place to support the **roof** of the active mining area; see also **retreat mining.**

round: Planned pattern of drill holes fired in sequence in **tunneling, shaft** sinking, or **stoping.**

royalty: A specific amount of money paid at regular intervals by the lessee or operator of an **exploration** or mining property to the owner of the ground, or the fee paid for the right to use a patented process.

rubbing surface: The total area (top, bottom, and sides) of an **airway.**

run of mine: Raw material as it exists in the **mine**; its average **grade** or quality.

safety fuse: A train of powder enclosed in cotton, jute yarn, or waterproofing compounds which burns at a uniform rate so it can fire a cap containing the detonation compound, which in turn sets off the explosive charge.

safety lamp: A lamp with steel-wire gauze covering every opening from the inside to the outside to prevent the passage of flame to any explosive gas it may encounter.

sample: A small portion taken from a rock or **mineral deposit** so that the **metal** content can be determined by **assay.**

sand: Grains of **sedimentary** material between .0625 and 2 millimeters in size.

sandstone: A **sedimentary** rock composed of **sand**-sized particles cemented together with calcium **carbonate,** silic, iron oxide, or another bonding element.

scab: See **strikebreaker.**

scaling: Removing loose rock from the **roof** or walls of a **mine,** usually with a bar.

schistosity: The tendency of a rock, especially a **metamorphic** one, that causes it to split into parallel layers; see also **cleavage.**

scoop: A battery- or diesel-powered piece of equipment on rubber tires designed to clean runways and haul supplies.

scrubber: See **flue-gas desulfurization.**

seal trench: A trench dug to disturb lateral, permeable layers of earth to allow runoff, then refilled with nonporous material to retain the water.

seam: A **stratum** or bed of **coal.**

secondary enrichment: Enrichment of a **vein** or **mineral deposit** by minerals that have been put into solution in one part of the vein or adjacent rocks and then redeposited in another.

sediment: Matter such as **sand** that settles to the bottom of a liquid, like the beds of rivers or bays.

sedimentary rocks: Rocks formed by compacting and cementing **sediment.**

selective mining: Obtaining a relatively high-**grade** mine product by expending resources in searching for and developing the separate strands of **ore.**

self-rescuer (SCSR): A small filtering device carried by an underground **miner** to provide immediate protection against **carbon monoxide** and smoke in case of a fire or explosion.

severance: The separation of a **mineral** interest from other land interests by grant or reservation; see also **mineral rights.**

Sevier mountain belt: The belt of mountain building in western North America that immediately preceded the Laramide and affected mostly the thin **sedimentary** cover on **basement** rocks.

shaft: A vertical or steeply inclined excavation, often entailing a transportation method, from the surface of the ground to the **underground workings.**

shale: A **sedimentary rock** composed of fine-grained particles of clay and silt consolidated and cemented to form a rock.

shear or shearing: The deformation of rocks by lateral movement along numerous parallel planes, generally resulting from pressure and producing such metamorphic effects as **cleavage** and **schistosity.**

shearer: A mining machine for **long-wall faces** that uses rotating action to peal the material from the face as it progresses along it.

shift: The number of hours or the part of any day worked.

shoot: That portion of a **vein** rich enough to be profitable, or a portion of a vein richer than that around it.

short-wall mining: An underground method where small areas (15 to 150 feet) are worked by a **continuous miner** in conjunction with **hydraulic roof** supports.

shot: A single explosive charge ignited in stone, **ore,** or **coal.**

shrinkage stoping: A method which uses part of the broken **ore** as a working platform and support for the walls of the **stope.**

siderite: Iron **carbonate** which has been roasted to drive off carbon dioxide and becomes sinter, which can be used in a **blast furnace;** see **sintering.**

siltstone: A rock of consolidated salt.

silver: (Ag) A white, **precious metal** often found with lead **ores** or in the oxidized zones of ore **deposits** and in hydrothermal **veins**.

single jacking: The action by one man of hand-drilling blasting holes by holding and turning the drill with one hand and hitting it with a sledge hammer with the other.

sinking: The process of driving a **shaft.**

sintering: Creating an agglomeration of larger particles by heating a mass of fine particles, such as lead **concentrates,** to just below the melting point.

skarn: Name for the **metamorphic rocks** surrounding an **igneous intrusive** where it comes in contact with a **limestone** or **dolomite** formation.

skip: A **car** used in a **shaft** or **slope** to move **coal, minerals,** or people.

slack: Small **coal;** the finest-sized soft coal, usually less than one inch in diameter.

slag: The **waste** product of the **smelting** process.

slate: A dense, fine-textured, **metamorphic rock** which has excellent parallel cleavage so that it breaks into thin plates or pencil-like shapes; any shale or slate accompanying coal.

slate bar: A long-handled tool to pry loose and hazardous material from **roof, face,** and **ribs.**

slickens: The debris or **tailings** deposited by **hydraulic mines** or **stamp mills.**

slickenside: A smooth, striated, polished surface produced on rock by friction.

slimes: Highly pulverized **ore** mixed with water to become a fine mud during **milling.**

slip fault: A steep **fault** where movement is horizontal; also called a **tear fault.**

slope mine: An **underground mine** with an opening that slopes up or down to the **seam** or **vein.**

sluicing: In **placer mining,** washing **ore** in large quantities in a ditch or trough.

slurry: Fine carbonaceous discharge from a **colliery** washing process.

smelting: The process of releasing **ore** from **gangue** by heating and melting.

solids: Minerals that have not been **undercut, sheared** out, or otherwise prepared for blasting.

sounding: Knocking on a **roof** to see whether it is safe to work under.

spad: A flat spike hammered into a wooden plug anchored in the mine ceiling from which a plumb line is suspended.

span: The horizontal distance between the side supports or solid **abutments** along the sides of a roadway.

specific gravity: The weight of a substance compared with the weight of an equal volume of water at four degrees Celsius.

spitter: The short **fuse** which ignites timed fuses in dynamite blasting.

square-set timbering: Wooden beams mortised and tenoned so that they fit together in cubes which can be stacked or placed on each other in cubes to support large **stopes;** invented in 1861 by Philip Diedsheimer.

squeeze: The settling, without breaking, of the **roof** and the gradual upheaval of the **floor** of a **mine** due to the weight of the overlying **strata.**

stamp mill: A device, or the building that houses it, to crush **ore** with a series of pestles (often called **stamps)** raised and lowered by a camshaft.

station: The end of a **level** that connects with a **shaft;** men and materials move to and from work areas through station(s).

steeply inclined: Deposits with a dip ranging from 0.7 to 1 rad (40 to 60 degrees).

stemming: The noncombustible material on top or in front of a charge or explosive.

step-out drilling: Holes drilled to intersect a mineralized horizon or structure along a **strike** or down **dip.**

stock: A small body of **intrusive igneous rock**, often with a generally circular outline.

stockpile: Broken **ore** heaped on the surface, pending treatment or shipment.

stoker: A device, once human but now a machine, to feed fuel into a boiler or furnace.

stope: An underground excavation from which **ore** has been extracted either above or below **mine level.**

stoping: The process of creating a **stope** by mining out the **ore.**

strata or stratum: A more or less homogenous layer of rock which identifies a geological group, system, or series.

stratigraphy: Strictly, the description of **bedded** rock sequences; used loosely, the sequence of bedded rocks in a particular area.

strato volcano: A volcanic cone with steep sides built up by successive layers of ash and lava flows.

strike: In mining, the direction of a **vein** or rock formation; in labor disputes, the refusal of workers to remain on the job as a measure to force employers to meet demands for salary or working conditions.

strikebreaker: A worker who refuses to join a **union,** or one who works despite a declared **strike;** often called a **scab.**

strike-slip fault: A steeply inclined **fault** with primarily horizontal displacement; also called a **tear fault.**

stringer: A narrow **vein** or irregular filament of a **mineral** or minerals traversing a rock mass.

stripping ratio: The amount of **overburden** that must be removed to gain access to a similar amount of **coal** or **mineral** material.

sublevel: A **level** or working horizon in a **mine** between main mining levels.

subsidence: The sinking, gradual or abrupt, of rock and soil layers into an **underground mine.**

sulfide: A compound of sulfur and some other element.

sump: The bottom of a **shaft** where superfluous water gathers.

sumping: Forcing the **cutter** bar of a machine into or under **coal.**

sump pump: A pump to remove superfluous water from the **bottom** of a **mine** or building.

surface mine: A **mine** where the **coal** or **ore** lies near the surface and can be extracted simply by removing the covering layers of rock and soil. See **open cut.**

supercontinents: An unusually large **continent** often consisting of numerous smaller ones sutured together.

suspension: Weaker **strata** hanging from stronger, overlying strata by means of **roof bolts.**

suture: A linear belt of deformed rocks that marks the location of a **continental** collision where one land fragment becomes attached to another.

syncline: A **fold** in **sedimentary** layers where the **strata** dip inward from both sides toward the axis.

table: A concentrating machine for separating (upgrading) heavy material **(ore)** from lighter **waste (gangue).** Particles of crushed low-**grade** ore and water pass across an inclined percussion (shaking) table, usually with longitudinal grooves **(riffles)** in its surface, agitated by side blows at right angles to the flow of the crushed ore and waste. Tables typically clean the **concentrate** produced by **jigs.**

tailgate: A subsidiary gate road to a **conveyor face**.

tailings: Material ejected from a **mill** after the recoverable valuable **mineral**s have been extracted.

tailings pond: A low-lying depression to confine **tailings;** its prime function is to allow enough time for heavy **metal**s to settle out or cyanide to be destroyed before water is discharged into the local watershed.

tar sands: Sandstone that contains some form of solid or almost solid hydrocarbon.

tear fault: See **strike-slip fault.**

tertiary: Lateral or panel openings in a **mine,** such as a **ramp** or **crosscut.**

theory of plate tectonics: The theory that holds that the **lithosphere** of the Earth consists of separate **plates** in motion in relation to one another.

thick-skinned mountain building: Mountain building that results from convergence of two **lithospheric plates** and deformation that extends beneath the **sedimentary** cover into the underlying **basement** rocks.

through steel: A system of dust collection from rock or roof drilling.

thrust fault: A gently inclined **fault** resulting from horizontal compressive forces and shortening where the upper block has moved up in relation to the lower one.

tidewater glaciers: Glacial ice that reached the shoreline due to ocean movement and is influenced by tides.

tillite: Rock formed by consolidating and cementing **glacial sediment,** usually unsorted in particle size and unstratified or poorly stratified.

timber: A collective term for underground wooden supports.

timbering: Placing timber supports in **mine workings** or **shafts** for protection against falls from the **roof, face,** or **rib.**

tipple: The place where the mine **cars** are tipped and the load emptied.

Tommy Knocker: A ghost or demon of Welsh origin believed to create the knocking sounds often heard in **mines.**

ton: A short or net ton is 2,000 pounds; a long or British ton is 2,240 pounds; a metric ton is approximately 2,205 pounds.

top: See **roof.**

top lander: A person who works above ground, for example, handling **ore cars** on the surface of a **mine.**

tram: Moving, self-propelled mining equipment.

tramway: An aerial device run by a line and pulley to transport **ore cars** to the top of a **dump** where **waste** rock can be dropped.

transfer: A vertical or inclined connection between two or more **levels** used as an **ore pass.**

transfer point: The place in a handling system where bulk material is transferred between conveyances.

transform fault margin: Two **lithospheric plates** sliding laterally past one another.

trend: The horizontal direction of a linear geological feature (for example, an **ore** zone), measured from true north.

trip: A train of **mine cars.**

triple junction: A point where three tectonic **plates** contact each other.

troy ounce: Unit of weight measurement used for all precious metals; the familiar 16-ounce avoirdupois pound equals 14.583 troy ounces.

tunnel: A horizontal or nearly horizontal underground passage, often an **adit** or entrance to a **mine.**

ultimate analysis: Precise chemical determination of the elements and compounds in **coal** or **ore.**

undercut, undermine, or underhole: To cut below the **coal face** or **ore vein** by chipping away the material beneath; also called **caving.**

underground or deep mine: A **mine** usually located several hundred feet below the Earth's surface

underground station: An enlargement of an **entry, drift,** or **level** at a **shaft** where **cages** stop to receive and discharge **cars,** personnel, and material.

union: An association of employees organized to bargain collectively with employers.

United Mine Workers of America: A **miners' union** organized in Ohio in 1890 and still active in many areas, including Utah.

universal coal cutter: A cutting machine designed to make both horizontal and **shearing** cuts in a **coal face**.

unpatented mining claim: A parcel of property located on federal land whose paramount title remains with the government.

upcast shaft: A **shaft** through which air leaves the **mine.**

uranyl ion: The soluble or ionic form of uranium, consisting of uranium combined with oxygen with a net charge of plus two.

valuation: An appraisal that estimates the value or worth of a **vein** or property.

vein: A mineralized zone developed more or less regularly in length, width, and depth, which clearly separates it from neighboring rock.

ventilation: Directed flow of fresh and returned air along underground passageways.

void: Pore space or other reopenings in rock.

volcanic activity: Active eruption of lava and other products from a **volcano.**

volcanic arc: Generally linear zone of **volcanic activity** related to a descending slab of **lithosphere.**

volcanic ash: Fine-grained particles erupted from a **volcano.**

volcanic edifice: A structure composed of **volcanic ash,** solidified lava, and small **intrusions** of **igneous rock** with a broad base and narrower, often-cone-shaped summit.

volcano genic: A term describing the volcanic origin of mineralization.

volcanic rocks: Igneous rocks that result from relatively rapid cooling, crystallization, and solidification of molten rock on the surface of the Earth.

volcano: A vent or conduit in the **crust** of the Earth through which melted lava, steam, and other gases are distributed over the surface.

vug: A small cavity in a rock, frequently lined with well-formed crystals.

wall rocks: Rock units on either side of an **ore body;** see **hanging wall** and **footwall.**

washing: Separating undesirable materials from **coal** based on differences in densities; pyritic sulfur, or sulfur combined with iron, is heavier and sinks in water; coal is lighter and floats.

Washoe-pan process: A method for reducing **silver ore** using both a **stamp mill** and the ***cazo* process.**

waste: Barren rock in a **mine,** or rocks or **minerals** which must be removed but have no value.

water gauge (standard U-tube): Instrument that measures differential pressure in inches of water.

weight: Fracturing and lowering of the **roof strata** at the **face** as a result of mining.

white damp: See **carbon monoxide.**

widowmaker: The name **miners** applied to a compressed air drill.

width: The thickness of a **lode** measured at right angles to the **dip.**

winning: The excavation, loading, and removal of **coal** or **ore** from the ground.

winze: An internal opening, vertical or horizontal, that is similar to a **shaft** and often connects two **levels.**

wobbly: A member of the **Industrial Workers of the World.**

working: The creaking noises emitted by a **coal seam** being squeezed by **roof** and **floor** which often warns **miners** that additional support is needed.

working face: Any place in a **mine** where material is extracted during a particular cycle.

working place: The distance from the **outby** side of the last open **crosscut** to the **face.**

workings: The entire system of openings in a **mine** that exploit the potential to remove **ore**.

working section: That section of a **mine** from the **faces** to the point where **coal** or **ore** is loaded onto belts or rail **cars** to begin its trip to the outside.

zone of oxidation: The portion of an **ore body** that has been **oxidized,** usually in the upper portion of the ore zone.

Sources: William Parry, *All Veins, Lodes, and Ledges throughout Their Entire Depth: Geology and the Apex Law in Utah Mines;* Albert H. Fay, *A Glossary of the Mining and Mineral Industry;* T. H. Watkins, *Gold and Silver in the West,* 278–79; Web sites for the following: Corriente Enterprises, Due North Resources, M. I. M. Holdings Limited, MiningLife, Rich River Exploration, Ltd., Saskatchewan Centre for Soils Research, Sterling Mining Company, United States Department of Environmental Protection.

Notes

Preface

1. The two mining mascots were a fox named Copper and a bear named Coal; the third was a rabbit called Powder, honoring the snow that brings in tourist dollars.
2. For information on "An Act to Promote the Development of the Mining Resources of the United States," see United States Land Office, *Mining Districts By-Laws, 1872–1909*, Series 3651, http://archives.utah.gov/reference/xml/series/3651.html.
3. For information on current mining districts and mining operations, see the Utah Mining Association's Web site: http://www.utahmining.org.

1—Geology and Utah's Mineral Treasures
William T. Parry

1. References and sources for this chapter are as follows:
William Lee Stokes, *Geology of Utah*, 331; Lehi F. Hintze, *Geologic History of Utah*, 202; Bill Fiero, *Geology of the Great Basin*, 198; John H. Stewart, *Geology of Nevada*, 136; William R. Dickenson, "Paleozoic Plate Tectonics and the Evolution of the Cordilleran Continental Margin," in *Paleozoic Paleogeography of the Western United States*, ed. J. H. Steward, C. H. Stevens, and A. E. Fritsche, 137–55; William R. Dickenson, "Cenozoic Plate Tectonic Setting of the Cordilleran Region in the United States," in *Cenozoic Paleogeography of the Western United States*, ed. J. M. Armentrout, M. R. Cole, and H. TerBest, Jr., 1–13; Warren Hamilton, "Mesozoic Tectonics of the Western United States," in *Mesozoic Paleogeography of the Western United States*, ed. D. G. Howell and K. A. McDougall, 23–70; Peter J. Coney, "Tertiary Evolution of Cordilleran Metamorphic Core Complexes," in *Cenozoic Paleogeography of the Western United States*, 15–28; Bruce Bryant, "Evolution and Early Proterozoic History of the Margin of the Archean Continent in Utah," in *Metamorphism and Crustal Evolution of the Western United States*, ed. W. G. Ernst, 431–45; Ian W. D. Dalziel, "Earth before Pangea," *Scientific American:* 58–63; B. C. Burchfiel and L. H. Royden, "Antler Orogeny: A Mediterranean-Type Orogeny," *Geology:* 66–69; R. J. Spencer, H. P. Eugster, B. F. Jones, and S. L. Rettig, "Geochemistry of Great Salt Lake, Utah I: Hydrochemistry Since 1850," *Geochimica et Cosmochimica Acta:* 727–37; R. J. Spencer, H. P. Eugster, and B. F. Jones, "Geochemistry of Great Salt Lake, Utah II: Pleistocene-Holocene Evolution," *Geochimica et Cosmochimica Acta:* 739–47.

2—Generating Wealth from the Earth, 1847–2000
Thomas G. Alexander

1. On the mining industry of Utah and its development, see Leonard J. Arrington, "Abundance from the Earth: The Beginnings of Commercial Mining in Utah," *Utah Historical Quarterly*: 192–219; Rossiter W. Raymond, *Statistics of Mines and Mining . . . for the Year[s] 1869 through 1871*, 4 vols.; Rodman W. Paul, *Mining Frontiers of the Far West, 1848–1880*; Rossiter W. Raymond, *Mineral Resources of the States and Territories West of the Rocky Mountains*; Rossiter W. Raymond, *Statistics of Mines and Mining in the States and Territories West of the Rocky Mountains*, 1870–75 and 1877; U.S. Geological Survey (1882–1923), succeeded by U.S. Bureau of Mines (1924–34), *Mineral Resources of the United States*, succeeded by U.S. Bureau of Mines (1933–present), *Minerals Yearbook*; Mark Wyman, "Industrial Revolution in the West: Hard-Rock Miners and the New Technology," *Western Historical Quarterly*: 39–57.
2. Leonard J. Arrington, *Great Basin Kingdom: An Economic History of the Latter-day Saints, 1830–1900*, 391, 345–46, 122–29; J. Kenneth Davies, *Mormon Gold: The Story of California's Mormon Argonauts*; Thomas G. Alexander, "From Dearth to Deluge: Utah's Coal Industry," *Utah Historical Quarterly*: 235–36.
3. On the iron-mining industry, see Gustive O. Larson, "Bulwark of the Kingdom: Utah's Iron and Steel Industry," *Utah Historical Quarterly*: 248–61.
4. Arrington, *Great Basin Kingdom*, 46.
5. Utah Economic and Business Review, *Measures of Economic Changes in Utah, 1847–1947*, 70.
6. Miles P. Romney, "Utah's Cinderella Minerals: The Nonmetallics," *Utah Historical Quarterly*: 221.
7. Arrington, *Great Basin Kingdom*, 391–92, 511 nn. 48–52.
8. *Measures of Economic Changes in Utah*, 70.
9. Ibid.
10. Ibid., 77.
11. Alexander, "From Dearth to Deluge," 240; Edward W. Parker, "Coal Cutting Machinery," *Cassier's Magazine*: 291–93.
12. On the introduction of new technology and its benefits and hazards, see Wyman, "Industrial Revolution in the West," 39–57 and Clark C. Spence, *Mining Engineers and the American West: The Lace-Boot Brigade, 1849–1933*.
13. *Measures of Economic Changes in Utah*, 68.
14. U.S. Bureau of Mines, *Minerals Yearbook*, 1945, 457.
15. Raymond, *Statistics of Mines and Mining West of the Rocky Mountains*, 1874, 256.
16. On the development and introduction of this technology, see Wyman, "Industrial Revolution in the West," 39–57.
17. For the development of these facilities in Big and Little Cottonwood Canyons, see Charles L. Keller, *The Lady in the Ore Bucket: A History of Settlement and Industry in the Tri-Canyon Area of the Wasatch Mountains*, 183–224.
18. This story is taken from Keller, 183–84.
19. Ibid., 187–89.
20. Ibid., 190–94.
21. U.S. Geological Survey, *Mineral Resources*, 1907, Part I, "Metallic Products," 440–41.
22. U.S. Geological Survey, *Mineral Resources*, 1922, Part I, "Metals," 378.
23. Arrington, "Abundance from the Earth," 212–14. For a contemporary investigation of the development of mines at Park City, see John M. Boutwell, *Geology and Ore Deposits of the Park City District, Utah*.
24. U.S. Geological Survey, *Mineral Resources*, 1914, Part I, "Metals," 726–27.
25. The following is based on Leonard J. Arrington and Gary B. Hansen, "*The Richest Hole on Earth*": *A History of the Bingham Copper Mine*, 19–56.

26. On the importance of eastern capital to the development of western mining and other enterprises, see William G. Robbins, *Colony and Empire: The Capitalist Transformation of the American West.*
27. Arrington and Hansen, *Richest Hole*, 52, 71.
28. Ibid., 70.
29. U.S. Geological Survey, *Mineral Resources*, 1915, Part I, "Metals," 394.
30. U.S. Bureau of Mines, *Mineral Resources*, 1925, Part I, "Metals," 414.
31. For the statement on gold, see U.S. Bureau of Mines, *Mineral Resources*, 1926, Part I, "Metals," 427. The section on silver reads, "Most of the silver was recovered from the smelting of crude ore and concentrates; no silver bullion or precipitates were marketed" (ibid., 473).
32. Arrington and Hansen, *Richest Hole*, 75.
33. On mining in some of these areas, see Arrington, "Abundance from the Earth," 192–219; Elroy Nelson, "The Mineral Industry: A Foundation of Utah's Economy," *Utah Historical Quarterly*: 179–91; Clark C. Spence, *British Investments and the American Mining Frontier, 1860–1891*; George A. Thompson and Fraser Buck, *Treasure Mountain Home: Park City Revisited*; Philip F. Notarianni, *Faith, Hope & Prosperity: The Tintic Mining District*; Paul Dean Proctor and Morris A. Shirts, *Silver, Sinners and Saints: A History of Old Silver Reef, Utah.*
34. John E. Lamborn and Charles S. Peterson, "The Substance of the Land: Agriculture v. Industry in the Smelter Cases of 1904 and 1906," *Utah Historical Quarterly*: 308–25.
35. U.S. Geological Survey, *Mineral Resources*, 1908, Part I, "Metallic Products," 550; U.S. Geological Survey, *Mineral Resources*, 1909, Part 1, "Metals," 462.
36. The information on the two explosions comes from Allan Kent Powell, "Scofield Mine Disaster," in *Utah History Encyclopedia*, ed. Allan Kent Powell, 491; and Janeen Arnold Costa, "Castle Gate Mine Disaster," in *Utah History Encyclopedia*, 77.
37. Utah Industrial Commission, *Report of the Industrial Commission of Utah, Period July 1, 1918 to June 30, 1920*, 290.
38. For a discussion of these rules, see Wyman, "Industrial Revolution in the West," 46–47.
39. Utah Industrial Commission, *Report, 1918 to 1920*, 273.
40. Ibid., 252.
41. Utah Bureau of Immigration, Labor, and Statistics, *Second Report of the State Bureau of Immigration, Labor and Statistics for the Years 1913–1914*, 220.
42. See, for instance, *Mrs. Wm. E. Sharpe v. Iron Blossom Consolidated* Mng. Co, *et al.* Claim No. 33 and Supplement to Decision in Claim No. 33, in Utah Industrial Commission, *Report of the Industrial Commission of Utah, Period July 1, 1917–June 30, 1918*, 68–72.
43. Utah Industrial Commission, *Report, 1918 to 1920*, 49–62.
44. Ibid., 271–72.
45. M. Henry Robison, "A Brief Economic History of Utah's Coal Industry," *Utah Economic and Business Review*: 7.
46. This information is based on an interview with Greg Fredde, president of the Utah Mining Association, 10 March 2003.
47. Utah Mining Association, "Utah Mining Facts: 'A Vital Part of Utah's Future.'" Percentage of payroll from mining in 2000: Emery (34.2), Carbon (23.3), Uintah (23.1), Duchesne (17.2), Sevier (59.7). Percentage of total tax base: Daggett (52.3), San Juan (44.6), Carbon (30.3), Uintah (29.5), Morgan (28.7), Sevier (28.2), Duchesne (23.8), Juab (22.4).
48. Robison, "A Brief Economic History," 6.
49. R. L. Bon and R. W. Gloyn, "2001 Summary of Mineral Activity in Utah":8, http://www.geology.utah.gov/utahgeo/rockmineral/activity/min2001.pdf.
50. Garth Mangum and MacLeans Geo-JaJa, "The Prospects for Utah Coal," *Utah Economic and Business Review*: 9, 14.
51. Herman W. Sheffer and William C. Henkes, "The Mineral Industry of Utah," in *Minerals Yearbook*, published by U.S. Bureau of Mines, 1965, vol. 3, *Area Reports, Domestic*, 809.

52. Robison, "A Brief Economic History," 8; American Coal Foundation, "How Coal Is Produced" in *Coal: An Ancient Gift Serving Modern Man*, http://www.ket.org/Trips/Coal/AGSMM/agsmmproduced.html.
53. Mike Gorrell, "Taking Its Lumps: Coal Mining in Utah Feels Pressure from Weak Demand, Low Prices," *Salt Lake Tribune*, 16 February 2003, E-3.
54. Paul Luff, "The Mineral Industry of Utah," in *Minerals Yearbook*, published by U.S. Bureau of Mines, 1952, vol. 3, *Area Reports*, 905.
55. Gorrell, "Taking Its Lumps," E-1; William H. Kerns, F. J. Kelly, and D. H. Mullen, "The Mineral Industry of Utah," *Minerals Yearbook*, U.S. Bureau of Mines, 1959, vol. 3, *Area Reports*, 1066.
56. Kerns, Kelly, and Mullen, "The Mineral Industry of Utah," 3:997; Merwin H. Howes, "The Mineral Industry of Utah," in *Minerals Yearbook*, published by U.S. Bureau of Mines, 1963, vol. 3, *Area Reports, Domestic*, 1080; Gorrell, "Taking Its Lumps," E-1, E-3; Bon and Gloyn, "2001 Summary of Mineral Activity in Utah": 9.
57. State of Utah, *Economic Report to the Governor* 71, 88, 92, 93, http//www.governor.state.ut.us/dea/publications/01erg/TofC.pdf.
58. Luff, "The Mineral Industry of Utah," 3:888–89.
59. Frank J. Kelly, William H. Kerns, and Breck Parker, in "The Mineral Industry of Utah," *Minerals Yearbook*, published by U.S. Bureau of Mines, 1956, vol. 3, *Area Reports*, 1183–84.
60. Lorraine B. Burgin, "The Mineral Industry of Utah," in *Minerals Yearbook*, published by U.S. Bureau of Mines, 1985, vol. 2, *Area Reports, Domestic*, 561.
61. Luff, "The Mineral Industry of Utah," 3:889–90, 905–6.
62. William H. Kerns, F. J. Kelly, and D. H. Mullen, "The Mineral Industry of Utah," in *Minerals Yearbook*, published by U.S. Bureau of Mines, 1960, vol. 3, *Area Reports*, 1033; Stephen R. Wilson, "The Mineral Industry of Utah," in *Minerals Yearbook*, published by U.S. Bureau of Mines, 1974, vol. 2, *Area Reports*, 719.
63. Paul Luff, "The Mineral Industry of Utah," in *Minerals Yearbook*, published by U.S. Bureau of Mines, 1953, vol. 3, *Area Reports*, 993.
64. Eileen K. Peterson and Robert W. Gloyn, "The Mineral Industry of Utah," in *Minerals Yearbook*, published by U.S. Bureau of Mines, 1992, vol. 2, *Area Reports, Domestic*, 541.
65. Kerns, Kelly, and Mullen, "The Mineral Industry of Utah," 1959, 3:998.
66. Merwin H. Howes, "The Mineral Industry of Utah," in *Minerals Yearbook*, published by U.S. Bureau of Mines, 1962, vol. 3, *Area Reports*, 1056–57; Howes, "The Mineral Industry of Utah," 1963, 3:1084.
67. Stephen R. Wilson and William C. Henkes, "The Mineral Industry of Utah," in *Minerals Yearbook*, published by U.S. Bureau of Mines, 1969, vol. 3, *Area Reports, Domestic*, 741; Wilson, "The Mineral Industry of Utah," 2:715.
68. Greg Fredde, interview with the author.
69. Douglas H. Hileman and William C. Henkes, "The Mineral Industry of Utah," in *Minerals Yearbook*, published by U.S. Bureau of Mines, 1967, vol. 3, *Area Reports, Domestic*, 787.
70. Francis C. Mitko, "The Mineral Industry of Utah," in *Minerals Yearbook*, published by U.S. Bureau of Mines, 1971, vol. 3, *Area Reports, Domestic*, 726.
71. Lorraine B. Burgin, "The Mineral Industry of Utah," 2:556–57.
72. Howes, "The Mineral Industry of Utah," 1962, 3:1062.
73. Sheffer and Henkes, "The Mineral Industry of Utah," 3:808–9.
74. Peterson and Gloyn, "The Mineral Industry of Utah," 2:545.
75. Lorraine B. Burgin, "The Mineral Industry of Utah," in *Minerals Yearbook*, published by U.S. Bureau of Mines, 1986, vol. 2, *Area Reports, Domestic*, 488, 591.
76. Michael N. Greeley and Robert W. Gloyn, "The Mineral Industry of Utah," in *Minerals Yearbook*, published by U.S. Bureau of Mines, 1989, vol. 2, *Area Reports, Domestic*, 476.
77. Lorraine B. Burgin, "The Mineral Industry of Utah," in *Minerals Yearbook*, published by U.S. Bureau of Mines, 1984, vol. 2, *Area Reports, Domestic*, 592.

78. Burgin, "The Mineral Industry of Utah," 1986, 2:485.
79. "The Mineral Industry of Utah," in *Minerals Yearbook*, published by U.S. Bureau of Mines, 1995, "Area Reports, Domestic," 274.
80. Peterson and Gloyn, "The Mineral Industry of Utah," 2:544.
81. U.S. Environmental Protection Agency, Region 8, Kennecott South, http://www.epa.gov/region08/superfund/sites/ut/kennes.
82. Wilson, "The Mineral Industry of Utah," 3:711; William A. McKinney, "The Mineral Industry of Utah," in *Minerals Yearbook*, published by U.S. Bureau of Mines, 1977, vol. 2, *Area Reports, Domestic*, 587; "The Noranda Smelting Process," http://www.norsmelt.com/norandaprocess.html.
83. Burgin, "The Mineral Industry of Utah," 1985, 2:557.
84. "The Mineral Industry of Utah," 274.
85. Greeley and Gloyn, "The Mineral Industry of Utah," 2:476.
86. Rodney E. Jeske and Robert W. Gloyn, "The Mineral Industry of Utah," in *Minerals Yearbook*, published by U.S. Bureau of Mines, 1991, vol. 2, *Area Reports, Domestic*, 518.
87. Greg Fredde, interview with the author.
88. Herman W. Sheffer and William C. Henkes, "The Mineral Industry of Utah," in *Minerals Yearbook*, published by U.S. Bureau of Mines, 1966, vol 3, *Area Reports, Domestic*, 769, 775.
89. Sheffer and Henkes, "The Mineral Industry of Utah," 1966, 3:769, 775.
90. Burgin, "The Mineral Industry of Utah," 1984, 2:591.
91. Burgin, "The Mineral Industry of Utah," 1985, 2:561.
92. Bureau of Labor Statistics, *Fatal Work Place Injuries in 1993: A Collection of Data and Analysis*, Report 891 (Washington, D.C.: Bureau of Labor Statistics, June 1995), 114; idem., *Fatal Workplace Injuries in 1995: A Collection of Data and Analysis*, Report 913 (Washington, D.C.: Bureau of Labor Statistics, April 1997), 84; idem., *Fatal Workplace Injuries in 2000: A Collection of Data and Analysis*, Report 961 (Washington, D.C.: Bureau of Labor Statistics, September 2002), 97.
93. *USA Today*, 23 January 2003.

3—General Patrick Edward Connor, Father of Utah Mining
Brigham D. Madsen

1. This article is taken almost entirely verbatim from Brigham D. Madsen, *Glory Hunter: A Biography of Patrick Edward Connor*. The material on Connor as a mining entrepreneur in Utah has been selected from the book with enough attention paid to his other major achievements as a military commander and political leader to give the reader an overview of his entire career. Patrick E. Connor spent most of his life developing mines in Utah Territory and well deserves the title, "The Father of Utah Mining."
2. Samuel Bowles, *Across the Continent*, 93; Anna Viola Lewis, "The Development of Mining in Utah," 40, 48.
3. Kate B. Carter, comp., *Our Pioneer Heritage*, vol. 7, 69; Leonard J. Arrington, *Great Basin Kingdom: An Economic History of the Latter-day Saints, 1830–1900*, 72–73 (reprint edition); Leonard J. Arrington, "Abundance from the Earth: The Beginnings of Commercial Mining in Utah," *Utah Historical Quarterly*: 194.
4. Of the many accounts of the Bingham discovery, Arrington's is the best: "Abundance from the Earth," 196, 199; see also John R. Murphy, *The Mineral Resources of the Territory of Utah*, 1–2; for a personal interview with Connor concerning the find, see Thomas B. H. Stenhouse, *The Rocky Mountain Saints*, 713–14; Hubert Howe Bancroft, *History of Utah: 1540–1886*, 741; William Fox, "Patrick Edward Connor: 'Father' of Utah mining," 45–48; Edward W. Tullidge, *History of Salt Lake City*, 697–98; Robert G. Raymer, "Early Mining in Utah," *Pacific Historical Review*: 81–88; Clarence King, *Statistics and Technology of the Precious Metals*, 407–8; Donald T. Schmidt, "Early Mining in Utah," 15–16.

5. Church of Jesus Christ of Latter-day Saints, *Journal History*, 10 December 1863.
6. Arrington, "Abundance from the Earth," 199.
7. Church of Jesus Christ of Latter-day Saints, *Journal of Discourses*, vol. 10, 254–55.
8. William Clayton, *Letterbooks*, vol. 7, reel 16, 28 November 1863.
9. Raymer, "Early Mining in Utah," 84; U.S. Geological Survey, *Contributions to Economic Geology*; King, *Statistics and Technology of the Precious Metals*, 419; Murphy, *Mineral Resources of the Territory of Utah*, iii, 1–2.
10. Fox, "Patrick Edward Connor," 58–59; Andrew Love Neff, *History of Utah*, 637–42.
11. U.S. War Department, *The War of the Rebellion: A Compilation of the Official Records of the Union and the Confederate Armies*, series 1, vol. 50, pt. 2, 721–52.
12. *Union Vedette*, 2 March 1864.
13. *Union Vedette*, 18 February 1864; Clayton, *Letterbooks*, reel 16; 3, 15 March 1864.
14. *Union Vedette*, 18 February; 17, 26 March 1864; Fox, "Patrick Edward Connor," 72.
15. Ibid.
16. *War of the Rebellion*, 845–46; Fox, "Patrick Edward Connor," 62–66.
17. *War of the Rebellion*, 887.
18. John M. Bourne, "Early Mining in Southwestern Utah and Southeastern Nevada, 1864–1873: The Meadow Valley, Pahranagat, and Pioche Mining Rushes," 12–23. Bourne's excellent thesis describes the development of this mining area. See also *Union Vedette*, 4 May; 10 June; 11 July; 16 August 1864.
19. Young to Bunker, 6 February 1864, quoted in Bourne, "Early Mining," 23.
20. Bourne, "Early Mining," 24–30.
21. Ibid., 31–35; Don Ashbaugh, *Nevada's Turbulent Yesterday*, 24.
22. *War of the Rebellion*, 355–803; *Union Vedette*, 8 July 1864.
23. Bourne, "Early Mining," 37–39.
24. *Union Vedette*, 1 July 1864.
25. High Council Meeting Minutes, 11 June 1864, Southern Utah Mission, St. George, Utah, 28–29, quoted in Bourne, "Early Mining," 40.
26. High Council Meeting Minutes, 40–54; *Union Vedette*, 26 September 1864; John M. Townley, *Conquered Provinces: Nevada Moves Southeast, 1864–1871*, 8.
27. Bourne, "Early Mining," 41.
28. Fred B. Rogers, *Soldiers of the Overland*, 114–15; Edward W. Tullidge, *Histories*, 2:76; B. S. Butler et al., *The Ore Deposits of Utah*, 362–63; Fox, "Patrick Edward Connor," 74; *Union Vedette*, 22 April 1864.
29. Fox, "Patrick Edward Connor," 72.
30. *Union Vedette*, 16 April 1864.
31. *Union Vedette*, 16, 23 April; 27 May; 9 June; 13, 16 July; 29 August 1864.
32. Dean R. Hodson, "The Origin of Non-Mormon Settlements in Utah, 1847–1896," 39–48; Janet Cook, "Stockton—Small Utah Town—Exciting History," *Sons of Utah Pioneers News*: 22; Rogers, *Soldiers of the Overland*, 116; Eugene E. Campbell, "The M-Factors in Tooele's History," *Utah Historical Quarterly*: 277–78; Stockton Bicentennial History Committee, *Brief History of Stockton, Utah*, 8, 97.
33. *Union Vedette*, 13 July 1864.
34. *Union Vedette*, 16 September; 19, 22 October 1864.
35. Arrington, "Abundance from the Earth," 204.
36. Ibid.
37. *Union Vedette*, 29 August; 13, 30 September; 15, 22 October 1864; *The War of the Rebellion*, 966.
38. Arrington, "Abundance from the Earth," 204.
39. *Union Vedette*, 22 October; 22 December 1864.
40. Fox, "Patrick Edward Connor," 109–10.
41. Tullidge, *History of Salt Lake City*, 228–29.

42. *Union Vedette*, 29 September; 6 November; 5 December 1865.
43. *Union Vedette*, 11, 30 June; 4, 16, 18 July 1866; *Rocky Mountain News* (Denver, Colorado), 7 July 1866.
44. *Millennial Star*: 605–6.
45. Samuel Bowles, *Our New West*, 229.
46. For a record of some of Connor's travels to Rush Valley, see *Union Vedette*, 24 July; 1, 7 August; 6 September; 28 December 1866.
47. Fox, "Patrick Edward Connor," 81–82, 110–14.
48. Grenville M. Dodge, "Personal Biography of Major General Grenville Mellon Dodge, 1831 to 1870," 2:569–70.
49. Daniel S. Tuttle, *Missionary to the Mountain West*, 366 (1987 reprint).
50. *Union Vedette*, 8, 10 January 1867; G. Owens, comp., *Salt Lake City Directory* (1867), 45; Hodson, "Origin of Non-Mormon Settlements," 42; Tullidge, *Histories*, 2:76.
51. Catherine V. Waite, *The Mormon Prophet and His Harem*, 280.
52. Hodson, "Origin of Non-Mormon Settlements," 110.
53. U.S. Congress, *A Report upon the Mineral Resources of the States and Territories West of the Rocky Mountains*, 130.
54. Dodge, "Personal Biography," 3:599–600.
55. P. E. Connor to E. M. Stanton, Salt Lake City, 29 April 1867.
56. U.S. Congress, *Mineral Resources of the States and Territories*, 484; see also *Journal History*, 19 February 1868.
57. *Stockton Independent*, 14 November 1868; *Mining and Scientific Press*, 28 November 1868; Rogers, *Soldiers of the Overland*, 251, 273; Brigham D. Madsen, *Corinne: The Gentile Capital of Utah*, 155.
58. U.S. Congress, *A Report upon the Mineral Resources of the States and Territories West of the Rocky Mountains*, 321.
59. Tullidge, *Histories*, 2:76.
60. Fox, "Patrick Edward Connor," 111.
61. *Utah Reporter*, 17 May 1870. Connor's Silver King Mine was in Rush Valley and should not be confused with a mine with the same name in Park City.
62. *Mormon Tribune*, 13 August 1870.
63. *Utah Reporter*, 6 October 1870.
64. *Salt Lake Tribune*, 13 August 1870.
65. *Utah Reporter*, 4, 28 October; 26 November 1870; William Clayton, *Letterbooks*, vol. 4, 580, 611, 646.
66. *Utah Reporter*, 8 March; 21, 28 May; 17 June 1870.
67. Madsen, *Glory Hunter: A Biography of Patrick Edward Connor*, 189, 271.
68. William Mulder and A. Russel Mortensen, eds., *Among the Mormons*, 378–79.
69. John Hanson Beadle, *The Undeveloped West*, 328; Clayton, *Letterbooks*, vol. 7, reel 15, 1 March; 27 July; 23 October 1872.
70. U.S. Congress, *A Report upon the Mineral Resources of the States and Territories West of the Rocky Mountains*, 218.
71. Murphy, *Mineral Resources of the Territory of Utah*, 20; *Salt Lake Tribune*, 31 October 1871.
72. *Corinne Reporter*, 19 April 1871.
73. Clayton, *Letterbooks*, reel 15, 16 August 1871.
74. B. A. M. Froiseth, *New Sectional Mining Map of Utah*, plat of City of Stockton; *Salt Lake Review*, 6 November 1871.
75. *Salt Lake Review*, 31 October 1871.
76. *Corinne Reporter*, 19 April; 16 June 1871; *Salt Lake Tribune*, 18 April 1870.
77. Raymond, *Statistics of Mines and Mining*, 220.
78. Fox, "Patrick Edward Connor," 111.
79. *Salt Lake Review*, 13 January 1872.

80. U.S. Congress, *Conditions of Mining Industry—Utah*, 255.
81. *Salt Lake Tribune*, 14 March; 18 April 1872; *Utah Mining Journal*, 6, 9 November 1872.
82. Aird G. Merkley, ed., *Monuments to Courage: A History of Beaver County*, 259–60.
83. The list included Flora No. 3 Ledge & Company, Flora No. 1 Ledge & Company, Merva Lode, Dora Lode, Hector Lode, Floral Tunnel Right, Connor Ledge & Co., Federal Chief Lode, Federal Chief No. 2, Federal Chief No. 3, Federal Chief No. 4, Erina Lode, Brittania Lode No. 2, and the Brittania Tunnel. The dates of filing ranged from 6 January to 17 June 1872. Beaver County Recorder's Office, North Star Mining District, Book A, 41, 42, 68, 135, 181; Book B, 11; Star Mining District, April 1871–January 1873, Book B, 215, 259, 263–65, 292, 344–45.
84. *Salt Lake Review*, 30 January 1872; *Salt Lake Tribune*, 21, 26 June 1872.
85. *Salt Lake Tribune*, 10 July 1872; *Corinne Reporter*, 8 July 1872.
86. Second Judicial District Court, Beaver County, Utah, Minute Book No. 3, 45–49.
87. Beaver County Recorder's Office, Book No. 1 of Notices, 7, 20–22,
88. *Salt Lake Tribune*, 15 June; 12 July 1872; *Utah Mining Journal*, 24 June; 8 July; 9, 27 August 1872; *Pioche Record*, 28 September 1872.
89. The chief silver mines at North Star were the Monahan, Gallagher, Keep, Shamrock, Last Discovery, Belfast, Aurora, and Montana. In the West District, Connor, Lighthall and Company were working the "valuable and extensive" Temperance and Medusa group of mines, whose ores assayed 80 dollars in silver to the ton. There were 300 tons of ore on the ground awaiting the construction of a smelter. The Flora, in the same district, and owned by Connor and Gallagher, had ores ranging in value from $40 to $700 to the ton in silver and 150 tons of ore on the dump awaiting processing. *Salt Lake Tribune*, 12 March 1873.
90. *Salt Lake Tribune*, 17, 19 August; 11 September 1873.
91. *M. Livingston vs. P. Edw. Connor*, Third District Court, Salt Lake City, Utah (1873).
92. *P. Edward Connor vs. Robert J. Goldring et al.*, Third District Court, Salt Lake City, Utah, 25 September 1873.
93. Salt Lake County Recorder's Office, Salt Lake County Abstracts, Book A2, Block 69, Plat A, 69, 202, 210, 221, 242, 265; Edward L. Sloan, ed., *Gazetteer of Utah and Salt Lake City Directory* (1874), 204.
94. *Utah Mining Gazette*, 10 February 1874.
95. *Utah Mining Gazette*, 17 January 1874.
96. Second Judicial District Court, Beaver County, Utah, Minute Book No. 1, 7 September 1874 to 17 September 1877, 18–19, 32–34, 41, 68–71; Minute Book No. 2, 16 October 1865 to 24 November 1879, 329, 331–32, 339–40, 355–56. These records are in the Beaver County recorder's office.
97. *Salt Lake Tribune*, 25 August 1874.
98. *Salt Lake Tribune*, 3 September, 3 November, 9 December 1874.
99. *Salt Lake Tribune*, 5 September 1874.
100. *Salt Lake Tribune*, 26 April; 18 June; 8, 28 November; 5, 9 December 1874; *Utah Mining Gazette*, 13 June 1874; *Mining and Scientific Press*, 26 December 1874.
101. *Amos Woodward et al. vs. P. E. Connor et al.*, Third District Court, Salt Lake City, Utah (1874).
102. *Steven F. Nuckolls vs. P. Ed. Connor et al.*, Third District Court, Salt Lake City, Utah (1874).
103. Beaver County Recorder's Office, "Real Estate in Beaver Co.," Index to Books A, B, C, and D (1857–75).
104. *Salt Lake Tribune*, 22 October 1875.
105. *Salt Lake Tribune*, 1 January 1875.
106. *Salt Lake Tribune*, 29 June 1875.
107. *P. Edward [sic] Connor vs. Enos A. Wall, John W. Johnson, Charles Reed and H. W. Lawrence*, Third District Court, Salt Lake City, Utah, 3 August 1875; and *P. Ewd. [sic] Connor vs. Hiram S. Jacobs*, Third District Court, Salt Lake City, Utah, 27 October 1876.
108. *P. Ewd. [sic] Connor vs. Hiram S. Jacobs*, Third District Court, Salt Lake City, Utah (1876).
109. *Salt Lake Tribune*, 3, 22 June 1876.

110. *Salt Lake Tribune*, 3 October; 5, 10 December 1878.
111. *Eureka Sentinel*, 18, 21 January; 8 May; 25, 28 October 1879; *Salt Lake Herald*, 28 June 1879; *Salt Lake Tribune*, 11 May; 27 June 1879.
112. *Salt Lake Tribune*, 22 January; 20 April; 8 May; 27 June; 21 August 1879; *Salt Lake Herald*, 22 January; 19 April; 8 May; 21 August; 4 September 1879; *Eureka Sentinel*, 24, 25 April; 7, 8 May; 12 August; 22, 23, 25, 28 October; 19, 20 December 1879.
113. U.S. Congress, *Report of the Governor of Utah*, (1880), 457.
114. Fox, "Patrick Edward Connor," 88–89.
115. *Salt Lake Tribune*, 1 January 1879.
116. King, *Statistics and Technology of the Precious Metals*, 447; *Salt Lake Tribune*, 8 March; 11 May; 4 September; 8 November 1879.
117. *Leonard S. Osgood vs. P. Edw. Connor*, Third District Court, Salt Lake City, Utah (1880); *Salt Lake Tribune*, 3 September 1880. Another lawsuit of 16 July 1880 in the Third District Court, *John S. Barrett and Oscar V. Walker vs. P. Edward Connor*, was dismissed under the statute of limitations because the suit was not filed within the prescribed four years allowed.
118. *Salt Lake Tribune*, 1, 24 January; 24 February; 25 July; 18 August; 13, 16 October 1880; *Western Mining Gazetteer*, 18 August; 22 September; 3 November 1880; *Engineering and Mining Journal*, 21 August; 21 November 1880; Daughters of Utah Pioneers, *History of Tooele County*, 343.
119. *Western Mining Gazetteer*, 18 August 1880; *Salt Lake Tribune*, 13, 16 October 1880.
120. *Salt Lake Tribune*, 13 October 1880; Fox, "Patrick Edward Connor," 89.
121. *Salt Lake Tribune*, 20 October 1880.
122. *Eureka Sentinel*, 20 October 1880; *Salt Lake Tribune*, 1 January; 25 July 1880.
123. Fox, "Patrick Edward Connor," 89–90. Connor, "one of the original locators of mines in Beaver County, after an absence of seven years, paid a visit to the Star and Frisco districts" on 31 August 1880. *Salt Lake Tribune*, 2 September 1880.
124. *Salt Lake Tribune*, 3, 4, 7 July 1880; Brigham H. Roberts, *A Comprehensive History of the Church of Jesus Christ of Latter-day Saints*, 5:625–26.
125. *Charles E. Mitchner and John R. Kelly vs. Great Basin Mining and Smelting Company and P. Edward Connor*, Third District Court, Salt Lake City, Utah (1881).
126. *Lawrence Bethune vs. P. Edward Connor*, Third District Court, Salt Lake City, Utah (1881).
127. *Salt Lake Tribune*, 4 January; 19 March 1881.
128. O. J. Hollister, *Resources and Attractions of Utah*, 31; *Salt Lake Tribune*, 4 January; 19 March 1881; *Western Mining Gazetteer*, 29 January 1881.
129. Fox, "Patrick Edward Connor," 90; Edward W. Tullidge, "The Mines of Utah," *Tullidge Quarterly Magazine*: 189–90.
130. John Codman, *The Round Trip*, 184–85.
131. *Salt Lake Tribune*, 20 February; 19 June; 17 July 1881.
132. *P. Edward Connor vs. Great Basin Mining and Smelting Company*, Third District Court, Salt Lake City, Utah (1882).
133. *The Utah Commercial*. Article was quoted in the *Eureka Sentinel*, 11 July 1882.
134. *Salt Lake Herald*, 24 April 1887.
135. U.S. Congress, *Report of the Governor of Utah*, (1887), 910; *The Engineering and Mining Journal*, 17 September 1887, 211.
136. *Salt Lake Tribune*, 30 March 1887.
137. Tooele County Recorder's Office, Deed Record, Book HH, 620.
138. *Salt Lake Tribune*, 1, 22, 28 January 1891.
139. Tooele County Recorder's Office, Deed Record, Book KK, 86–88; Book MM, 254.
140. *Salt Lake Tribune*, 17, 18, 19, 20, 21 December 1891; *Salt Lake Herald*, 19, 20, 22 December 1891; *Deseret News*, 19, 20 December 1891; *San Mateo Times Gazette*, 26 December 1891.
141. Salt Lake County, Probate Court Records.

142. Salt Lake County Clerk's Office, Utah, Estates, Books M1, N1, 01, 1891–94 and Misc. Dates, 73–80, 387–90, 479.
143. *Millennial Star*: 606.

4—The Stories They Tell
Carma Wadley

1. T. H. Watkins, *Gold and Silver in the West*, 16.
2. John Greenway, ed., *Folklore of the Great West: Selections from Eighty-three Years of the Journal of American Folklore*, 298.
3. Ibid.
4. Carma Wadley, "Pay Dirt," Ron Kunz, interview with the author, September 1997, in *Deseret News*, 21 September 1997.
5. Kate B. Carter, comp., "Mining and Railroad Ghost Towns," 172–73. Daughters of Utah Pioneers, lesson for November 1970, comp. Kate B. Carter.
6. Greenway, 290.
7. Ibid., 292.
8. Kate B. Carter, comp., "Mormon Folklore," 589. Daughters of Utah Pioneers, lesson for May 1964, comp. Kate B. Carter.
9. Kate B. Carter, *Heart Throbs of the West*, 10:134. The Three Nephites are mentioned in the Book of Mormon (3 Nephi 28) as requesting, like John the Beloved in the Bible, to remain on the earth to help people until Christ returns.
10. Greenway, 294.
11. Ibid.
12. Carter, "Mormon Folklore," 585–86.
13. Gale R. Rhoades and Kerry Ross Boren, *Footprints in the Wilderness: A History of the Lost Rhoades Mine*, jacket blurb.
14. Kerry Ross Boren and Lisa Lee Boren, *The Gold of Carre-Shinob: The Final Chapter in the Mystery of the Lost Rhoades Mines, Seven Lost Cities and Montezuma's Treasure*, xii.
15. Carter, "Mining and Railroad Ghost Towns,"170.
16. Carter, "Mormon Folklore," 582–84.
17. Colleen Whitley, "La Platta: Where Mines and Men Collided," *Utah Magazine*: 26; Robin Tippets, "Ghosts of La Platta," *Ogden Standard-Examiner*, 2 December 1983.
18. Whitley, "La Platta," 26.
19. Arnold Irvine, "The People Who Made Kennecott," *Deseret News*, 27 March 1985, C3.
20. "1913 Manhunt Remembered," *Salt Lake Tribune*, 21 November 2002.
21. *Through Our Eyes: 150 Years of History as Seen Through the Eye of the Writers and Editors of the* Deseret News, 82.
22. Ed Lion, UPI news story, 12 August 1977, *Deseret News* files.
23. Dean Nolan and Fred Thompson, *Joe Hill: IWW Songwriter*, 5. For more information on Joe Hill, see Philip S. Foner, *The Case of Joe Hill*; Philip S. Foner, *The Letters of Joe Hill*; Gibbs M. Smith, *Joe Hill*.
24. Lion, UPI news story.
25. Carter, "Mining and Railroad Ghost Towns," 161–62.

5—Saline Minerals
J. Wallace Gwynn

1. J. L. Clark, "History of Utah's Salt Industry, 1847–1970" is a well-documented reference detailing the history of Utah's salt industry. Unless otherwise noted, the majority of the information for

this section on the history of salt production from Great Salt Lake comes from this reference with permission from the author.

2. J. L. Clark and Norman Helgren, "History and Technology of Salt Production from Great Salt Lake," in *Great Salt Lake: A Scientific, Historical, and Economic Overview*, ed. J. W. Gwynn, 203–17.
3. Danny Bauer, "History of the Cargill (Formerly AKZO) Salt Company from Early Pioneer Times to the Present," in *Great Salt Lake: An Overview of Change*, ed. J. W. Gwynn, 217–19.
4. Ibid.
5. Nathan Tuttle and James Huizingh, "A Brief History of the Morton Salt Company," in *Great Salt Lake: An Overview of Change*, 213–16.
6. Ibid.
7. Ibid.
8. David Butts, "IMC Kalium Ogden Corporation: Extraction of Non-metals from Great Salt Lake," in *Great Salt Lake: An Overview of Change*, 227–33.
9. Utah Mining Association, "IMC Kalium Sold—Changes Name," 2001, http//www.utahmining.org/01dec.htm (accessed 2 April 2002).
10. Corey Milne, Great Salt Lake Minerals, personal communication with the author in April 2002.
11. G. T. Tripp, "Production of Magnesium from the Great Salt Lake," in *Great Salt Lake: An Overview of Change*, 221–25.
12. Ibid.
13. L. H. Austin, "Problems and Management Alternatives Related to the Selection and Construction of the West Desert Pumping Project," in *Great Salt Lake: An Overview of Change*, 303–12.
14. Tripp, "Production of Magnesium," 221–25.
15. Ibid.
16. Ibid.
17. Butts, "IMC Kalium Ogden Corporation," 227–33.
18. Utah Mining Association, "ICM Kalium Sold."
19. Butts, "IMC Kalium Ogden Corporation," 227–33.
20. C. D. Anderson, and Val Anderson, "Nutritional Enterprises on Great Salt Lake: North Shore Limited Partnership and Mineral Resources International," in *Great Salt Lake: An Overview of Change*, 235–41.
21. J. W. Gwynn, "History of Potash Production from the Salduro Salt Marsh (Bonneville Salt Flats), Tooele County," *Survey Notes*: 1–18.
22. J. W. Gwynn, "History of Potash Production from the Salduro Salt Marsh (Bonneville Salt Flats), Tooele County," in *Great Salt Lake: An Overview of Change*, 421–22.
23. Unless otherwise noted, the information for this section on the history of salt production from east of Nephi in Juab County comes from J. L. Clark, "History of Utah's Salt Industry," with permission from the author.
24. I. J. Witkind, M. P. Weiss, and T. L. Brown, "Geologic Map of the Manti 30' x 60' Quadrangle, Carbon, Emery, Juab, Sanpete, and Sevier Counties, Utah," 1987.
25. Sadie Greenlaw, "A Brief History of Juab County," http://utahreach.usu.edu/juab/visitor/history.htm (accessed 13 May 2002).
26. I. J. Witkind et al.; I. J. Witkind and M. P. Weiss, "Geologic Map of the Nephi 30' x 60' Quadrangle, Carbon, Emery, Juab, Sanpete, Utah, and Wasatch Counties, Utah," 1991.
27. Unless otherwise noted, information for the history of salt production from the Redmond area comes from three sources: J. L. Clark, "History of Utah's Salt Industry"; R. M. Young et. al., *Sevier County, Utah: Past to Present* (some sections by Verle Peterson and Milo and Neal Bosshardt); or Neal Bosshardt, "Redmond Minerals, Inc."

28. R. J. Hite, "Salt Deposits of the Paradox Basin, Southeast Utah and Southwest Colorado," in *Saline Deposits*, ed. R. B. Mattox, 319–30; R. J. Hite, "Potash Deposits in the Gibson Dome Area, Southeast Utah."
29. Robert Evans and K. O. Linn, "Fold Relationships within Evaporites of the Cane Creek Anticline, Utah," in *Third Symposium on Salt*, ed. J. L. Rau and L. F. Dellwig, 1:286–97.
30. This section on the history of potash production from the Cane Creek area is based on information provided by Rick York, general manager of Moab Salt LLC, on 15 August 2001 unless specified otherwise.
31. H. S. Kerr, "An Analysis of Utah's Potash Industry."
32. Evans and Linn, "Fold Relationships within Evaporites."
33. Untitled Utah Geological Survey report; no date or author.
34. Margie Phillips, "Cane Creek Mine Solution Mining Project, Moab Potash Operations, Texasgulf, Inc.," 261–62.
35. J. W. Gwynn, "Mining Conditions and Problems Encountered within the Texasgulf-Cane Creek Potash Mine near Moab, Utah, and Potential Mining Conditions at Gibson Dome, Utah"; Mine Safety and Health Administration, historical data on mine disasters in the United States, http://www.msha.gov/mshainfo/factshet/mshafct8.htm (accessed 26 June 2001).
36. Daniel Jackson, "Solution Mining Pumps New Life into Cane Creek Potash Mine."
37. Information for this section on the assessment and development of Sevier Lake comes from J. W. Gwynn, "Assessment and Development of Saline Resources of Sevier Lake, Millard County, Utah," in *Energy and Mineral Resources of Utah*, ed. M. L. Allison, 125–34.
38. Unless otherwise noted, information for this section on the Preuss salt zone comes from J. W. Gwynn, "The Saline Resources of Utah," *Survey Notes*: 21–26.
39. G. R. Mansfield, *Geography, Geology, and Mineral Resources of Part of Southeastern Idaho*, 338.
40. Bryce Tripp, Utah Geological Survey, personal communication with the author in 2002.
41. R. A. Bishop, "Whitney Canyon-Carter Creek Gas Field, Southwest Wyoming," 591–99.
42. P. O. Maher, "The Geology of the Pineview Field Area, Summit County, Utah," 345–50.
43. Gwynn, "The Saline Resources of Utah"; G. C. Mitchell, "Stratigraphy and Regional Implications of the Argonaut Energy No. 1 Federal, Millard County, Utah," 503–14.
42. Dyni, 1970.
43. J. G. Gwynn, "Utilization of Lisbon Oil Well Field Brine, San Juan County, Utah."

6—Coal Industry
Allan Kent Powell

1. Leonard Arrington writes, "By the fall of 1851 virtually every man in Utah territory who understood the working of coal or iron had been sent to Parowan." Leonard J. Arrington, "Planning an Iron Industry for Utah 1851–1859," *Huntington Library Quarterly*: 242.
2. Ronald G. Watt, *A History of Carbon County*, 41–42.
3. Leonard J. Arrington, "Utah's Coal Road in an Age of Unregulated Competition," *Utah Historical Quarterly*.
4. State of Utah, *Report of the State Coal Mine Inspector*, 1896, 37.
5. State of Utah, *Report of the State Coal Mine Inspector*, 1901, 21
6. Nancy Taniguchi, "The Denver and Rio Grande Western Railway," in *Utah History Encyclopedia*, ed. Allan Kent Powell, 135.
7. For a full account of this story, see Nancy J. Taniguchi, *Necessary Fraud: Progressive Reform and Utah Coal*.
8. For accounts of the disaster, see James W. Dilley, *History of the Scofield Mine Disaster*; Allan Kent Powell, "Tragedy at Scofield," *Utah Historical Quarterly*: 182–94; Craig Fuller, "Finns and the Winter Quarters Mine Disaster," *Utah Historical Quarterly*: 123–39; and Nancy J. Taniguchi, "An

Explosive Lesson: Gomer Thomas, Safety, and the Winter Quarters Mine Disaster," *Utah Historical Quarterly*: 140–57.

9. Allan Kent Powell, *The Next Time We Strike: Labor in the Utah Coal Fields 1900–1933*, 141–52; Helen Zeese Papanikolas, "Toil and Rage in a New Land: The Greek Immigrants in Utah," *Utah Historical Quarterly*: 176–81; Janeen Arnold Costa, "A Struggle for Survival and Identity: Families in the Aftermath of the Castle Gate Mine Disaster," *Utah Historical Quarterly*: 279–92; Michael Katsanevas, "The Emerging Social Worker and the Distribution of the Castle Gate Relief Fund," *Utah Historical Quarterly*: 241–54; and Philip F. Notarianni, "Hecatomb at Castle Gate, Utah, March 8, 1924," *Utah Historical Quarterly*, 63–74.
10. For statistics on the number of miners who lost their lives in the coal mines, see Fred Civish, *The Sunnyside War*. While the book is a very interesting and readable fictional account of the 1922 coal miners strike in Sunnyside, the appendix includes a list of names, dates, and mines for 1,383 coal miners who lost their lives in Utah mines from 1896 to the present. The author notes that the compilation of this list is a work in progress because more and more names will probably be found in obscure records—especially those killed before Utah became a state in 1896. Civish also includes in his list of twentieth-century casualties one poignant and instructive entry—"The Unknown Miner Many Dates Many Mines." It is also impossible to list the thousands of miners whose lives were cut short either by suffering accidents inside the mines or developing respiratory diseases such as black lung from years of breathing coal dust.
11. J. Eldon Dorman, *Confessions of a Coal Camp Doctor*, 11–12.
12. Fuller, "Finns and the Winter Quarter Mine Disaster," 123–39.
13. Philip F. Notarianni, "Italianita in Utah," in *The Peoples of Utah*, ed. Helen Zeese Papanikolas, 310.
14. *Eastern Utah Advocate*, 3 December 1903.
15. Philip F. Notarianni, "The Italian Immigrant in Utah: Nativism (1900–1925)," 54–77.
16. Joseph Stipanovich, "Falcons in Flight: The Yugoslavs," in *The Peoples of Utah*, 363–83.
17. Helen Zeese Papanikolas, "The Exiled Greeks," in *The Peoples of Utah*, 409–13.
18. Watt, *History of Carbon County*, 246–51.
19. Helen Zeese Papanikolas and Alice Kasai, "Japanese Life in Utah," in *The Peoples of Utah*, 336–37.
20. *Millennial Star*: 174. See also Sheelwant Bapurao Pawar, "An Environmental Study of the Development of the Utah Labor Movement: 1860–1935," 102.
21. For a more detailed history of this strike, see Powell, *The Next Time We Strike*, 37–50.
22. *Salt Lake Herald*, 16 February 1901.
23. For a more detailed account of the 1903–4 strike, see Powell, *The Next Time We Strike*, 51-80. For an account of the role of the Utah National Guard in the strike, see Richard C. Roberts, *Legacy: The History of the Utah National Guard from the Nauvoo Legion Era to Enduring Freedom*, 47–55.
24. For accounts of Mother Jones, see *Autobiography of Mother Jones*, ed. Mary Field Barton; Dale Fetherling, *Mother Jones the Miners' Angel*; and Elliot J. Gorn, *Mother Jones: The Most Dangerous Woman in America*.
25. Powell, *The Next Time We Strike*, 88–89.
26. Ibid., 91–93.
27. For two excellent accounts of the Ludlow massacre, see George S. McGovern and Leonard F. Guttridge, *The Great Coalfield War;* and Zeese Papanikolas, *Buried Unsung: Louis Tikas and the Ludlow Massacre*.
28. Powell, *The Next Time We Strike*, 105–14.
29. For accounts of the 1922 strike, see Powell, *The Next Time We Strike* 121–40, Helen Zeese Papanikolas, "The Carbon County Strike of 1922," in "Toil and Rage in a New Land," 166–75; and Roberts, *Legacy: A History of the Utah National Guard*, 158–72.
30. Copies of reports submitted by agents of the Globe Company are in the Spring Canyon Area Coal Company Records, MS 252, box 3, folder 4. See Powell, *The Next Time We Strike*, 156–59.

31. For an account of the strike and the conflict between the two unions, see Helen Zeese Papanikolas, "Unionism, Communism, and the Great Depression: The Carbon County Coal Strike of 1933," *Utah Historical Quarterly*: 254–300; and Powell, *The Next Time We Strike*, 165–94.
32. James B. Allen, *The Company Town in the American West*, ix.
33. Ibid., x.
34. Ibid.
35. Watt, *History of Carbon County*, 183–99. These are the pages of chapter nine entitled, "The Coal Camps."
36. Wayne L. Balle, "'I Owe My Soul': An Architectural and Social History of Kenilworth, Utah." *Utah Historical Quarterly*: 250–78.
37. State of Utah, *Report of the Industrial Commission, 1925–26*, 37.
38. For excellent accounts of coal company doctors, see Dorman, *Confessions of a Coal Camp Doctor*, 11–12; and Troy Madsen, "The Company Doctor: Promoting Stability in Eastern Utah Mining Towns," *Utah Historical Quarterly*: 139–56.
39. Utah Geological Survey and Department of Natural Resources, *2001 Annual Review and Forecast of Utah Coal Production and Distribution*, 1. Coal consumption is broken down into 12.48 million tons for power plants within Utah, 7.42 million tons for domestic power plants outside the state, and an additional 2.4 million tons exported to Pacific Rim countries.
40. Ibid. It should be noted that while production increased dramatically from 1982 to 2001, the price per ton the coal companies received dropped significantly from $29.42 in 1982 to $17.76 in 2001.
41. Ibid., 4.
42. Ibid., 3.

7—Uranium Boom
Raye C. Ringholz

1. John Wesley Powell, *The Exploration of the Colorado River and Its Canyons*, 206. This book is a replica of the original *Canyons of the Colorado*.
2. Darroll P. Young, "Economic Impact of the Uranium Industry in San Juan County, Part One," *Blue Mountain Shadows* vol. 16 (winter 1995–96): 6.
3. William E. Ford in *Dana's Manual of Minerology*, quoted in Young, "Economic Impact of the Uranium Industry," 4.
4. Quotation attributed to Merwin Shumway in Leland Shumway, "Cottonwood Vanadium Mill 1937 Garbutt-Kimmerle Venture," *Blue Mountain Shadows*: 50.
5. Terrence R. Fehner and F. G. Gosling, "Origins of the Nevada Test Site," in *The Birth of the Nuclear Age*.
6. Young, "Economic Impact of the Uranium Industry," 6.
7. W. P. Huleatt, Scott W. Hazen, Jr., and William M. Traver, Jr., "Exploration of Vanadium Region of Western Colorado and Eastern Utah."
8. Bronson qtd. in Daroll P. Young, "Economic Impact of the Uranium Industry in San Juan County," 10.
9. Holger Albrethsen, Jr., and Frank E. McGinley, "Summary History of Domestic Uranium Procurement under U.S.A.E.C. Contracts, Final Report."
10. Raye Ringholz, *Uranium Frenzy: Boom and Bust on the Colorado Plateau*, 77.
11. A. W. Knoerr, "Can Uranium Mining Pay?" *Engineering and Mining Journal*: 68.
12. *Review Journal* (Las Vegas), 28 January 1951.
13. *Times Independent* (Moab, Utah), 3 January 1952.
14. R. P. Fischer, "Federal Exploration for Carnotite Ore," *Engineering and Mining Journal*: 67.
15. Knoerr, "Can Uranium Mining Pay?" 68.

16. Maxine Newell, *Charlie Steen's Mi Vida;* interviews with the Steens by the author, 1968.
17. Raye C. Ringholz, "Bonanza at Big Indian," in *Uranium Frenzy: Saga of the Nuclear West*, 61–62.
18. Howard Balsley (speech before the American Mining Congress on 23 September 1952).
19. Ringholz, "Bonanza at Big Indian," 76; Pick's quote comes from Peter Wyden, "Greenhorn's Trail to Uranium Riches, Part 2," *St. Louis Post-Dispatch*, 30 June 1953.
20. Report of Court Proceedings, *John N. Begay et al., plaintiffs, vs. the United States of America, defendant*, Arizona U.S. District Court, Civil No. 80-982, 3 August 1983.
21. Duncan Holaday, memo to Henry N. Doyle, 11 April 1949.
22. Henry N. Doyle, "Survey of Uranium Mines in Navajo Reservation," 14–17 November 1949; 11–12 January 1950. Ringholz, *Uranium Frenzy*, 49–51.
23. Interview with Wallace Bennett.
24. The Soviets resumed their nuclear tests in 1961, and in 1963 the Limited Test Ban Treaty authorized underground testing only.
25. United States Department of Energy, "Statistical Data of the Uranium Industry" http://www.energy.gov/engine/basicSearch.do

8—Beryllium Mining
Debra Wagner

1. Sources for the information on bertandite mining come from the files of the Brush Resources Company, P.O. Box 815, Delta, Utah 84624. The information in those files came from Leland J. Davis, geologist, retired; Jack C. Valiquette, plant manager, retired; and John R. Wagner, mine supervisor.

9—Iron County
Janet Seegmiller

1. York Jones was a major contributor to this chapter on iron mining. Jones is a mining engineer who worked more than 40 years in the ore bodies of Iron County. His research extends through four major time periods of mining and other historical subjects. He and his wife, Evelyn Kunz Jones, have coauthored three books about the pioneers and government of Cedar City. Jones provided information for Graham D. MacDonald III's book, *The Magnet: Iron Ore in Iron County Utah*.
2. Kenneth L. Cook, *Magnetic Surveys in the Iron Springs District, Iron County, Utah*, 2; W. E. Young, *Iron Deposits in Iron County, Utah*, 1.
3. Some resource books which explain the geology of southern Utah are these: William Lee Stokes, *Geology of Utah;* Herbert E. Gregory, *Geology of Eastern Iron County, Utah;* and Halka Chronic, *Roadside Geology of Utah*.
4. MacDonald, *The Magnet*, 1–2; United States Steel Corporation, *The Making, Shaping and Treating of Steel*.
5. Rick Fish, "The Southern Expedition of Parley P. Pratt, 1849–50," 5–17, 87–88; "Charles C. Rich Diary" in LeRoy R. Hafen and Ann W. Hafen, eds., *Journals of Forty-Niners—Salt Lake to Los Angeles*, vol. 2 of *Far West and the Rockies*, 184; S. George Ellsworth, *The Journals of Addison Pratt*, 385.
6. William B. and Donna T. Smart, *Over the Rim: The Parley P. Pratt Exploring Expedition to Southern Utah, 1849–50*, 179.
7. "Journal of George A. Smith, President of the Iron County Mission," vol. 2 (28 April, 3 May, 6 May 1851).
8. *Deseret News*, 25 June 1852.
9. Details of the difficulties and different companies are discussed in Morris A. Shirts and Kathryn H. Shirts, *A Trial Furnace: Southern Utah's Iron Mission*.
10. Erastus Snow, letter to the editor, *Deseret News*, 25 December 1852.

11. The iron ore from which iron and steel are made is an oxide, a compound of iron (Fe) and oxygen (O). Common forms—hematite (Fe_2O_3) and magnetite (Fe_3O_4)—are found in natural deposits. To produce iron, oxygen atoms are separated from iron ore by reduction, usually by causing the oxygen to react with carbon, hydrogen, or carbon monoxide, leaving the iron free as a metal. In steelmaking, the ore is smelted, producing pig iron and slag, which contain the oxidized and unreduced substances. In the 1850s an iron furnace master was like a chef with a mental file of iron recipes, and he improvised as he worked, adding a dash of one ingredient or another. The resulting pig iron varied in quality but was generally usable. In Cedar City, the British method of using coke was tried, even though it was relatively uncommon in the United States. Coke was coal reduced in covered piles or burned in closed kilns. Ironworkers also used charcoal made by burning wood in kilns, as was done in the beehive kiln at Irontown. Exhaustible forests limited the use of charcoal.

    The Deseret Iron Company used a simple blast furnace, charged or filled with ore, fuel, and lime. Alternating layers of ingredients were added in measured lots when flames broke through the previous ones. Compressed air injected into the furnace by tuyeres placed on either side or in back formed the blast and made the furnace burn hotter. Air was compressed by water or steam power. Molten iron sank through the charge to the bottom of the furnace and collected in a pool. Slag drained off the top continually, but at intervals the furnace operators broke out a clay plug at the base of the hearth and drained molten iron into sand molds, forming iron "pigs" or bars, hence the term *pig iron*. Processes such as melting the pig iron in a cupola or puddling furnace further refined it and allowed it to be shaped by hammering for "wrought" or "bar" iron. Castings produced useful articles such as hand irons, cooking pots, tools, machinery parts, or wagon wheels.
12. *Deseret News*, 16 October 1852.
13. *Millennial Star*, 6 January 1855, 2.
14. The bell called the people of Cedar City together for church services, funerals, dances and plays; fires, floods, and other dangers; and all community celebrations. It is now at the Iron Mission State Park in Cedar City.
15. *Journal of Discourses*, vol. 2, 27 May 1855, 281–82.
16. When the men from Iron County reached the mines, no miners were there. The settlers had brought no picks or shovels but were not willing to return empty handed. They looked for ore to load in their wagons and located a rock slide which looked like lead ore. The deposit was far up a hillside, and they had no way to carry the ore. Undaunted, they took off their buckskin trousers, tied up the waists, filled them with ore, slung one leg over each shoulder, and carried the load down the hill. The lead was forged into bullets in readiness for the invading army. William R. Palmer, "History of Iron County," William R. Palmer Collection, box 22, 1922.
17. Brigham Young to Isaac C. Haight, 8 October 1858, Brigham Young's letterbook, MS f219, #8, p. 433.
18. William R. Palmer, *Forgotten Chapters of History: A Series of Talks Given over Radio Station KSUB, 1951–1955*. Transcription from audiotapes, no. 153, 1978, 1.
19. John Lee Jones, "John Lee Jones as a Missionary"; The Iron Mission Park Commission, "Building the Iron Mission Park in Cedar City, a Proposal to the Union Pacific Railroad," 42.
20. Morris Shirts, "The Demise of the Deseret Iron Company" (address given at Mormon History Association Annual Meeting, 3 May 1986), 18.
21. ElRoy Smith Jones, *John Pidding Jones, His Ancestors and Descendants*, 10–11; Ivan Jones, "The Iron Works of the John P. Jones & Sons Company, Founded 1874, Johnson's Springs [Enoch], Utah," typescript in author's possession, 1995. In 1994 a monument was dedicated by Jones's descendants at the site of his blast furnace and foundry in Enoch.
22. This national historical site was given to the State of Utah by the Cedar City chapter of the Sons of Utah Pioneers and is maintained by the Iron Mission State Park, Utah State Parks and Recreation.

23. John C. Cutler, who later became governor of Utah, was married to Thomas Taylor's daughter.
24. Brent D. Corcoran, "'My Father's Business': Thomas Taylor and Mormon Frontier Enterprise," *Dialogue:* 111–12; Leonard J. Arrington, "Iron Manufacturing in Southern Utah in the Early 1880s: The Iron Manufacturing Company of Utah," *Bulletin of the Business Historical Society:* 3. These articles discuss the complicated business dealings among Thomas Taylor, the Mormon Church, and the iron mines, although details seem contradictory or perhaps not fully known.
25. Corcoran, "My Father's Business," 113–14.
26. Ibid., 124, 131–32.
27. *Salt Lake Herald,* 6 October 1886; quoted in Arrington, "Iron Manufacture in Southern Utah in the Early 1880s," 165.
28. MacDonald, *The Magnet,* 13–15. The Utah Iron Ore and Steel Corporation also built a small steel plant in Midvale, Utah, in 1915. Since it survived on government contracts during World War I, the plant closed when the war and the contracts ended.
29. Quoted in MacDonald, *The Magnet,* 16.
30. *Iron County Record,* 20 June 1946.
31. MacDonald, *The Magnet,* 31–32.
32. *Iron County Record,* 5 October, 17 November 1949; 5, 26 June 1952; 31 July 1952; 7 August 1952; 5 July 1956; 10 July, 12 November 1959. Also *Encyclopedia International,* 1966, 17: 263–64; MacDonald, *The Magnet,* 42.
33. *Iron County Record,* 17 May 1951.
34. York F. Jones, "History of Mining in Iron County," unpublished manuscript, Chapter V, "Iron Mountain Mining, 1935–1970," table titled "Iron Ore Production, 1923–1968."
35. MacDonald, *The Magnet,* 47.
36. When the Blowout pit closed in 1968 at a depth of 625 feet, 7,168,047 tons of hard magnetite ore had been mined with an average iron content of 60 percent iron (Fe).
37. York F. Jones, BHP-Utah International operations manager, interview with the author, June 1994.
38. LaMar G. Jensen, Iron County Treasurer, to York Jones of Utah International, Inc., 23 December 1975, copy in York F. Jones, "History of Mining in Iron County," Chapter V.
39. "Historical Information for Iron County Mines" (report prepared by Roy Benson, manager of mining, Geneva Steel, for this history, 6 January 1995; tax information from Merna H. Mitchell).
40. "Geneva Steel Has Six Weeks to Find a Buyer or Financing," *Salt Lake Tribune,* 16 March 2002; "Geneva Steel Fails to Beat the Bankruptcy Deadline," *Salt Lake Tribune,* 17 November 2002; "Geneva Facing Bleak Future," *Daily Universe,* 24 October 2002; "Geneva Steel LLC," *AISE Steel News,* 29 March 2003; Dave Anderton, "Sale Ok'd of Geneva Property." *Deseret Morning News,* 29 July 2004, E-1.
41. Keera Ward, "Mine to open near city," *University Journal,* 3 October 2005, 3. Also available at http://www.palladonmining.com. Meg Cady, "Palladon Ready for Iron," *The Spectrum* 29 May 2006, 1.
42. Grant Tucker, formerly of Cedar City, contributed to this section, with additional information provided by Clemont Adams. Grant Tucker, "Notes on Iron County Coal Mining," in author's possession, 29 August 1994; Gregory, *Geology of Eastern Utah,* 145–50.
43. Henry Lunt, "Journal," 13 August–2 September 1852. York Jones and Morris Shirts located these early mining sites while researching the Iron Mission and Henry Lunt (per a conversation with York Jones, 7 September 1994).
44. Tucker, "Notes on Iron County Coal Mining," 1.
45. Paul Averitt, *Geology and Coal Resources of the Cedar Mountain Quadrangle, Iron County Utah,* 54.
46. Ibid., 60.
47. William C. Adams, "History of Coal Mining in and around Cedar City," interview by Clemont B. Adams, 7 July 1965, typescript in author's possession.

48. *Iron County Record*, 28 November 1913, 1.
49. This may have been the same two-story cabin used by Francis Webster, Henry Lunt, and Christopher Arthur as a hideout during the polygamy raids in 1887. It must have been renovated if it was the same structure. Pictures of the Corry Hotel/boardinghouse at the Iron County Coal Company mine taken in 1918 show a large frame structure with glass-paned windows.
50. L. W. Macfarlane, *Dr. Mac: The Man, His Land, and His People*, 223–24 (2d edition); *Salt Lake Herald Republican*, 13 November 1916, 10; *Iron County Record*, 18 November 1913; 8, 22 February; 12 April; 26 July; 15 November 1918.
51. William C. Adams interview, 7.
52. The height of the face of the coal mine is stated as 11 feet in the *Iron County Record*, 22 July 1937, and 15 feet in the *Iron County Record*, 14 October 1937. Grant Tucker, son of Guy C. Tucker, says the face was opened up to 15 feet, with two clay seams in the coal seam, one about 18 inches from the roof, and the other about 24 inches from the floor. Tucker, "Notes on Iron County Coal Mining," 1; "Dr. A. L. Graff Locates Long Lost Coal Mine Of High Coking Qualities," *Iron County Record*, 22 July 1937, 1. Averitt in *Geology and Coal Resources of the Cedar Mountain Quadrangle*, 59, places the coke ovens near the Old Kanarraville Mine. However, the ovens are adjacent to the Graff Kleen Koal Mine, which dates the opening of the mine at this site to the 1880s.
53. Gregory, *Geology of Eastern Iron County, Utah*, 145–48.
54. Tucker, "Notes on Iron County Coal Mining," 1–2.
55. Ibid., 2.
56. "Air Cleaning, Diesel Haulage Move Koal Kreek Ahead," *Coal Age*: 72–75.
57. Averitt, *Geology and Coal Resources of the Cedar Mountain Quadrangle*, 60–61.
58. Edward H. Hahne is the major contributor to the section on silver mining. He was general manager at the Escalante Silver Mine.
59. Placer means there were nuggets large enough to be found through panning or washing for gold in streams or, more likely, in washes that were wet in the spring.
60. "Sheriff's Sale," *Iron County Record*, 10 September 1904.
61. *Gold Guidebook for Nevada and Utah*, 104.
62. *Stateline Oracle*, 28 November 1903 (microfilm available at Sherratt Library); *Iron County Record*, 9, 30 January 1903; 13 February 1904.
63. *Iron County Record*, 3 March 1905.
64. *Iron County Record*, 30 April 1909; 17 February 1911.
65. The name Deer Lodge, from Deerlodge Canyon in eastern Lincoln County, Nevada, is also associated with this district.
66. *Gold Guidebook*, 57.
67. *Iron County Record*, 11 January 1918.
68. *Gold Guidebook*, 57; also information given to the author by Dr. Blair Maxfield, 8 September 1995.
69. Joseph Fish, "History of Enterprise," typescript, 131.
70. E. H. Hahne, "History of the Mine as I Remember It," typescript, c. 1996, copy in the author's possession.
71. E. H. Hahne, "History of the Mine"; Bruce Lee, "After 2 Years, $30 Million, Mine Producing Silver," *Salt Lake Tribune*, 11 January 1982, B-8.
72. E. H. Hahne to State of Utah Natural Resources, 21 February 1986, copy in author's possession.
73. Hecla Mining Company Annual Report, 1987.
74. *Iron County Record*, 18, 25 December 1929; 18 January 1930.
75. *Iron County Record*, 10 July 1903, 4; 30 January 1904, 1; 2 April 1904, 1. Further mining was done during World War I. *Iron County Record*, 25 October 1918.
76. *Iron County Record*, 11 Mar 1910.

77. *Iron County Record*, 24 October 1918.
78. Macfarlane, *Dr. Mac*, 204–6; Elroy Nelson, *Utah's Economic Patterns*, 182.
79. Averitt, *Geology and Coal Resources of the Cedar Mountain Quadrangle*, 64–65.

10—Bingham Canyon
Bruce D. Whitehead and Robert E. Rampton

1. Leonard. J. Arrington and Gary B. Hansen, *The Richest Hole on Earth: A History of the Bingham Copper Mine*, 11.
2. The following story, including quotations, comes from W. W. Gardner, interview with Heber J. Hart, in *Kennescope* (a company magazine of Kennecott's Utah Copper Division, November 1974), 2; Kennecott Copper Corporation Archives.
3. Connor was still a colonel and commander at Fort Douglas in 1863. He was appointed major general of the Utah militia in 1870. T. A. Rickard, *The Utah Copper Enterprise*, 15. See also chapter three of this volume and Brigham D. Madsen, *Glory Hunter: A Biography of Patrick Edward Connor.*
4. Those present were Archibald Gardner; George B. and Alex Ogilvie; Hugh O'Donneel; M. C. Lewis; Dr. Robert K. Reid, surgeon at Fort Douglas; Col. Charles Jeffrey Sprague, paymaster at Fort Douglas; Samuel Egbert, farmer and stockman in West Jordan; Neil Anderson, Swedish immigrant working in Bingham Canyon; Patrick Edward Connor; Richard Colter Drum; along with William A. Hickman, General Edward McGarry, Captain Daniel McLean, and Colonel Robert Pollock, officers at Fort Douglas; H. O. Pratt, telegraph operator; John Hardcastle; Alex, Henry, and Thomas Bexsted; James Briniger; James Finnerty; G. W. Carleton; M. J. Jenkins; H. O Pratt; Robert Pollack; David McLean; and H. B. Eldred. Gardner, interview with Hart, 3; Lynne R. Bailey, *Old Reliable*, 17.
5. The name West Mountain was selected as the English translation of the Indian word *Oquirrh*.
6. Arrington and Hansen, *Richest Hole on Earth*, 12.
7. Rickard, *Utah Copper Enterprise*, 16.
8. Beatrice Spendlove, "History of Bingham Canyon, Utah," 8.
9. Ibid.
10. Bailey, *Old Reliable*, 69–70.
11. Ibid., 51.
12. Arrington and Hansen, *Richest Hole on Earth*, 11–12.
13. T. A. Rickard, *A History of American Mining*, 191. Colonel Wall's military title was given to him by his friends.
14. Rickard, *Utah Copper Enterprise*, 17.
15. A. B. Parsons, *The Porphyry Coppers*, 50.
16. Spendlove, "History of Bingham Canyon," 13.
17. *Engineering and Mining Journal*, (9 July 1898), 67.
18. Arrington and Hansen, *Richest Hole on Earth*, 18.
19. Untitled manuscript, Kennecott Copper Corporation Archives, c. 1957.
20. *Salt Lake Tribune*, 1 January 1899.
21. *Deseret News*, 8 February 1905. A discussion of emissions and farming appears in Michael A. Church, "Smoke Farming: Smelting and Agricultural Reform in Utah, 1900–1945," *Utah Historical Quarterly*: 196–218.
22. *Deseret News*, 5, 14, 15 November 1906.
23. Spendlove, "History of Bingham Canyon," 25.
24. Jackling earned the rank of colonel by his service to Colorado Governor J. H. Peabody with the Colorado National Guard from 1903 to 1904 and service to Utah Governor William Spry with the Utah National Guard from 1909 to 1913. Consequently, he is frequently called Colonel Jackling.

25. Bailey, *Old Reliable*, 42.
26. Ibid.
27. Untitled manuscript, Kennecott Copper Corporation Archives, c. 1957.
28. Ibid.
29. Arrington and Hansen, *Richest Hole on Earth*, 37.
30. Ibid.; italics in original.
31. Ibid.
32. Parsons, *Porphyry Coppers*, 68–69.
33. Ibid.
34. Bailey, *Old Reliable*, 53.
35. Arrington and Hansen, *Richest Hole on Earth*, 52.
36. Bailey, *Old Reliable*, 32, 46.
37. David B. Morris, "Digging Out: Kennecott Resurfaces in an Era of Global Competition" manuscript, 1993, 13, Kennecott Copper Corporation Archives.
38. Parsons, *Porphyry Coppers*, 79.
39. Arrington and Hansen, *Richest Hole on Earth*, 64.
40. Ibid., 64–67; Parsons, *Porphyry Coppers*, 50.
41. Bailey, *Old Reliable*, 63.
42. Helen Z. Papanikolas, "Life and Labor among the Immigrants of Bingham Canyon," *Utah Historical Quarterly*: 290.
43. Ibid., 292.
44. Bailey, *Old Reliable*, 98.
45. "The People Who Made Kennecott," *Deseret News*, 27 March 1985, C1. The IWW was reputed to have socialist, anarchist, or communist ties and was seen as among the most violent of the emerging labor organizations.
46. A *padrone*, literally "master" or "boss" in Italian, was a contractor who provided laborers for a business or industry, usually, as in the case of Skliris, for a fee paid by each worker.
47. Papanikolas, "Life and Labor among the Immigrants," 295.
48. Ibid., 296.
49. Ibid.
50. Ibid., 307; "The People Who Made Kennecott," C1.
51. Bailey, *Old Reliable*, 103.
52. Ibid.
53. Ibid., 104.
54. Papanikolas,"Life and Labor among the Immigrants," 298.
55. *Salt Lake Tribune*, 19 September 1912, quoted in Papanikolas, "Life and Labor among the Immigrants," 300–301.
56. Ibid.
57. Bailey, *Old Reliable*, 107–8.
58. Ibid., 103.
59. Papanikolas, "Life and Labor among the Immigrants," 302–3.
60. Bailey, *Old Reliable*, 103.
61. "Outline of History of Bingham Canyon and Kennecott Utah Copper," 2, Kennecott Copper Corporation Archives.
62. Draft of script for Kennecott informational video, 26 August 2003, in possession of W. S. Adamson and Associates, Inc.
63. Spendlove, "History of Bingham Canyon," 64.
64. Harvey O'Connor, *The Guggenheims: The Making of an American Dynasty*, 352.
65. Arrington and Hansen, *Richest Hole on Earth*, 68.

66. Ibid.
67. Telluride Power Company document, Kennecott Copper Corporation Archives.
68. Bailey, *Old Reliable*, 83.
69. Ibid., 172.
70. Ibid., 154–57
71. Spendlove, "History of Bingham Canyon," 153–54.
72. Ibid., 154–55.
73. Ibid., 149–53.
74. Helen Z. Papanikolas, "Georgia Lathouris Mageras: Magerou, the Greek Midwife," *Utah Historical Quarterly*: (Fall 1965); reprinted in Colleen Whitley, ed., *Worth Their Salt, Too: More Notable but Often Unnoted Women of Utah*, 159–70.
75. Alta Miller, "A Short Sketch of My Life," in *Worth Their Salt, Too*, 145–46.
76. Floralee Millsaps, "Ada Duhigg: Angel of Bingham Canyon," in *Worth Their Salt, Too*, 156.
77. Ibid., 159.
78. Parsons, *Porphyry Coppers*, 80, quoted in Arrington and Hansen, *Richest Hole on Earth*, 70.
79. Arrington and Hansen, *Richest Hole on Earth*, 70.
80. "World's Biggest Artificial Hole," *Literary Digest*, 17.
81. Utah Copper Company, "Annual Report," 1933, 7.
82. Kennecott Copper Corporation, "Annual Report," 1934, 5.
83. Kennecott Copper Corporation, "Annual Report," 1935, 6.
84. Bailey, *Old Reliable*, 164.
85. Kennecott Copper Corporation, "Annual Report," 1936, 5.
86. Ibid., 14.
87. Kennecott Copper Corporation, "Annual Report," 1942, 6.
88. Kennecott Copper Corporation, "Annual Report," 1947, 5.
89. Arrington and Hansen, *Richest Hole on Earth*, 77.
90. Ibid.
91. Bailey, *Old Reliable*, 172.
92. Rosie the Riveter was a popular icon of women working in men's jobs. She appeared on posters, in newspapers, and eventually on T-shirts.
93. Morris, "Digging Out," 21.
94. Kennecott Copper Corporation, "Annual Report," 1945, 2.
95. Jackling received still other honors. On 19 April 1955, Brigadier General Maxwell E. Rich, the Utah adjutant general, by Special Order promoted Colonel Jackling to the honorary rank of brigadier general in the Utah National Guard: "In recognition of outstanding and meritorious service rendered the Utah National Guard, the State of Utah, and the United States of America during a long and distinguished career which has included two World Wars, and in appreciation of technological and managerial contributions of important significance to the State of Utah." State of Utah, Military Department, Office of the Adjutant General, Special Orders 11, 19 April 1955.
96. "Daniel Jackling, Engineer, Is Dead," *New York Times*, 15 March 1956.
97. Morris, "Digging Out," 20.
98. J. P. O'Keefe, general manager, Utah Copper Division, in *Kennescope* (March–April 1963), 2, Kennecott Copper Corporation Archives.
99. Ibid., 4–5.
100. Bailey, *Old Reliable*, 176–79.
101. Morris, "Digging Out," 22–23.
102. Ibid., 34.
103. Ibid.
104. Louis J. Cononelos and Philip F. Notarianni, "Kennecott Corporation," 2, Kennecott Copper Corporation Internal Archives.

105. Ibid.
106. Draft of script for Kennecott informational video "A Story about Kennecott Utah Copper," 26 August 2003; data in Kennecott files.
107. Kennecott Utah Copper Charitable Foundation, "Annual Report," 2003. An interesting addition to the Visitors Center occurred in 2004, when the Theater Candy Company donated a panorama of four separate photographs taken in 1950 by Hal Romel. Originally in black and white, the pictures were hand tinted and on display originally in a theater lobby.

11 — Silver Reef and Southwestern Utah's Shifting Frontier
W. Paul Reeve

1. *Salt Lake Tribune*, "Southern Utah" 18 August, "Southern Utah" 19 December 1875; "Bonanza City" 13 February, "Bonanza City" 5 April 1876. For other reports from the region during the same time period, see ibid., "Bonanza City" 24 March, "Harrisburg Disaster" 25 March; "Harrisburg Disaster" 1 April, "Southern Utah" 2 May 1876.
2. *Pioche Daily Record* (Pioche, Nevada), 17 April 1873. For a more detailed account of these events and a broader context for mining activity in southwestern Utah and southeastern Nevada, see W. Paul Reeve, "Mormons, Miners, and Southern Paiutes: Making Space on the Nineteenth-Century Western Frontier," chapters 1–3.
3. Irving Telling, "History of William Haynes Hamblin," in Irving Telling, *A Preliminary Study of the History of Ramah, New Mexico*.
4. The version of the initial discovery here is gleaned from a report written by Captain Hempstead and published in the "Editorial Notes—Discovery, Location etc., of the Panacka Lead," *Daily Union Vedette* (Salt Lake City), 2 July 1864. Hempstead was at the claims in 1864 and likely learned the information in his report firsthand. In any case his is the most detailed and a chronologically close retelling of the first location by Hamblin with Moroni as guide. There is, however, another version of the discovery in a letter from Edward Bunker, LDS bishop at Santa Clara, to Brigham Young (20 January 1864, Brigham Young Collection, office files, 1832–78, microfilm, reel 40, box 29, folder 17). According to Bunker, the Paiutes had been trying to persuade Hamblin "to go with them to a lead mine as they said the Mormons wanted lead. Last fall he [Hamblin] consented to go with them. He found the mine situated about 12 miles from Meadow Valley lying about northwest from here and about 120 miles distant. He braught [*sic*] some of the ore home with him." Bunker also mentions Hamblin giving a gun to a Paiute, not as inducement to show him the place but as incentive to keep the spot secret, especially from a group of California prospectors then searching for wealth in the area.

    It is difficult to know which version is more accurate. I have relied upon Hempstead's account because it seems more plausible that Hamblin would be twisting Moroni's arm to show him the ore, rather than the other way around. Given Hamblin's two-year search for gold in California, it is difficult to imagine him resisting Indian enticements to find wealth nearby. Bunker's telling is perhaps tailored to Brigham Young as audience because it makes Hamblin a reluctant participant instead of an active prospector, much more in line with Young's general policy against mining. See also the testimony of William Pulsipher in the *Raymond and Ely vs. Hermes* mining case and Judge Pitzer's closing argument as a lawyer for the Hermes Company in that case, *Pioche Daily Record*, 29 March; 12, 24 April 1873.
5. The Bunker group consisted of William Pulsipher, William Hamblin, Alsen Hamblin, Daniel C. Cill, Andrew Gibbons, Benjamin Brown Crow, Jeremiah Leavitt, A. Chamberlain, and the county surveyor, Israel Ivins. See Daniel Bonelli to George A. Smith, 30 April 1864, Brigham Young Collection, office files, 1832–78, microfilm, reel 40, box 29, folder 17; Bunker to Young, 20 January 1864; John M. Bourne, "Early Mining in Southwestern Utah and Southeastern Nevada, 1864–1873: The Meadow Valley, Pahranagat, and Pioche Mining Rushes," 22–23.

6. Brigham Young to Edward Bunker, 6 February 1864, Brigham Young Collection, outgoing correspondence, book 7, p. 194.
7. Ibid.
8. For more on Connor, see Brigham D. Madsen, *Glory Hunter: A Biography of Patrick Edward Connor* and chapter three of this volume.
9. See Reeve, "Mormons, Miners, and Southern Paiutes," 30–54, for a more detailed account of these events.
10. Bonelli to Smith, 30 April 1864; Panaca Ward, Uvada Stake, manuscript history and historical reports, microfilm, LR 6708, series 2.
11. P. Edw. Connor, Brigadier-General, Commanding, to Captain David J. Berry, 30 April 1864, in U.S. War Department, *The War of the Rebellion: A Compilation of the Official Records of the Union and Confederate Armies*, series 1, vol. 50, pt. 2, 845; Micajah G. Lewis to Captain David J. Berry, 13 May 1864, in U.S. War Department, *War of the Rebellion*, 845.
12. Reeve, "Mormons, Miners, and Southern Paiutes," 41.
13. *Daily Union Vedette*, 8 July 1864.
14. Erastus Snow to Brigham Young, 19 June 1864, Brigham Young Collection, office files, 1832–78, microfilm, reel 55, box 42, folder 18.
15. "Our Notes Continued—Snowstorms and Birch Stakes—The Mines," *Daily Union Vedette*, 1 July 1864. Italics in original.
16. "Our Southern Notes Resumed—The St. George Party and its Cache," *Daily Union Vedette*, 8 July 1864, "The Meadow Valley Mines," *Daily Union Vedette*, 11 July 1864.
17. *Pioche Daily Record*, "The Great Mining Suit," 28, 30 March; 1 April 1873; Panaca Ward manuscript history.
18. *Pioche Daily Record*, 28, 29 March 1873.
19. *Annual Report of the State Mineralogist of the State of Nevada for 1866*, 64; *Daily Union Vedette*, 31 January 1866; Church of Jesus Christ of Latter-day Saints, *Journal History*, 6 July 1866; Bourne, "Early Mining," 55–56.
20. *Annual Report of the State Mineralogist*, 64; *American Journal of Mining, Milling, Oil-Boring, Geology, Mineralogy, Metallurgy, etc.* 1 (12 May 1866): 100; *Journal History*, 6 July 1866, 2–3.
21. *Daily Union Vedette*, 18 November 1865. Sale resigned his post as recorder in October 1865. *Annual Report of the State Mineralogist* notes that the number of mining locations on record for Pahranagat "probably reaches one thousand" (64).
22. Thomas C. W. Sale to O. H. Irish, 4 May 1865, Letters Received by the Office of Indian Affairs, Utah Superintendency, 1863–65, microfilm roll #901; James W. Hulse, *The Silver State: Nevada's Heritage Reinterpreted*, 53 (2d edition).
23. *The Mining and Scientific Press* 13 (21 July 1866): 38–39; 13 (8 September 1866): 151; *American Journal of Mining*: 100.
24. *Daily Union Vedette*, "From South Utah, Silver Regions," 1 November 1865; "The Pahranagat Silver District," 31 January 1866.
25. *Daily Union Vedette*, "Mr. Editor," 30 May 1866.
26. James W. Hulse, "Boom and Bust Government in Lincoln County, Nevada, 1866–1909," *Nevada Historical Society Quarterly*: 78–79 n. 5.
27. The house debate over the bill is contained in *Congressional Globe*, part 3, 2368–70. To trace the bill (Senate Bill 155) through the Senate and House, see *Congressional Globe*, part 1, 645; part 2, 1386, 1401, 1535; part 3, 2358, 2377, 2381. The bill also added a portion of northwestern Arizona Territory to Nevada, which was ratified by the Nevada Legislature in 1867. See Donald Bufkin, "The Lost County of Pah-Ute," *Arizoniana: The Journal of Arizona History*: 7; and John M. Townley, *Conquered Provinces: Nevada Moves Southeast, 1864–1871*.
28. Rossiter W. Raymond, *Statistics of Mines and Mining in the States and Territories West of the Rocky Mountains* (first annual report of U.S. Commissioner of Mining Statistics), 1868, pp. 114–15.

29. Mel Gorman, "Chronicle of a Silver Mine: The Meadow Valley Mining Company of Pioche," *Nevada Historical Society Quarterly*: 71; *The Raymond and Ely vs. The Kentucky Mining Co. Judge Beatty's Decision*, 4.
30. Reeve, "Mormons, Miners, and Southern Paitues," 44–46.
31. *Journal History*, 25 November 1872, 1.
32. Orson Welcome Huntsman, "Diary of Orson W. Huntsman," typescript, 53–54, 74, 81, 82, 94, 96, 110, 111.
33. Paul Dean Proctor and Morris A Shirts, *Silver, Sinners and Saints: A History of Old Silver Reef, Utah*, 47–49.
34. Mark A. Pendleton, "Memories of Silver Reef," *Utah Historical Quarterly*: 99–118; Proctor and Shirts, *Silver, Sinners and Saints*, 26. See also chapter four of this book for the full story.
35. Pendleton, "Memories of Silver Reef," 99–118; Proctor and Shirts, *Silver, Sinners and Saints*, 26.
36. *Silver Reef Miner*, 29 October 1878, quoted in Alfred Bleak Stucki, "A Historical Study of Silver Reef: Southern Utah Mining Town," 12.
37. Stucki, "Historical Study of Silver Reef," 13; Proctor and Shirts, *Silver, Sinners and Saints*, 29.
38. "Southern Utah," *Salt Lake Tribune*, 2 May 1876.
39. James G. Bleak, "Annals of the Southern Utah Mission," typescript, special collections, B:173–75, 179–82; see also B:183–97; 419–23.
40. *Journal History*, 5 June 1870, 6.
41. Stucki, "Historical Study of Silver Reef," 13–18; Proctor and Shirts, *Silver, Sinners and Saints*, 27–34.
42. Stucki, "Historical Study of Silver Reef," 18; Proctor and Shirts, *Silver, Sinners and Saints*, 35–36; Ferris recounted his involvement at Silver Reef in a letter to the editor of *Mines and Methods*, April 1920. The quote here is from that letter, as reprinted in Proctor and Shirts, *Silver, Sinners and Saints*, 35–36.
43. Stucki, "Historical Study of Silver Reef," 18; Proctor and Shirts, *Silver, Sinners and Saints*, 37.
44. Brian F. Hahn, "Walker Brothers," in *Utah History Encyclopedia*, ed. Allan Kent Powell, 616–17.
45. Proctor and Shirts, *Silver, Sinners and Saints*, 37–38; Stucki, "Historical Study of Silver Reef," 18–19.
46. Stucki, "Historical Study of Silver Reef," 18–20; Proctor and Shirts, *Silver, Sinners and Saints*, 37–38.
47. *Salt Lake Tribune*, 19 December 1875; Stucki, "Historical Study of Silver Reef," 20–21; Proctor and Shirts, *Silver, Sinners and Saints*, 38–41.
48. *Salt Lake Tribune*, 1 April 1876; Proctor and Shirts, *Silver, Sinners and Saints*, 39.
49. Stucki, "Historical Study of Silver Reef," 21–22.
50. *Salt Lake Tribune*, 25 March 1876, quoted in Stucki, "Historical Study of Silver Reef," 24.
51. "Bonanza City," *Salt Lake Tribune*, 5 April 1876.
52. Stucki, "Historical Study of Silver Reef," 25; Proctor and Shirts, *Silver, Sinners and Saints*, 41–43.
53. Reeve, "Mormon, Miners, and Southern Paiutes," 208.
54. For reports of merchants and miners moving from Pioche to Silver Reef, see the *Pioche Weekly Record*, 19 May 1877; 6 April 1878; Proctor and Shirts, *Silver, Sinners and Saints*, 47–49; Stucki, "Historical Study of Silver Reef," 31–39.
55. Mark A. Pendleton, "Naming Silver Reef," *Utah Historical Quarterly*: 29–31; Stucki, "Historical Study of Silver Reef," 26–28; Proctor and Shirts, *Silver, Sinners and Saints*, 27.
56. Pendleton, "Naming Silver Reef"; "Correspondence," *Pioche Weekly Record*, 13 April 1878.
57. Proctor and Shirts, *Silver, Sinners and Saints*, chapter twelve.
58. Stucki, "Historical Study of Silver Reef," 121–22.
59. Proctor and Shirts, *Silver, Sinners and Saints*, 171–75; Stucki, "Historical Study of Silver Reef," 88–89.
60. Proctor and Shirts, *Silver, Sinners and Saints*, 47, 49, 175–81; Stucki, "Historical Study of Silver Reef," 117–18.

61. Stucki, "Historical Study of Silver Reef," 119–20; Proctor and Shirts, *Silver, Sinners and Saints*, 181–87.
62. Proctor and Shirts, *Silver, Sinners and Saints*, 181–87.
63. Ibid., 44–45, 150–51, 187–88; Stucki, "Historical Study of Silver Reef," 63, 102; Doris F. Salmon, "Enos A. Wall," in *Utah History Encyclopedia*, 617–18.
64. Proctor and Shirts, *Silver, Sinners and Saints*, 188–91.
65. Stucki, "Historical Study of Silver Reef," 37; the Wells Fargo building has been restored and currently serves as a museum and art gallery; see Proctor and Shirts, *Silver, Sinners and Saints*, 202.
66. Stucki, "Historical Study of Silver Reef," 42–43; Proctor and Shirts, *Silver, Sinners and Saints*, 123–31.
67. Proctor and Shirts, *Silver, Sinners and Saints*, 113–19; Douglas D. Alder and Karl F. Brooks, *A History of Washington County: From Isolation to Destination*, 86.
68. Nels Anderson, *Desert Saints: The Mormon Frontier in Utah*, 428–32.
69. Proctor and Shirts, *Silver, Sinners and Saints*, 113–18.
70. Stucki, "Historical Study of Silver Reef," 43–46; *Silver Reef Miner*, 29 October 1879; 24 March 1882, quoted in Stucki, "Historical Study of Silver Reef," 44–45.
71. Stucki, "Historical Study of Silver Reef," 46–50; Robert J. Dwyer, "Pioneer Bishop: Lawrence Scanlan, 1843–1915," *Utah Historical Quarterly*: 135–58.
72. W. Paul Reeve, "In 1879 a Mormon Choir Sang for a Catholic Mass in St. George," *The History Blazer*, March 1995; Alder and Brooks, *History of Washington County*,115–16; Stucki, "Historical Study of Silver Reef," 47–48; Francis J. Weber, "Catholicism among the Mormons, 1875–79," *Utah Historical Quarterly*: 141–48; Dwyer, "Pioneer Bishop."
73. Reeve, "In 1879 a Mormon Choir Sang for a Catholic Mass in St. George."
74. Ibid.
75. Nancy Perkins, "Religious Friends Relive Historic Moment," *LDS Church News*, 5.
76. "Hurrah for Leeds!" *Salt Lake Tribune*, 22 November 1876; Stucki, "Historical Study of Silver Reef," 69–71.
77. Stucki, "Historical Study of Silver Reef," 79–85.
78. See Reeve, "Mormons, Miners, and Southern Paiutes," 162–66, for examples of Young's speeches aimed at Pioche.
79. Church of Jesus Christ of Latter-day Saints, *Journal of Discourses*, 1 January 1877, 18:305. For additional evidence of Young's anti-mining stance, especially as it applied to southern Utah Saints, see Reeve, "Mormons, Miners, and Southern Paiutes," 150–79.
80. Alder and Brooks, *History of Washington County*, 114–15; Stucki, "Historical Study of Silver Reef," 74–76.
81. Stucki, "Historical Study of Silver Reef," 52–53.
82. Proctor and Shirts, *Silver, Sinners and Saints*, 149–52; Stucki, "Historical Study of Silver Reef," 93–94; Anderson, *Desert Saints*, 432–33.
83. Ibid.
84. Stucki, "Historical Study of Silver Reef," 94–95.
85. *Silver Reef Miner*, 5 February 1881, quoted in Stucki, "Historical Study of Silver Reef," 95.
86. "Letter from Silver Reef," *Salt Lake Tribune*, 9 February 1881.
87. Ibid. See also the following articles from the *Salt Lake Tribune*: "The Silver Reef Trouble" and "Silver Reef," 4 February 1881; "Silver Reef Matters," 9 February 1881; "Silver Reef Strike," 8 February 1881; "The Silver Reef Row," 10 March 1881; and "Silver Reef Affairs," 19 March 1881.
88. *Pioche Weekly Record*, 5 March 1881; Stucki, "Historical Study of Silver Reef," 96; Proctor and Shirts, *Silver, Sinners and Saints*, 153.
89. *Pioche Weekly Record*, 12 February 1881.
90. Ibid., 5 March 1881; see also Proctor and Shirts, *Silver, Sinners and Saints*, 152–53.

91. Stucki, "Historical Study of Silver Reef," 96–97; Proctor and Shirts, *Silver, Sinners and Saints*, 153–54; Pendleton, "Memories of Silver Reef."
92. Stucki, "Historical Study of Silver Reef," 98; Proctor and Shirts, *Silver, Sinners and Saints*, 154–55.
93. Stucki, "Historical Study of Silver Reef," 99–104; Proctor and Shirts, *Silver, Sinners and Saints*, 169–70.
94. Stucki, "Historical Study of Silver Reef," 104–6; Alder and Brooks, *History of Washington County*, 114–15.
95. Alder and Brooks, *History of Washington County*, 116–17.
96. This account of Washington County's oil industry is taken from W. Paul Reeve, "Rocks That Burned Led to Oil Discoveries in Southwestern Utah," *The History Blazer*, April 1995. Sources include Hazel Bradshaw, ed., *Under Dixie Sun*, 278–79; Harvey Bassler and J. B. Reeside, Jr., "Oil Prospects in Washington County, Utah," in *Contributions to Economic Geology*, 93–97; James Jepson, Jr., *Memories and Experiences of James Jepson, Jr.*, ed. Eta Holdaway Spendlove, 23–24; *Washington County News*, 31 January 1929; "Big Stills Shipped to 'Dixie' to Be Used in Refining Utah's First Oil Output," *Salt Lake Tribune*, 25 June 1924; and Osmond L. Harline, "Utah's Black Gold: The Petroleum Industry," *Utah Historical Quarterly*: 291–311.

12—Alta, the Cottonwoods, and American Fork
Laurence P. James and James E. Fell, Jr.

1. This work began when Laurence P. James interviewed many old mining people about their work in the Cottonwood-American Fork area. Their gift of time to answer innumerable questions about the Wasatch region was priceless. So, too, was the opportunity to work with Charles Keller of Salt Lake City, Utah, and Richard Winslow of the Public Library in Portsmouth, New Hampshire; their help is especially appreciated. Both authors wish to thank the many enthusiastic historians and chroniclers who have provided their time and data. Finally, both authors wish to thank Corwin Grueble of the University of Colorado, Denver, for his computer expertise.
2. Brigham H. Madsen, *Glory Hunter: A Biography of Patrick Edward Connor*, 1–30; Rossiter W. Raymond, *Statistics of Mines and Mining in the States and Territories West of the Rocky Mountains*, 1870, p. 168; Charles L. Keller, *The Lady in the Ore Bucket: A History of Settlement and Industry in the Tri-Canyon Area of the Wasatch Mountains*, 127–36. For more information on Patrick Connor, see chapter three.
3. F. C. Calkins and B. S. Butler, *Geology and Ore Deposits of the Cottonwood and American Fork Area, Utah*, 72; Laurence P. James, *Geology, Ore Deposits, and History of the Big Cottonwood Mining District, Salt Lake County, Utah*, 31.
4. Keller, *Lady in the Ore Bucket*, 120–25; John R. Murphy, *The Mineral Resources of the Territory of Utah*, 74.
5. Keller, *Lady in the Ore Bucket*, 127–36; U.S. Congress, *Report of the Emma Mine Investigation*.
6. U.S. Congress, *Emma Mine Investigation*; Keller, *Lady in the Ore Bucket*, 125–26.
7. James F. Day, testimony in *Emma Mine Investigation*.
8. Clark C. Spence, *British Investments and the American Mining Frontier: 1860–1901*, 139–45.
9. L. E. Chittenden, *The Emma Mine: A Statement of the Facts Connected with the Emma Mine, Its Sale to Emma Silver Mining Co., Ltd. of London, and Its Subsequent History and Present Condition*.
10. Spence, *British Investments*, 139–45; testimony of James F. Day in U.S. Congress, *Emma Mine Investigation*.
11. Rossiter W. Raymond, *Statistics of Mines and Mining in the Territories West of the Rocky Mountains*, 1874, p. 260; Calkins and Butler, *Geology and Ore Deposits*, 77–79.
12. Issac F. Marcosson, *Anaconda*; Jonathan Bliss, *Merchants and Miners in Utah: The Walker Brothers and Their Bank*, 162–76.

13. Spence, *British Investments*, 140–48; *Engineering and Mining Journal*, vols. 17 (1874) to 26 (1878).
14. Raymond, *Statistics of Mines and Mining*, 1874, p. 260; Calkins and Butler, *Geology and Ore Deposits*, 77–79.
15. Quoted in Spence, *British Investments*, 143. *Emma Mine Investigation*, 126.
16. Spence, 140–48; testimony of Silliman in U.S. Congress, *Emma Mine Investigation*.
17. U.S. Congress, *Emma Mine Investigation*, 45; W. Turrentine Jackson, "The Infamous Emma Mine: A British Interest in the Little Cottonwood District, Utah Territory," *Utah Historical Quarterly*: 339–62; W. Turrentine Jackson, "British Impact on the Utah Mining Industry," *Utah Historical Quarterly*, 347–75.
18. Spence, *British Investments*, 159–62.
19. U.S. Congress, *Emma Mine Investigation*; Spence, 144–82.
20. Keller, *Lady in the Ore Bucket, 146*.
21. Bureau of Land Management, General Records, 1796–1981, section 49.3.2, http://www.archives.gov/research/guide-fed-records/groups/049.html#49.3.2. Metis Culture, 1850–1854, http://www.agt.net/public/dgarneau/metis43.htm.
22. Murphy, *Mineral Resources of the Territory*, 32; D. B. Huntley, "Mining Industries of Utah," in Appendix I of "Statistics and Technology of the Precious Metals," in U.S. Bureau of Census, *Census of Population*, 13:444.
23. Rossiter W. Raymond, *Statistics of Mines and Mining in the States and Territories West of the Rocky Mountains*, 1873, p. 352.
24. Huntley, "Mining Industries of Utah," 13:444.
25. Keller, *Lady in the Ore Bucket*, 179.
26. Lawrence Goodwin, *The Populist and the Free Silver Movement*.
27. Laurence P. James, "George Tyng's Last Enterprise: A Prominent Texan and a Rich Mine in Utah," *Journal of the West*: 429–37; Huntley, "Mining Industries of Utah," 13:444.
28. Spence, *British Investments*, 178–82; W. Turrentine Jackson, *The Enterprising Scot: Investing in the American West after 1873*.
29. Nels H. Johnson to Laurence P. James, various letters in 1963 and 1970, Laurence P. James papers, in possession of the author, Golden, Colorado.
30. Charles L. Keller, "James T. Monk: The Snow King of the Wasatch," *Utah Historical Quarterly*: 139–58.
31. Gerald M. McDonough, *The Hogles*. George A. Thompson and Fraser Buck, *Treasure Mountain Home*, passim.
32. Charles Tyng, *Before the Wind: Tyng Family Memoir, 1808–1933*, ed. Susan Fels; James, "George Tyng's Last Enterprise," 429–37.
33. *Salt Lake Mining Review*, various issues, 1904; Laurence P. James, "Metalliferous Recent Sediments? Mineralogical and Historical Approaches to Utah Mill Tailings," in M.L. Allison, ed., *Energy and Mineral Resources of Utah*, 57–73.
34. J. Cecil Alter, *Early Utah Journalism; Salt Lake Mining Review*, 30 November 1903, p. 13 ff.
35. Richard Knight to Laurence P. James, various letters in 1976 and 1977, James papers; Jesse William Knight, *Jesse Knight Family*.
36. James, *Geology, Ore Deposits, and History of Big Cottonwood*; Keller, 214–17.
37. B. S. Butler, G. F. Loughlin, V. C. Heikes et al., *The Ore Deposits of Utah*; Calkins and Butler, *Geology and Ore Deposits*, 75–85.
38. Butler, Loughlin, and Heinkes et al., *Ore Deposits of Utah*; Calkins and Butler, *Geology and Ore Deposits*; Laurence P. James and Henry O. Whiteside, "Promoting the Alta Tunnel: The Rise and Fall of F. V. Bodfish," *Journal of the West*: 89–102; John Marshall with Zeke Zanoni, *Mining the Hard Rock in the Silverton San Juans*, 145.

39. A. B. Thomas to Laurence P. James, 1966; Cesar Ibanyez to James, interview in James papers.
40. R. F. Marvin to Laurence P. James, 1996, letters in James papers; *Western Mineral Survey* (Salt Lake City), 1926–1932.
41. Keller, *Lady in the Ore Bucket*; R. F. Marvin to Laurence P. James, 1967, letters in James papers.
42. E. H. Newman to Laurence P. James, 1984; S. Sargis to James, 1973, interviews in James papers.
43. See W. T. Parry, chapter one of this volume; Laurence P. James, "Big and Little Cottonwood (Alta) Mining Districts, Salt Lake County," in Robert W. Gloyn, ed. *The Mining Camps of Utah*, forthcoming.
44. Marvin to James, 1969 letters; J. W. Wade to Laurence P. James, 1963; J. J. Beeson to Laurence P. James, 1974, interviews in James papers.
45. E. G. Despain to Laurence P. James, 1974, interviews in James papers. Keller, *Lady in the Ore Bucket*, 194–98.
46. J. I. Kasteler to Laurence P. James, 1976, interviews In James papers.
47. Duane Shrontz, *Alta: A People's Story*, 41 (1989 edition); Alexis Kelner, *Skiing in Utah: A History*.
48. Kelner, *Skiing in Utah*.

13—Park City
Hal Compton and David Hampshire

1. M. B. Kildale, "Geology and Mineralogy of the Park City District," *Bulletin of the Mineralogical Society of Utah*: 5.
2. Miriam H. Bugden, *Geology and Scenery of the Central Wasatch Range, Salt Lake and Summit Counties, Utah*, 8.
3. David Hampshire, Martha Sonntag Bradley, and Allen Roberts, *A History of Summit County*, 83–84.
4. John M. Boutwell, *Geology and Ore Deposits of the Park City District, Utah*, 19.
5. Kate B. Carter, *Our Pioneer Heritage*, 7:117. For more on Connor's involvement with Utah mining, see chapter three of this work and Brigham D. Madsen, *Glory Hunter: A Biography of Patrick Edward Connor*.
6. Boutwell, *Geology and Ore Deposits of Park City*, 19.
7. U.S. Bureau of Census, Census Reports, vol. 1, *Twelfth Census of the United States taken in the year 1900*. Population, Part 1. Washington, D.C., U.S. Census Office, 1901, 644 and 790.
8. George A. Thompson and Fraser Buck, *Treasure Mountain Home: Park City Revisited*, 15 (1993 edition).
9. *Park Record*, 6 February 1931.
10. Carter, *Our Pioneer Heritage*, 7:121.
11. Thompson and Buck, *Treasure Mountain Home*, 5–8.
12. Ibid., 11.
13. E-mail from Marsac descendent Virginia Douglas, 28 September 2001, copy in possession of Hal Compton.
14. Thompson and Buck, *Treasure Mountain Home*, 11–13.
15. *Park Record*, 7 January 1905.
16. *Park Record*, 16 March 1917.
17. *Park Record*, 13 January 1938.
18. Boutwell, *Geology and Ore Deposits of Park City*, 136.
19. Robin Moench, "The Cornish Pump," *Park City Lodestar*: 14–16.
20. W. P. Hardesty, "The Drainage Works of the Ontario Silver Mine," *Engineering News*: 440.
21. "Death of David Keith." *Park Record*, 19 April 1918.
22. O. N. Malmquist, *The First 100 Years*, 181.

23. Kent Sheldon Larsen, "Life of Thomas Kearns," 10–11. "The Passing of David Keith," *Salt Lake Mining Review*, 30 April 1918, 27–28.
24. Thompson and Buck, *Treasure Mountain Home*, 35.
25. Boutwell, *Geology and Ore Deposits of Park City*, 145.
26. *Park Record*, 28 October 1927.
27. Margaret D. Lester, *Brigham Street*, 104.
28. Boutwell, *Geology and Ore Deposits of Park City*, 25; Hardesty, "The Drainage Works of the Ontario Silver Mine," 440–43.
29. Thompson and Buck, *Treasure Mountain Home*, 12.
30. Stephen L. Carr and Robert W. Edwards, *Utah Ghost Rails*, 104–6.
31. *Park Record*, 28 July 1894, 13 August 1904.
32. Thompson and Buck, *Treasure Mountain Home*, 49–52.
33. Judy Dykman and Colleen Whitley, *The Silver Queen*, 8.
34. *Salt Lake Mining Review*, 30 November 1913.
35. David Hampshire, "Remembering Park City's Great Fire," *Utah Historical Quarterly*: 225–42.
36. *Salt Lake Herald*, 20 June 1898, p. 4.
37. *Salt Lake Tribune*, 20 June 1898.
38. Thompson and Buck, *Treasure Mountain Home*, 61–63.
39. Ibid., 39–40.
40. Ibid., 64.
41. *Park Record*, 6 June 1908, 17 February 1911, 24 March 1933; Cheryl Livingston, "Mother Rachel Urban, Park City's Leading Madam," in *Worth Their Salt: Notable but Often Unnoted Women of Utah*, ed. Colleen Whitley, 122–29. Livingston contends that the brothels were located on Main Street until 1907. However, there are references in the *Park Record* to "fast houses" in Deer Valley as early as 1884.
42. Glenwood Cemetery, National Register of Historic Places Registration Form, Julie Osborne and Hal Compton, May 1996, Utah State Historic Preservation Office..
43. Sheelwant B. Pawar, "An Environmental Study of the Development of the Utah Labor Movement: 1860–1935," 92–103.
44. Ibid., 173.
45. *Park Record*, 1 October 1904.
46. *Salt Lake Mining Review*, 30 January 1915; Thompson and Buck, *Treasure Mountain Home*, 91–92. In spite of their similar names and proximity to one another, the Silver King (which became the Silver King Coalition) and the Silver King Consolidated (King Con) mines were, at this time, separate entities with different owners.
47. Thompson and Buck, *Treasure Mountain Home*, 57.
48. *Salt Lake Mining Review*, 30 August 1916.
49. *Park Record*, 29 October; 5, 12 November 1915.
50. Kenneth Lawson, "The Mines Saw Red," *Park City Lodestar*: 50–54.
51. *Salt Lake Tribune*, 7 May 1919.
52. *Park Record*, 9 May 1919.
53. John Ervin Brinley, Jr., "The Western Federation of Miners," 172.
54. *Salt Lake Mining Review*, 15 May 1919; Lawson, "The Mines Saw Red," 50–54.
55. Mike Ivers papers, MS 370, folders 2, 7, and 8. This collection includes a series of reports to the Silver King Coalition Mining Company, written between 27 May and 6 August 1919, which carefully document the movements of the strikers. The reporter, someone identified only as "Opr. #240," clearly had the confidence of the strike organizers.
56. *Salt Lake Tribune*, 22 June 1919.
57. *Park Record*, 18 July 1919.

58. *Park Record*, 7 July 1916; *Salt Lake Mining Review*, 30 October 1916, 15 January 1917.
59. *Salt Lake Mining Review*, 30 March 1922, 15 January 1923.
60. *Park Record*, 15 June 1917; *Salt Lake Mining Review*, 15 August 1917.
61. "Funeral of George W. Lambourne," *Salt Lake Tribune*, 3 August 1935, front page.
62. *Salt Lake Mining Review*, 15 February, 15 April, 15 December 1921.
63. *Salt Lake Mining Review*, 15 January 1922, p. 1.
64. *Salt Lake Mining Review*, 30 March 1922, p. 17.
65. *Salt Lake Mining Review*, 30 October, 30 December 1922.
66. *Salt Lake Mining Review*, 15 January 1923, p. 39.
67. *Salt Lake Mining Review*, 30 December 1923, p. 17.
68. *Salt Lake Mining Review*, 30 January, 15 March 1927.
69. *Park Record*, 24 August 1928, 29 March 1929, 22 August 1930.
70. *Park Record*, 31 January, 4 April 1930.
71. *Park Record*, 20 February 1931, 1 May 1931, 19 February 1932.
72. *Park Record*, 6 March, 1 May, 5 June 1931; 18 March 1932, 30 December 1932, 24 February 1933.
73. James Ivers, interview with David Hampshire, 13 September 1997, transcript in possession of the author. Ivers held several positions at the Silver King during the 1930s and 1940s, including chief engineer. He moved to northern Michigan in 1950 but returned to Park City in January 1965 as president of United Park City Mines. Ivers was the third generation of a Park City family to play an important role in the Silver King. His grandfather, also named James Ivers, had been a partner of Keith and Kearns in the Silver King. His father, with the same name, was the general manager of the Silver King Coalition from 1935 to 1952 and president from 1952 to 1953.
74. *Park Record*, 15 December 1932, p. 1, 30 June 1933, p. 1, 24 November, 1933.
75. *Park Record*, 22, 29 December 1933; 5 January, 9 February, 24 August 1934.
76. *Park Record*, 12 October 1934; 8 February 1935, p. 1.
77. *Park Record*, letter to the editor, 3 May 1935, p. 1.
78. *Park Record*, 12 July 1935, p. 1.
79. *Park Record*, 8 February 1935, 4 March 1943, 24 February 1949.
80. *Park Record*, 13 March 1936, p. 1.
81. *Deseret News*, 12 October 1936.
82. *Deseret News*, 3, 9, 11 December 1936.
83. *Salt Lake Tribune*, 13 December 1936, p. 1 and p. 10A.
84. *Park Record*, 17 December 1936.
85. James Ivers interview.
86. *Park Record*, 31 March 1938.
87. *Park Record*, 28 April, 5 May 1938.
88. *Park Record*, 6 June, p. 1, 12 December 1940; 14 August 1941.
89. *Park Record*, 4 March, p. 1, 16 September 1943; 8 February 1945.
90. *Park Record*, 26 December 1946, 18 March 1948.
91. *Park Record*, 29 April, 24 June 24, 22 July 1948; 28 April 1949.
92. *Salt Lake Tribune*, 22, 30 June 1949, p. 19.
93. *Salt Lake Tribune*, 10 September 1949; *Park Record*, 7 September 1950.
94. *Summit County Bee*, 22 January 1953, p. 5.
95. *Salt Lake Tribune*, 10 May 1953, p. B17.
96. Summit County Commission minutes, 8 March 1954, pp. 388–89.
97. *Summit County Bee*, 6 September 1956.
98. *Salt Lake Tribune*, 31 August, 18 September 1962; 8, 22 December 1963, p. 1B.
99. *Summit County Bee*, 7 December 1961; 5 April 1962.

100. *Salt Lake Tribune*, 17 February, 18 November 1972; 19 April 1973.
101. James Ivers interview.
102. William J. Grismer, memo to L. J. Randall and W. H. Love, 16 September 1966, Clark L. Wilson papers, box 16, folder 1.
103. *Salt Lake Tribune*, 14 April, 15 July 1970.
104. *Park Record*, 19 January 1978; *Salt Lake Tribune*, 29 March 1978.
105. *Salt Lake Tribune*, 21 August 1979, 16 April 1982; *Park Record*, 28 May 1981, p. 4A.
106. *Salt Lake Tribune*, 4 November 1999.

14—Tintic Mining District
Philip F. Notarianni

1. Waldemar Lindgren and G. F. Loughlin, eds., *Geology and Ore Deposits of the Tintic Mining District, Utah*, 15 (see pp. 77–90 for a discussion of faulting); Hal T. Morris, "General Geology of the East Tintic Mountains, Utah," in *Geology of the East Tintic Mountains and Ore Deposits of the Tintic Mining District*, vol. 12 of *Guidebook to the Geology of Utah*, 1.
2. *Eureka Reporter*, 15 February 1918, 1.
3. *Deseret News*, 27 February, 5 March 1856; Peter Gottfredson, *Indian Depredations in Utah*, 100–107 (2d edition); *Eureka Reporter*, 25 October 1963, 3.
4. Papers at the Juab County Recorder's Office, Nephi, Utah. For a printed copy see *Eureka Reporter*, 25 October 1963, 10.
5. See, for example, Ray Allen Billington, *Westward Expansion: A History of the American Frontier*, 618–19 (3d edition).
6. *Salt Lake Mining Review*, 15 September 1899, 5.
7. C. Eric Stoher, *Bonanza Victorian, Architecture and Society in Colorado Mining Towns*, 10–17.
8. *Eureka Reporter*, 29 August 1929, 2.
9. V. C. Heikes, "History of Mining and Metallurgy in the Tintic District," in Lindgrin and Loughlin, (eds.), *Geology and Ore Deposits*, 105.
10. Ibid.
11. Ibid. An excellent source of mining terminology appears in Albert K. Fay, *A Glossary of the Mining and Mineral Industry*. Also see the glossary in this volume.
12. Clarence Reeder, Jr., "The History of Utah's Railroads, 1869–1883," 360.
13. Heikes, "History of Mining and Metallurgy," 105. This site at Ironton apparently was south and west of the present Tintic Junction, just west of Eureka.
14. Reeder, "History of Utah's Railroads," 360–63. It is interesting to note that Jay Gould, the eastern financier with interests in Carbon County, had stock in the Utah Southern Railroad extension. The Utah Southern and Utah Southern extension consolidated under the Union Pacific as part of the Utah Central Railway in July 1881.
15. Heikes, "History of Mining and Metallurgy," 105; *Salt Lake Tribune*, 1 January 1890, 7.
16. *Salt Lake Tribune*, 1 January 1892, 24; Heikes, "History of Mining and Metallurgy," 106. For the agreement between the Tintic Range Railway and the D&RGW, see "Important Contracts," Denver + Rio Grande Rialroad, Davis Yard, Salt Lake City, Utah, 1903. Copy in author's possession.
17. Heikes, "History of Mining and Metallurgy," 114.
18. Ibid.
19. Ibid., 115.
20. Ibid., 115–16; *Salt Lake Tribune*, 1 January 1880, 7.
21. *Salt Lake Tribune*, 1 January 1892, 24.
22. *Western Mining Gazetteer*, 8 January 1881, 2.
23. Pearl D. Wilson, June McNulty, and David Hampshire, *A History of Juab County*, 101.

24. *Salt Lake Tribune*, 1 January 1880, 7.
25. *Salt Lake Tribune*, 1 January 1887, 6.
26. *Salt Lake Tribune*, 1 January 1889, 6.
27. *Salt Lake Tribune*, 1 January 1890, 6. The report also noted that wood was fast disappearing and coal would be the new fuel. Mines in Carbon County would eventually figure in supplying that coal.
28. Ibid.
29. *Salt Lake Tribune*, 1 January 1891, 2.
30. *Salt Lake Tribune*, 1 January 1892, 24.
31. Ibid.
32. *Salt Lake Tribune*, 1 January 1893, 27.
33. *Western Mining Gazetteer*, 8 January 1881, 2.
34. *Salt Lake Mining Review*, 30 May 1899, 7; 30 December 1899, 6; 30 August 1900, 10; 15 October 1900, 9; and 15 May 1909, 22 (article entitled, "The Railroads and Smelters").
35. *Salt Lake Mining Review*, 15 September 1900, 8.
36. See the following: *Salt Lake Tribune*, 1 January 1897, 2; 1 January 1898, 17; Gary F. Reese, "Uncle Jesse: The Story of Jesse Knight"; J. William Knight *The Jesse Knight Family: Jesse Knight, His Forebears and Family*; Kay Harris, *The Towns of Tintic*, 157–67; Alice P. McCune, *History of Juab County*, 229–31, 237–41; *Salt Lake Mining Review*, 15 June 1909, 17; Knight Investment Company, Papers, MS 278.
37. See *Eureka Reporter*, 9 July 1909, 1; 11 October 1912, 1; and "The Beginning of Tintic Mining District," unpublished manuscript.
38. For a review of the Chinese in Utah, see Don C. Conley, "The Pioneer Chinese of Utah," in Helen Zeese Papanikolas, *The Peoples of Utah*, 251–77.
39. *Utah State Gazetteer and Business Directory*, 1892–93.
40. Craig Fuller, interview with the author, Salt Lake City, Utah, 26 March 1980. For a complete discussion of this influx, see Philip Taylor, *The Distant Magnet: European Emigration to the U.S.A.*; Sanborn maps, Eureka, 1898, 1908; and A. William Hoglund, "No Land for Finns: Critics and Reformers View the Rural Exodus from Finland to America between the 1880s and World War I," in *The Finnish Experience in the Western Great Lakes Region: New Perspectives*, ed. Michael G. Kami, Matti E. Kaups, Douglas J. Ollila, Jr., 36–54; *Utah Gazetteer*, 1892–93.
41. U.S. Census Bureau, *Thirteenth Census of the United States*, 1910, 3:884, U.S. Census Bureau, *Fourteenth Census of the United States*, 1920, 3:1040; and U.S. Census Bureau, *Fifteenth Census of the United States*, 1930, 3:1105.
42. *Eureka Reporter*, 30 January 1903, 8.
43. *Eureka Reporter*, 11 October 1912, 1.
44. Article reprinted in the *Eureka Reporter*, 28 June 1912, 8.
45. Reverend Louis J. Fries, *One Hundred and Fifty Years of Catholicity in Utah*, 95–97.
46. McCune, *History of Juab County*, 196–97.
47. Henry Martin Merkel, *History of Methodism in Utah*, 151–53.
48. *Tintic Miner* (Eureka, Utah), 1 November 1895 (unpaginated).
49. Richard E. Lingenfelter, *The Hardrock Miners: A History of the Mining Labor Movement in the American West, 1863–1893*, 216.
50. In 1886 the miners had formed a local union and affiliated with the Knights of Labor. See Johnathan Ezra Garlock, "A Structural Analysis of the Knights of Labor: A Prolegomenon to the History of the Producing Classes," 411.
51. *Salt Lake Tribune*, 22 February 1893, 5.
52. *Salt Lake Tribune*, 7 March 1893, 5; 8 March 1893, 7.
53. *Salt Lake Tribune*, 28 February 1893, 7; Lingenfelter, *Hardrock Miners*, 217.
54. *Salt Lake Tribune*, 8 March 1893, 7.

55. *Salt Lake Tribune*, 11 March 1893, 5; 8 March 1893, 5.
56. *Salt Lake Tribune*, 14 March 1893, 5; 15 March 1893, 8; 16 March 1893, 8; Eureka City Criminal Justice Docket Ledger, 1893, Tintic Mining Museum, Eureka, Utah.
57. *Salt Lake Tribune*, 30 March 1893, 5; 4 April 1893, 5; and 5 April 1893, 6.
58. Allan Kent Powell, "Mormon Influence on the Unionization of Eastern Utah Coal Miners, 1903–33," *Journal of Mormon History*, 92.
59. *Salt Lake Tribune*, 20 March 1893, 6.
60. *Salt Lake Tribune*, 28 March 1893, 5.
61. *Salt Lake Tribune*, 29 April 1893, 6.
62. *Deseret News*, 29 April 1893, 4.
63. *Salt Lake Tribune*, 6 May 1893, 7.
64. *Deseret News*, 15 March 1893, 3; *Salt Lake Tribune*, 17 March 1893, 5; 1 April 1893, 5; 2 April 1893, 5; 3 April 1893, 5; and 6 June 1893, 8.
65. *Salt Lake Tribune*, 11 June 1893, 3; 14 June 1893, 3; and 30 June 1893, 8.
66. *Salt Lake Tribune*, 7 April 1893, 7.
67. In the *Salt Lake Tribune*, 14 June 1893, 3, Eureka union men charged that the Beck people had blown up the houses and blamed the union to create a public outcry. The outcry did not materialize.
68. *Salt Lake Tribune*, 1 January 1895, 26.
69. Melvin Dubofsky, "The Origins of Western Working Class Radicalism, 1890–1905," *Labor History:* 136–37.
70. For more information on labor relations in the mining industry generally, see Vernon H. Jensen, *Heritage of Conflict: Labor Relations in the Non-ferrous Metals Industry Up to 1930*.
71. *Salt Lake Tribune*, 1 January 1895, 26–27; Heikes, "History of Mining and Metallurgy," 116; Harris, *Towns of Tintic*, 134.
72. *Salt Lake Mining Review*, 15 July 1899, 5.
73. *Salt Lake Tribune*, 1 January 1899, 17.
74. *Salt Lake Tribune*, 1 January 1895, 26–27.
75. *Salt Lake Tribune*, 1 January 1896, 21; 1 January 1897, 20; 1 January 1898, 17.
76. *Salt Lake Mining Review*, 15 September 1899, 5.
77. Heikes, "History of Mining and Metallurgy," 108; Don Maguire, *Utah's Great Mining Districts*, 121.
78. Maguire, *Utah's Great Mining Districts*, 22.
79. Harris, *Towns of Tintic*, 119; McCune, *History of Juab County*, 201–3.
80. Eureka City Minute Book, Book I, 1; Eureka City Criminal Justice Docket Ledger, 1893. In one incident, the mayor, Hugo Deprezin, was arrested in a house of prostitution and, when questioned about his presence there, remarked that he had been to the dentist and received a shot of Novocain, and hence had no knowledge of his actions.
81. *Salt Lake Tribune*, 1 January 1898, 17; *Eureka Reporter*, 25 July 1929, 1.
82. *Eureka Reporter*, 5 December 1902, 8; 14 May 1909, 12; 6 March 1903, 1; 20 November 1903, 8; 20 March 1903, 8.
83. *Eureka Reporter*, 13 March 1903, 1, 8; 13 November 1903, 8; 19 February 1904, 5; 16 January 1903, 8.
84. *Eureka Reporter*, 4 June 1909, 3; 23 July 1909, 1.
85. Heikes, "History of Mining and Metallurgy," 106–8.
86. Chief Consolidated Mining Company Meeting Minutes, vol. 1, 16 February 1909–December 1922; *Salt Lake Mining Review*, 15 March 1909, 30; *Eureka Reporter*, 19 March 1909, 8; 30 July 1909, 1; 28 January 1910, 1; Abstracts of Title, Eureka, Utah, Juab County.
87. Heikes, "History of Mining and Metallurgy, " 107–8.
88. See the production table in Federal Emergency Relief Administration, *The Significance to the Rural Relief Problem of Economic Fluctuations in the Tintic Metal Mining Region in Utah*, 50.
89. Raymond D. Steele, *Goshen Valley History*, 208–11.

90. *Eureka Reporter*, 18 September 1914, 1.
91. *Eureka Reporter*, 23 February 1917, 1.
92. *Eureka Reporter*, 11 October 1918, 1; 18 October 1918, 1; 25 October 1918, 1; 1 November 1918, l; Jack Lucas, interview with the author, Eureka, Utah, 15 August 1976.
93. Federal Emergency Relief Administration, *Rural Relief Problem*, 50.
94. *Eureka Reporter*, 20 November 1925, 1; 8 October 1926, 1; 17 February 1928, 7. See "Annual Report of the RGW Railway Company to the Stockholders for the Year Ending June 30, 1892," p. 28, box 43.
95. Federal Emergency Relief Administration, *Rural Relief Problem*, 189–94.
96. *Eureka Reporter*, 22 February 1940, 1.
97. Wilson, McNulty, and Hampshire, *History of Juab County*, 224.
98. *Eureka Reporter*, 6 June 1941, quoted in Wilson, McNulty, and Hampshire, *History of Juab County*, 228.
99. Wilson, McNulty, and Hampshire, *History of Juab County*, 230
100. Ibid., 232.
101. Ibid., 237.
102. Philip F. Notarianni, *Faith, Hope and Prosperity: The Tintic Mining District*.

15—San Francisco Mining District
Martha S. Bradley

1. United States Land Office, "Mining Districts By-Laws, 1872–1909," Series 3651, http://archives.utah.gov/reference/xml/series/3651.html.
2. U.S. Congress, *Statutes at Large, Treaties, and Proclamations, of the United States of America*, 1872, vol. 17, chapter 152.
3. For more on Brigham Young's recommendations about mining and the earliest mining explorations in the state, see chapter three of this volume
4. Miriam B. Murphy, "When the Fabulous Horn Silver Mine Caved In," *History Blazer:* http://historytogo.utah.gov/hornmine.html
5. Joe Smith, interview with Rosemary Davies, 7 October 1974, Milford, Utah, copy available at the Utah State Historical Society, Salt Lake City.
6. Martha S. Bradley. *A History of Beaver County*, 112.
7. Aird G. Merkley, ed., *Monuments to Courage: A History of Beaver County*, 255.
8. *Deseret News*, 30 June 1880.
9. Alton Smith, interview with Rosemary Davies, 30 September 1978, Milford, Utah, copy available at the Utah State Historical Society, Salt Lake City.
10. Evan Patterson, "Summary of Beaver County Minutes, 1856–1883," copy in the possession of the author.
11. Frank Robertson, quoted in the *Deseret News*, 8 September 1969. Robertson, along with Beth Kay Harris, authored *Boom Towns of the Great Basin*.
12. Fred Hewitt, Letters.
13. Philip F. Notarianni, "The Frisco Charcoal Kilns," *Utah Historical Quarterly* 50 (winter 1982): 40–46.
14. Fred Hewitt, Letters.
15. *Utah Mining Gazette*, 25 July 1874, 381.
16. Merkley, *Monuments to Courage*, 243.
17. *Salt Lake Tribune*, 1 July 1877.
18. Notarianni, "The Frisco Charcoal Kilns," 40–46.
19. Ibid.
20. John Franklin Tolton, "Memories of the Life of John Franklin Tolton," typescript.

21. Martha S. Bradley, 189.
22. *Weekly Press*, 14 October 1910.
23. *Richfield Reaper*, 30 April 1937.
24. "Rich Ore Strike on Indian Creek," *Beaver Press*, 3 July 1931.
25. "Huge Body of Lead Ore Located in West Mt. Range," *Beaver Press*, 21 August 1931.
26. "Beaver's Gold Fields," *Beaver Press*, 10 July 1931.
27. "Great Future Is Seen for Fortuna Plan Development," *Beaver Press*, 21 August 1931.
28. "Fortuna Mining Company Preparing to Ship Ore from Property," *Beaver Press*, 25 September 1931.
29. "Prospectors Rush Work in Fortuna Plan Development," *Beaver Press*, 26 February 1931.
30. "Mining Notes," *Beaver Press*, 17 July 1931.
31. "Interest in Mining Being Evidenced," *Beaver Press*, 20 September 1932.
32. "Plans Being Made to Reopen Sulphur Mines in This Company," *Beaver Press*, 25 September 1931.
33. "Milford Men Developing New Mine in the Bradshaw District," *Beaver Press*, 4 September 1931.
34. "New Prospect Being Developed by Horn Silver," *Beaver Press*, 4 September 1931.
35. "Sulphurdale Mines Begin Operations; Machinery Arrives," *Beaver Press*, 10 June 1932.
36. "Reopening of Mines," *Beaver Press*, 21 July 1933.
37. "Work Starts on Custom Rod Ore Mill in Beaver," *Beaver Press*, 9 September 1932.
38. "Installation of Process Ore Mill Underway in Beaver," *Beaver Press*, 16 September 1932.
39. "Ore from Fortuna Mine Gives Assay of $38.60 per Ton," *Beaver Press*, 4 November 1932.
40. "Gold Mine Sells for $250,000," *Beaver Press*, 5 April 1935.
41. "Important Mining Deal Consummated in Beaver," *Beaver Press*, 10 April 1936.
42. "Car of Ore to Be Shipped from the Sheep Rock Mine," *Beaver Press*, 9 October 1936.
43. "Star District Mining Activities," *Beaver Press*, 5 April 1935.
44. "Rich Ore Strike," *Beaver Press*, 3 May 1935.
45. *Beaver Press*, 28 June 1935.
46. Ibid.
47. "Quadmetals Will Resume Operations in Near Future," *Beaver Press*, 6 December 1935.
48. "San Francisco District Is Alive with Mining Activity," *Beaver Press*, 22 October 1937.
49. "Smelter Official Is Much Impressed with Frisco Strike," *Beaver Press*, 12 November 1937.
50. "Ray Barton Heads Bonanza Company," *Beaver Press*, 19 November 1937.
51. "New Mining Company Receives Charter," *Beaver Press*, 7 January 1938.
52. "Horn Silver Hits Heavy Production," *Beaver Press*, 7 March 1940.
53. Ibid.
54. "Scheelite Adds Millions to Value of Old Hickory Mine; Mill Will Be Installed," *Beaver Press*, 5 December 1940.
55. "Tungsten Discovered in West Granite Mining District of Beaver," *Beaver Press*, 26 February 1942.
56. "Work Resumed on Tungsten Claims in West Mountains," *Beaver Press*, 6 August 1942.
57. "Beaver County's Tungsten Developments Move Along," *Beaver Press*, 3 March 1944.
58. "Startling Discovery Made in Old Forgotten Mine Tunnel," *Beaver Press*, 4 June 1943.
59. "Beaver County Tungsten Mines Are Making Surprising Showing," *The Beaver Press*, 3 March 1944.
60. "Beaver County's Minerals," *Beaver Press*, 26 May 1944.
61. "West Mountain District Attracts Eastern Capital," *Beaver Press*, 6 November 1943.
62. "Penn-Utah Mining Company Is Formed to Develop Claims," *Beaver Press*, 27 October 1944.
63. "Beaver County Producer Again Active," *Milford News*, 9 August 1945.
64. "A Miner's Report of the Milford District Metal and Mineral Production during the War Years 1943–1945," *Milford News*, 25 October 1945.
65. "Milford Is Becoming Mining Capital of Southern Utah," *Beaver Press*, 25 January 1946.

66. "Metal Producers to Construct Mill to Work Low-Grade Ores," *Milford News*, 21 August 1947.
67. "Horn Silver Shuts Down; Lincoln Still at Work," *Milford News*, 21 August 1947.
68. "Horn Silver to Resume Production Sept. 15; Mill Equipment on Way," *Milford News*, 4 September 1947.
69. "Mill for Horn Silver Nearing Completion," *Milford News*, 12 February 1948.
70. "Metal Producers Mill Adds Swing Shifts; Working Horn Silver Dump and Pit Ores," *Milford News*, 23 July 1948.
71. According to the executive summary produced to secure the original loan to purchase the rights, "The Company's properties include about 3,794 acres of patented mining claims either owned, under contract to be acquired, or leased, plus unpatented mining claims, State of Utah mineral leases, and 133.45 acres of fee lands in and near Milford. The total property holdings controlled or under contract aggregate about 40,532 gross acres, including substantial overlap of certain properties. The approximate total of net mineral acres controlled or under contract by the Company is about 37,700 net acres." "Executive Summary," Mining, 0396 WUC Copper Mine, Utah, $8-Loan, http://www.help-finance.com/download/0396-Eng.doc

16—Uinta Basin
John Barton

1. Many Utahns are unaware that nearly half of the Uinta Basin is in Colorado. Since the eastern rim is formed by the Rocky Mountains, Steamboat Springs, Meeker, and surrounding areas are part of the Uinta Basin. There are two accepted spellings for *Uinta*. "Uinta" is generally used for natural features such as the Uinta Basin or the Uinta Mountains, whereas "Uintah" is used for human institutions such as Uintah County and the Uintah Utes. Though not known to all basin residents or consistently used, this distinction explains the difference in spelling.
2. The Uintas have several peaks over 12,000 feet, and King's Peak is the highest in the state at 13,528 feet above sea level. King's Peak was named after Clarence King, early director of the U.S. Geological Survey. See John W. Vancott, *Utah Place Names*, 214.
3. Doris K. Burton, *A History of Uintah County*, 186.
4. For more information on Spanish mining and Indian folktales on the subject, see Gale R. Rhoades and Kerry Ross Boren, *Footprints in the Wilderness: A History of the Lost Rhodes Mines;* see also Gale R. Rhoades, *The Lost Gold of the Uintah's: The Rest of the Story*. Neither of these books is considered a credible source by most historians; however, they contain the most complete details of the many stories and folktales about the lost Rhoades Mines and other gold finds in the Uinta Mountains. See also chapter four of this book.
5. Jerry D. Spangler, *Paradigms and Perspectives: A Class I Overview of Cultural Resources in the Uinta Basin and Tavaputs Plateau*, 772.
6. At the start of the atomic era after World War II, uranium exploration peaked. Between 1949 and 1958, 161 tons of uranium ore was produced from several mines. The total yield was 648 pounds of U308 at .20 percent and 395 pounds of V205 at 0.16 percent. Uranium mining in the Uinta Basin was too limited in quantity and quality to continue, however. See Thomas D. Fouch et al., eds., *Hydrocarbon and Mineral Resources of the Uinta Basin, Utah and Colorado*.
7. See the "The Dredge Christened," *Vernal Express*, 23 October 1908.
8. Robert G. Pruitt, Jr., *The Mineral Resources of Uintah County*, 89. *Vernal Express* 25 July 1913. Later attempts also met with failure.
9. Ibid. See also Burton, *History of Uintah County*, 135.
10. Burton, *History of Uintah County*, 89–90. The Silver King Mine in the Uinta Basin should not be confused with the more famous and profitable mine in Park City.
11. Ibid., 90.
12. Ibid., 139.

13. *Roosevelt Standard*, 18 April 1917.
14. *Roosevelt Standard*, 24 September 1919, 10 March 1920, 23 March 1921.
15. Burton, *History of Uintah County*, 140.
16. *Beehive History*: 23; see also George H. Hansen and H. C. Scoville, comps., *Drilling Records for Oil and Gas in Utah*, Bulletin 50 (February 1955).
17. By 1978 the Altamont/Bluebell field was producing 33,607 barrels of oil daily, which amounts to 39 percent of the oil output from the state; see Wayne L. Walquist, ed., *Atlas of Utah*, 211. This percentage dropped as drilling in the 1980s extended to additional new oil fields in Utah.
18. Utah is not usually thought of as an oil state, and the Uinta Basin is the exception. Additional limited drilling has occurred in several other counties, including San Juan, Emery, and Carbon.
19. Walquist, *Atlas of Utah*, 110.
20. Ibid. The 42 percent figure is an average. The percentages change somewhat from region to region throughout the state's oil-producing areas from year to year.
21. *Beehive History*: 23; Walquist, *Atlas of Utah*, 211.
22. *Vernal Express*, 28 September 1994. Local production did drop significantly with the oil glut. In 1984, 37,902,000 barrels of oil were produced in the state, and in 1988, the output was 33,017,000 barrels. However, the value of that oil was significantly different. The 1984 oil was valued at $1,031,313,000, while 1988 production, although only 4.88 million barrels different, was worth only $470,492,000—less than half due to the decline in oil prices per barrel.
23. *Vernal Express*, 28 September 1994. Drilling costs are higher in Utah than any other onshore state except Alaska and Louisiana.
24. Ibid.
25. The gross taxable sales for Duchesne County demonstrate the extent of the recession. The $134,586,446 gross taxable sales in 1984 marked the peak of the county's growth. By 1988 the figure had fallen to $71,468,095, a 53 percent decrease. See Statistical Abstract of Utah, 1990, table 8.
26. *Utah House Journal*, 1990, 121; see also *Utah Code Unannotated*, Title 59, chapter 5, paragraphs 101 and 102; Beverly Ann Evans, interview with the author, 5 December 1994.
27. Stephen Speckman, "Shale oil—now? Company says $40 per barrel production is possible in Utah," *Deseret Morning News*, 2 June 2006, A1.
28. M. Dune Picard, ed., *Geology and Energy Resources, Uinta Basin of Utah*, 227.
29. Coal Mine Basin mines included the Farmers, Weeks, Wardle, Stringham, Pack-Allen, and Hartle Mines.
30. Pruitt, *Mineral Resources of Uintah County*, 67.
31. "Coal Mining in Ashley Valley," *Vernal Express*, 13 December 1928.
32. Carl Gardner, interview with the author, Vernal, Utah, 19 August 2002. Carl Gardner at the time of the interview had just turned 100 but could remember his experiences during the Depression well.
33. Geologists refer to gilsonite as *asphaltite*; another less-used name is *uintaite*.
34. Pruitt, *Mineral Resources of Uintah County*, 7.
35. Doris K. Burton, "The Strip," *The Outlaw Trail Journal*: 2–11.
36. Burton, *History of Uintah County*, 132.
37. John D. Barton, *A History of Duchesne County*, 79.
38. Charles William Smith, *From Then until Now*, 796.
39. *Daily Sentinel*, 18 January 1905; see also Burton, *History of Uintah County*, 93.
40. Marie Kaczmarek, "Ghost Town on the Old Uintah Railway," *The Outlaw Trail Journal*, (Winter 1993): 15–20.
41. Pruitt, *Mineral Resources of Uintah County*, 47.
42. Earl White, vice president of operations, American Gilsonite Company, interview with the author, Bonanza, Utah, 28 January 2003.

43. Burton, *History of Uintah County*, 99.
44. For commercial mining of gilsonite, a vein must be at least five feet wide. Of the estimated 30 million tons of total gilsonite in Utah, much of it is in veins too small to justify mining.
45. Barton, *History of Duchesne County*, 128.
46. Burton, *History of Uintah County*, 147.
47. *Vernal Express*, 27 November 1991.

Resources and Bibliography

Abbreviations in This Bibliography

BYU Brigham Young University, Provo, Utah
DUP Daughters of Utah Pioneers, Salt Lake City, Utah
GPO Government Printing Office, Washington, D.C.
LDS Church of Jesus Christ of Latter-day Saints
SLC Salt Lake City
SUU Southern Utah University, Cedar City, Utah
UGA Utah Geological Association
UGMS Utah Geological and Mineral(ogical) Survey, Salt Lake City, Utah
UGS Utah Geological Survey, Salt Lake City, Utah
UHQ Utah Historical Quarterly
USGS United States Geological Survey
USHS Utah State Historical Society, Salt Lake City, Utah
U of U University of Utah, Salt Lake City, Utah
USU Utah State University, Logan, Utah

Archives and Organizations

The following maintain information on Utah mining:

Brigham Young University, Provo, Utah, Harold B. Lee Library and J. Reuben Clark Law Library

Oral-history interviews with miners and mining families, 1973–79; papers and records from miners, their families, and companies; field notes; pamphlets from companies and unions; correspondence; diaries; financial records; maps; photographs; scrapbooks; union bylaws and statements; analyses of legal decisions; Utah Stock Certificate Collection, 1874–1913. http://www.lib.byu.edu.
Federal Emergency Relief Administration. "The Significance to the Rural Relief Problem of Economic Fluctuations in the Tintic Metal Mining Region in Utah." 1934.
Fish, Joseph. "History of Enterprise." Typescript.
Huntsman, Orson W. Diary of Orson W. Huntsman. Typescript.
Knight Investment Company. Papers, 1885–1953. MS 278.
Schmidt, Donald T. "Early Mining in Utah." Paper submitted to History 697, BYU, 12 March 1959.
Spring Canyon Area Coal Company Records, 1913–54.

Tunks, Phil S. "Rocky Mountain Giant: Utah Construction and Mining Co. Has Turned into a Global Enterprise." *Barron's* 40, no. 9 (June 1960). Reprinted by Utah Construction and Mining Co.
United Mine Workers of America, Spring Canyon Local Union 6210. Constitution, Bylaws, and Rules of Order. Spring Canyon, Utah. 1935.
Utah Fuel Company. Papers, 1890–1919.

Church of Jesus Christ of Latter-day Saints, LDS Church Archives and Church Historical Department, SLC

Histories, minutes, diaries, maps, pictures, and documents related to LDS and Utah history. Many of these are being made accessible through the website. http://lds.org/churchhistory/library.
High Council Meeting Minutes, Southern Utah Mission, St. George, Utah.
Journal History. 1830–present. See especially 1863 and 1866.
Journal of Discourses. 26 vols. Liverpool, England: LDS, 1854–86.
Millennial Star. Vol. 28 (1866); vol. 45 (1883); also 6 January 1855.
Panaca Ward, Uvada Stake manuscript history and historical records, microfilm.
Smith, George A. "Journal of George A. Smith, President of the Iron County Mission."
Telling, Irving. A *Preliminary Study of the History of Ramah, New Mexico*. N.p., n.d. Copy.
Young, Brigham. Brigham Young Collection, including Letterbook and Letterpress Books 1863–64.

Iron Mission State Park, Cedar City, Utah

Information on the Mormon Iron Mission and its development in Iron County; displays of horse-drawn vehicles used from 1850 to 1920; collection of Native American and Mormon pioneer artifacts. http://www.stateparks.utah.gov/park_pages/iron.htm.

Kennecott Utah Copper Corporation Archives, SLC, Utah.

Records of the Kennecott Company, contracts, photographs, maps, reports, correspondence. http://www.kennecott.com.
Gardner, W. W. Interview with Heber J. Hart. *Kennescope*, November 1974.
Kennecott Copper Corporation Annual Reports, 1936, 1942, 1945, 1947.
Morris, David B. "Digging Out: Kennecott Resurfaces in an Era of Global Compeition." Manuscript, 1993.
O'Keefe, J. P., General Manager, Utah Copper Division, *Kennescope*, March–April 1963.
Utah Copper Company. Annual reports, 1933–35.

Murray Historical Museum, Murray City Offices, Murray, Utah

Information on smelters and smokestacks in Murray: blueprints and plans, photographs; records of ethnic groups who worked in the smelters; scale models. http://www.murray.ut.us/cityhall/heritageweb.

Park City Museum, Park City, Utah

History of the Park City Mining District and Miners Hospital; maps, photographs, scrapbooks; details of mines. http://www.parkcityhistory.org

Southern Utah University, Cedar City, Utah, Gerald R. Sherratt Library

Information on the Iron Mining District, including the Iron Mission, iron mines, and mining; manuscripts, diaries, and papers of early settlers; photographs of the district from the 1920s through the 1980s. http://www.li.suu.edu.
Iron Mission Park Commission. "Building the Iron Mission Park in Cedar City, a Proposal to the Union Pacific Railroad." N.d.
Jones, Ivan. "The Iron Works of the John P. Jones & Sons Company, Founded 1874, Johnson's Springs [Enoch], Utah." Typescript, 1995.

Jones, John Lee. "John Lee Jones as a Missionary."
Jones, York F. "History of Mining in Iron County." York Jones Collection.
Lunt, Henry. "Journal."
Palmer, William R. "History of Iron County." William R. Palmer Collection, 1922.
Union Pacific Railroad. "Building the Iron Mission Park in Cedar City." C. 1970.

Tintic Mining Museum, Eureka, Utah

Records of the Tintic Mining District, including manuscripts, artifacts, and photographs. http://www.juabtravel.com/euraka.htm.
"The Beginning of Tintic Mining District." Manuscript.

University of Utah, SLC, J. Willard Marriott Library, S. J. Quinney Law Library and American West Center

The Utah Mining Project, an ongoing effort to collect stories, both written and oral, from miners, mine owners, and family members, including a guide to the collection; correspondence; financial records, maps, mining company records and reports; papers from mine owners, supervisors, engineers, and inventors; photographs; scrapbooks; surveys; union bylaws and statements. http://www.utah.edu/libraries_computing/index.html.
Bleak, James G. "Annals of the Southern Utah Mission." 2 vols, A and B. Typescript.
Coal Index, 1860–1946.
Ivers, Mike. Papers. MS 370.
Uranium Mining Oral History Project, tape and transcript, 1970.
Utah Fuel Company. Papers, 1889–1933.
Wilson, Clark L. Papers.

Utah Geological Survey, SLC

Previously known as the Utah Geological and Mineralogical Survey and then as the Utah Geological and Mineral Survey. Geologic information in the form of maps, surveys, and publications for government, industries, and individuals. http://www.ugs.utah.gov.
Bosshardt, Neal. "Redmond Minerals, Inc." Unpublished report, 2002.
Gwynn, J. G. "Utilization of Lisbon Oil Well Field Brine, San Juan County, Utah." Unpublished report to UGS: 1–78.

Utah Mining Association, SLC

Association of mining companies and suppliers for mining since 1915. Information and news articles on economics, employment, laws, and current needs and trends in mining. http://www.utahmining.org.

Utah State Archives, SLC

Researcher's guide to mining resources throughout the state; inventories of collections; all county records, including incorporations of mining districts, claims, minutes of meetings, and incorporations of individual mining companies, presently being filmed and indexed. http://www.archives.utah.gov.

Utah State Historical Society, SLC

Mining records from throughout the state, including maps, photographs, and pamphlets from companies, miners, and government entities; contracts, reports, prospectuses; safety rules and laws; scrapbooks. http://www.history.utah.gov.
Connor, P. E. Letter to E. M. Stanton, Salt Lake City, 19 (29) April 1867. A1987.
Hewitt, Fred. Letters.
Tolton, John Franklin. "Memories of the Life of John Franklin Tolton." Typescript.

Utah State University, Merrill Library, Logan, Utah

Companies' pamphlets and records; diaries; equipment catalogs; maps; individual and company correspondence and papers; photographs; scrapbooks; union bylaws and statements. http://www.library.usu.edu.

Palmer, William R. *Forgotten Chapters of History: A Series of Talks Given over Radio Station KSUB, 1951–1955*. Transcription from audiotapes. No. 153, 1978.

Newspapers

Most Utah newspapers, including many cited in this book, are currently being digitally recorded, reproduced, and indexed by person, place, and topic at http://www.digitalnewspapers.org.

The following newspapers are cited in this work:

The Beaver Press

(BYU) *Daily Universe*

Deseret Morning News

Deseret News

Iron County Record

Milford News

New York Times

(Ogden) *Standard-Examiner*

Pioche Record

St. Louis Post-Dispatch

Salt Lake Tribune

(SUU) *University Journal*

Vernal Express

Western Mining Gazetteer

Unpublished Works

Agreement between Tintic Range Railway and the Denver and Rio Grande Western. "Important Contracts," 1903. Copy in possession of Philip Notorianni, USHS.

Albrethsen, Holger, Jr., and Frank E. McGinley. "Summary History of Domestic Uranium Procurement under U.S.A.E.C. Contracts." Final report to U.S. Department of Energy. October 1982.

"Annual Report of the RGW Railway Company to the Stockholders for the Year Ending June 30, 1892." Box 43. Copy in possession of Philip Notorianni, USHS.

Chief Consolidated Mining Company Meeting Minutes, vol. 1, 16 February 1909–December 1922. Chief Consolidated Mining Company, Eureka, Utah.

Dodge, Grenville M. "Personal Biography of Major General Grenville Mellon Dodge, 1831 to 1870." 3 vols. Manuscript. Des Moines: Iowa State Historical Department.

Doyle, Henry N. "Survey of Uranium Mines in Navajo Reservation." 14–17 November 1949, 11–12 January 1950. Unpublished document in author's possession.

Hecla Mining Company Annual Report, 1987. Copy in possession of Janet Seegmiller, SUU.

Hite, R. J. "Potash Deposits in the Gibson Dome Area, Southeast Utah," Open file report 82-1067. USGS, 1982.

Hite, R. J., and S. W. Lohman. "Geologic Appraisal of Paradox Basin Salt Deposits for Waste Emplacement." Open file report 4339-6. USGS, 1973.

Huleatt, W. P., Scott W. Hazen, Jr., and William M. Traver, Jr. "Exploration of Vanadium Region of Western Colorado and Eastern Utah." Report to U.S. Department of the Interior and Bureau of Mines, September 1946.

Published Works

"Air Cleaning, Diesel Haulage Move Koal Kreek Ahead." *Coal Age*, February 1962, 72–75.
Alder, Douglas D., and Karl F. Brooks. *A History of Washington County: From Isolation to Destination*. SLC: USHS and Washington County Commission, 1996.
Alexander, Thomas G. "From Dearth to Deluge: Utah's Coal Industry." *UHQ* 31, no. 3 (Summer 1963): 235–36.
———. *Utah: The Right Place, The Official Centennial History*. SLC: Gibbs Smith Publisher, 1995.
Alexander, Thomas G., and John F. Bluth. *The Twentieth Century American West: Contributions to an Understanding*. Provo: Charles Redd Center for Western Studies, BYU, 1983.
Allen, James B. "The Company Town: A Passing Phase of Utah's Industrial Development." *UHQ* 34, no. 2 (Spring 1966): 138–60.
———. *The Company Town in the American West*. Norman: University of Oklahoma Press, 1966.
Allgaier, Frederick K. *Surface Subsidence over Longwall Panels in the Western United States: Final Results at the Deer Creek Mine, Utah*. Pittsburgh: U.S. Bureau of Mines, 1988.
Allison, M. L., ed. *Energy and Mineral Resources of Utah*. UGA Publication 18. SLC: UGA, 1991.
Alter, J. Cecil. *Early Utah Journalism*. SLC: USHS, 1938.
———. *Utah Since Statehood, Historical and Biographical*. Chicago: S. J. Clarke Publishing, 1919.
Alverson, Bruce. "The Limits of Power: Comstock Litigation, 1859–1864." *Nevada Historical Quarterly* 43, no. 1 (Spring 2000): 74–99.
American Coal Foundation. "How Coal Is Produced." *Coal: An Ancient Gift Serving Modern Man*. http://www.ket.org/Trips/Coal/AGSMM/agsmmproduced.html.
American Mining Congress, Utah Chapter. *Legislation of 1919 of Interest to Miners and Mining Men of America*. SLC: Arrow Press, 1919.
America's Successful Men of Affairs: An Encyclopedia of Contemporaneous Biography, edited by Henry Hall. 2 vols. New York: New York Tribune, 1895–96.
Amundson, Michael A. *Yellowcake Towns: Uranium Mining Communities in the American West*. Boulder: University Press of Colorado, 2002.
"Anaconda Will Develop Carr Fork Copper Deposit in Utah's Oquirrh Mountains." *Engineering and Mining Journal* 175, no. 9 (September 1974): 21.
Anderson, George. *Emma Silver Mining Company, Limited*. London: William Brown and Company, 1872.
Anderson, Nels. *Desert Saints: The Mormon Frontier in Utah*. Chicago: University of Chicago Press, 1942.
Annual Report of the State Mineralogist of the State of Nevada for 1866. Carson City, NV: Joseph E. Eckley, State Printer, 1867.
Armentrout, J. M., M. R. Cole, and H. TerBest, Jr. *Cenozoic Paleogeography of the Western United States*. Bakersfield, CA: Pacific Coast Paleography Symposium, 1979.
Arrington, Leonard J. "Abundance from the Earth: The Beginnings of Commercial Mining in Utah." *UHQ* 31, no. 3 (Summer 1963): 192–219.
———. *The Changing Economic Structure of the Mountain West, 1850–1950*. USU Monograph Series 10. Logan: USU Press, 1963.
———. *Great Basin Kingdom: An Economic History of the Latter-day Saints, 1830–1900*. Cambridge, MA: Harvard University Press, 1958. Reprint, Lincoln: University of Nebraska Press, 1966.
———. "Iron Manufacturing in Southern Utah in the Early 1880s: The Iron Manufacturing Company of Utah." *Bulletin of the Business Historical Society* 25, no. 3 (September 1951): 149–68.
———. "Planning an Iron Industry for Utah 1851–1859." *Huntington Library Quarterly* 21 (May 1958): 237–60.
———. *Utah's Audacious Stockman, Charlie Redd*. Logan: USU Press, 1995.
———. "Utah's Coal Road in an Age of Unregulated Competition." *UHQ* 23 (January 1955): 35–63.
Arrington, Leonard J., and Gary B. Hansen, *"The Richest Hole on Earth": A History of the Bingham Copper Mine*. USU Monograph Series 9. Logan: USU Press, 1963.

Arrington, Leonard J., and John R. Alley, Jr. *Harold F. Silver: Western Inventor, Businessman, and Civic Leader*. Logan: USU Press, 1992.

Arrington, Leonard J., and Thomas G. Alexander. *A Dependent Commonwealth: Utah's Economy from Statehood to the Great Depression*. Provo: BYU Press, 1974.

Ashbaugh, Don. *Nevada's Turbulent Yesterday*. Los Angeles: Westernlore Press, 1963.

Atherton, Lewis. "The Mining Promoter in the Trans-Mississippi West." *Western Historical Quarterly* 1 (January 1970): 35–50.

———. "Structure and Balance in Western Mining History." *Huntington Library Quarterly* 30 (November 1966): 55–84.

Atkin, Claude F. *Mining, Smelting and Railroading in Tooele County*. Tooele, UT: Tooele County Historical Society, 1986.

Atkinson, W. W., Jr., ed. *Guidebook to Geology of the Oquirrh Mountains and Regional Setting of the Bingham Mining District, Utah*. Abstracts of talks at the annual field conference, Sept. 24–25, 1976. SLC: UGA, 1976.

Averitt, Paul. *Geology and Coal Resources of the Cedar Mountain Quadrangle, Iron County, Utah*. USGS Professional Paper 389. Washington, D.C.: GPO, 1962.

Bailey, Cristina. "The Legendary Gold Mines of the Uintas." *Outlaw Trail Journal* (Winter 2003): 11–26.

Bailey, Lynne R. *Old Reliable*. Tucson, AZ: Westernlore Press, 1966.

———. *The Search for Lopez: Utah's Greatest Manhunt*. Tucson, AZ: Westernlore Press, 1990.

———. *Shaft Furnaces and Beehive Kilns: A History of Smelting in the Far West, 1863–1900*. Tucson, AZ: Westernlore Press, 2002.

Baisley, Howard W. *The Continuing Plight of the Smaller Uranium Producer*. Denver: Colorado Mining Association, 1950.

Ball, Russell H. *Uranium Resources in Utah: An Assessment*. Provo: Center for Business and Economic Research, BYU, 1977.

Balle, Wayne L. "I Owe My Soul: An Architectural and Social History of Kenilworth, Utah." *UHQ* 56, no. 3 (Summer 1988): 250–78.

Bancroft, Hubert Howe. *History of Utah: 1540–1886*. San Francisco: History Company, 1890.

Barrett, G. W. "Colonel E. A. Wall: Mines, Miners, and Mormons." *Idaho Yesterdays* 15 (Spring 1971): 3–11.

———. "Enos Andrew Wall, Mine Superintendent and Inventor." *Idaho Yesterdays* 15 (Spring 1971): 24–31.

"Bankruptcies, August–December 2001." *Timesizing*, 31 March 2003. http://www.timesizing.com.

Barton, John D. *A History of Duchesne County*. SLC: USHS, 1998.

Barton, Mary Field, ed. *Autobiography of Mother Jones*. Chicago, 1925.

Bate, Kerry William. "Iron City: Mormon Mining Town." *UHQ* 50, no. 1 (Winter 1982): 47–58.

Beadle, John Hanson. "The Silver Mountains of Utah." *Harper's* 53, no. 317 (October 1876): 641-651.

———. *The Undeveloped West*. Philadelphia, 1873.

———. *Western Wild, and the Men Who Redeem Them: An Authentic Narrative, Embracing an Account of Seven Years Travel and Adventure in the Far West*. Detroit: J. C. Chilton Publishing, 1882.

Beal, Wilma C. My *Story of Silver Reef*. N.p.: Heritage Press, 1987.

Beard, H. R., I. L. Nicholas, and D. C. Seidel. *Absorption of Radium and Thorium from Wyoming and Utah Uranium Mill Tailings Solutions*. Washington, D.C.: U.S. Bureau of Mines, 1979.

Beck, William O. "The Journeys of a Victorian Jason: Moreton Frewen's Western American Mining Investments, 1890–1900." *Journal of the West* 11, no. 3 (July 1972): 513–30.

Beeson, J. J. Report on the Property of the Emma Silver Mines Company in Little Cottonwood Mining District, Alta, Utah. SLC: Arrow Press, 1919.

Bellus, Donald P. "Iron Mining, Healthy Utah Industry." *Mining Engineering* 15 (November 1963): 34.

Bersticker, A. C. *Symposium on Salt*. Cleveland: Northern Ohio Geological Society, Inc., 1962.

"Bill May End Mining Threats to Bryce." *National Parks* 64, nos. 5–6 (May–June 1990): 12.

Billington, Ray Allen. *Westward Expansion: A History of the American Frontier*. 3d ed. New York: Macmillan, 1967.

Bingham and Eastern Copper Mining Company. *The Kernel in the Bushel of Chaff*. New Haven, CT: Bingham and Eastern Copper Mining Company, 1900.

Biographical Record of Salt Lake City and Vicinity. Chicago: National Historical Record, 1902.

Bishop, R. A. "Whitney Canyon-Carter Creek Gas Field, Southwest Wyoming." In *Geologic Studies of the Cordilleran Thrust Belt*. Vol. 2. Edited by Richard Blake Powers. Denver: Rocky Mountain Association of Geologists Symposium, 1982.

Bituminous Operators' Special Committee. *The Campaign of the United Mine Workers of America against the Non-Union Mines of Utah in Aid of Its 1922 Nation-Wide Strike*. Washington, D.C.: Press of Ramsdell, n.d.

Bliss, Jonathan. *Merchants and Miners in Utah: The Walker Brothers and Their Bank*. SLC: Western Epics, 1983.

Bon, Roger L., and R. W. Gloyn. "2001 Summary of Mineral Activity in Utah." http://www.geology.utah.gov/utahgeo/rockmineral/activity/min2001.pdf

Boren, Kerry Ross, and Lisa Lee Boren. *The Gold of Carre-Shinob: The Final Chapter in the Mystery of the Lost Rhoades Mines, Seven Lost Cities, and Montezuma's Treasure*. Springville, UT: Bonneville Books, 1998.

———. *The Utah Gold Rush: The Lost Rhoades Mine and the Hathenbruch Legacy*. Springville, UT: Council Press, 2002.

Boutwell, John M. *Copper Deposits at Bingham, Utah*. Washington, D.C.: GPO, 1933.

———. *Geology and Ore Deposits of the Park City District, Utah*. Washington, D.C.: GPO, 1912.

Bowles, Samuel. *Across the Continent*. Springfield, MA, 1866.

———. *Our New West*. New York, 1869.

Bradley, Martha S. *A History of Beaver County*. SLC: USHS and Beaver County Commission, 1989.

Bradshaw, Hazel, ed. *Under Dixie Sun*. Panguitch, UT: DUP, Washington County Chapter, 1950.

Bronder, Leonard D. *Taxation of Coal Mining: Review with Recommendations*. Denver: Western Governors Conference Regional Energy Policy Office, 1976.

Brown, Clarice J. *Someone Should Remember: Frisco, Utah Horn Silver Mine*. Beaver, UT: privately published, 1996.

Brüyn, Kathleen. *Uranium Country*. Boulder: University Press of Colorado, 1955.

Buchanan, Frederick Stewart, ed. *A Good Time Coming: Mormon Letters to Scotland*. SLC: U of U Press, 1988.

Bufkin, Donald. "The Lost County of Pah-Ute." *Arizoniana: The Journal of Arizona History* 5 (Summer 1964): 7.

Bugden, Miriam H. *Geology and Scenery of the Central Wasatch Range, Salt Lake and Summit Counties, Utah*. SLC: UGS, 1991.

Bullock, Kenneth C. *Minerals and Mineral Localities of Utah*. SLC: UGMS, 1981.

Bunnell, Mark D., and Theodore W. Taylor. *Contributions to Economic Geology in Utah, 1986*. SLC: UGMS, 1987.

Burchfiel, B. C., and L. H. Royden. "Antler Orogeny: A Mediterranean-type Orogeny." *Geology* 19 (1991): 66–69.

Burton, Doris K. *A History of Uintah County*. SLC: USHS, 1997.

———. "The Strip." *The Outlaw Trail Journal* (Summer 1997): 2–11.

Butler, B. S. "Relation of Ore Deposits to Thrust Faults in the Central Wasatch Region, Utah." *Economic Geology* 14 (1919).

Butler, B. S., and V. C. Heikes. *A Reconnaissance of the Cottonwood-American Fork Mining Region, Utah*. USGS Bulletin 620. Washington, D.C.: GPO, 1915.

Butler, B. S., G. F. Loughlin, V. C. Heikes, et al. *The Ore Deposits of Utah*. USGS Professional Paper 111. Washington, D.C.: GPO, 1920.

Cady, Meg. "Palladon Ready for Iron," *The Spectrum* 29 May 2006:1.

Calkins, F. C., and B. S. Butler. *Geology and Ore Deposits of the Cottonwood and American Fork Area, Utah.* USGS Professional Paper 201. Washington, D.C.: GPO, 1943.

Campbell, Eugene E. "The M-Factors in Tooele's History." *UHQ* 51, no. 3 (Summer 1983): 277–78.

Cannon, Donald Q. "Angus M. Cannon: Frustrated Mormon Miner." *UHQ* 57, no. 1 (Winter 1989): 36–45.

Carr, Stephen L., and Robert W. Edwards. *Utah Ghost Rails.* SLC: Western Epics, 1989.

Carroll, John A., ed. *Reflections of Western Historians.* Papers of the seventh annual conference of the Western History Association, 1967. Tucson: University of Arizona Press, 1969.

Carter, Kate B. *Heart Throbs of the West.* 12 vols. SLC: DUP, 1939–51.

———. "Mining and Railroad Ghost Towns." DUP Lessons for 1970. SLC: DUP, 1970.

———. "Mormon Folklore." Lesson for May 1964. SLC: DUP, 1964.

———. *The Story of Mining in Utah.* SLC: DUP, 1963.

———, comp. *Our Pioneer Heritage.* 20 vols. SLC: DUP, 1958–77.

Chittenden, L. E., *The Emma Mine: A Statement of the Facts Connected with the Emma Mine, Its Sale to Emma Silver Mining Co., Ltd. of London, and Its Subsequent History and Present Condition.* Court brief. New York: L. B. H. Tyrrel, n.d (c. 1877).

Chronic, Halka. *Roadside Geology of Utah.* Missoula, MT: Mountain Press Publishing, 1990.

Church, Michael A. "Smoke Farming: Smelting and Agricultural Reform in Utah, 1900–1945." *UHQ* 72, no. 3 (Summer 2004): 196–218.

Civish, Fred. *The Sunnyside War.* Springville, UT: Bonneville Books, 2003.

Clayton, James E. *Mining Institute: What It Has Done and What It Proposes.* SLC, 1884.

Clayton, William. *Letterbooks.* 9 vols. Bancroft Library, Berkeley, California.

Clifton, Julie. *Federal Coal Leases, Central Utah, 1985.* Washington, D.C.: GPO, 1986.

"Coal Firm Renews Plans to Strip-Mine near Bryce." *National Parks* 60, no. 11 (November–December 1986): 38–39.

Codman, John. *The Round Trip.* New York: G. P. Putman's Sons, 1881.

Cook, Douglas R. *Geology of the Bingham Mining District and Northern Oquirrh Mountains.* SLC: UGMS, 1961.

Cook, Janet. "Stockton—Small Utah Town—Exciting History," *Sons of Utah Pioneers News* 3 (March 1958).

Cook, Kenneth L. *Magnetic Surveys in the Iron Springs District, Iron County, Utah.* R.I. no. 4586. Washington, D.C.: U.S. Bureau of Mines, 1950.

Cope, Henry Norris. *American Mining Code.* Washington, D.C.: GPO, 1899.

Corcoran, Brent D. "'My Father's Busines': Thomas Taylor and Mormon Frontier Enterprise." *Dialogue* 28, no. 1 (Spring 1995): 105–41.

Costa, Janeen Arnold. "A Struggle for Survival and Identity: Families in the Aftermath of the Castle Gate Mine Disaster," *UHQ* 56, no. 3 (Summer 1988): 279–92.

Crampton, Frank A. *Deep Enough: A Working Stiff in the Western Mine Camps.* Norman: University of Oklahoma Press, 1982.

Crane, G. W. *The Tintic Mining District of Utah.* SLC: American Institute of Mining Engineers, Utah Section, 1915.

Crump, Scott. *Copperton.* SLC: Publishers Press, 1978.

Cunningham, Frances Blackham. *Driving Tour: Selected Abandoned Coal Mine Sites, Castle Coal Country, Carbon and Emery Counties.* Helper, UT: City of Helper, 1990.

Daggett, Ellsworth. *Circular to Applicants for Mineral Surveys.* SLC: Surveyor General's Office, 1890.

Dalziel, Ian W. D. "Earth before Pangea." *Scientific American*, January 1995, 58–63.

Danford, Harry Edmund. *Builders of the West: Explorers, Trappers, Miners, and Settlers West of the Mississippi.* New York: Vantage Press, 1959.

Daughters of Utah Pioneers. *Centennial Echoes from Carbon County.* Compiled by Thursey Jessen Reynolds. UT: DUP, Carbon County Chapter, 1948.

———. *History of Tooele County*. SLC: DUP, 1961.

Davies, J. Kenneth. *Mormon Gold: The Story of California's Mormon Argonauts*. SLC: Olympus, 1984.

Dilly, James W. *History of the Scofield Mine Disaster*. Provo, UT: Skelton Company, 1900.

Doelling, Helmut H. *Bibliography of Utah Radioactive Occurrences*. SLC: UGMS, 1983.

Dorigatti, Barbara Thompson. *History of Beaver County*. SLC: DUP, 2002.

Dorman, J. Eldon. *Confessions of a Coal Camp Doctor*. Price, UT: Peczuh Printing, 1995.

Dotson, John L. "Duel in the Sun." *Newsweek* 27 October 1975: 10.

Dubofsky, Melvin. "The Origins of Western Working Class Radicalism, 1890–1905." *Labor History* 7 (Spring 1966): 136–37.

Dunn, Richard P. *Utah's Federal Mineral Lease Revenues, Uses and Potential: A Report to the 42nd Legislature*. SLC: Office of Legislative Research, 1976.

Dwyer, Robert J. "Pioneer Bishop: Lawrence Scanlan, 1843–1915." *UHQ* 20 (April 1952): 135–58.

Dykman, Judy, and Colleen Whitley. *The Silver Queen*. Logan: USU Press, 1998.

Edgerton, Mary Wright. *A Governor's Wife on the Mining Frontier: The Letters of Mary Edgerton from Montana, 1863–1865*. Edited by James L. Thane, Jr. SLC: Tanner Trust Fund, U of U Library, 1976.

Ellsworth, George S. *The Journals of Addison Pratt*. SLC: U of U Press, 1990.

Elston, D. P., and E. M. Shoemaker. "Salt Anticlines of the Paradox Basin, Colorado and Utah." In *Symposium on Salt*. Edited by A. C. Bersticker. Cleveland: The Northern Ohio Geological Society, Inc, 1962, 131–46.

Emmons, Samuel F., and Arnold Hague. *Descriptive Geology, U.S. Geological Exploration, 40th Parallel*. Vol. 2. Washington, D.C.: GPO, 1877.

Enchanted Wilderness Association. *Will Power Plants Destroy the Enchantment?* SLC, 1972.

Encyclopedia International. New York: Grolier, 1966.

Erickson, R. L. "Safety at the Arthur and Magna Mills of the Utah Copper Company." *Mining Congress Journal* 29, no. 7 (July 1943): 16–19.

Ernst, W. G., ed. *Metamorphism and Crustal Evolution of the Western United States*. Englewood Cliffs, NJ: Prentice-Hall, 1988.

Evans, Charles A. *A Study of Nuclear Power Plants and Their Effect on the Coal Industry of Utah*. SLC: U of U Press 1965.

Everett, Floyd D. *Mining and Mineral Operations in the Rocky Mountain States: A Visitor's Guide*. Washington, D.C.: U.S. Bureau of Mines, 1977.

"Executive Summary," Mining, 0396 WUC Copper Mine, Utah, $8-Loan http://www.help-finance.com/download/0396-Eng.doc.

Farlaino, Frank G. *A Sunnyside Utah History*. UT, 1991.

Fay, Albert K. *A Glossary of the Mining and Mineral Industry*. U.S. Bureau of Mines Bulletin 95. Washington, D.C.: GPO, 1947.

Fehner, Terrence R., and F. G. Gosling. *The Birth of the Nuclear Age*. Washington, D.C.: U.S. Department of Energy, 2000.

Fell, James E., Jr. *Ores to Metals: The Rocky Mountain Smelting Industry*. Lincoln: University of Nebraska Press, 1979.

Fetherling, Dale. *Mother Jones the Miners' Angel*. Carbondale: Southern Illinois University Press, 1974.

Fiero, Bill. *Geology of the Great Basin*. Reno: University of Nevada Press, 1986.

"First Apprenticeship Program for Hardrock Miners Launched in Utah." *Engineering and Mining Journal* 178, no. 6 (June 1977).

Fischer, R. P. "Federal Exploration for Carnotite Ore." *Engineering and Mining Journal*. (December 1949).

Fitch, Franklyn Y. *The Life, Travels and Adventures of an American Wanderer*. New York: John W. Lovell, c. 1883.

Fjelsted, Boyd L. "Utah's Gross State Product for the Years 1963–1986." *Utah Economic and Business Review* 48 (July–August 1988): 1–15.

———. "Utah's Gross State Product for the Years 1977–1989." *Utah Economic and Business Review* 51 (December 1991): 1–15.

Foner, Philip S. *The Case of Joe Hill*. New York: International Publishers, 1965.

———, ed. *The Letters of Joe Hill*. New York: Oak Publications, 1965.

Forster, C. B., K. D. Solomon, and L. P. James. "Ownership of Mine-Tunnel Discharge." *Ground Water* 38, no. 4 (2000): 487-496.

Fouch, Thomas D., C. Vito, and F. Nuccio, eds. *Hydrocarbon and Mineral Resources of the Uinta Basin, Utah and Colorado*. UGA Guidebook 20. Field Symposium SLC: UGA, 1992.

Fries, Reverend Louis J. *One Hundred and Fifty Years of Catholicity in Utah*. SLC: Intermountain Catholic Press, 1927.

Fuller, Craig. "Finns and the Winter Quarters Mine Disaster." *UHQ* 70, no. 2 (Spring 2002): 123–39.

"Geneva Steel LLC," *AIST Steel News* 29 March 2003. http://www.steelnews.com/companies/chapter11/geneva_steel.htm.

Gifford, Gerald F., Don D. Dwyer, and Brian E. Norton. *Bibliography of Literature Pertinent to Mining Reclamation in Arid and Semi-Arid Environments*. Logan: USU Press, 1972.

Gloyn, Robert W., ed. *The Mining Camps of Utah*. SLC: UGA and UGS, forthcoming.

Gold Guidebook for Nevada and Utah. Dana Point, CA: Minobras Mining Services, 1981.

Gold. Mercur District, Utah. The California Groups of Gold Mining Properties. St. Paul: Banning Press, 1900.

Goodrich, Mary Joanna Dern. *Life in Mercur: The Recollections of a Young Girl in a Great Basin Mining Town at the Turn of the Century*. Edited by Newell G. Bringhurst. Reno: Nevada Historical Society, 1978.

Goodwin, Charles C. *As I Remember Them*. SLC: Commercial Club, 1913.

Goodwin, Lawrence. *The Populist and the Free Silver Movement*. American History Documentary Series. Durham, NC: Duke University Press, 1995.

Gordon, Mitchell. "New Stage of Development, Utah Construction and Mining Co. Has Struck It Rich." *Barron's* 48 (10 June 1968): 9.

Gorman, Mel. "Chronicle of a Silver Mine: The Meadow Valley Mining Company of Pioche." *Nevada Historical Society Quarterly* 29 (Summer 1986): 71.

Gorn, Elliot J. *Mother Jones: The Most Dangerous Woman in America*. New York: Hill and Wang, 2001.

Gottfredson, Peter. *Indian Depredations in Utah*. 2d ed. SLC: Marlin G. Christensen, 1969.

Goulding, Harry. "The Navajos Hunt Big Game . . .Uranium." *Popular Mechanics*, June 1950, 89.

Greene, J. Thomas. *A Legal Guide for the Uranium Prospector*. SLC: U of U College of Mines and Mineral Industries and UGMS, 1954.

Greenlaw, Sadie. "A Brief History of Juab County [Utah]." http://www.utahreach.usu.edu/juab/visitor/history.

Greenway, John, ed. *Folklore of the Great West: Selections from Eighty-three Years of the Journal of American Folklore*. Palo Alto, CA: American West Publishing, 1969.

Greever, William S. *The Bonanza West: The Story of the Western Mining Rushes, 1848–1900*. Norman: University of Oklahoma Press, 1963.

Gregory, Herbert E. *Geology of Eastern Iron County, Utah*. USGS Bulletin 37. Washington, D.C.: GPO, 1950.

Gwynn, J. W. *Great Salt Lake: An Overview of Change*. UGS Special Publication. SLC: UGS, 2002.

———. "History of Potash Production from the Salduro Salt Marsh, Tooele County." *Survey Notes* (UGMS) 28, no. 2 (1966): 1–18.

———. "Mining Conditions and Problems Encountered within the Texasgulf-Cane Creek Potash Mine near Moab, Utah, and Potential Mining Condition at Gibson Dome, Utah." UGMS Report 184. SLC: UGMS, 1984.

———. "The Saline Resources of Utah." *Survey Notes* (UGMS) 23, no. 3 (1989): 21–26.

———, ed. *Great Salt Lake: A Scientific, Historical, and Economic Overview.* UGMS Bulletin 116. SLC: UGMS, 1980.

Hachman, Frank C., Craig Bigler, and Douglas C. W. Kirk. *Utah Coal: Market Potential and Economic Impact.* SLC: Bureau of Economic and Business Research, U of U, 1968.

Hafen, LeRoy R., and Ann W. Hafen, eds. *Journals of the Forty-Niners—Salt Lake to Los Angeles.* Vol. 2 of *Far West and the Rockies.* Glendale, CA: Arthur H. Clark, 1954.

Hampshire, David. "Remembering Park City's Great Fire." *UHQ* 66, no. 3 (Summer 1998): 225–42.

Hampshire, David, Martha Sonntage Bradley, and Allen Roberts. *A History of Summit County.* SLC: USHS, 1998.

Hannah, Richard. *A Case Study of Underground Coal Mining Productivity in Utah.* N.p., 1981.

Hannah, Richard, and Garth L. Mangum. *The Coal Industry and Its Industrial Relations.* SLC: Olympus Publishing Company, 1985.

Hansen, Gary B. "Industry of Destiny: Copper in Utah." *UHQ* 31, no. 3 (Summer 1963): 161–67.

Hansen, George H., and H. C. Scoville, comps. *Drilling Records for Oil and Gas in Utah.* UGMS Bulletin 50. SLC: UGMS, 1955.

Hardesty, W. P. "The Drainage Works of the Ontario Silver Mine." *Engineering News* 32 (November 1894).

Harline, Osmond L. "Utah's Black Gold: The Petroleum Industry." *UHQ* 31, no. 3 (Summer 1963): 291–311.

Harris, Kay. *The Towns of Tintic.* Denver: Sage Books, 1961.

Harrow, Edward C. *The Gold Rush Overland Journal of Edward C. Harrow.* Austin, TX: Michael Vinson, 1993.

Hauck, F. R. *Cultural Resource Evaluation in South Central Utah, 1977–1978.* SLC: U.S. Bureau of Land Management, 1979.

Herring, Dean F. *From Lode to Dust: The Birth and Death of Kimberly, Utah.* Eugene, OR: privately published, 1989.

Heubach, Emil. *Guide to the Gold Fields of Colorado, New Mexico, Utah and Arizona.* Chicago: Chicago, Burlington and Quincy Railroad, 1880.

Hintze, Lehi F. *Geologic History of Utah.* Provo: BYU Press, 1988.

Hoglund, A. William. "No Land for Finns: Critics and Reformers View the Rural Exodus from Finland to America between the 1880s and World War I." In *The Finnish Experience in the Western Great Lakes Region: New Perspectives.* Edited by Michael G. Karni, Matti E. Kaups, and Douglas J. Ollila, Jr. Turku, Finland: Institute for Migration, 1975: 36-54.

Holbrook, Stewart H. *The Rocky Mountain Revolution.* New York: Henry Holt and Co., 1956.

Hollister, O. J. *Gold and Silver Mining in Utah.* Dallas, TX: American Institute of Mining Engineers, 1887.

———. *Resources and Attractions of Utah.* SLC, 1882.

Hooker, W. A., and E. M. March. *The Horn Silver Mine.* New York: K. Tompkins, 1879.

Howell, D. G., and K. A. McDougall. *Mesozoic Paleogeography of the Western United States.* Los Angeles: Pacific Section of the Society of Economic Paleontologists and Mineralogists, 1978.

Hulse, James W. "Boom and Bust Government in Lincoln County, Nevada, 1866–1909." *Nevada Historical Quarterly* 1, no. 2 (November 1957): 65-80.

———. *The Silver State: Nevada's Heritage Reinterpreted.* 2d ed. Reno: University of Nevada Press, 1998.

Huntley, D. B. "Mining Industries of Utah." In Appendix I of "Statistics and Technology of the Precious Metals." In U.S. Census Bureau, *Tenth Census of the Population,* 1880. Vol. 13.

Husband, Michael B. "History's Greatest Metal Hunt: The Uranium Boom on the Colorado Plateau." *Journal of the West* 21, no. 4 (October 1982): 17–23.

IHS Energy Group. Electronic oil-well database for Rich and Summit Counties, Utah. In *PI/Dwight Plus.* Englewood, CO: IHS Energy Group, 2002. CD-ROM.

"IMC Kalium Sold—Changes Name." Utah Mining Association Web site. http://www.utahmining.org (accessed 2001).

Jackson, Daniel. "Solution Mining Pumps New Life into Cane Creek Potash Mine." Reprinted from *Engineering and Mining Journal*, vol. 174, 7, 59-69. New York: McGraw-Hill, 1973.

Jackson, Samuel B. *200 Trails to Gold: A Guide to Promising Old Mines and Hidden Lodes throughout the West*. Garden City, NY: Doubleday, 1976.

Jackson, W. Turrentine. "British Impact on the Utah Mining Industry." *UHQ* 31, no. 3 (Summer 1963): 347–75.

———. *The Enterprising Scot: Investing in the American West after 1873*. Edinburgh, Scotland: Edinburgh University Press, 1968.

———. "The Infamous Emma Mine: A British Interest in the Little Cottonwood District, Utah Territory." *UHQ* 23, no. 4 (October 1955): 339–62.

James, Laurence P. *Geology, Ore Deposits, and History of the Big Cottonwood Mining District, Salt Lake County, Utah*. UGMS Bulletin 114 SLC: UGMS, 1979.

———. "George Tyng's Last Enterprise: A Prominent Texan and a Rich Mine in Utah." *Journal of the West* 7 no. 3 (July 1969): 429–37.

———. "Sulfide Ore Deposits Related to Thrust Faults in Northern Utah." In *Overthrust Belt of Utah*. SLC: UGA, 1982: 91-100.

———. "The Tintic Mining District." *Survey Notes* (UGMS) 18 (1984): 1, 4–13, 18.

James, Laurence P., and E. H. McKee. "Silver-Lead-Zinc Ores Related to Possible Laramide Plutonism near Alta, Salt Lake County, Utah." *Economic Geology* 80 no. 2: 497-504.

James, Laurence P., and S. C. Taylor. "Strong-Minded Women: Desdemona Stott Beeson and Other Hard Rock Mining Entrepreneurs." *UHQ* 46 no. 2 (Spring 1978): 136-150.

James, Laurence P., and Henry O. Whiteside. "Promoting the Alta Tunnel: The Rise and Fall of F. V. Bodfish," *Journal of the West* 20 (1981): 89–102.

Jameson, W. C. *Buried Treasures of the Rocky Mountains: Legends of Lost Mines, Train Robbery Gold, Caves of Forgotten Riches, and Indian Burial Silver*. Little Rock, AR: August House, 1993.

Jensen, Gordon F., and Boyd L. Fjelsted. *Utah's Potential in the Manufacturing or Servicing of Mining Equipment*. SLC: Utah Engineering Experiment Station, 1976.

Jensen, Vernon H. *Heritage of Conflict: Labor Relations in the Non-ferrous Metals Industry Up to 1930*. New York: Greenwood Press, 1968.

Jepson, James, Jr. *Memories and Experiences of James Jepson, Jr.* Edited by Eta Holdaway Spendlove. N.p., 1944.

Jones, ElRoy Smith. *John Pidding Jones, His Ancestors and Descendants*. SLC: American Press, 1977.

Josephy, A. M. "Murder of the Southwest." *Audubon* 13, no. 4 (July 1971): 52.

Justesen, Osmon. *Early Day Iron Mining in Utah*. SLC, 1918.

Kaczmarek, Marie. "Ghost Town on the Old Uintah Railway." *The Outlaw Trail Journal* (Winter 1993): 15–20.

Kalt, William D. *Awake the Copper Ghosts!: The History of Banner Mining Company and the Treasure of Twin Buttes*. Tucson, AZ: Banner Mining, 1968.

Kasteler, J. I., and J. H. Hild. *Silver, Copper, and Bismuth Deposits of South Hecla Mine, Alta,Salt Lake County, Utah*. U.S. Bureau of Mines Report of Investigations 4170. Washington, D.C.: GPO, 1948.

Katsanevas, Michael. "The Emerging Social Worker and the Distribution of the Castle Gate Relief Fund." *UHQ* 50, no. 3 (Summer 1982): 241–54.

Keller, Charles L. *The Lady in the Ore Bucket: A History of Settlement and Industry in the Tri-Canyon Area of the Wasatch Mountains*. SLC: U of U Press, 2001.

———. "James T. Monk: The Snow King of the Wasatch." *UHQ* 66 no. 2 (Spring 1998): 139–58.

Kelly, Keith A. *The Impact of a Proposed Severance Tax on the Utah Underground Coal Industry*. SLC: Utah State Tax Commission, 1982.

Kelner, Alexis. *Skiing in Utah, A History*. SLC: Kelner, 1980.

"Kennecott Copper." *Forbes* (15 April 1972): 32.

Kildale, M. B. "Geology and Mineralogy of the Park City District." *Bulletin of the Mineralogical Society of Utah* 8, no. 2 (1956): 5-11.

Killian, James F. *The History of the Deseret Coal Mine*. Provo: BYU, 1966.

King, Clarence. *Statistics and Technology of the Precious Metals*. Washington, D.C.: GPO, 1885.

Knight, Amberly. "Hot Rocks Make Big Waves: The Impact of the Uranium Boom on Moab, Utah 1948–57." *UHQ* 69, no. 1 (Winter 2001): 29–45.

Knight, Jesse William. *The Jesse Knight Family: Jesse Knight, His Forebears and Family*. SLC: Deseret News Press, 1940.

Knoerr, A. W. "Can Uranium Mining Pay?" *Engineering and Mining Journal* (December 1949).

Knudsen, G. W. "Safety at the Utah Copper Mine." *Mining Congress Journal*, March 1944, 35–37.

Kopp, R. S., and R. E. Cohenour, eds. *Cenozoic Geology of Western Utah*. UGA Publication 16; papers from the annual symposium and field conference in 1987. SLC: UGA, 1987.

Kraut, Ogden. *Relief Mine II: Through Others' Eyes*. SLC: Pioneer Press, 1998.

Lamborn, John E., and Charles S. Peterson. "The Substance of the Land: Agriculture v. Industry in the Smelter Cases of 1904 and 1906." *UHQ* 53, no. 4 (Fall 1985): 308–25.

Lang, Herbert H. "Uranium Also Had Its 'Forty-Niners'." *Journal of the West* 1, no. 2 (October 1962): 161.

Lankford, William T., ed. *The Making, Shaping and Treating of Steel*. Pittsburgh, PA: Association of Iron and Steel Engineers, 1985.

Larson, Gustive O. "Bulwark of the Kingdom: Utah's Iron and Steel Industry." *UHQ* 31, no. 3 (Summer 1963): 248–61.

———. "Where Mormons Found a Mountain of Iron." *Deseret Magazine*, May 1953, 11–14.

Lawson, Kenneth. "The Mines Saw Red." *Park City Lodestar* 21, no. 2 (Summer 1998): 50–54.

Lester, Margaret D. *Brigham Street*. SLC: USHS, 1979.

Lewis, Daniel Rich. *La Plata, 1891–1893: Boom, Bust, and Controversy*. SLC: USHS, 1982.

Lewis, Robert Strong. *Condensed Mining Handbook of Utah*. SLC: U of U School of Mines, 1945.

———. *Mining, the Priceless Heritage*. Third Annual Frederick William Reynolds Memorial Lecture. SLC: U of U Extension Division, 1939.

Lewis, Robert Strong, and Kirac Eray. *Bibliography on Open-Cut Mining*. Bulletin of the State School of Mines 23. SLC: U of U School of Mines, 1945

Leydet, Francois. "A Nation's Quandary: Coal vs. Parklands." *National Geographic* 158 (December 1980): 776–803.

Lindgren, Waldemar, and G. F. Loughlin, eds. *Geology and Ore Deposits of the Tintic Mining District, Utah*. Washington, D.C.: GPO, 1919.

Lindstrom, Philip. "Experience at the Radon Uranium Mine." *Mining and Engineering*, December 1964, 56.

Lingenfelter, Richard E. *The Hardrock Miners: A History of the Mining Labor Movement in the American West, 1863–1893*. Berkeley: University of California Press, 1974.

"Long Way from Utah." *Time* 92 (20 September 1968): 100.

Look, A. B. *1,000 million Years on the Colorado Plateau, Land of Uranium*. Denver: Bell Publications, 1955.

Luce, Willard. "They're Drilling for Uranium Now." *Westways* 46 (January 1954): 2–3.

Lunt, Henry. *Life of Henry Lunt and Family Together with a Portion of His Diary*. Provo: BYU, 1955. Microform, Provo, UT: BYU Photography Studio, 1960.

MacDonald, Graham D., III. *The Magnet: Iron Ore in Iron County Utah*. Cedar City, UT: privately published, 1990.

Macfarlane, L. W. *Dr. Mac: The Man, His Land, and His People*. SLC: U of U Press, 1956. Reprint, Cedar City, UT: Southern Utah State College Press, 1985. Page references are to the 1985 edition.

Madsen, Brigham D. *Corinne: The Gentile Capital of Utah*. SLC: USHS, 1980.

———. *Glory Hunter: A Biography of Patrick Edward Connor*. SLC: U of U Press, 1990.

Madsen, Troy. "The Company Doctor: Promoting Stability in Eastern Utah Mining Towns." *UHQ* 68, no. 2 (Spring 2000): 139–56.

———. "Medicine and the Mines: The Company Doctor as a Mediator within the Eastern Utah Coal Camps." *Theatean* 28 (1999): 103–37.

Maguire, Don. *Utah's Great Mining Districts*. SLC: Rio Grande Western Railway, 1899.

Maher, P. O. "The Geology of the Pineview Field Area, Summit County, Utah." In *Geology of the Cordilleran Hingeline*. Edited by J. Gilmore Hill. Denver: Rocky Mountain Association of Geologists Symposium, 1976.

Majors, Alexander. *Seventy Years on the Frontier*. Columbus, OH: Long's College Book Co., 1950.

Malmquist, O. N. *The First 100 Years*. SLC: USHS, 1971.

Mangum, Garth, and Geo-Jala MacLeans. "The Prospects for Utah Coal." *Utah Economic and Business Review* 46 nos. 7–8 (July–August 1986): 1–15.

"Many Fingers, Many Pies." *Forbes* 100 (15 July 1967): 22.

Marcosson, Isaac F. *Anaconda*. New York: Dodd, Mead, 1957.

Mariger, Marietta M. *Saga of Three Towns: Harrisburg, Leeds, Silver Reef*. Panguitch, UT: Garfield County News, 1951.

Marshall, John, with Zeke Zanoni. *Mining the Hard Rock in the Silverton San Juans*. Silverton, CO: Simpler Way Book Company, 1996.

McAuliffe, Eugene. *Early Coal Mining in the West, Beginning with 1868*. New York: Newcomers Society of England, American Branch, 1948.

McCormick, John S. *Silver in the Beehive State*. SLC: USHS, 1988.

McCourt, Tom. *White Canyon: Remembering the Little Town at the Bottom of Lake Powell*. Price, UT: South-paw Publications, 2003.

McCune, Alice P. *History of Juab County*. Springville, UT: DUP, Juab County Chapter, 1947.

McDonald, John. "Dig Marriner Eccles' Company: Its Digs." *Fortune*, January 1961, 86.

McDonough, Gerald M. *The Hogles*. SLC: McMurrin-Henriksen, Western Epics, 1988.

McGovern, George S., and Leonard F. Guttridge. *The Great Coalfield War*. Boston: Houghton Mifflin, 1972.

McGregor, Joseph K., and Carl Abston. *Photographs of Historical Mining Operations in Colorado and Utah from the U. S. Geological Survey Library*. Reston, VA: USGS, 1994.

McPhee, William Miller. *The Trail of the Leprechaun: Early History of a Utah Mining Camp*. Hicksville, NY: Exposition Press, 1977.

Mellinger, Philip J. *Race and Labor in Western Copper: The Fight for Equality 1896–1918*. Tucson: University of Arizona Press, 1995.

Merkel, Henry Martin. *History of Methodism in Utah*. Colorado Springs: Dentan Printing, 1938.

Merkley, Aird G., ed. *Monuments to Courage: A History of Beaver County*. Milford, UT: DUP of Beaver County and Stephen A. Williams of *Milford News*, 1948.

Metis Culture, 1850-1854, http://www.agt.net/public/dgarneau/metis43.htm.

Mitchell, G. C. "Stratigraphy and Regional Implications of the Argonaut Energy No. 1 Federal, Millard County, Utah." In *Basin and Range Symposium and Great Basin Field Conference*. Edited by Gary W. Newman and Harry D. Denver: Goode. Rocky Mountain Association of Geologists and UGA, 1979, 503–14.

Mitchell, G. C. and R. E. McDonald. "Subsurface Tertiary Strata, Origin, Depositional Model and Hydrocarbon Exploration Potential of the Sevier Desert Basin, West Central Utah." In *Cenozoic Geology of Western Utah*. Edited by R. S. Kopp and R. E. Cohenour. UGA Publication, 16. SLC: UGA, 1987, 533–56.

Moench, Robin. "The Cornish Pump." *Park City Lodestar* 10, no. 1 (Winter 1987): 14–16.

Moon, Samuel. *Tall Sheep: Harry Goulding, Monument Valley Trader*. Norman: University of Oklahoma Press, 1992.

Morgan, Nicholas G. *Daniel Cowan Jackling and Utah Copper*. SLC, 1968.

Mulder, William, and A. Russell Mortensen, eds. *Among the Mormons*. New York: Alfred A. Knopf, 1958.

Murphy, John R. *The Mineral Resources of the Territory of Utah*. San Francisco: A. L. Bancroft, 1872.

Murphy, Miriam B. "When the Fabulous Horn Silver Mine Caved In." *History Blazer*, January 1996. http://www.historytogo.utah.gov/hornmine.html.

Navin, Thomas R. *Copper Mining and Management*. Tucson: University of Arizona Press, 1978.

Neff, Andrew Love. *History of Utah*. SLC, 1904.

Nelson, Elroy. "The Mineral Industry: A Foundation of Utah's Economy." *UHQ* 31, no. 3 (Summer 1963): 179–91.

———. *Utah's Economic Patterns*. SLC: U of U Press, 1956.

Newell, Maxine. *Charlie Steen's Mi Vida*. N.p.: privately published, 1976. Reprinted, Moab, Utah: Moab's Printing Place, 1992.

Niebur, Jay E. *Arthur Redman Wilfley*. Louisville: Colorado Historical Society, 1982.

Nolan, Dean, and Fred Thompson. *Joe Hill: IWW Songwriter*. Chicago: Industrial Workers of the World, Chicago General Membership Branch, n.d.

"The Noranda Smelting Process." http://www.norsmelt.com/norandaprocess.html.

Notarianni, Philip F. *Carbon County, Eastern Utah's Industrialized Island*. SLC: USHS, 1981.

———. *Faith, Hope and Prosperity: The Tintic Mining District*. Eureka, UT: Tintic Historical Society, 1982.

———. "The Frisco Charcoal Kilns." *UHQ* 50, no. 1 (Winter 1982): 40–58.

———. "Hecatomb at Castle Gate, Utah, March 8, 1924." *UHQ*, 70, no. 1 (Winter 2002): 63–74.

———. *The Symbol of an Era: Bullion Beck & Champion Mining Company Headframe and the Tintic Mining Company*. SLC: Utah Division of Oil, Gas and Mining, 1989.

"Not So Nice for Bryce." *Rocky Mountain Magazine*, March 1980, 10.

O'Conner, Harvey. *The Guggenheims: The Making of an American Dynasty*, New York: Covici, Friede, 1937.

O'Donnell, J. C. (Buck). *The Good Old Days*. SLC: Shaft and Development Machines Company, 1968.

Palmer, W. H. *Home of the Dividend Payers*. SLC: Tribune-Reporter Printing, 1910.

Papanikolas, Helen Zeese. "Georgia Lathouris Mageras: Magerou, the Greek Midwife." *UHQ* 38, no. 2 (Spring 1970): 50-60.

———."Life and Labor among the Immigrants of Bingham Canyon." *UHQ* 33, no. 4 (Fall 1965): 289–315.

———. *The Peoples of Utah*. SLC: USHS, 1976.

———. "Toil and Rage in a New Land: The Greek Immigrants in Utah." *UHQ* 38, no. 2 (Spring 1970): 100–203.

———. "Unionism, Communism, and the Great Depression: The Carbon County Coal Strike of 1933." *UHQ* 41, no. 3 (Summer 1973): 254–300.

Papanikolas, Zeese. *Buried Unsung: Louis Tikas and the Ludlow Massacre*. SLC: U of U, 1982.

Parker, Edward W. "Coal Cutting Machinery." *Cassier's Magazine* 22 (July 1902): 291–93.

Parry, William T. *All Veins, Lodes, and Ledges throughout Their Entire Depth: Geology and the Apex Law in Utah Mines*. SLC: U of U Press, 2004.

———. "Geology of Utah Mining Camps."

Parry, William T., Craig B. Forster, D. Kip Solomon, and Laurence P. James. "Ownership of Mine-Tunnel Discharge." *Ground Water* 38 (2000): 487–96.

Parsons, A. B. *The Porphyry Coppers*. New York: American Institute of Mining and Metallurgical Engineers, 1933.

Patterson, Edna B., Louise A. Ulph, and Victor Goodwin. *Nevada's Northeast Frontier*. Sparks, NV: Western Printing and Publishing Company, 1969.

Paul, Rodman W. *Mining Frontiers of the Far West, 1848–1880*. Rev. ed. Albuquerque: University of New Mexico Press, 2001.

Pendleton, Mark A. "Memories of Silver Reef." *UHQ* 4 (October 1930): 99–118. Also on a CD, USHS, *The Utah History Suite*.

———. "Naming Silver Reef." *UHQ* 5 no. 1 (January 1932): 29–31.

Pepper, Choral. *Western Treasure Tales*. Niwot: University Press of Colorado, 1998.

Perkins, Nancy. "Religious Friends Relive Historic Moment." *LDS Church News*, 22 May 2004, 5.

Perry, J. H., G. N. Aul, and Joseph Cervik. *Methane Drainage Study in the Sunnyside Coalbed, Utah*. Washington, D.C.: U.S. Bureau of Mines, 1978.

Peterson, Richard H. *Bonanza Rich: Lifestyles of the Western Mining Entrepreneurs*. Moscow: University of Idaho Press, 1991.

Phillips, Margie. "Cane Creek Mine Solution Mining Project, Moab Potash Operations, Texasgulf, Inc." In *Canyonlands Country: A Guidebook of the Four Corners Geological Society*. Edited by James F. Fassett and Sherman A. Wengerd. Four Corners Geological Society, 1975.

Picard, M. Dune, ed. *Geology and Energy Resources, Uinta Basin of Utah*. SLC: UGA, 1985.

Pitchard, George E. *A Utah Railroad Scrapbook*. SLC: privately published, 1987.

Pope, Elizabeth. *The Richest Town in the U.S.A.* N.p., 1956.

Porath, Joseph H. *The Town That Copper Killed*. Austin, TX: Western Publications, 1971.

Powell, Allan Kent. "Labor's Fight for Recognition in the Western Coalfields." *Journal of the West* 25, no. 2 (April 1986): 20–26.

———."Mormon Influence on the Unionization of Eastern Utah Coal Miners, 1903–33." *Journal of Mormon History* 4 (1977): 91–100.

———. *The Next Time We Strike: Labor in the Utah Coal Fields 1900–1933*. Logan: USU Press, 1985.

———. "Tragedy at Scofield." *UHQ* 41, no. 2 (Spring 1972): 182–94.

———, ed. *Utah History Encyclopedia*. SLC: U of U Press, 1994.

Powell, John Wesley. *Canyons of the Colorado*. New York: Floyd & Vincent, 1895. Reprinted as *The Exploration of the Colorado River and Its Canyons*. New York: Dover Publications, 1961. Page reference is to the 1961 edition.

Proctor, Paul Dean, and Morris A. Shirts. *Silver, Sinners and Saints: A History of Old Silver Reef, Utah*. N.p.: Paulmar, 1991.

Project on Government Oversight. *NRC Sells Environment Down the River: Radiation Flows Unchecked into the Colorado River*. N.p., 1999.

Pruitt, Robert G., Jr. *The Mineral Resources of Uintah County*. UGMS Bulletin 71. SLC: UGMS, 1961.

Rau, J. L., and L. F. Dellwig, eds. *Third Symposium on Salt*. Vol. 1. Cleveland: Northern Ohio Geological Society, 1970.

Rayback, Joseph G. *A History of American Labor*. New York: Free Press, 1966.

Raymer, Robert G. "Early Mining in Utah." *Pacific Historical Review* 8 (1939): 81–88.

Raymond, Rossiter W. *Mineral Resources of the States and Territories West of the Rocky Mountains*. Washington, D.C.: GPO, 1869.

———. *The Mines of the West: A Report to the Secretary of the Treasury*. New York: J. B. Ford, 1869.

———. *Report on the Mineral Resources of the United States and Territories*. First annual report, 1868. Washington D.C.: GPO, 1869.

———. *Statistics of Mines and Mining . . . for the Year[s] 1869 through 1871*. 4 vols. Washington D.C.: GPO, 1870–72.

———. *Statistics of Mines and Mining in the States and Territories West of the Rocky Mountains*. First annual report of U.S. Commissioner of Mining Statistics. 40th Cong., 3d sess., 1868. House Ex. Doc. 54: 114–15.

———. *Statistics of Mines and Mining in the States and Territories West of the Rocky Mountains*. Washington, D.C.: GPO, 1870, 1872–75, 1877; New York: J. B. Ford, 1871.

Reeve, W. Paul. *A Century of Enterprise: The History of Enterprise, Utah, 1896–1996*. Enterprise, UT: City of Enterprise, 1996.

———. "In 1879 a Mormon Choir Sang for a Catholic Mass in St. George." *The History Blazer*, March 1995.

———. "Rocks That Burned Led to Oil Discoveries in Southwestern Utah." *The History Blazer*, April 1995.

Reinhart, Herman Francis. *The Golden Frontier: The Recollections of Herman Francis Reinhart, 1851–1869*. Austin: University of Texas, 1962.

"Removing a Mountain of Copper." *Collier's*, 1 April 1916, 16.

Reps, John William. *Cities of the American West: A History of Frontier Urban Planning*. Princeton, NJ: Princeton University Press, 1979.

Rhoades, Gale R. *The Lost Gold of the Uintah's: The Rest of the Story*. Duchesne, UT: Benzoil Oil Corporation, 1995.

Rhoades, Gale R., and Kerry Ross Boren. *Footprints in the Wilderness: A History of the Lost Rhoades Mines*. SLC: Publishers Press, 1971. Reprints, SLC: Dream Garden Press, 1980, 1984. Page reference is to the 1984 edition.

Rice, George Graham (Jacob Herzig). *My Adventures with Your Money*. New York: Ridgway Company, 1911.

Rickard, T. A. *History of American Mining*. New York: McGraw-Hill, 1932.

———. *The Utah Copper Enterprise*. San Francisco: Abbott Press,1919.

Ringholz, Raye C. *Uranium Frenzy: Boom and Bust on the Colorado Plateau*. New York: W. W. Norton, 1989.

———. *Uranium Frenzy: Saga of the Nuclear West*. Logan: USU Press, 2002.

Robbins, William G. *Colony and Empire: The Capitalist Transformation of the American West*. Lawrence: University Press of Kansas, 1994.

Roberts, Brigham H. *A Comprehensive History of the Church of Jesus Christ of Latter-day Saints*. 5 vols. SLC: LDS, 1930.

———. *The Autobiography of Brigham H. Roberts*. Edited by Gary James Bergera. SLC: Signature Books, 1990.

Roberts, Richard C. *Legacy: The History of the Utah National Guard from the Nauvoo Legion Era to Enduring Freedom*. SLC: National Guard Association of Utah, 2003.

Robertson, Frank C., and Beth Kay Harris. *Boom Towns of the Great Basin*. Denver: Sage Books, 1962.

Robertson, Ruth Winder. *This Is Alta*. SLC: Wheelwright Lithography, 1972.

Robison, M. Henry. "A Brief Economic History of Utah's Coal Industry." *Utah Economic and Business Review* 37 (April 1977): 1-15.

Rogers, Fred B. *Soldiers of the Overland*. San Francisco, 1938.

Rolker, Charles M. *The Silver Sandstone District of Utah*. N.p., 1880.

Rollins, Vance W. *Boom and Bust: Uranium Stocks*. SLC: U of U Press, 1964.

Romney, Miles P. "Utah's Cinderella Minerals: The Nonmetallics." *UHQ* 31, no. 3 (Summer 1963): 220–34.

Rumsey, Judson S. *A Digest of the Decisions of the Supreme Court of Utah: Reported in Volumes 1 to 36 Inclusive*. Chicago: Callaghan and Co., 1912.

Rush Valley Silver Mining Association. Boston: Wright and Potter, 1865.

Salt Lake City Chamber of Commerce. *Mineral Wealth of Utah*. SLC: Salt Lake City Chamber of Commerce, 1926.

———. *The Mining Industry of Utah*. SLC: Salt Lake City Chamber of Commerce, 1941.

———. *Utah and Its Mineral Wealth*. SLC: Salt Lake City Chamber of Commerce, 1930.

———. *What Mining Means to Utah: A Compilation of Papers Presented under the Auspices of the Mining Committee of the Chamber of Commerce of Salt Lake City during 1927–1928*. SLC: Salt Lake City Chamber of Commerce, 1929.

Salt Lake City Directory. Compiled by G. Owens. SLC, 1867.

Salt Lake Stock and Mining Exchange. *Mercur, Utah's Johannesburg*. SLC: W. M. Wantland, 1899.

Schindler, Harold. "Utah: Producers Brace for 300% Jump in Output by 1980s." *Coal Age* 79 (May 1974): 94.

Schumacher, Otto L. *Federal Land Status in the Overthrust Belt of Idaho, Montana, Utah, and Wyoming*. Washington, D.C.: U.S. Bureau of Mines, 1979.

Seamons, Marian Larsen. *More Precious than Gold*. Eureka, UT: privately published, 1990.

Sessions, Gene A., and Sterling D. Sessions. *Utah International: A Biography of a Business*. Ogden, UT: Weber State University, 2002.

Shaffer, Steven. *Of Men and Gold: The History and Evidence of Spanish Gold Mines in the West*. Spanish Fork, UT: privately published, 1994.

Sharrock, Floyd W. *1960 Excavations, Glen Canyon Area*. SLC: U of U Press, 1961.

Shepardson, John W. "Utah [Coal Industry]." *Coal Age* 78, no. 5 (April 1973): 159.

Shirts, Morris A. "The Demise of the Deseret Iron Company." *UHQ* 56 no. 1 (Winter 1988): 23–35.

Shirts, Morris A., and Kathryn H. Shirts. *A Trial Furnace: Southern Utah's Iron Mission*. Provo: BYU Press, 2001.

Shoebotham, H. M. *Anaconda: Life of Marcus Daly, the Copper King*. Harrisburg, PA: Stackpole Company, 1956.

Shrontz, Duane. *Alta, Utah: A People's Story*. SLC: Salt Lake Winter Sports Association, 1989. Reprint, Alta, UT: Two Doors Press, 2002. Page reference is to the 1989 edition.

Shumway, Leland. "Cottonwood Vanadium Mill 1937 Garbutt-Kimmerle Venture." *Blue Mountain Shadows* 16 (Winter 1995): 50.

Sloan, Edward L., ed. *Gazetteer of Utah and Salt Lake City Directory*. SLC: Salt Lake Herald Publishing Company, 1874.

Smart, William B., and Donna T. Smart. *Over the Rim: The Parley P. Pratt Exploring Expedition to Southern Utah, 1849–50*. Logan: USU Press, 1999.

Smith, Charles William. *From Then until Now*. Roosevelt, UT: Ink Spot Printing Company, 1982.

Smith, Dwight LaVern. "Hoskaninni: A Gold Mining Venture in Glen Canyon." *El Palacio* 69 (Summer 1962): 77–84.

Smith, Gibbs M. *Joe Hill*. SLC: G. M. Smith, 1969.

Smith, Wilbur H. *Bingham of West Mountain Mining District: Comprehensive List of References*. SLC: Marriott Library, U of U, 1977.

Sorgenfrei, Robert. "'A Fortune Awaits Enterprise Here': The Best Mining Expedition to the Grand Canyon." *Journal of the West* 40, no. 4 (Winter 1998): 437–62.

Spangler, Jerry D. *Paradigms and Perspectives: A Class I Overview of Cultural Resources in the Uinta Basin and Tavaputs Plateau*. SLC: Bureau of Land Management, 1995.

Spence, Clark C. *British Investments and the American Mining Frontier, 1860–1891*. Ithaca, NY: Cornell University Press, 1958.

———. *Mining Engineers and the American West: The Lace-Boot Brigade, 1849–1933*. New Haven, CT: Yale University Press, 1970.

Spencer, R. J., H. P. Eugster, and B. F. Jones. "Geochemistry of Great Salt Lake, Utah II: Pleistocene-Holocene Evolution." *Geochimica et Cosmochimica Acta* 49 (1985): 739–47.

Spencer, R. J., H. P. Eugster, B. F. Jones, and S. L. Rettig. "Geochemistry of Great Salt Lake, Utah I: Hydrochemistry since 1850." *Geochimica et Cosmochimica Acta* 49 (1985): 727–37.

Spicer, Wells. *Handbook of Ophir Mining District*. SLC: Utah Mining Gazette Steam Book and Job Press, 1874.

Spurr, Josiah Edward. *Economic Geology of the Mercur Mining District, Utah*. Washington, D.C.: GPO, 1895.

Stanton, Robert Brewster, C. Gregory Crampton, and Dwight L. Smith. *The Hoskaninni Papers: Mining in Glen Canyon, 1877–1902*. SLC: U of U Press, 1961.

Steele, Raymond D. *Goshen Valley History*. N.p., 1960.

Stenhouse, Thomas B. H. *The Rocky Mountain Saints*. London, 1874.

Steward, J. H., C. H. Stevens, and A. E. Fritsche, eds. *Paleozoic Paleogeography of the Western United States*. Los Angeles: Pacific Section, Society of Economic Paleontologists and Mineralogists, 1977.

Stewart, John H. *Geology of Nevada*. Reno: Nevada Bureau of Mines and Geology, 1980.

Stockton Bicentennial History Committee. *Brief History of Stockton, Utah*. Tooele, Utah, 1976.

Stoher, Eric. *Bonanza Victorian, Architecture and Society in Colorado Mining Towns*. Albuquerque: University of New Mexico Press, 1975.

Stokes, William Lee. *Geology of Utah*. SLC: Utah Museum of Natural History and UGMS, 1986.

"The Stormont Mines, Silver Reef, Utah." *Engineering and Mining Journal* 29 (17 January 1880): 45–46.

Taniguchi, Nancy J. "An Explosive Lesson: Gomer Thomas, Safety, and the Winter Quarters Mine Disaster." *UHQ* 70 no. 2 (Spring 2002): 140–57.

———. *Necessary Fraud: Progressive Reform and Utah Coal*. Norman: University of Oklahoma Press, 1996.

Taylor, Philip. *The Distant Magnet: European Emigration to the U.S.A.* New York: Harper and Row, 1971.

Taylor, Raymond W., and Samuel W. Taylor. *Uranium Fever; or, No Talk under 1 Million*. New York: Macmillan, 1970.

Thayer, William Makepeace. *Marvels of the New West*. Norwich, CT: Henry Bill Publishing, 1888.

Tholen, Faye Farnsworth. "Vipont, Utah: A Lost and Almost Forgotten Ghost Town." *UHQ* 71, no. 3 (Summer 2003): 215–32.

Thompsen, George A. *Faded Footprints: The Lost Rhoades Mine and Other Hidden Treasures of the Uintahs*. SLC: Roaming the West, 1991. Reprint, SLC: Dream Garden Press, 1996.

———. "The Mormon Wizard [Jessie Knight]." *Frontier Times* 43, no. 3 (April–May 1969): 26.

Thompson, George A., and Fraser Buck. *Treasure Mountain Home: Park City Revisited*. SLC: Dream Garden Press, 1981; reprint, 1993. Page references are to the 1993 edition.

Through Our Eyes: 150 Years of History as Seen through the Eyes of the Writers and Editors of the Deseret News. SLC: Deseret News Publishing Company, 1999.

Tilby, Wilma B. "Jakob Brand's Register of Dutchtown, Utah's Lost German Mining Colony." *UHQ* 62, no. 1 (Winter 1994): 53–70.

Timmins, William M. *The Copper Strike and Collective Bargaining*. Chicago: Commerce Clearing House, 1970.

Todd, Arthur C. *The Cornish Miner in America: The Contribution to the Mining History of the United Sates by Emigrant Cornish Miners—The Men Called Cousin Jacks*. Spokane, WA: Arthur H. Clark, 1995.

Tooele County Historical Society. *Mining, Smelting, and Railroading in Tooele County*. Tooele, UT: Tooele County Historical Society, 1986.

Toone, Bessie Berry. *Nuggets from Mammoth*. UT, 1966.

Townley, John M. *Conquered Provinces: Nevada Moves Southeast, 1864–1871*. Charles Redd Monographs in Western History 2. Provo: BYU Press, 1973.

Tullidge, Edward W. *Histories*. 2 vols. SLC: Juvenile Instructor, 1889.

———. *History of Salt Lake City*. SLC, 1886.

———. "The Mines of Utah." *Tullidges Quarterly Magazine* 1, no. 2 (January 1881): 189–90.

Turner, Frederick Jackson. *The Frontier in American History*. New York: Henry Holt, 1920.

Turner, Orson. *History of Sunnyside, Utah: An Oral History Interview*. Logan: USU Press, 1974.

Tuttle, Daniel S. *Missionary to the Mountain West*. 1906. Reproduced, SLC: U of U Press, 1987.

Tyng, Charles. *Before the Wind: Tyng Family Memoir, 1808–1933*. Edited by Susan Fels. New York: Viking Press, 1999.

United Mine Workers of America. "Agreement between the United Mine Workers of America and Utah Coal Operators: Effective April 1, 1939." Cheyenne: *Wyoming Labor Journal* (1937).

United Mine Workers of America, District 22. "Constitution of District Number 22, United Mine Workers of America." Cheyenne: *Wyoming Labor Journal* (1934).

University of Utah Bureau of Economic and Business Research. "Measures of Economic Changes in Utah, 1847–1947." *Utah Economic and Business Review* 7 (December 1947): 68-69.

University of Utah, Department of Mining and Fuels Engineering. *Applied Research and Evaluation of Process Concepts for Liquefaction and Gasification of Western Coals*. Washington, D.C.: Department of Energy, 1976.

Utah Environmental and Agricultural Consultants. *Environmental Setting, Impact, Mitigation and Recommendations for a Proposed Products Pipeline between Lisbon Valley, Utah and Parachute Creek, Colorado*. Denver: Colony Development Operation, 1973.

Utah, Her Cities, Towns and Resources. Chicago: Manly and Litteral, 1891–92.

Utah Mining Association. *A Report of the Uranium Committee of the Utah Mining Association on Cost of Finding, Mining and Milling Uranium Ore in Utah*. SLC: Utah Mining Association, 1961.

———. *Utah Mining Facts: A Vital Part of Utah's Future*. SLC: Utah Mining Association, 1955.

———. *Utah's Mining Industry: An Historical, Operational, and Economic Review of Utah's Mining Industry*. SLC: Utah Mining Association, 1955, 1959, 1967.

Utah State Gazetteer and Business Directory. SLC: Stenhouse and Company, 1892–93.

Utah State Historical Society. *Utah. . . Treasure House of the Nation: Century of Mining, 1863–1963*. SLC: USHS, 1963.

Valle, Doris. *Looking Back around the Hat: A History of Mexican Hat*. Mexican Hat, UT: privately published, 1986.

Vancott, John W. *Utah Place Names*. SLC: U of U Press, 1990.

Vine, James D. *Reconnaissance during 1952 from Uranium-Bearing Carbonaceous Rocks in Parts of Colorado, Utah, Idaho, and Wyoming*. Washington, D.C.: USGS, 1953.

Von Linden, Julius. "The Newly-Discovered Oil Fields of Utah." *Illustrated American* 16 (November 1895): 624.

Waite, Catherine V. *The Mormon Prophet and His Harem*. Cambridge, MA, 1866.

Walquist, Wayne L., ed. *Atlas of Utah*. Weber State College: BYU Press, 1981.

Warren, Cecil W., comp. *Index to the Salt Lake Mining Review, 1899–1928*. SLC: UGMS, 1971.

Warren, Henry L. J. *The Story of the Tintic Mining District, Utah*. SLC: Rio Grande Western Railway, 1897.

Watkins, T. H. *Gold and Silver in the West*. Palo Alto, CA: American West Publishing, 1971.

Watt, Ronald. *A History of Carbon County*. SLC: USHS and Carbon County Commission, 1997.

Watts, A. C. *Coal and the Coal Industry: What It Means to Utah*. SLC: Paragon Printing, 1917.

Weber, Francis J. "Catholicism among the Mormons, 1875–79." *UHQ* 44, no 2 (Spring 1976): 141–48.

Weed, Walter Harvey. *The Copper Mines of the United States in 1905*. N.p., 1906.

Wegemann, Carroll H. *The Coalville Coal Field, Utah*. Washington, D.C.: GPO, 1915.

Weiss, Norm. *Helldorados, Ghosts, and Camps of the Old Southwest*. Caldwell, ID: Caxton Printers, 1977.

"Western Coal Mining as a Way of Life: An Oral History of the Colorado Coal Miners to 1914." *Journal of the West* 24, no. 3 (July 1985): 10.

Westwood, Dick. "Howard W. Balsley, Dean of Uranium Miners and Civic Leader of Moab." *UHQ* 59, no. 4 (Fall 1991): 395-406.

White, David, and G. H. Ashley. *Contributions to Economic Geology (Short Papers and Preliminary Reports), 1919: Part II. Metal Fuels*. Washington, D.C.: GPO, 1920.

Whitley, Colleen. "La Platta: Where Men and Mines Collided." *Utah Magazine*, January– February 1976, 25–26.

———, ed. *Worth Their Salt: Notable but Often Unnoted Women of Utah*. Logan: USU Press, 1996.

———, ed. *Worth Their Salt, Too: More Notable but Often Unnoted Women of Utah*. Logan: USU Press, 2000.

Whitney, Orson F. *History of Utah*. 4 vols. SLC: George Q. Cannon and Sons, 1892–1904.

Who Was Who in America. Historical Volume, 1607–1896. Rev. ed. Chicago: A. N. Marquis, 1967.

Who Was Who in America. 1897–1941. Chicago: A. N. Marquis, 1943.
Wilkerson, Christine. *Utah Gold: History, Placers, and Recreational Regulations.* SLC: UGS, 1997.
Wilkins, Thurman. *Clarence King.* 2d ed. New York: McMillan,1988.
Williams, Burton J. "Mormons, Mining and the Golden Trumpet of Moroni." *Midwest Quarterly* 8, no. 1 (October 1966): 67.
Wilson, Calvert. *Wilson's Mining Laws.* Los Angeles: Press of Baumgardt Publishing, 1917.
Wilson, Pearl D., June McNulty, and David Hampshire. *A History of Juab County.* SLC: USHS and Juab County Commission, 1990.
Wistensen, Martin J. *Socio-Economic Impact of the Kaiparowits Mining Operation.* Provo: Center for Business and Economic Research, BYU, 1974.
"World's Biggest Artificial Hole." *Literary Digest,* 119 (6 April 1935): 17.
Wyman, Mark. "Industrial Revolution in the West: Hard-Rock Miners and the New Technology." *Western Historical Quarterly* 5 (January 1974): 39–57.
Young, Darroll P. "Economic Impact of the Uranium Industry in San Juan County, Part One." *Blue Mountain Shadows* 16 (Winter 1955): 10.
Young, R. M., et al. *Sevier County, Utah: Past to Present.* N.p., 1998.
Young, W. E. *Iron Deposits in Iron County Utah.* Report of Investigations, no. 4076. Washington, D.C.: U.S. Bureau of Mines, 1947.

Publications and Reports of Government Agencies

United States

Army Corps of Engineers. *Final Environmental Impact Statement for the Kennecott Tailings Modernization Project, Magna, Utah.* 1995.
Bureau of Census. *Census of the Population, Characteristics of the Population, United States.* 1870–2000.
———. Statistical Abstract of the United States. 2001.
Bureau of Indian Affairs. *Responses to Comments Received from Public Review of the Fall 1979 Draft of Uranium Development in the San Juan Basin Region.* 1980.
Bureau of Land Management. *At the Blackbird Mine: Putting Micro-Organisms to Work on Reclamation.* 1976.
———. *Decision Record Finding of No Significant Impact for Sodium Leasing in the Green River Basin of Southwestern Wyoming.* 1996.
———. *Energy Mineral Rehabilitation Inventory and Analysis: Summary Report, Alton Coal Field, Kane County, Utah.* 1975
———. *Environmental Impact Statement and Proposed Resource Management Plan for the Green River Resource Area, Rock Springs, Wyoming.* 1996.
———. *Environmental Impact Statement on the Federal Oil Shale Management Program.* 1983.
———. *Escalante, Paria, Zion Planning United Management Framework Plan Summary.* 1980.
———. General Records, 1796-1981, section 49.3.2, http://www.archives.gov/research/guide-fed-records/groups/049.html#49.3.2.
———. *Index to Mining Districts, State of Utah.* Map. 1977.
———. *Inland Resources Monument Butte-Myton Bench Waterflood Environmental Assessment.* 1997.
———. "Moon Lake Power Plant—Units 1 and 2: Environmental Impact Statement Preparation Plan." Preliminary draft. 1980.
———. *Price River Resource Area Management Framework Plan: Summary.* 1984.
———. *Proposed Alunite Project.* 1977.
———. *Record of Decision: Final San Juan River Regional Coal Environmental Impact Statement.* 1987.
———. *Regulations Pertaining to Coal Management, Federally Owned Coal: 43 CFR 3400.* 1979.
———. *San Juan Basin Cumulative Overview and Comment Letters.* 1983.

———. *Snyder Oil Corporation Horseshoe Bend Waterflood Project Environmental Assessment*. 1997.

———. *Uinta–Southwestern Utah Coal Region: Round Two Draft Environmental Impact Statement*. 1983.

———. *Uinta–Southwestern Utah Regional Coal Environmental Impact Statement*. 1981.

———. *Uinta–Southwestern Utah Coal Study Region*. 1982.

Bureau of Mines. *In Situ Leach Mining*. Proceedings of the Bureau of Mines Technology Transfer Seminars, Phoenix, AZ, April 4, and Salt Lake City, UT, April 6, 1989.

———. *Mineral Resources of the United States*. 1924–34.

———. *Minerals Yearbook*. 1933–present.

———. *Mining and Mineral Operations in the Rocky Mountain States: A Visitor Guide*. 1977.

Bureau of Reclamation, *Environmental Statement: WESCO Gasification Project and Expansion of Navajo Mine by Utah International Inc., San Juan County, New Mexico*. 1974.

Congress. *Congressional Globe*. 39th Cong., 1st sess., 1867, part 1: 645; part 2: 1386, 1401, 1535; part 3: 2358, 2368–70, 2377, 2381.

———. House. *Conditions of Mining Industry—Utah*. Executive Document, 42nd Congress, 3rd Session, 1567.

———. House. *Monumental Abuse: The Clinton Administration's Campaign of Misinformation in the Establishment of the Grand Staircase-Escalante National Monument: Report of the Committee on Resources, House of Representatives, Together with Additional Views*. GPO, 1998.

———. House. *Report of the Emma Mine Investigation*. 44th Cong., 1st sess., 1875–76, Rep. 579, serial #1711.

———. House. *Report of the Governor of Utah*. 46th Cong., 2d sess., 1880, Exec. Doc. 1, serial #1911.

———. House. *Report of the Governor of Utah*. 50th Cong., 1st sess., 1887, Exec. Doc. 1, serial #2541.

———. House. *A Report upon the Mineral Resources of the States and Territories West of the Rocky Mountains*. 39th Cong., 2d sess., 1867, Exec. Doc. 29, serial #2189.

———. House. *Sagebrush Rebellion: Impacts on Energy and Minerals: Oversight Hearing before the Subcommittee on Mines and Mining of the Committee on Interior and Insular Affairs, House of Representatives, Ninety-sixth congress, Second Session, on Sagebrush Rebellion, Impacts on Energy and Minerals: Hearing Held in Salt Lake City, Utah, November 22, 1980*. 96th Cong., 2d sess., 1980.

———. Senate. *Health Impact of Low-Level Radiation, 1979: Joint Hearing before the Subcommittee on Health and Scientific Research of the Committee on Labor and Human Resources and the Committee on the Judiciary*. 96th Congress, first session, June 19, 1979.

———. *A Staff Report on the Wilberg Mine Disaster of 1984, Orangeville, Utah*..

———. *Statutes at Large, Treaties, and Proclamations, of the United States of America*. 1872. Vol. 17.

Department of Energy. "Statistical Data of the Uranium Industry." GJO-100 (83). 1983.

Department of the Interior. *Development of Coal Resources in Central Utah: Environmental Statement*. 1979.

Environmental Protection Agency, Region 8, Kennecott South. http://www.epa.gov/region08/superfund/sites/ut/kennes.

———. "Field Demonstration of Permeable Reactive Barriers to Remove Dissolved Uranium from groundwater, Fry Canyon, Utah, September 1997 through September 1998." Interim report. 2000.

Forest Service. *Forest Planning Unit: Coal Unsuitability Study*. 1980.

———. *The Impacts Associated with Energy Developments in Carbon and Emery Counties, Utah*. 1975.

———. *Record of Decision, Gardner Canyon Gypsum Mine*. 1995.

———. *Revegetation Potential of Surface-Mineable Coal Lands in the Interior West*. 1979.

———. *South Twin Lode Mining and Development Proposal: Final Environmental Impact Statement*. 1990.

———. *Uinta National Forest Gardner Canyon Gypsum Mine Proposal: Final Environmental Impact Statement*. 1994.

———. *User Guide to Engineering: Mining and Reclamation in the West*. 1979

———. *User Guide to Soils: Mining and Reclamation in the West*. 1979

———. *User Guide to Vegetation: Mining and Reclamation in the West*. 1979

General Land Office. "Mining Districts By-Laws, 1872–1909." Series 3651. http://archives.utah.gov/reference/xml/series/3641.html

Geological Survey. *Contributions to Economic Geology*. 1903.

———. "Cottonwood Special Quadrangle." Map; scale 1:25,000. 1936.

———. *Development of Coal Resources in Southern Utah*. 1979.

———. *Gold Placers in Utah: A Compilation*. 1966.

———. *Mineral Resources of the United States*. 1882–1923.

Handbook of Federal and Utah State Laws on Energy/Mineral Resource Development: A Cooperative Effort between State and Federal Agencies. SLC, 1975.

Mine Safety and Health Administration. Historical Data on Mine Disasters in the United States. http://www.msha.gov/mshainfo/factshet/mshafct8.htm (accessed 26 June 2001).

National Archives. Letters Received by the Office of Indian Affairs, Utah Superintendency, 1863–65. Microfilm roll #901.

National Park Service. *Prospector, Cowhand, and Sodbuster: Historic Places Associated with the Mining, Ranching, and Farming Frontiers in the Trans-Mississippi West*. 1967.

Nuclear Regulatory Commission. *Environmental Impact Statement Related to Reclamation of the Uranium Mill Tailings at the Atlas Site, Moab, Utah*. 1999.

———. *Environmental Statement Related to the Plateau Resources Limited Shooterin Canyon Uranium Project* (Garfield County, Utah). 1979.

Office of Surface Mining Reclamation and Enforcement. *Southern Utah Petition Evaluation: Draft 522 SMCRA Evaluation and Environmental Statement OSM-EIS-4*. 1980.

War Department. *The War of the Rebellion: A Compilation of the Official Records of the Union and the Confederate Armies*. Series 1, vol. 50, pt. 2, serial #3584. Washington, D.C.: GPO, 1897.

State of Utah

Bureau of Immigration, Labor, and Statistics. *Second Report of the State Bureau of Immigration, Labor and Statistics for the Years 1913–1914*. 1915.

Department of Employment Security. See Department of Workforce Services.

Department of Natural Resources. *Abandoned Mine Lands in Utah: Hazards and Environmental Problems in Selected Mining Districts*. 1991.

———. *Mining Utah's Heritage: The Story of Mining in Utah*. 1997.

———. *Surface Mining Reclamation and Enforcement Provisions for Coal*. 1978.

Department of Workforce Services. "Industry Profile, Mining." http://www.wi.dws.state.ut.us/regions/mining.pdf.

———. Economic Data and Analysis Unit. *Annual Report of Labor Market Information*. 2000.

———. Job Service and the College of Eastern Utah. *Utah's Coal Mining Industry: Perspectives on Its History and Potential Emphasizing the Need to Meet Coal Industry Training Requirements*. 1976.

Division of Oil, Gas, and Mining. *Final Report on Selected Research Projects Leading to the Development of Utah Coal, Tar Sands and Oil Shale*. 1978.

———. *Mining Utah's Heritage: The Story of Mining in Utah*. 1992.

———. *Monthly Oil and Gas Production Report*. 1975–present.

———. *Utah Abandoned Mine Reclamation Program*. 1982.

———. Utah Directory of Mining and Manufacturing. 1976.

———. Utah Oil and Gas Activity, January through June 1989. 1989.

———. Utah's Coal Mining Industry: Perspectives on Its History and Potential. 1976.

———. *Your Guide to the Utah Division of Oil, Gas and Mining*. 1997.

Geological [and Mineralogical] Survey. *2001 Annual Review and Forecast of Utah Coal Production and Distribution*. 2003.

———. Geology of Big Cottonwood Mining District. Compiled By Max D. Crittenden, Jr., and Laurence P. James. 1978.

———. *Geology of the East Tintic Mountains and Ore Deposits of the Tintic Mining District*. Vol 12 of *Guidebook to the Geology of Utah*. 1957.

———. *Gold Placers in Utah: A Compilation*. 1966.

Governor's Office of Planning and Development. Andalex Resources and the Proposed Smoky Hollow Mine: A Fiscal Impact Analysis and Economic Overview. 1993.

———. *Economic Report to the Governor*. http://www.governor.utah.gov/dea Published annually.

Industrial Commission. *General Safety Orders Covering Coal Mining Operations in the State of Utah*. Revised periodically.

———. *Report of the Industrial Commission of Utah, Period July 1, 1917 to June 30, 1918*. SLC: F. W. Gardner, n.d.

———. *Report of the Industrial Commission of Utah, Period July 1, 1918 to June 30, 1920*. Kaysville, UT: Island Printing Company, n.d.

———. *Report of the Industrial Commission, 1925–26*.1926.

Military Department, Office of the Adjutant General, Special Orders 11, 19 April 1955.

Office of Education. Mineral Lease Revenues and Utah's Public Education System. 1984.

Office of Energy and Resource Planning. *Annual Review and Forecast of Utah Coal Production and Distribution*. Published annually.

Office of the Legislative Auditor General. A *Performance Audit of the Division of State Lands*. Published annually.

———. A *Survey of Oil and Gas Auditing*. 1987.

The Oil and Gas Conservation Act the Mined Land Reclamation Act and the General Rules and Regulations and Rules of Practice and Procedure. 1955.

Report of the State Coal Mine Inspector. 1896, 1901.

Secretary of State. Mines and Mining Laws: State of Utah. 1913–present.

State Bar. *Oil and Gas: Practice and Procedure before the Utah Board of Oil, Gas and Mining: A Nuts and Bolts Approach for the Practioner*. 1984.

State Board for Vocational Education. *Vocational Training for War Production Workers: Course of Study for Metal Mining Occupations*. WPT-U 200-36. 1943.

Statistical Abstract of Utah. 1990.

Utah Code Unannotated, Title 59, chapter 5.

Utah House Journal. 1990

Counties and Cities of Utah

Beaver County, Utah
- Recorder's Office
  - Book No. 1 of Notices
  - Index to Books A, B, C, and D.
  - North Star Mining District, Book A, Book B.
  - "Real Estate in Beaver Co."
  - Star Mining District, Book B.
- Second Judicial District Court
  - Minute Book No. 1.
  - Minute Book No. 2.
  - Minute Book No. 3.

Eureka, Utah
  - Abstracts of Title. Eureka, Utah, Juab County, Juab County Courthouse, Nephi, Utah.
  - Eureka City Criminal Justice Docket Ledger, 1893. Eureka City Hall
  - Eureka City Minute Book 1. 1893.

Salt Lake County, Utah
- Clerk's Office
  - Estate Books M1, N1, O1. 1891–94.
  - Probate Court Records.

Recorder's Office
Salt Lake County Abstracts, Book A2.
Summit County, Utah
Summit County Commission minutes, 8 March 1954. Summit County Courthouse, Coalville, Utah.
Tooele County, Utah
Recorder's Office
Deed Record Books HH, KK, MM.

Dissertations and Theses

Andberg, Donald B. "Principles and Practices of Reduction of Sand and Gravel Excavation Visual Impacts in the Wasatch Front Foothills of Central Utah." Master's thesis, USU, 1981.

Bamberger, Clarence Greenwalde. "Report on the Daly West Mining Co., Park City, Utah." Master's thesis, Cornell University, 1908.

Bourne, John Michael. "Early Mining in Southwestern Utah and Southeastern Nevada, 1864–1873: The Meadow Valley, Pahranagat, and Pioche Mining Rushes." Master's thesis, U of U, 1973.

Bringhurst, Newell G. "The Mining Career of George H. Dern." Master's thesis, U of U, 1967.

Brinley, John Ervin, Jr. "The Western Federation of Miners." PhD diss., U of U, 1972.

Clark, J. L. "History of Utah's Salt Industry, 1847–1970." Master's thesis, BYU, 1971.

Condon, David DeLancey. "A Preliminary Study of the Social and Economic Geography of Utah with Special Emphasis on the Tintic Mining District." Master's thesis, BYU, 1934.

Ercanbrack, Dennis L. "A History of the American Fork Mining District, 1870–1920." Master's thesis, USU, 1970.

Fish, Rick. "The Southern Expedition of Parley P. Pratt, 1849–50." Master's thesis, BYU, 1992.

Fox, William. "Patrick Edward Connor: 'Father' of Utah Mining." Master's thesis, BYU, 1966.

Garlock, Johnathan Ezra. "A Structural Analysis of the Knights of Labor: A Prolegomenon to the History of the Producing Classes." PhD diss., University of Rochester, 1974.

Gibbs, Lynda Broadbent. "The History of the United Mine Workers of America in Carbon County, Utah up to 1933." Master's thesis, BYU, 1968.

Hansen, Gary B. "A Business History of the Copper Industry of Utah." Master's thesis. USU, 1963.

Hendrickson, A. LaMar. "An Economic Study of Coal Mine Taxation in Utah." Master's thesis, USU, 1941.

Hodson, Dean R. "The Origin of Non-Mormon Settlements in Utah, 1847–1896." PhD diss., Michigan State University, 1971.

Hughes, Charles E. "The Development of the Smelting Industry in the Central Salt Lake Valley Communities of Midvale, Murray and Sandy Prior to 1900." Master's thesis, BYU, 1990.

Kerr, H. S. "An Analysis of Utah's Potash Industry." Master's thesis, U of U, 1965.

Larsen, Kent Sheldon. "Life of Thomas Kearns." Master's thesis, U of U, 1964.

Lees, Jon Eugene. "Utah's Uranium Boom and Penny Stocks." Master's thesis, U of U, 1974.

Lewis, Anna Viola. "The Development of Mining in Utah." Master's thesis, U of U, 1941.

Milner, Michael Joseph. "The 1967–1968 Copper Strike, Selected Issues and Implications." Master's thesis, U of U, 1969.

Notarianni, Philip F. "The Italian Immigrant in Utah: Nativism (1900–1925)." Master's thesis, U of U, 1972.

Olson, Emmett K. "Mining Methods of the Park Utah Consolidated Mines Company." Bachelor's thesis, U of U, 1950.

Pawar, Sheelwant B. "An Environmental Study of the Development of the Utah Labor Movement: 1860–1935." PhD diss., U of U, 1968.

Payne, Robert L. "The Effect of the Uranium and Petroleum Industries upon the Economy of San Juan County, Utah." Master's thesis, BYU, 1964.

Pelton, Roger T. "Utah Mining."Project for the degree of mining engineer, Columbia University, 1903. Also available on *The Utah History Suite*, CD, USHS.

Powell, Alan Kent. 'Labor at the Beginning of the Twentieth Century: The Carbon County, Utah Coal Fields, 1900 to 1905." Master's thesis, U of U, 1972.

Quackenbush, Stanley Fulton. "Utah Business Corporations, 1847–1895." Master's thesis, University of Illinois at Urbana, 1971.

Randall, Deborah Lyn. "Park City, Utah: An Architectural History of Mining Town Housing, 1869 to 1907." Master's thesis, U of U, 1985.

Reeder, Clarence, Jr., "The History of Utah's Railroads, 1869–1883." PhD diss., U of U, 1970.

Reeder, Ray M. *The Mormon Trail: A History of the Salt Lake to Los Angeles Route to 1869*. PhD diss., BYU, 1966.

Reese, Gary F. "'Uncle Jesse': The Story of Jesse Knight." Master's thesis, BYU, 1961.

Reeve, W. Paul. "Mormons, Miners, and Southern Paiutes: Making Space on the Nineteenth-Century Western Frontier." PhD diss., U of U, 2002.

Shumway, Gary Lee. "The Development of the Uranium Industry in San Juan County, Utah." Master's thesis, BYU, 1964.

Spendlove, Beatrice. "History of Bingham Canyon, Utah." Master's thesis, U of U, 1937.

Stringham, Bronson F. "Mineralization in the West Tintic Mining District, Utah." PhD diss., Columbia University, 1941.

Stuckenschneider, Victor H. "A Study of the Impact of Mining on Utah's Economy during the Great Depression." Master's thesis, U of U, 1959.

Stucki, Alfred Bleak. "A Historical Study of Silver Reef: Southern Utah Mining Town." Master's thesis, BYU, 1966.

Tribe, Patrick Riley. "Park City, Utah: An Analysis of the Factors Causing a Decline of Mining, and of the Factors Which Have Contributed to Its Development as a Recreational Area." Master's thesis, BYU, 1967.

Wegg, David Spencer. "Bingham Mining District, Utah." Master's thesis, U of U, 1915.

Westfield, James. "Coal Mine Explosions in Utah." Bachelor's thesis, U of U, 1935.

Young, Arthur. "Annie Laurie Gold Mines." Bachelor's thesis, U of U, 1937.

Law Cases

John N. Begay et al., plaintiffs, vs. the United States of America, defendant. Report of court proceedings in Arizona U.S. District Court, August 3, 1983, Civil No. 80-982.

"The Raymond and Ely vs. The Kentucky Mining Co." Judge Beatty's Decision. Pioche, NV: Record Publishing, 1873. Copy at the Nevada Historical Society, Reno, Nevada.

Salt Lake City Third District Court

Amos Woodward et al. vs. P. E. Connor, 1874.

Charles E. Mitchner and John R. Kelly vs. Great Basin Mining and Smelting Company and P. Edward Connor, 1880.

John S. Barrett and Oscar V. Walker vs. P. Edward Connor, 1880.

Lawrence Bethune vs. P. Edward Connor, 1881.

Leonard S. Osgood vs. P. Edw. Connor, 1880.

M. Livingston vs. P. Edw. Connor, 1873.

P. Edward Connor vs. Enos A. Wall, John W. Johnson, Charles Reed and H. W. Lawrence, 1875.

P. Edward Connor vs. Great Basin Mining and Smelting Company, 1882.

P. Ewd. [sic] *Connor vs. Hiram S. Jacobs*, 1876.

P. Edward Connor vs. Robert J. Goldring, et al., 1873.

Steven F. Nuckolls vs. P. Ed. Connor et al., 1874.

Maps

Baker, A. A., F. C. Calkins, M. D. Crittenden, Jr., and C. S. Bromfield. "Geologic Map of the Brighton Quadrangle, Utah." USGS Map GQ-534. Washington, D.C.: GPO, 1966.

Bassler, Harvey, and J. B. Reeside, Jr. "Oil Prospects in Washington County, Utah." In *Contributions to Economic Geology*, USGS Bulletin 726. Washington, D.C.: GPO, 1922.

Bon, Roger L., and Sharon Wakefield. *Large Mine Permits in Utah*. SLC: UGS, 1998–2002.

———. *Small Mine Permits in Utah*. SLC: UGS, 1999–2002.

Bromfield, C. S., and L. L. Patten. "Mineral Resources of the Lone Peak Wilderness Area, Utah and Salt Lake Counties, Utah." *U.S. Geological Survey Bulletin*, 1981.

Bryant, Bruce. "Geologic Map of the Salt Lake City 30' x 60' Quadrangle. North-Central Utah, and Uintah County, Wyoming." USGS Miscellaneous Investigations Series, Map 1, 1990. 1994.

Collier, Frances L. "Claim Map of Cottonwood-American Fork Area, Utah." Washington, D.C.: USGS, 1943.

Dyni, J. R. "Sodium carbonate resources of the Green River Formation in Utah, Colorado, and Wyoming." U.S. Geological Survey Open-File Report 1996: 96–729.

Dyni, J. R., Hite, R. J., and Raup, O. B. "Lacustrine deposits of bromine-bearing halite, Green River Formation northwestern Colorado." In *Third Symposium on Salt*, vol. 1. Edited by J. L. Rau, and L. F. Dellwig. Cleveland: The Northern Ohio Geological Society, Inc., 1970: 166–79.

Froiseth, B. A. M. "New Sectional Mining Map of Utah." SLC, 1871.

Mansfield, G. R. *Geography, Geology, and Mineral Resources of Part of Southeastern Idaho*. USGS Professional Paper 152. Washington, D.C.: GPO, 1927.

Mattox, R. B. *Saline Deposits*. Special Paper 88. New York: Geological Society of America, 1968.

Owen, Charles Mostyn. *West Mountain Mining District*. Denver: Rio Grande Western Railway Company, 1900.

Sanborn maps of Eureka, 1898, 1908.

West, Joseph A. "West's New Sectional and Topographical Map of Utah." N.p., 1885.

Witkind, I. J., M. P. Weiss, and T. L. Brown. "Geologic map of the Manti 30' x 60' quadrangle, Carbon, Emery, Juab, Sanpete, and Sevier Counties, Utah." Miscellaneous Investigations Series, Map I-1631, scale 1:100,000. USGS, 1987.

Witkind, I. J., and M. P. Weiss. "Geologic map of the Nephi 30' x 60' quadrangle, Carbon, Emery, Juab, Sanpete, Utah, and Wasatch Counties, Utah." USGS Miscellaneous Investigations Series, Map I-1937, scale 1:100,000, 1991.

Wong, George. "Preliminary Map of the Mining Districts of the Basin and Range Area of Utah." Reston, VA: USGS, 1981.

Videos

Atomic Stampede. Produced by Ken Verdoia. SLC: KUED, 1994.

Copper Canyon, American Dream. Produced by Colleen Casto. SLC: KUED, 2003.

Fire in the Hole: Mine Wars of the West. SLC: KUED, 2000.

Joe Hill. Produced by Ken Verdoia. SLC: KUED, 1998.

Tintic Revisited. Provo, UT: KBYU, 1995.

Treasure House: The Utah Mining Story. Bountiful, UT: Groberg Communications, 1995.

About the Authors

Utah native Thomas G. Alexander earned a PhD in American history from the University of California at Berkeley. He is currently the Lemuel Hardison Redd, Jr., Emeritus Professor of Western American History at Brigham Young University. He specializes in Utah history, western history, environmental history, and Mormon history and is the author of more than 120 articles and the author, co-author, editor, or co-editor of 22 books and monographs including *A Clash of Interests: Interior Department and Mountain West*; *Mormons and Gentiles: A History of Salt Lake City* (with James B. Allen); *Utah, the Right Place: The Official Centennial History*; *The Rise of Multiple-Use Management in the Mountain West: A History of Region 4 of the Forest Service*; *Grace and Grandeur: A History of Salt Lake City*; *Things in Heaven and Earth: The Life and Times of Wilford Woodruff, A Mormon Prophet*; and *Mormonism in Transition: A History of the Latter-day Saints, 1890–1930*. Winner of numerous awards, he is a fellow of the Utah State Historical Society and of the Utah Academy of Sciences, Arts and Letters and former chair of the Utah Board of State History. He has served as president of the American Historical Association–Pacific Coast Branch, the Mormon History Association, and the Utah Academy of Sciences, Arts, and Letters. He is formerly chair of the Utah Humanities Council Advisory Board and president of Phi Alpha Theta. Alexander is also a member of the council of the Western History Association and a member of the Organization of American Historians and the American Society for Environmental History.

John Barton grew up in the Uinta Basin. His master's thesis at Brigham Young University was "Antoine Robidoux and the Fur Trade of the Uintah Basin." He currently serves as a senior lecturer of history for the Uintah Basin Campus of Utah State University, where he has taught history since 1988. He was named as the Teacher of the Year (1999) at USU-UB and has received recognition as an outstanding teacher on the CON/NET system of distance education. In 2004 he received the Extension Award for "Innovative Practices in Teaching." Barton is a recipient of a Merit Award from the Utah Humanities Council for the Outlaw Trail History Project and a Western History Research Fellowship from the Redd Center of Western Studies at Brigham Young University and was named by Governor Olene Walker to serve on the Utah State History Board. He is the author of numerous articles and reviews as well as three books: *A History of Duchesne County; From Tabernacle to Temple: The Story of the Vernal Utah Temple;* and *Buckskin Entrepreneur: Antoine Robidoux and the Fur Trade in the Uintah Basin, 1824–1844.* In 1991 Barton became the founding editor of *The Outlaw Trail Journal,* and he is a popular lecturer on topics ranging from outlaws and mountain men to Ute culture and perspective.

A native of Salt Lake City, MARTHA SONNTAG BRADLEY-EVANS earned her bachelor's degree from the University of Utah, her master's from Brigham Young University, where she has also taught, and a PhD from the University of Utah, where she is currently associate professor of architectural history and the director of the Honors Program. She has contributed articles to various books and journals including *Utah Historical Quarterly, Sunstone, Dialogue: A Journal of Mormon Thought,* and the *Journal of Mormon History.* Bradley is the author of several books including *Four Zinas: A Story of Mothers and Daughters on the Mormon Frontier; Kidnapped from the Land: The Government Raids on the Short Creek Polygamists;* and *Pedestals and Podiums: Utah Women, Religious Authority, and Equal Rights.* She also wrote the histories of Beaver and Kane Counties for the Utah Centennial County History Series. *A History of Beaver County* provided the foundation for her chapter in this volume.

HAL COMPTON is a Utah native who holds a BS degree in communications from the University of Utah. He served as a second lieutenant in the United States Army Signal Corps in Korea. He worked for CBS Radio and Television in Hollywood, California, and in public relations for GTE in California. Compton became interested in mining while exploring the gold country of northern California, and in 1987 he retired to Park City, Utah, where he currently serves as the research historian for the Park City Historical Society and Museum. In 1998 he was named the Museum Volunteer of the Year by the Utah Museum Association. The Park City mayor and city council have also designated him Park City's historian laureate. He is a member of the board of directors of the Park City Historical Society and Museum and president of the Glenwood Cemetery Association. Compton works in the Park City Museum and Visitor Information Center and gives historic walking tours of Park City's Main Street. For several years he conducted underground tours for the Silver Mine Adventure in the Ontario Mine.

JAMES E. FELL, JR., earned a BA in chemistry from Colby College and both an MA and a PhD in American history from the University of Colorado in Boulder. He has taught at several universities, including Regis University School for Professional Studies; Metropolitan State College of Denver; Colorado School of Mines; University of California, San Diego; Colby College; and Arapahoe Community College, and is currently adjunct associate professor of history at the University of Colorado at Denver. He has also served as an editor for *Arizona and the West* (now the *Journal of the Southwest*), *Business History Review*, and Cordillera Press. Fell was a Harvard-Newcomen Fellow in business history in the Harvard University Graduate School of Business Administration and manager of communications for United Banks of Colorado (now a division of Wells Fargo Banks), a chemist for the Eastman Kodak Company, and a research historian for Historic Preservation and Exhibits for the Colorado Historical Society. He is the author of numerous journal and encyclopedia articles and four books: *Ores to Metals: The Rocky Mountain Smelting Industry; Arthur Redman Wilfley: Miner, Inventor, and Entrepreneur,* in collaboration with Jay E. Niebur; *Aurora: Gateway to the Rockies*, with Steven F. Mehls and Carol J. Drake; *Mining the Summit: Colorado's Ten Mile District, 1860–1960*, with Stanley Dempsey. He also served as coeditor with P. D. Nicolaou and G. D. Xydous of *Book of Proceedings: 5th International Mining History Congress*.

A native of Centerville, Utah, J. Wallace Gwynn received BS and PhD degrees in mineralogy from the University of Utah. Since graduating in 1970, he has spent most of his professional career working with the Great Salt Lake. For the last 30 years he has worked as a saline-minerals geologist with the Utah Geological Survey. Gwynn is the author of numerous publications dealing with Utah's natural resources. Published topics include bituminous sandstones of the P. R. Springs area, oil-well brines of the Uinta and Paradox basins, subsurface brines of Sevier Lake, low-temperature geothermal resources along the Wasatch Front, and the brines and mineral resources of Great Salt Lake. He has served on the Great Salt Lake Technical Team and the Bonneville Salt Flats Technical Review Committee and he enjoys working with students on science-fair projects.

David Hampshire is the director of communication programs for Questar Corporation. A Park City resident, he is coauthor of the books *A History of Juab County* and *A History of Summit County* for the Utah Centennial County History Series and of *No Western Parallel: The Story of the Questar Corporation* and has written many articles on local history and preservation for *Utah Historical Quarterly*, *Utah Preservation*, and *Beehive History*.

Laurence P. James earned a bachelor's degree with honors from Stanford and a PhD in geology and geochemistry from Pennsylvania State University. His wide experience ranges from teaching at universities in the United States and Asia to research in government geological organizations to working for private mineral companies in this and several other countries. He has published and lectured widely on geology and the impacts of ore discoveries on economy and history. His major works in mining history include articles in *Journal of the West*, *Utah Historical Quarterly* and the *Journal of Geochemical Exploration*. Along with James E. Fell, he is writing a study of metal mining and entrepreneurship in the western Wasatch Mountains of Utah. James currently works with NewWest Gold Corporation providing expertise in mining property evaluation, regional exploration concepts, and mapping geology in the eastern Great Basin.

A Utah native, BRIGHAM D. MADSEN holds a bachelor's degree from the University of Utah and a master's and PhD from the University of California, Berkeley. He is currently a professor emeritus from the University of Utah, where he served as chair of the History Department, director of libraries, academic and administrative vice president, and dean of the Division of Continuing Education. He is the author or editor of many articles and several books including *The Bannock of Idaho; The Now Generation: Student Essays on Social Change in the Sixties; North to Montana!: Jehus, Bullwhackers, and Mule Skinners on the Montana Trail; The Lemhi: Sacajawea's People; The Northern Shoshoni; Corinne: The Gentile Capital of Utah; Gold Rush Sojourners in Great Salt Lake City, 1849 and 1850; The Shoshoni Frontier and the Bear River Massacre; Chief Pocatello, the "White Plume"; Exploring the Great Salt Lake: The Stansbury Expedition of 1849–50*; and his autobiography, *Against the Grain: Memoirs of a Western Historian*.

PHILIP F. NOTARIANNI is currently the director of the Utah State Historical Society/ Division of State History, where he has worked for the past 29 years. He also serves as a lecturer in the Ethnic Studies Program at the University of Utah. A resident of Magna, Utah, he received BS and MA degrees in history from the University of Utah, an MA degree in history from the University of Minnesota, and a PhD from the University of Utah in 1980. He continues to teach the Peoples of Utah class at the University of Utah that analyzes the ethnic and cultural diversity of the state. He has served as a member of the temporary faculty in cultural anthropology for the University of Calabria in Cosenza, Italy. Notarianni has published various books and articles, as author or editor, including *The Avenues of Salt Lake City; Carbon County: Eastern Utah's Industrialized Island; Faith, Hope, and Prosperity: The Tintic Mining District*; "Italianita in Utah: The Immigrant Experience," in *The Peoples of Utah*, ed. Helen Z. Papanikolas; "Places of Origin: Calabresi in Carbon County, Utah," in *Old Ties, New Attachments: Italian-American Folklife in the West*, ed. David A. Taylor and John Alexander Williams; "Italians in Utah" and "Frank Bonacci," in *Italian American History and Culture: An Encyclopedia*, ed. Salvatore J. LaGumina, Frank J. Cavaioli, Salvatore Primeggia, and Joseph A. Varacalli; and "Ethnic Folklore Studies," in *Folklore in Utah*, ed. David Stanley.

WILLIAM T. PARRY was born, raised, and educated in Utah. After receiving BS, MS, and PhD degrees from the University of Utah, he worked for Shell Oil Company in Texas and then taught at Texas Tech University. He subsequently joined the faculty of the University of Utah, first in mining and geological engineering then in geology and geophysics. Parry received a number of awards for his teaching, including commendations from the Geology Department and College of Mines, a University Distinguished Teaching Award and a Presidential Teaching Scholar Award. He has published numerous technical papers dealing with mineralogy, chemistry, and ore deposits from Africa to Alaska. The geology of Utah is a favorite subject. He is now professor emeritus and continues to teach and write about Utah geology.

Allan Kent Powell was born in Price, Utah, and grew up in nearby Huntington. He earned his BA, MA, and PhD degrees in history at the University of Utah. He has worked at the Utah State Historical Society for more than 30 years, where he is currently the public history coordinator and editor of the *Utah Historical Quarterly*. He is also an associate instructor of history at Westminster College and has participated in the Utah Humanities Council Speakers Bureau and other programs for many years. Much of Powell's career has been involved in researching and writing history. His books include *The Next Time We Strike: Labor in Utah's Coal Fields 1900–1933; Splinters of a Nation: German Prisoners of War in Utah; Utah Remembers World War II*; and *The Utah Guide: A Travel Guide to the State*. His major editing projects have included *The Utah History Encyclopedia*, and, as general editor, the Utah Centennial County History Series, which was completed in 1999 and includes a book-length history of each of Utah's 29 counties.

Robert E. Rampton is a native of Bountiful, Utah, and was educated in Davis County schools. After military service, he attended the University of Utah and earned a BFA degree with emphasis in journalism and radio and television broadcasting in 1950. He is a former staff writer and photographer for the *Salt Lake Tribune* and served as western regional correspondent for McGraw-Hill publications. Rampton's professional career includes service in the public relations departments of United States Steel Corporation, Westinghouse Air Brake Company, and Ketchum Public Relations in Pittsburgh, Pennsylvania, and with the Evans Group in Salt Lake City. He has written extensively for technical and trade publications serving the coal and metal mining, processing, and manufacturing industries. He emerged from retirement and joined W. S. Adamson and Associates in 1998. In 2004 he retired for the third time.

W. Paul Reeve holds a BA and an MA from Brigham Young University and a PhD from the University of Utah. His doctoral dissertation received "best" awards from both the University of Utah's History Department and Brigham Young University's Joseph Fielding Smith Institute for LDS History. He has published extensively in journals and wrote the centennial history of Enterprise, Utah. Reeve has also presented papers at numerous conferences, ranging from the Western History Association to the Conference on Expanding the Interdisciplinary Conversation. In 2006, University of Illinois Press will publish a revision of his dissertation under the title

Making Space on the Western Frontier: Mormons, Miners, and Southern Paiutes. Reeve is currently an assistant professor of history at the University of Utah.

Popular author RAYE C. RINGHOLZ graduated from the University of Utah with a BA in English and history and has published many books on the West including *Paradise Paved: The Challenge of Growth in the New West; Little Town Blues: Voices from the Changing West; The Wilderness Handbook; Diggings and Doings in Park City; Walking through Historic Park City; Park City Trails; On Belay! The Life of Legendary Mountaineer Paul Petzoldt; Barrier of Salt: The Story of the Great Salt Lake; Guidebook to Canyonlands Country*; and *Uranium Frenzy: Saga of the Nuclear West*, from which she drew her chapter for this book. She has also published dozens of articles in venues ranging from *Salt Lake Magazine* and the *Salt Lake Tribune* to *Ford Times* and *Field and Stream*. In addition to being a prolific author, she has served as an officer or board member for many civic and academic organizations, including the Kimball Art Center, Writers at Work, the Park City Historical Society and the Park City Miner's Day Parade, the Wilderness Education Association, and the Junior League of Salt Lake City. Ringholz has been awarded the Paul Petzoldt Award from the Wilderness Education Association and both the Executive Committee Award and the Board of Directors Award from the Wilderness Education Association.

JANET BURTON SEEGMILLER holds a BA from the University of Utah in journalism and an MA in information resources and library sciences from the University of Arizona. She is currently associate professor of library media and special collections librarian at the Sherratt Library at Southern Utah University. Her publications include "Walter K. Granfer: A Friend to Labor, Industry, and the Unfortunate and Aged," *Utah Historical Quarterly; A History of Iron County: Community above Self*, a Utah Centennial County History; and *Be Kind to the Poor: The Life Story of Robert Taylor Burton*. She has made scholarly presentations and read papers to several groups including the Association of College and Research Libraries, Mormon History Association, Utah Library Association, Communal Studies Association, Iron County Sesquicentennial Lecture Series, and Dixie Heritage Lecture Series. Seegmiller is on the board of editors for both *Mormon Historical Studies* and the *Journal of Mormon History* and is a member of the American Library Association, Utah Library Association, and the Conference of Intermountain Archivists. She has served as president of the Iron Mission Historical Society, board member for the Iron County Museum Foundation, and council member, conference chair, and newsletter editor for the Mormon History Association.

A native of Providence, Utah, CARMA WADLEY earned a bachelor's degree from Utah State University and a master of arts in communications from Brigham Young University. She has worked for the *Deseret Morning News* as a writer and feature editor. Currently a senior writer, she frequently contributes articles and photographs dealing with her extensive travels. Several of the stories she covered for the *News* dealt with mining, including some of the industry's greatest disasters. In meeting and talking with miners, Wadley became fascinated with their stories and tales and has collected mining folklore for years.

Delta, Utah, native Debra Wagner started working with Brush Resources in 1986 and has moved through the company serving as a clerk, an administrative secretary, and administrative assistant. Currently she is the company's human relations manager. She is known throughout the company for her high standards and competence. Her duties over the years have ranged from working with customers and employees to organizing company materials to creating the history section of the company's Web site, which she expanded to form this chapter.

Bruce D. Whitehead, a native of Statesville, North Carolina, is president and owner of W. S. Adamson and Associates, a pioneer public relations firm in Salt Lake City. The company was organized in 1950 by the late W. S. "Bill" Adamson to serve the public relations and public affairs needs of Kennecott Copper Corporation. The firm still provides public communications services to Kennecott Utah Copper and other Kennecott companies. Whitehead holds a BA degree in communications from Quincy University. He joined Adamson in 1966 after 20 years as an on-the-air personality; as producer and director with Illinois, Idaho, and Utah radio and television broadcasters; and of service with the Armed Forces Radio Service in Austria. His career encompasses 39 years of managing and supervising public relations, public affairs, and general communications programs for a variety of local, regional, national, and international clients. He acquired ownership of Adamson and Associates in 1989.

Born and raised in Ogden, Utah, Colleen Whitley holds degrees from Weber (then) College, the University of Utah, and Brigham Young University. She has taught students in every grade level from elementary school through graduate school, worked with Job Corps, and helped to open one of the first alternative high schools in the state. She recently retired from teaching for the English and Honors Departments of Brigham Young University and has lectured for the Utah Humanities Council Speakers' Bureau. Widely published, she has written poetry, fiction, and newspaper, journal, and magazine articles, edited *Worth Their Salt: Notable but Often Unnoted Women of Utah* and *Worth Their Salt, Too: More Notable but Often Unnoted Women of Utah*. She also edited and wrote several chapters for *Brigham Young's Homes* and coauthored *The Silver Queen: Her Royal Highness Suzanne Bransford Emery Holmes Delitch Engalitcheff, 1859–1942* with Judy Dykman.

Index

Italic indicates an illustration.

A

Absaraka mine, 162
Act to Promote the Development of Mining Resources, 359
Adams, David, 385
Adams, David B., 199
Adams, Don, 160
Adams, Joe, 160
Adams, Orson, 257, 258
Adams, William C., 212, 213
Adams and Kiesel, 107
Adams group, 152
Adamson, Ephriam, 335, 336
adularia, 215
AEC. *See* Atomic Energy Commission
AFL. *See* American Federation of Labor
Africa, 15, 65, 231, 374
AKZO Salt/AKZO Nobel, 112
Alaska, 19, 24, 33, 235, 236, 241, 383
Albion Basin, 282, 292
Albion Range, 11
Albuquerque, New Mexico, 217
Alder, Douglas, 265, 267
Alladin Uranium Company, 161
Allen, Colonel, 268, 269
Allen, James B., 138
Alliance Tunnel, 324
Alpha Mine, 155
Alta, Utah, 22, 41, 42, 74, 92, 93, 238, 259, 272–74, 280–85, 287–97, *303*, *306*
Alta and Hecla Mining Company, 295
Alta fault, 7, 28
Alta Independent, 288
Alta Mining and Development Company, 281
Alta Mining District, 82, 272–*317*
Alta Quincy Mine, 290. *See also* Quincy Mine
Alta Scenic Railway, 296
Alta Tunnel and Transportation Company, 289
Alta United Mines Company, 290
Altamont, 34, 383–85
aluminum, 37, 170, 219, 398, 400
AMAX, Inc., 55, 111–14
Amazon Mine, 85
American Emma Company, 280
American Federation of Labor (AFL), 207, 209
American Flag Mine, 322
American Fork Canyon, 12, 35, 44, 272, 282
American Fork Mining District, 27, 272–*317*, *273*, *274*
American Mining Congress, 154
American Orsa Inc., 121
American Salt, 111–13
American Smelting and Refining Company (ASARCO), 43, 147, *185*, 224, 225, 286, 287, 292, 296 340, 374
Anaconda Copper Company, 163, 240, 293, 340, 443
Anchor Mine, 322, 324
Ancient Order of Foresters of America, 349
Ancient Order of United Workmen, 328, 349
Anderson (Chairman), 279
Anderson, Dewey, 161
Anderson, Harold, 237
Anderson, Hartley, 115
Anderson Creek, *317*, 379, 382
Andes Mountains, 15
Andean-type mountain range, 16, 33
Andrews, James, 257
Aneth oil field, 34, 53
Angel of Bingham Canyon, 239
Antelope Range, 217, 218
Antelope Springs, 218
anthracite, 134, 396
anticline, 6, *7*, 9, 28, 29, 32, 33, 145, 147, 152, 318, 393, 401
Antler Mountains, *10*, 12, 13, 15, 25
Apache Mine, 91, 162
apex, 369, 393
apex law, 417
Apex mining camp, 225
Apollo Management, L. P., 113, 115
Arab oil embargo, 383
Arapien, 34
Arapien Shale, 101, 119
Arapien Valley, 120
Archean Period, *10*, 11
Arches National Park, 34
ARCO Oil Company, 54, 57, 125
Arctic, 16, 290
Arentz, Sam, 217
Argenta Mining and Smelting Company, 64, 285, 286
Argentina, 18
Argo, Colorado, 343
Argonaut Mine, 76, 79, 80, 125, 429
Arinosa, Utah, 118
Arizona, 5, 6, 11, 15, 22, 25, 62, 121, 146, 152, 230, 241, 254, 255, 323, 432, 440
Arkansas, 201, 227
Army Corps of Engineers, 146
Arrow Mine, 162, 164
Arrow Uranium, 164
Arrowhead Mine, 217
arsenic, 155, 156, 225
Arthur mill, 43, 240, 241, 246
AS&R. *See* American Smelting and Refining Company
ASARCO. *See* American Smelting and Refining Company
Ashley, Delos, 255
Ashley Creek, 390
Ashley Field, 382
Ashley Valley, 386, 454
Asphalt Ridge, 382, 386
Atkins, M., 370
Atkins Bar, 365
Atkinson, Utah, 329
Atlantic City Mine, 209
Atlas Company, 162
Atlas Engineering Works, *303*
Atlas Mine, 162
Atlas Plant, 54
Atomic Age, 163, 453
Atomic Energy Commission (AEC), 53, 147–50, 152–54, 156–60, 162–64
Atomic Mine, 162
atomic research, 146
atomic tests, 219

atomic weapons, 52, 147
Aulbach, Adam, 64
Australia, 11, 247, 265, 284
Austria, Austrians, 131, 132, 231, 347, 348

B

Baby McKee shaft, *303*
Bald Mountain, 321, 329
Baldwin, N., 61
Bale, Dr. W. F., *180*
Balkans, 131, 132
Ballard, M. Russell, 267
Balsley, Howard, 145, 154, 432
Baltimore, Maryland, 223, 343
Bannack, Montana, 59
Baptists, 349
Barbee, August, 259
Barbee, William Tecumseh, 250, 256, 258–60, 267, 268
Barbee and Walker Mine and Mill, 262, 265, 268, 269
Barber Asphalt Corporation, 389
Basic Manufacturing and Technology, 210
Bates, Grace, 206
Batie, Ralph, 157
Baxter, H. Henry, 277, 278
Beadle, John Hansen, 68, 69, 424
Bear Lake, 124
Bear River, 23, 58, 67
Bear River Massacre, 58
Beattie, W. J., 349
Beauman Building, 346
Beaver Copper Mine, 374
Beaver County, Utah, 42, 44, 70, 71, 213, 218, 219, 257, 269, 270, 359–77
Beaver River, 71, 366, 371
Beaver River Power Company, 371
Beck, John, 349
Becqueral, Henri, 144
beehive kilns, 203, *313*, 369
Beggs mill, 329
Behren's Trench, 115
Belcher Mine, 368
Bellorphan Mine, 282, 291
Bells Canyon, 24
Bemont, G., 144
Bentley, Jim, 152
Bently, Richard I., 257
Benton, Marshall, 350
Bergin Mine, 358
Berry, David J., 61
bertrandite, 166–68
Beryl Junction, 216
beryllia ceramic, 170
beryllium, 37, 53, 157, 159, 166–70
Besame Mucho Mine, 152
Bestelmeyer, John, 374
Beta Mine, 155
Big Bonanza Mine, 291, 390
Big Buck Mine, 152, 160, 162
Big Cottonwood Canyon, 11, 12, 22, 27, 28, 35, 272, 273, 277, 285, 286, 288–90, 295, 297, *303–5*, 319, 332
Big Cottonwood formation, 9
Big Cottonwood Mining District, 12, 22, *272–317*, *274–75*
Big Four mill, 329
Big Fourteen Mine, 216
Big Indian Wash, 151, 152, 161
Big Project Mine, 373
Big Rock Candy Mountain, 148
Bikini Atoll, 147, 157
Bingham, Utah, 3, *10*, 14, 21, 22, 27, 29, 30, 35, 36, 42, 43, *184*, *185–87*, *188*, *190*, 220–49, 264, 272, 292, 293, 335
Bingham and Garfield Railroad, 240
Bingham Canyon, 11, *26*, 54, 59, 60, 91, *184–87*, *190*, 220–49, 286, 377, 436–38
Bingham Gold and Copper Company, 229
Bingham Junction, 229
bismuth, 156, 294, 362
Bissell, E., 374
Black and Diamond Mine, *316*
Black Hawk Mine and mining camp, 129, 138, 206
Black Jack Mine, 162
Black Sea, 29
Blackhawk formation, 32, 33
Bladen, Thomas, 199
Blair, M., 203
Blanding, Utah, 145
Blood, Henry H., 335, 336, 386
Blowout Pit, *182*, 207, 209, 434
Blue Bell Mine and Plant, 43, *312*
Blue Diamond Material Company, 342, 355
Blue Ledge Mining District, 320
Blue Lizard Mine, 148
Blue Rock Mine, 342, 355
Blue Rock Club, 355
Bluebell dolomite, *10*, 31
Bluebell oil field, 34, 383, 384
Bodfish, Fredrick Valentine, 289, 444
Bogen Mine, 329
Bohunk, 139, 348
Bolivia, 18
Bonacci, Frank, 136
bonanza, 78, 144, 156, 256, 259–62, 284, 291, 321, 332, 339, 374, 386, 389, 390, 394
Bonanza City, 260, 261, *314*, *315*, 439, 441
Bonanza Flat, 319, 321
Bonneville Limited, 118
Bonneville Potash Corporation, 118
Bonneville Salt Flats, 101, 102, 115–18, 428
Book Cliffs, 32, 33
Bookcliff Mountains, 378, 386
Booth, Thomas, 119
Boren, Kerry Ross, 89, 427
Borland, A., 261
boron, 101
borrasca, 282, 284, 285, 289, 394
Bosshardt, LaMar, 120, 121
Bosshardt, Milo, 120, 121
Boston, Massachusetts, 74, 77, 78, 229, 230, 232
Boston Consolidated Mining Company, 42, 43, 225, 229, 230, 232
Boston Exchange, 75
Bostwick, F. E., 351
Bothwell, J. R., 263
Boutwell, J. M., 318
Bowen, Henry, 215
Bowles, Samuel, 65, 422, 424
Bowman, Ralph, 162
Bowman, Ray, 162
Box, Thomas, 252, 254
Box Elder County, Utah, 42
Boy Scouts, 239
BP. *See* British Petroleum
BPC. *See* Bonneville Potash Corporation
Braden Copper Company, 241
Bradley, George O., 43, 359, 445, 451, 452
Bradshaw District, 373, 452
Braine Fissure, 287
Braine Silas, 160, 250, 273, 287
Bransford, Susanna, 326
Brewer, W. Y., 152, 387
Brickyard Mine, 226
Brigham City, Utah, 107
Brigham Street, 323, 324
Brighton, 274, 297, 332
British Empire, 200
British Isles, 199, 231
British Petroleum (BP), 247, 248
Broadwater mill, 329
bromine, 101
Bronson, Fletcher, 147, 148
Bronson, Grant, 147, 148
Brooke Smelter, 75
Brooks, Karl, 265
Broomcamp, John M., 374
Broughton Mine, 76, 79, 80
Brown, Amanda, 281, 296
Brownies, 239
Bruhn, F., 159, 160
Bruner, James P., 274
Brush Creek, 382, 390
Brush Resources, 166
Brush Wellman, 53, 166
Brant, J. R., 216, 237
Buckeye Reef, 259, 263
Buckeye shaft, 262, 268
Buckhorn shaft, 373
Buel, George, 216
Bulgarians, 231
Bullet Mine, 87
Bullion Beck Mine, *313*, 342, 343, 345, 346, 348–53

Bullion Coalition mining camp, *312*
Bunker, Edward, 61, 252, 439, 440
Bureau of Mines, 147, 293, 419–22
Burford, Blanton W., 149
Burke, James, 210, 215
Burke pits, 210
Burleigh, Charles, 41
Burmester, Utah, 107, 108, 113
Bush, George W., 84, 160, 385
Butte, Montana, 102, 259, 277, 284
Byram, A., 363

C

Cactus Mine, 364, 371
California, 4, 11, 18, 38, 42, 51, 53, 59, 62–64, 66–69, 72, 75, 118, 132, 135, 154, 198, 200, 205–7, 210, 214, 220, 223, 226, 232, 245, 251, 254–56, 258, 261, 262, 264, 273, 275, 297, 321, 323, 329, 335, 341, 343, 367, 373, 374, 376, 383, 405
California gold rush, 222
California Volunteers, 58, 63, 64, 222
Call, C. J., 110
Cameron, Hal, 161
Cameron, J. C., 369
Camomile, L. E., 332
Camp Douglas, 58, 59, 64–66, 319
Camp Floyd, 63, 223, 344
Camp Floyd Railroad, 223
Camp Kearny, California, 135
Camp Relief, 63
Campbell, A. G., 363, 365
Campbell, Andrew, 63
Campfire Girls, 239
Canada, 18, 24, 52, 121, 146, 164, 231, 265, 323
Cane Creek, 122, 165
Cane Creek Mine, 122
Cannon, Angus, 90, 204, 349
Cannon, George Q., 204, 349
Canyon Fuel Company, 52
Canyon Range, 6, 342
Capel Salt Company, 117
Carbide lamps, 212
Carbon County, Utah, 3, 38, 126–38, 141, 205, 211, 232
Carbon Fuel Company, 205
carbonate, 11, 13, 14, 27, 101, 276, 282, 344, 356, 370, 372, 395, 397, 399, 404, 411, 412
Carbonate Mine, 366, 374
Cardiff, Wales, 288
Cardiff Fork, 28, 288, *303*
Cardiff Mine, 28, 289, 290, 292, 294, *304*
Cardiff Mining and Milling Company, 288
Cargill Salt, 112, 428
Carlisle, Henry C., 292
Carmel, 19
carnotite, 144–46, 148, 149, 152, 153, 155, 431
Carr, Joe, 54, 60, 270
Carr Fork (mining camp), *189*, 223–25, 229, 231, 245
Carr Fork Bridge, 245
Carter, Joseph, 216
Carter, Julian, 72
Carter, W. H., 65
Carter Oil, 383
Cassidy, Butch, 388
Cassidy, Ed, 108
Cassidy, John, 261
Castle Dale, Utah, 127
Castle Gate, Utah, 44, 46, 128–31, 133, 134, 138, 232
Castle Gate Mine, 44, 46, 128–31, 133, 134, 138, 232
Castle Valley, 127–29, 138, *174*, *175*
Cathedral of the Madeleine, 324
Catholic Church, 266, 346, 355
Catholic Diocese of Salt Lake, 267
Cave Mine, 370
CCC. *See* Civilian Conservation Corps
Cedar Canyon, 200, 204, 212, 214, 218
Cedar City, Utah, 32, 38, 61, 126, *183*, 197–219
Cedar Mercantile Company, 206
Cedar Mountain, 7, 215, 218, 434–36
Cedar Plaster Company, 218
Cedar Valley, 199, 214, 236, 342
cement, 37, 117, 209, 218, 219, 402
Cenozoic Era, *10*, 20, 25, 418
Centennial-Eureka Mine, 342, 356
Center Creek, 199
Central City, 274, 275, 280, 281
Central Pacific Railroad, 67
Centurion Mines, 358
CF&I. *See* Colorado Fuel and Iron Company
Chalk Creek, 126
Chambers, R. C., *308*, 321–24
Champion Silver Mine and Mining Company, 343, 345, 367
Channing, R. H., 234
Chapter 11 bankruptcy, 211
Chase Brass and Copper, 241
Chemical Salt Production Company, 109
Cherry Creek, 344
Chesapeake Mine, 210
Chevron Oil Company, 385, 389–91
Cheyenne, Wyoming, 5, 25
Chicago, Illinois, 92, 146, 218, 290, 372, 374, 376
Chicago Silver Mining, 73
Chief Consolidated Mine and Mining Company, 31, 43, 342, 355, 357, 358
China, 131, 149, 231, 320
China Bridge, 320
China Joe, 266
Chinatown, 265, 320
Chinese, 131, 231, 264–66, 320, 328, 347, 348, 351
Chinese Exclusion Act, 266, 347
Chinle formation, *10*, 32, 34–36, 152, 153
Chino Mines, 230, 244
Chipita Wells oil deposit, 386
Chisholm, Robert B., 274–77
chloride, 23, 101, 102, 104, 113–15, 124, 125, 217, 218, 250, 260, 409
Chloride Point Mine, 72, 79, 80
Chloride Products Incorporated (CPI), 118
Christy Mining and Milling, 262
chrome, 37
Chuar group, 33
Chugach Mountains, 24
Chung, Sing, 347
Church of Jesus Christ of Latter-day Saints (Mormon), 37, 39, 59–64, 66–70, 73, 74, 76, 77, 80, 86, 101, 102, 105, 106, 108, 119, 120, 126, 129–31, 135, 139, 160, 198, 199, 203, 204, 217, 220, 250–59, 265–73, 277, 285, 286–88, 297, 318–20, 342, 344, 347, 348–52, 355, 362, 365, 366,
CIO. *See* Congress of Industrial Organizations
Circle Cliffs, 155, 165
Cisco, Utah, 152
Citicorp, 211
City of Zion, 365
Civil War, 39, 58, 80, 273
Civilian Conservation Corps (CCC), 206
Clark, Joseph, 63
Clark, W. C., 263
Clarke, William B., 107
Clarkson Smelter, 345
Clay, Ben, 222
Clayton, William, 59, 60, 68, 69
Clayton Peak, 22
Clay's bar, 222
Clean Air Act, 54, 396
Clear Creek mining camp, 128, 130, 131, 133, 138
Clearfield, Utah, 110
Clemson, George W., 371
Cleveland, Ohio, 270, 291
Climax Molybdenum, 163
Clive, Utah, 54
Clover Valley, Utah, 252, 254
coal, 3, *10*, 19, 32, 33, 36–40, 44–47, 50–53, 55, 57, 65, 77, 85, 126–42, 152, *171–78*, *183*, 197, 199–205, 207, 211–15, 232, 267, 323, 347, 366, 370, 386, 387, 392, 394–412, 414, 416
Coal Canyon, 127
Coal City mining camp, 138
Coal Creek, 199, 202, 212, 214

Coal Creek Canyon, 199
Coal Mine Basin, 386, 387, 454
coal miners, 44, 47, 52, 126, 130, 132–38, 140, 200
Coalbed, Utah, 126
Coalville, Utah, 33, 38, 126
cobalt, 156
Codman, John, 77, 426
Coeur d'Alene Mine Owner's Association, 349
Cohen, Hartwig, 227
Cold War, 56, 64, 149
Collins, A., 351
Collins Chairlift, 296
Collins Gulch, 92, 93, 282
Colorado, 13–15, 19, 25, 33–35, 38, 42, 43, 53, 121, 128, 131, 134, 135, 149–51, 154, 159, *179*, 219, 226–28, 232, 233, 235, 263, 275, 277, 284, 286, 295, 296, 343, 380, 382, 383, 386, 388, 389
Colorado Fuel and Iron Company, 204, 206, 207, 209, 210
Colorado Plateau, 34, 142, 144, 145, 147, 149, 150, 154, 157, 161, 162, 164, 211, 378
Colorado Raw Materials, 157
Colorado River, 54, 123, 147
Colorado Springs, Colorado, 227
Columbia Mine, 129, 139, 205–9
Columbia River, 22
Columbia Steel Company, 129, 205–7
Columbian Iron and Mining Company, 206–8
Columbus Mine, 289, 291
Columbus Consolidated Mining Company, 287–89, 303
Columbus Day, 238, 239
Comer, G. W., 352
Comet Mine, 364
Commercial mining camp, 364
Commonwealth Mine, 382
communist, 137, 330, 337, 437
company stores, 46, 134, 136, 138, 139, *174*, 202, 333, 336, 350, 352
company towns, 130, 138–40, *176*, 225
Compass Mineral Group, 113
Comstock Lode (Nevada), 273, 275–77, 290
Comstock Mine (Utah), 207, 209–11, 255, 268, 269, 297, 323, 329, 335, 341
Congar, O. H., 255, 273, 274
conglomerate, 11, 14, 25, 32, 33, 144, 151, 155, 397
Congo, 144, 146
Congress, 71, 137, 146, 147, 154, 160, 252, 255, 256, 281, 283, 359, 387
Congress of Industrial Organizations (CIO), 207
Connellsville mining camp, 128
Connor, Johanna, 59, 60, 66, 95
Connor, Kate, 60, 67, 68, 96
Connor, Maurice J., 79
Connor, Patrick Edward, 58–80, 94, 220, 222, 252, 253, 272, 275, 297, 319
Connor, Patrick Edward, Jr., 80
Connor Springs, 75
Consolidated Fuel Company, 129
Consolidated Implement Company, 371
Construction Trade Unions, 207
Consumers Mutual Coal Company, 129, 130, 136
Continental Alta Mining Company, 288, 290, 306
Continental Bank, 161
Continental Mines and Smelters Company, 42
Continental Oil, 383
Continental Uranium Company, 148
continuous miner, 52, 397, 399, 402, 404, 412
Controlled Metals, 164
Cook, Frank, 118
Cooke, Jay, 363
Coombs, Jack, 161
Cooper, Joe, 147, 148
Coors, Adolphson, 387
copper, 3, *10*, 12, 20, 22, 24, 27, 29, 30, 35, 37, 39, 40, 42–44, 46, 53, 55, 71, 90, 112, 132, 147, 148, 170, *184–94*, 216, *221–36*, 239–49, 258, 262, 264, 286, 288–90, 293, 297, 318, 334, 336, 342, 344, 345, 354, 358, 362, 371–77, 379, 382, 393, 395, 405, 407
Copper Belt Railroad, 223, 228
Copper Chief Mine, 364
Copper Gulch, 364
Copper Heights mining camp, 225
Copper Hill, *185*, *188*, *193*, 229, 230, 240
Copper Zone Company, 218
Copperfield mining camp, 225, 231
Copperton, Utah, 43, 54, *193*, 229, 244, 246–48
Copperton concentrator, *193*, 246–48
Coral Sea, 11
Corinne, Utah, 68, 278, 424, 425
Cornish, 270, 323–25, 347, 397, 445
Cornish pump, *307*, 324, 325, 397, 445
Cornwall, 82, 322
Corry, A. R., 216
Corry, Andrew, 212, 357
Corry Hotel, 213, 435
Corry mine, *183*, 213
Cortez, 81
Coschina, Joseph M., 259
Cosmopolitan Restaurant, *302*
Cottonwoods canyons, 272–*317*
Cottonwood Creek, 41, 127, 212, 280, 287
Cottonwood Wash, 165
Cougar Spar Mine, 217, 218
Coursa, James D., 74
Courthouse Rock, 34, 346
Cousin Jacks, 231
Cowboy Mine, *314*, 389
CPI. *See* Chloride Products Incorporated
craton, 5, 25, 398
Creole Mine, 215, 329
Crescent Bar, 365
Crescent Eagle Well, 122
Crescent Mine and Mill, 322, 325, 329
Crescent Tramway, 325
cretaceous, 6, 7, *10*, 19–21, 25, 28, 32, 33, 36, 211
Cricket Mountains, 359
Cripple Creek, Colorado, 227, 286
Crismon Mammoth Mine, 345
Croatian, 132
Cromer Brokerage Company, 162
Crosby, George H., 256, 268
Crouch, George W., 70
Crow, Benjamin Brown, 439
Crowther, Henry M., 42
Crystal Creek, 362
Crystal Peak, 124
Crystal White Salt Company, 108, 109
Cub Scouts, 239
Cullen, Matthew, 363
Curie, Marie, 144, 157
Curie, Pierre, 144, 157
Cutler, John C., 203, 434
Czechoslovakia, 144, 156, 159

D

D. George Harris and Associates, 112, 113
D&RGW. *See* Denver & Rio Grande Western Railroad
Dacotah Mine, 281
Daggett County, Utah, 50, 420
Dailey, Mike, 329
Dalley, Parley, 213
Dalton, William, 279
Daly, John J., 323, 324, 329, 331
Daly, Marcus William, 29, 43, 327, 332
Daly Mine and Mill, 321, 324, 332
Daly-Judge Mine, 324, 329
Daly-West Mining Company, 29, 43, 322, 324, 325, 326, 327, 331
Dark, Charles W., 350
Daughters of Rebekah, 355
Davenport locomotive, 229
Davenport Mine, 276
Davis, Erwin, 277
Davis, L. D., 75
Davis County, Utah, 53

Davis horse, *304*
Davis Mine, 214
Day, James F., 274, 275
De Courcy, 74
De Lamar, Joseph R., 226, 227
Dedrichs, Joseph, 215
Deep Creek oil deposit, 5, 21, 22, 386
Deer Trail Mine, 15
Deer Valley, *308*, 327, 329, 333
Deerlodge Canyon, 215, 435
Defense Department, 357
Del Rio, Texas, 91
Delaware Mine, 73, 237
Delitch, Radovan, 326
Delta, Utah, 53, 101, 124, 166, 170
Delta Mine, 35, 53, 155, 156, 161
Demetrakopoulos, George, 135
Denmark, 120, 265, 320
Denver, Colorado, 54, 122, 123, 128, 154, 224
Denver and Rio Grande Western Railroad (D&RGW), 38, 122, 127–29, 131, 132, 228, 232, 236, 325, 333, 344, 389
Department of Defense, 166
Department of Energy, 54, *181*
Department of Health, 54
Department of Labor, 135
Dern, George, 333
Deseret Generation and Transmission, 386
Deseret Iron Company, 200, 201, 433
Deseret limestone, *10*, 28, 98, 110, *173*, *299*, *300*, *306*, *310*, *312*, *313*, 366
Deseret Livestock Company, 106, 108, 109
Deseret Salt Company, 107, 109, 110
Desert Mound, 205, 206, 208
Devil's Castle, 282
Devonian Period, *10*, 25, 28, 33
Dewey, Albert P., 71
Diamond, Utah, 343, 345
Diamond Crystal Salt, 111, 112
Diamond Mountain, 390
Diamond Salt, 106
Dinkeyville mining camp, 225
Dinosaur National Monument, 382
Dirty Devil River, 35, 154
"Dirty Harry" atomic explosion. *See* Shot Harry
disaster, 44, 46, 55, 82, 130, 131, 133, 141, 164, *173*, 218, 282, 326–38, 390
Dividend, Utah, 256, 279, 348, 356, 358
Dixie Junior College, 159
Dodge, Grenville, 65–67, 258
Dodge, W. H., 325
doghole miners, 148, 161
Dolores River, 14, 121, 149
Domtar, 111
Dorman, J. Eldon, 130
Dougan, J. L., 383
Douglass, Earl, 382
Dove Creek, Colorado, 150, 151
Dove Creek Mercantile Store, 151
Doyle, Henry, 158, 159, 432
Dragerton mining camp, 139
Dragon, Utah, *314*, 389
Dragon Wash, 386
Drake, Zeth, 216
Dream Mine, 86
Drum, Lt. Col., 52, 64, 405, 436
Dry Canyon, *97*
Dry Fork, 54, *190*
Dry Valley, 165
Dubofsky, Melvin, 352
Duchesne County, Utah, 50, 55, 125, 378, *380*, *381*, 382–85, 390
Duffin Mine, 264
Duggan, John, 349–51
Dugway, Utah, 30, 90
Duhigg, Ada, 239, 438
Dumke, Zeke, 162
Duncan, Chapman, 203
Duncan, Homer, 203
Duncan Mine, 207
Durango, Colorado, 146
Dyer Mine, *317*, 379
dynamite, 41, 121, 285, 351, 413

E

Eagle and Bluebell Plant, 43, *312*, 342
Eagles social club, 355
Early, W. J., 258
East Tintic District, 31, *312*, 342, 347
East Tintic Mountains, 21, 27
Echo, Utah, 325, 344
Echo Canyon, 25, 33, 126
Egypt, 383
Einstein, Albert, 145
Eisenbud, Merrill, 157
elaterite, 390
electric, 31, 39, 41–43, 51, 52, 75, 77, 98, 140, 154, 164, *189*, *191*, 206, 209–11, 236, 240, 244, 245, 287, 289, 375, 376, 394, 399, 401, 404
Elephant City Mine and mining camp, 70
Elk Horn Saloon, *301*
Elko, Nevada, 5, 12
Ellingsworth, R. E., 373
Elliott, H. R., 216
Ellsworth, E. W., 257, 432
Elmore, Ohio, 53
Ely, Nevada, 230, 254, 261, 439, 441
Ely Mill, 345
Ely Mining district, 256
Emery County, Utah, 50, 53, 55, 121, 127, 140, 141
Emery, Albion, 325
Emma Company, 277, 279, 280, 284
Emma Hill, 275, 276, 286
Emma Mine, 28, 92, 259, 277–90, 305
Emma Silver Mining Company, 277–90
Emma Silver Mining Company, Ltd., 278
Empire Canyon, 323, 324, 328
Employment Security Office, 338
Enewetak Atoll, 147
Engalitcheff, Nicholas, 326
England, 64, 82, 131, 200, 265, 278, 320, 322
English, 68, 69, 130, 131, 134, 200, 231, 232, 235, 239, 277, 284, 347
Enoch, Utah, 202, 433
Enterprise, Utah, 216, 217
Entrada sandstone, 7, 15
Environmental Protection Agency (EPA), 54, 55, 209, 210, 358
Episcopalians, 349
Equity Company, 383
erosion, 6, 12, 13, 15, 22, 24, 36, 147, 197
Escalante, Utah, 119
Escalante Desert, 219
Escalante Silver Mine, 216, 217, 435
Escalante Valley, 208
Etaugh, George, 79
Eureka, Nevada, 78
Eureka, Utah, 21, 27, 31, 36, 43, 67, 73, 77, 119, *312*, *313*, 335
Eureka City Mining Company, 355
Eureka Gulch, 343
Eureka Hill, 342, 343, 345, 346, 353
Eureka Home Dramatic Company, 355
Eureka Hotel, 71, 74, 346, 353
Eureka Lilly, 31
Eureka Tunnel Company, 77, 78
Evans, Beverly, 385
Evans, Morris R., 289, 292
Evening Star Mine, 150
Excelsior Mine, 210

F

Fairbanks, Avard, 244
Fairmont Lode, 80
Farmington, Utah, 11, 24, 27
Farmington Bay, 107
Farmlands Industries, 391
Fay mining camp, 215, 372
Federal Bureau of Investigation (FBI), 156
Federal Trade Commission, 109
Federal Uranium Company, 161
Ferlin, Carl Bernard, 291
Ferris, John S., 258, 259
Ferron Creek, 127
Ferry, Edward R., 320–22, 324
Ferry, William, 322, 329
Ferry Mansion, 322

Fife, Lou, 218
Fife, Otto, 218
Finland, 131, 231, 449
Finley, R. J., 372
Finns, 130, 131, 134, 231, 429, 430, 449
First National Bank of Utah, 277
Fischer, R. P., 150
Fish, Hamilton, 278
Fish Springs, Utah, 85
Fisher, Cyrus M., 278
Fitch, Walter, 355
Fitchville, Utah, 28, 355
Fitzgerald, J. Terrance, 267
Five Mile Pass, 31
Flagstaff Mine, 28, 34, 92, 276, 281, 283, 319–22
Flagstaff Mountain, 280
Flemetis, Steve, 135
Flora Mining and Smelting Company, 70
Floral Springs Water Company, 262
Florence & Raving Mining Company, 88
Flying Diamond Company, 164
Fontana, California, 51, 207
Fontana mill, 210
Foote, B., 119
Ford, William E., 144, 431
Forest City, Utah, 282, 283
Forest Service, 296
Fort Bridger, Wyoming, 64
Fort Douglas, 58, 60, 79, 135, 220, 253, 272
Fort Duchesne, *316*, 387, 388
Fortuna Mine, 372, 373, 452
France, 144, 231
Franklin, Utah, 60
Frazer and Chalmers Company, 353
Free and Accepted Masons, 349
Freed, David, 109
Freedom Mine, 148, 430
Freeman Gulch mining camp, 225
Freemasons, 328
Fremont, John C., 101
Frenchman's Flat, 149, 160
Frisco mining camp, 43, 44, 225, 359–76
Frisco Mining and Smelting, 370
Frisco Peak, 359, 362
Frog Valley. *See* Deer Valley
Frogtown mining camp, 225, 231
Froiseth's mining maps, 69
Frost, Burr, 199
Fruita, Colorado, 389
Fuller, Elijah, 257
Fundadoro Mine, 152

G

Galbraith, William W., 107
galena, 66, 67, 223, 275, 329, 337, 363, 372, 401
Gallagher, John P., 70, 71, 73, 425
Gamma Mine, 155
Gampion Mine, 366
Garbutt, Frank, 145, 431
Gardison limestone, 28
Gardner, Archibald, 59, 220
Gardner, Carl, 387
Gardner, Reed, 214
Gardner, W. W., 220, 329
Garfield, James A., 77, 121
Garfield County, Utah, 15, 33, 121
Garfield Electrolysis, 43
Garfield Smelter, 244
Garfield, Utah, 228, 229, 286, 292
gas, 33, 34, 37, 50, 51, 53, 55–57, 101, 102, 114, 124–26, 130, 147, 157, 168, *173*, 197, 211, 217, 225, 271, 327, 378, 382–85, 391, 395, 396, 399, 401, 404, 406, 411, 429, 454
Gashwiler, S. F., 261
Gay Deceiver Mine, 73
Geiger counter, 146, 150, 151, 153, 155, *179*, 219, 294
Gemini Mine, 342, 343, 346
Gemmell, De, 226, 227, 229, 234
General Conference, 200, 204
General Connor Tunnel, 76, 77, 79, 80
General Dodge Mine, 65
General Electric, 164
General Garfield Mine, 329
General Land Office, 359
Geneva Steel, 51, 207–11, 218, 357
gentile (non-Mormon), 61–63, 65, 66, 68–70, 77, 80, 204, 253, 258, 267, 344, 351
George Arthur Rice and Company Bank, 346, 353
Georgia, 226
Georgia Mine, 73
Germania smelter, 41, 285
Germans, 231
Germany, 117, 144, 156, 159, 231, 240, 242, 265
Gibbs, Lorin W., 109
Gibson (Trooper), 63
Gilbert Development, 210
Gillilan, J. B., 349
Gilson Mountains, 342, 387
gilsonite, 37, 53, *314–16*, 387–91
Girl Scouts, 239
Glen Allen Mine, 329
Glen Canyon Dam, 214
Glenco Canyon, 329
Glenwood Cemetery, 328
Godbe, William S., 69, 72, 109, 281, 370
Godiva Mine, 342, 346, 349
Golconda Mine, 15, 25
gold, *10*–12, 20, 22, 24, 27, 29–31, 37–40, 42, 43, 53, 58–61, 64, 65, 68, 69, 71, 74, 75, 79, 81, 85–89, 142, 144, *183*, 197, 198, 215–19, 222–25, 229, 231, 234, 241, 244–46, 248, 256, 258, 272, 275, 280, 282–85, 289, 291, 292, 294, 297, *317*–19, 323, 326, 333, 334, 337, 340, 342–44, 353, 354, 359, 362, 364, 371–76, 378, 379, 382, 393, 395, 396, 398, 399, 403, 405, 407–10, 417
Gold Dome Mine, 215
Gold Hill, 11, 323
Gold Mission, 58
Gold Springs Mining District, *183*, 216
Golden Gate mill, 227
Golden Porphyry Mine, 290
Golden Reef Mine, 366
Goldring, Robert J., 71
Gordon Creek, 132, 138
Goshen Valley, 286, 342, 344, 345, 450
Gottheimer, Albert, 278
Government Mine, 386
Grace, Idaho, 236
Graff, Arnold, 212–14
Graff Kleen Koal, 212
Graff Point, 212
Graff Tipple Road, 213
Grambo, Margaret, 302
Gramlich, J. W., 150
Grampian Mine, 363
Gran Quivira Mine, 81
Grand Canyon, 11, 15, 84
Grand County, Utah, 53, 147, 164
Grand Haven, Michigan, 321
Grand Junction, Colorado, 146, 150, 152, 154, 163, *314*, 388
Granite Mining District, 375, 452
Granite Mountain, 197, 205, 207
Grant, Albert, 278
Grant, Heber J., 217
Grant, Jedediah M., 318
Grant, Ulysses S., 278–80
Grantsville, Utah, 107, 108, 112
Grasselli zinc mill, 329
Graves, William H., 262
Great Basin, 6, 20, 22, 36, 63, 67, 68, 72, 74–79, 198, 202, 362, 367
Great Blue limestone, *10*, 30, 31
Great Depression, 38, 56, 136, 239, 240, 271, 291–93, 296, 333, 357, 431
Great Salt Lake, 23, 38, 39, 63, 67, 68, 96, 101, 102, *103*, 105, 107, 110–13, *114*, 115–17, *185*, 418
Great Salt Lake Minerals and Chemicals Corporation, 115
Great War. *See* World War I
Great Western Iron Mining, 203
Great Western Salt Company, 120
Greece, 131, 132, 231, 238
Greek Camp (mining camp), 225, 231
Greek Orthodox, 132
Greeks, 5, 130, 132, 135, 136, 139, 155, 225, 231–35, 238, 239, 347, 348, 430

Green, Earnest F., 213
Green, Leslie, 206
Green Monster Mine, 319
Green River, 13, 34, 125, 127, 165, 379
Greenwich Pharmaceutical, 164
Grizzly Gulch, 284, 288, 306
gross state product (GSP), 48–50, 56
Grouse Creek, 11
GSL. *See* Great Salt Lake Minerals and Chemicals Corporation
GSP. *See* Gross State Product
Guggenheim, Meyer, 43
Guggenheim, Simon, 56
Guggenheim family, 43, 230, 236, 286
Gulf of Alaska, 19, 33
Gulf of Mexico, 13, 16, 19, 33
Gulf Oil, 383
Gunnison Valley Salt Company, 120
Gusher, Utah, 383, 388
Gust, Leonard, 237
Gustaveson Oil Company, 271
Gwilliam Brothers Salt Company, 107
gypsum, 18, 19, 37, 119, 142, 197, 218, 219, 402

H

Haggin, James Ben Ali, 42, 56, 321
Hagglund, Joel, 91
Haight, Isaac, 202
halite, 119, 122, 402
Hall, Clarence H., 375
Hamblin, William, 61, 62, 251, 252, 439
Hamilton, Samuel, 257, 418
Hamlin Valley, 215
Hampton, Benjamin Y., 370
Hand, Wayland D., 87
Hanford, Washington, 146
Hanks, Ebenezer, 203
Hanks, Ephraim, 319
Hanksville, Utah, 154, 155
Hanson, George, 352, 424
Happy Jack Mine, 147, 148
Hardy, August, 268
Hardy Salt Company, 106, 109, 111
Harrington-Hickory Consolidated Mining Company, 365, 376
Harris, D. George, 112, 113
Harris, Fay, 372
Harris, Thomas, 372
Harrisburg Mining District, 257–61, 439
Harrison, Peter, 261
Harrison, Richard, 199
Harrison House, 264, 265
Harvey, Amelia D., 79
Hashimoto, Edward Daigoro, 132
Hatch, Ira, 254
Hatfield, William, 345, 346
Hawkes, Samuel, 363
Hawkeye Mine, 28, 329
Hay Canyon, 386
Hayes, Alfred, 92
Hayes, Dan, 151, 152, 160
Hazen, Scott W., 147
HBCH. *See* Highland Boy Community House
Hearst, George, 42, 56, 321, 322
Heaston, Al, 225, 237
Heaston Heights mining camp, 225
Heaton, Dan, 222
Heaton, Geneva, 206
Hebron, Utah, 256
Hecla Mining Company, 163, 217, 290, 295, 339, 340
Heikes, V. C., 42
Heines, John W., 281
Helper, Utah, 127, 129, 132, 134, 135, 137–39, 141, 347
Helper Railroad and Railroad Museum, 141
Hemby, Jim, 373
Hempstead Lode, 62, 65, 253, 439
Henderson, Nevada, 113
Hendrickson, W. H., 371
Henry Mountains, 165, 211
Hermosa group, 101, 121
Hewitt, Freed, 367, 451
Hiawatha Mine and mining camp, 129, 138, 139, 232
Hickory Mine, 365, 375, 376, 452
Hidden Splendor Mining Company, 35, 162, 163
Higbee, Samuel A., 216
Higgins, Charles C., 354
Highland Boy, 224, 225, 229, 231, 234, 237–39
Highland Boy Community House (HBCH), 238, 239
Highland Boy Mine, 224, 225, 234
Hiko, Nevada, 255
Hill Creek oil deposit, 386
Hill, Joe, 91, 98
Himalayas, 4
Hinchman, Charles, 263
Hines, Edward, 285
Hiroshima, Japan, 146
Hite, Utah, 53, 87, 148
Hitler, Adolph, 146
Hoffman, Charles, 261
Holaday, Duncan, *180*
Holden, Liberty, 291
Holden, Utah, 291
Hogle-Kearns Investment Group, 113
Holland, 231
Holmes, Edwin, 326
Holt, George A., 216
Holt, Henry D., 216
Holy Cross sisters, 266
Homansville Mill and Smelter, 345, 346
Homestake formation, 197
Honaker Trail, 33
Hooker Electro Chemical Company, 109
Hooper, William H., 256, 364
Hope Mine, 390
Horn Silver Mining Company, 43, 72, 250, 259, 362–64, 366, 367, 370–76
Horse Creek, 362
Hotel Utah, 232
House Range, 25
Housel and Hopkins Salt Company, 111
Houston, Texas, 154
Howland Mine, 287
Hoyt, Will L, 209
Hudson, William T., 154, 226
Huish, C. E., 354
Huleatt, W. P., 147
Humboldt County, Nevada, 262
Humboldt Mine, 79, 80
Humbug Mine, 85, 347
Humphrey, A. E., 390
Hunt, W. A., 348
Huntington, Utah, 161, 429
Huntington Canyon, 128, *172*
Huntington River, 127
Huntley, Dwight B., 284
Huntsman, Orson, 256
Huntsville, Utah, 90
Hurricane Cliffs, 211
Hurricane fault, 6, 8, 32
Hurricane Katrina, 385
Hussey, Warren, 277
Hyde, A. E., 349–51
Hyde, Hiram, 351

I

Idaho, 22, 38, 45, 60, 101, 117, 124, 209, 217, 226, 232, 235, 236, 339
igneous, 14–16, 19–22, 24–32, 36, 42, 197, 215, 318, 343, 372, 393, 397, 401, 403, 408, 412, 413, 416
Iliff, Thomas C., 348
Illinois, 198, 227, 265, 275, 276, 372
Illinois Tunnel Company, 279
IMC Global, 113, 115
IMC Kalium Ogden Corporation, 115
IMC Salt, 113
IMCU. *See* Iron Manufacturing Company of Utah
immigrants, 38, 60, 130–32, 134–36, 141, 142, 231, 232, 236, 239, 277, 284, 321, 323, 324, 347
Imperial Mine, 362
Independence Mine, 216
Independent Coal and Coke Company, 55, 129, *175–78*
Independent Order of Odd Fellows (IOOF), 328, 346, 349
Indian Creek, 165, 372, 452
Indian Queen Mine, 362
Indiana, 150, 263
Indianola, Utah, 25, 33

Industrial Workers of the World (IWW, wobblies), 91, 232, 235, 330, 403
Inland Crystal, 38, 105–8, 110, 120
Inland Salt, 38, 105
International Society of Mine Safety Professionals, 248
Interstate Venture Company, 218
Intrepid Mining, LLC, 123
IOOF. *See* Independent Order of Odd Fellows
Iowa, 239
Ireland, 200, 231, 265, 320
Irish, 231, 266, 270, 277, 323, 324, 347, 348, 352, 440
iron, *10*, 19, 37, 38, 42, 53, 55, 65, 66, 126, 129, 145, *182*, *183*, 197–217, 219, 229, 293, 296, 342, 346, 369, 370, 372, 379, 382, 392, 397, 405, 409, 411, 412, 417
Iron Blossom Mine, 342, 346
Iron City, Utah, 203, 204
Iron County, Utah, 19, 38, *182*, *183*, 197, 198–219
Iron County Coal Company, *183*, 213
iron curtain, 149
Iron Manufacturing Company of Utah (IMCU), 204
Iron Masters, 200
Iron Mission, 126, 202, 212, 432–34
Iron Mountain, *182*, 197, 203, 204, 206–8, 434
Iron Springs, *182*, 198, 203–7, 210, 211, 432
Iron Springs Gap, 205, 207
Ironton, Utah, 205–7, 344, 448
Irontown, Utah, 129, 203, 204, 433
Isom, Kate, 206
Italians, 130–32, 134, 135, 231, 235, 238, 239, 348, 430, 437
Italy, 131, 231, 265
Ivers, James, 325, 329, 333, 336, 340, 447
IWW. *See* Industrial Workers of the World

J

Jack Rabbit, Nevada, 108, 204
Jackling, Daniel Cowan, 43, 56, *184*, *192*, 226–30, 234, 235, 241, 244, 245, 438
Jackson, Thomas, 135
Jackson, Wyoming, 124
Jack, James, 38
Jacob, L., 120, 282
Jacobs, Hiram S., 73
Jacobs, Hyman, 261
Jacobs and Sultan's store, 265
Jacobs Smelter, 70, 74
Janney, Frank G., 43, 261
Jap Town/Japanese mining camp, 132, 139, 225, 231
Japan, 52, 131, 146, 207, 208, 242, 293
Japanese, 139, 207, 232, 235, 239, 347, 430
Jeffs, D. W., 374
Jenkins, Richard, 119
Jennie Mine, *183*, 216
Jeremy and Company, 105
Johnny Mine, 215
Johnson, H. T., 216
Johnson, J. W., 73
Johnson, John S., 284, 285
Johnson, Joseph E., 267
Johnson, Lyndon B., 339
Johnston, Albert Sidney, 63, 201
Johnston's army, 202
Joklik, G. Frank, 247
Jolly Jack Mine, 162
Jones, Allan E., 163, *182*
Jones, John Pidding, 202, 203
Jones, William, 63
Jones, York L., 210
Jordan, John, 216
Jordan Lode, 222
Jordan River, 23, 59, 67
Jordan Silver Mining Company, 60, 222
Jordan Ward House, 59, 222
Jordanelle Reservoir, 325
Josephine Mine, 87
Juab County, Utah, 42, 44, 50, 53, 85, 166, 203, 218, 342, 343, 346, 348, 355, 367
Judge, John, 65, 324, 325, 329, 331
Judge Building, 324
Judge Memorial High School, 324
Judge Mine and Mill, 29, *308*, 324, 439
Judge Mining and Smelting, 324, 331, 332
Judge Tunnel, 324, 341
Julia Dean Mine, 223
Jumbo Mine, 216
Jumboldt Mine, 76
Jupiter Mine, 329
Jurassic Period, 7, *10*, 15, *16*, *18*, 25, 34, 36, 101, 119, 124, 144, 197

K

Kaiparowits Plateau, 33, 57, 211
Kaiser Chemicals, 111
Kaiser Steel Company, 51, 207
Kanab, Utah, 211
Kanarra Mountain, 212, 213
Kanarraville, Utah, 214
Kane County, Utah, 33, 55, 211
Kansas City, Missouri, 105, 107, 227, 239
Kate Connor Gold and Silver Mining, 60
Kearns, Thomas, 56, 113, 286, 322, 323, 325, 326, 329, 331
Keetley, John, 28, 324, 331, 332
Keetley Drain Tunnel, 324, 331, 332, 339
Keiley, Denis, 348
Keith, David, 56, 323, 325, 329, 331, 445, 446
Keith, Minor Cooper, 290
Keller, Helen, 91
Kelly, Annie, 350
Kemple, John, 257, 258
Kenilworth Mine and mining camp, 129, 132, 135, 136, 138, *176*, *178*
Kennecott Copper Corporation, 247
Kennecott Research Center, 246
Kennecott Utah Copper, 53, 54–55, 57, 112, *189*, *192*, *193*, *221*, 235, 241–49, 293, 358
Kennecott Utah Copper Charitable Foundation, 249, 439
Kennecott Wire and Cable Company, 241
Kennedy, James, 320
Kennicott, Robert, 235, 236
Kentucky, 259, 265, 441
Kerr McGee, 113
Keystone Mine, 329, 346, 349, 353, 385
Kibbe, A. Payne, 162
Kiely, Dennis, 266
Kildale, M. B., 318
Kimball, Heber C, 318
Kimball, Heber P., 72
Kimberly, Utah, 27
Kimmerle, H. T., 145
King, Clarence, 284
King David Mine, 362, 371, 372, 374, 376
King Midas Mine, 162
Kingston Mine, 223
Kinner Mine, 263, 264
Kirkman limestone, 34
Klaproth, M. H., 144
Knickerbocker and Argenta Mining and Smelting, 64
Knight, Jesse, 42, 56, 85, 129, 286, 347, 352
Knight Investment Company, 288
Knights of Columbus, 355
Knights of Labor, 131, 133, 328, 351
Knights of Pythias, 349
Knights of Robert Emmet, 355
Knightsville, Utah, 347
Knoerr, A. W., 149–51
Knolls ponds, 113, 114
Knutsford Hotel, 228
Koal Kreek, 212, 214, 435
Kolasakis, Steve, 135
Kolob coalfield, 211, 213
Korean War, 52, 149
Koyle, John Hyrum, 85, 86
Kunz, Ron, 83, 427, 432

L

La Pierre House, 69
La Platta (ghost town), 90
LaBarge oil field, 124

labor unions, 133, 231, 232, 235, 243, 349
Ladies Aid Society, 355
Lake Bonneville, 22–24, 115, 117, 123
Lake Crystal Salt Company, 108, 111
Lake Huron, 22
Lake Mead, 160
Lake Michigan, 22
Lady of the Lake Mine, 320, 329
Lake Point, Utah, 67, 68, 106, 109–13
Lake Uinta, 378
Lambourne, George W., 331, 332, 447
Lambrides, Vasilios, 234
laramide, *10*, 20, 25, 28, 378, 404, 412
Lark, Utah, 30, 335
Lark limestone, 30
Larsen, Shorty, 161, 162
Las Vegas, Nevada, 38, 149, 160, 202, 431
LaSal Mining and Development, 163
LaSal Mountains, 34
Last Chance Mine, 72
Latey, John H., 72
Latham, Milton S., 262
Latuda Mine, 129, 138
Lawrence, Henry W., 280
Lawson, John, 72, 446
Lay, Elzy, 388
LDS. *See* Church of Jesus Christ of Latter-day Saints
lead, *10*, 12, 14, 15, 20, 22, 24, 27–32, 38–40, 42–44, 51, 53, 54, 60, 62–65, 72, 74–76, 79, 81, 90, 113, 134, 159, 163, 197, 199, 202, 215, 217–19, 223–25, 229, 234, 241, 257, 273–76, 282, 283, 286–89, 291, 294, 297, 318, 319, 321, 322, 329, 333–39, 342, 345, 349, 354, 355, 357–59, 362, 363, 366, 371–76, 379, 382, 393, 401, 403, 404, 407, 412, 433, 439, 452
Leadmine mining camp, 225
Leadville, Colorado, 42, 284
Leahy, Phil, 146
Leamaster, William, 216
Leany, William, 257
Lee, J. Bracken, 244
Leeds, Utah, 32, 251, 257–59, 261, 262, 264, 267, 268
Leeds Mining and Milling, 262, 264
Leeds Mining Company, 261
Lees, Samuel, 281
Lehi, Utah, 218, 418
Lehi and Tintic Railroad Company, 344
Leigh, Samuel F., 218
Leland, Warren, 64
Leonore Mine, 77
Leslie Salt Company, 109, 111–13
leukemia, 160
Lessing, Lewis, 375
Lewis, John L., 136, 137
Lewiston Peaks, 29
Lexico Chemical and Mining Company, 389
Leyson Mine, 212
Liberal Party, 68, 80
Liberty Fuel Company, 129
limestone, 7, *10*–12, 14, 15, 19, 27–35, 37, 119, 197, 200, 205, 215, 275, 292, 294, *305*, 318, 343, 363, 373, 395, 397, 404, 412
Lincoln County, Nevada, 203, 435, 440
Lincoln Mine, 374
Linda Mujer Mine, 152
Lindgren, Waldemar, *312*
Lindsay Hill Pit, *182*
Lindsey, David A., 167
Linmar Mine, 385
Lion Coal Company, *171*, *177*
Lion Creek, 150
Lisbon Uranium Company, 162
Lisbon Valley, 34, 165
lithium carbonate, 101
Lithium Corporation of America (Lithcoa), 114
lithospheric plates, 4, 5, 6, 9, 24, 397, 404, 406, 408, 415
Little Bell Mine, 332
Little Chief Mining Company, 355
Little Cottonwood Canyon, 9, 22, 24, 27, 28, 35, 41, 42, 70, 71, 74, 92, 272, 277–*317*
Little Cottonwood Creek, 41, 280, 287
Little Cottonwood Mining District, *274*, 277–*317*
Little Cottonwood Transportation Company, 290
Little Mountain, 115
Little Pinto Creek, 203
Little Salt Creek, 119
Little Salt Lake, 198, 199
Little Muddy Creek, 212
Little Wild Horse Mesa, 155
Little Willow, 27
Little Willow Creek, 290
Live Yankee and Bellorphan claims, 282, 291, 292
Liverpool, England, 322
Livingston, M., 71
Locke, G. E., 233
Logan Canyon, 85
London, England, 70, 247, 248, 278, 279, 284, 443
Lone Peak, 213
Lone Tree Mountain, *183*, 213
Long Tom Saloon, 365
Lopez, Raphael, 91
Los Alamos, New Mexico, 146
Los Angeles, California, 198, 205, 218, 324, 366, 371, 372
Los Angeles and Salt Lake Railroad, 228, 334
Lost Jack Write Mine, 87
Lost Rhoades Mine, 89, 427
Louder, James N., *301*
Louma, Abe, 131
Low, James, 366
Lowe, Jack, 376
Lowe Peak, 29
Loyal Order of Moose, 328
Lubbock, Henry, 262, 265, 268
Luce, Ben D., 352
Lucky Bill Mine, 320
Lucky, Ohio, 113
Lucky Strike Mine, 162
Lulu Mine, 375
Lund, Utah, 205
lung cancer, 156, 159, 160
Lunt, Henry, 435
Lutherans, 349
Luxor Capital Partners, 211
Lyman, E. Ray, 209
Lynndyl, Utah, 166
Lyon, James E., 275–79

M

Mabey, Charles R., 46
Macfarlane, John, 226
Macfarlane, Kenneth, 214
Macfarlane Mine, 212
MacNeill, Charles M., 43, 227, 228
Magcorp, 55, 114
Mageras, Georgia Lathouris, 238
Maggie Mines, 89, 259, 262
magma, 27, 197, 403, 405
Magna, Utah, 43, 53, 54, *187*, 228, 238, 240, 241, 247
Magna mill, 240, 241, 245, 246
magnesium, 23, 37, 53, 55, 101, 102, 104, 113–15, 117, 125, 395, 399, 428
Magnolia Trading Company, 128
Major Evans Gulch, 292
Malmborg, Charles H., *305*
Mammoth Basin, 343
Mammoth lode, 215
Mammoth Plaster and Cement Company, 218
Mancos shale, 34
Manhattan Project, 146, 147, 152
Manly, Lewis, 254
Manning Canyon, 25
Mao Tse-tung, 149
Maple Canyon, 212
Markagunt Plateau, 211
Markham Gulch mining camp, 225
Marks, Anna, 346
Marsac, Sophie de, 321–25
Marsac mill, 321–24
Marsden, Albert, 213
Marsden, L. N., 209
Marsh Creek, 22
Marshall, Charles, 68

Marshall, John A., 44, 226
Marshall Islands, 147
Marsing, June, 155
Marvin, Robert F., 295, 445
Mary Ellen Gulch, 282, 291, 292
Marysvale, Utah, 15, 56, 158
Mason, James, 321, 322
Masons, 349, 355
Mathews, W. W., 345
Matzatzal craton, 25
Maud Withey steam engine, 325
Maxfield, Blair, 216
Maxfield limestone, 27, 28, 286
Mayflower Mine, 28, *310*, 325, 329, 337, 339, 340
Maynard, Ed, 259
McCarry, E. A., 375
McCarty, B. F., 72
McCleve, Nancy Jane, 259
McCluskey, William, 254
McCormick, Bill, 151–54, 160, 162
McCornick, W. S., 287, 329, 353, 354
McCrystal and Company, 346
McDaniel, Council, 109
McDonald, J. H., 216
McGarry, E. A., 375, 376
McGarry shaft, 375
McHenry Canyon, 28, 321, 329
McHenry Mine and Mill, 320–22, 325, 329
McKay Mine, 276
McKean, James B., 73
McLaughlin, David, 321, 322
McLean, John, 128
McMullen, Brigham Y., 257
McNalley, 259, 263
Meadow Valley, 60–62, 251, 253, 256, 261
Means, Allen Hay, 292
Melich, Mitch, 154
Mendocino Junction, 20, 405
Mercur, Utah, 29–31, 35, 42, 43, 226, 227
Mesa Verde, 33, 34
Mesozoic Era, 6, *10*, 15, 25, 28, 34, 35, 378
Metal Producers Company, 376
Methodism /Methodist, 239, 348, 349, 252
Mexican Hat, Utah, 54
Mexicans, 91, 233, 348
Mexico, 13, 16, 19, 25, 33, 91, 146, 217, 226, 230, 241, 265, 290
Meyer's Hotel, 346
Mi Alma Mine, 152
Mi Amorcita Mine, 152
Mi Corazon Mine, 152
Mi Vida Mine, 152, 153, 160–63, 432
Michigan, 111, 290, 291, 369
Michigan Bunch, 321, 322
Michigan-Utah aerial tram, 306
Midas fault, 29, 162
Middle East, 385
Midvale, Utah, 43, 54, 225, 286, 287, 290, 294
Miles Number One oil well, 383
Milford, Utah, 15, 70, 359, 362–66, 370, 371, 373, 376, 377
Milford stamp mill, 365
Mill Canyon, 71
Millard County, Utah, 101, 123, 140, 218, 359, 366
Miller, Alta, 238, 239
Miller, Jacob, 282
Miller, William, 282
Miller Hill, 282
Miller Mining and Smelting Company, 282, 283, 287
Miller-Beggs mill, 329
Milner Spur, 205
Minard, F. H., 228
Mine Safety and Health Administration (MSHA), 167, 406, 408
Mineral Fork, 27, 285, 290
mineral resources, 3, 37, 65, 66, 71, 197
Mineral Resources International, 115, 428
Mineral X, 146
Miner's Day Parade, 328
Miners Hospital, 329
Miner's National Bank, 277
Miners Union, 137, 268, 269, 349
Minneapolis, Minnesota, 216
Minnesota, 154, 210, 216, 229, 281
Mississippi, 200
Missouri, 107, 109, 198, 226
Missouri School of Mines, 227
Moab, Utah, 3, 14, 34, 54, 87, 101, 122, 145, 147, 149, 150, 152, 154, 156, 161
Moab Drilling Company, 162
Moab fault, 7
Moab Salt, Inc., 123
Modena, Utah, 215, 216
Moenkopi formation, 15, 35
Moffat, Utah, 387, 388
Mohrland Mine, 129, 139, *174*
molybdenite, 43, 242, 244, 246
molybdenum, 22, 30, 37, 53, 163, 241, 242, 244, 246, 248, 294, 379
Monitor lode, 275, 276
Monk, James, 285
Montana, 38, 45, 51, 57, 59, 77, 79, 80, 102, 107, 131, 232, 257, 259, 263, 277, 284, 323, 383
Montello Salt Company, 117
Montezuma Creek, 145
Montezuma Mine, 121, 165, 276
Monticello, Utah, 53, 87, 145–48
Montreal Mine, 374
Montrose, Colorado, 121
Monument uplift, 25
Monument Valley, 158, 165
Moore, Charles, 334
Moore, Thomas, 71
Morehouse Canyon, 362
Morgan, Nicholas G., 50, 244, 420
Mormon. *See* Church of Jesus Christ of Latter-day Saints
Mormon Battalion, 258
Morning Star Mine, 150
Morris, Durham, 208
Morrison, John G., 91, 98
Morrison formation, *10*, 15, 25, 34, 36, 144, 147, 149–52
Morton Salt Company, 105, 106, 107, 108, 110, 112, 113, 120
Moscow Mine, 374, 376
Mother Jones, 134
Mount Panacker, 253
Mount Rainier, 27
Mount Timpanogos, 14, 18, 24
Mountain Cheaf Mine, 288, *304*
Mountain Lake Mining District, 273, 320
Mountain Lion Iron Project, 209, 211
Mountain Meadows, 62, 201
Mountain Mesa Uranium Company, 163
Moyer, Charles, 233, 234
MRI. *See* Mineral Resources International
MSHA. *See* Mine Safety and Health Administration
Mt. Calvary Cemetery, 323
Mt. Emmons, 385
Mt. Olivet Cemetery, 323, 324
Muddy River, 35, 155
Mujer Sin Verquenza Mine, 152
Murphy, John R., 69
Murphy, Metalliferous, 84, 257
Murphy, Miriam, 363
Murray, Eli, 371
Murray, Utah, 41, 44, 91, 225, 238, 286, 287, 291, 292, 294
Murtha, Harry, 374
Mutual Improvement Association, 355
Mutual Metal Mines, 290
My Adventures with Your Money, 290
Myers, W. D., 353
Myrin, Alarik, 385
Myton Bench oil deposit, 386

N

Nagasaki, Japan, 146
Nagler, Robert, 280, 281
Naildriver Mine, 329
National Bituminous Code, 137
National Bulk Carriers, 109, 112
National Cancer Institute, 159
National Coal Company, 129
National Industrial Recovery Act (NIRA), 137
National Lead, 113, 163
National Miners Union (NMU), 137

Native Americans, 96, 101, 120, 239
natural gas, 33, 37, 51, 53, 126, 378, 382, 391, 401
Naturita, Colorado, 146
Navajos, 15, *17*, 35, 142, 158
Nebo Salt Manufacturing Company, 119
Nebraska, 121
Needles, California, 62
Nelson, A. H., 107
Nelson, Eliza, 328
Nelson Company, 107
Nephi, Utah, 18, 101, 119
Nevada, 5, 11–16, 18, 20, 25, 38, 45, 61, 67, 71, 73, 74, 77, 78, 84, 113, 117, 118, 149, 159, 160, *183*, 203, 204, 215, 216, 219, 226, 230, 232, 241, 244, 251, 252, 254–57, 262, 268, 269, 271, 273, 275, 277, 292, 321, 323, 343, 344, 363, 367, 371, 418
Nevada Consolidated Copper Corporation, 239, 241
Nevada Mines, 244
New Deal, 137
New Harmony coal field, 211
New Haven, Connecticut, 278
New Jersey, 216
New Jersey Zinc, 163
New Mexico, 25, 146, 217, 230, 241, 439
New Park Mining Company, 337–39
New York, 64, 74, 114, 157, 216, 236, 262, 263, 265, 275, 277–79, 282, 286, 290
New York and Utah Mining and Prospecting Company, 273, 287
New York Central Railroad, 277
New York City, New York, 42, 64
New York Mine, 329
New York Mining Company, 329, 364, 371
Newfoundland Mountains, 25
Newfoundland Pond, 113
Newhouse, Samuel, 42, 43, 137, 224, 225, 230, 329, 364, 371
Newhouse Hotel, 137
Newman, Ershel, 292
Niagara mining camp, 225
Nichols, Bert, 373
Nielson, Bud, 161
Nile River, 15, 59
Nims, Fredrick, 321, 322
NIRA. *See* National Industrial Recovery Act
NL Industries, 113
NMU. *See* National Miners Union
Nobel, Alfred, 41
Nobel Corporation, 112
Noble Furnace, 201
Noranda Mines Limited, 55, 247, 340, 341
Norem, Bill, 340
Normal fault, 6–8, 407
Norsk Hydro, 113
North American Salt Company, 112, 113
North Carolina, 114, 263
North Lily Mining Company, 358
North Point, 105
North San Rafael, 165
North Shore Limited, 110, 115, 428
North Star District, 70–73
North Star Mine, 274, 276
Northland Mine, 325
Norway, 113, 265, 320
Norwegian, 295
Nova Scotia, 231, 323
Nuckolls, Steven F., 72
Nucla, Colorado, 159
Nugget sand deposit, 28, 34, 222, 223
Nunn, Lucien L., 42

O

O. K. Mine, 376
Oak Leaf Mine, 255, 273
Oak Ridge, Tennessee, 146
Occupational Safety and Health Administration (OSHA), 209
Odlum, Floyd B., 156, 162
Ofer Mine, 215
Ogden, Richard, 268
Ogden, Utah, 111, 115, 117, 129
Ogilvie, George B., 59, 220, 222
Ohio, 53, 113, 233, 247, 265, 270, 416
oil, 33, 37, 41, 43, 51, 53, 55, 57, 64, 75, 111, 122, 124, 125, 150, 151, 161, 163, 168, 216, 219, 225, 247, 250, 271, 378, 382–86, 388, 389, 391, 395, 401, 404
Oklahoma, 356, 383
Oklahoma slope, 356
Olcott, William J., 229
Old Hickory Mine, 374
Old Maggie mill, 262
Old Spanish Trail, 198
Oliver, B. P., 79, 80
Oliver, W. L., 261
Ontario, Canada, 323
Ontario Canyon, *307*, 322, 328, 329
Ontario Gulch, 321
Ontario Mine, 42, 83, *307–9*, 321–27, 330–32, 337, 341
Ontario Mining Company, 215, 321, 331
Ontario rocks, 22, 29
OPEC. *See* Organization of Arab Petroleum Countries
Open pit, 54, 168, *193*, 206, 407
Ophir anticline, 29, 30, 32
Ophir Canyon, 29, 31
Ophir Hill Mine, 29, 30
Ophir Mining District, 67, 69, 72, 73, 259
Ophir, Utah, *10*, 12, 27, 29–32, 69, 80, 89, 215, 259
Oquirrh Basin, 13, 14, 18, 25, 28, 29
Oquirrh formation, 29, 34
Oquirrh Mining, 70
Oquirrh Mountains, 5, 20, 29, 30, 42, 54, 59, 60, 63, 67, 220, 222, 228, 342
Order of the Eastern Star, 355
Ordovician, *10*, 27, 362
Oreana, Nevada, 73
Oregon Short Line, 344
Orem, Utah, 51, 207, 294
Organization of Petroleum Exporting Countries (OPEC), 383, 384
Osborn, L. D., 72
Osgood, Leonard S., 74
OSHA. *See* Occupational Safety and Health Administration
Outokumpu furnace, 55
Overland Park, Kansas, 113
Oxford Saloon, 365
O'Connor, William, 296
O'Keefe, J. P., 245
O'Laurie, Dan, 154
O'Loughlin, Matthew, 268, 270

P

P. R. Spring oil deposit, 386
Pack-Allen Mine, 386
Packard, Miland O., 128
padrone, 132, 232, 235, 407, 437
Pahranagat Mining District, 254–56, 271, 275
Pahvant Mountains, 6, 8
Paiute, 251–54, 439–42
Paleozoic Era, 6, 7, *10–12*, 25, 27, 31, 32, 35, 167, 418
Palladon Venture, Ltd., 211
Pan Hellenic Grocery Store, 232
Panaca, Nevada, 61, 62, 253, 254, 440
Panama Canal, 207, 240
Pangea, 13, 15, 29, 418
Panic of 1873, 283
Panic of 1893, 286, 349
Panic of 1907, 129, 288
Paradox Basin, 13, 14, 25, 29, 33, 34, 102, *116*, 121, 122, 125, 165, 429
Paradox Valley, 14
Park City, Utah, 12, 15, 21, 22, 27–29, 34–36, 42–44, 83, 91, 120, 273, 285, 287, 289, *307–11*, 318–41, 344, 382, 390
Park City Community Center, 328
Park City Mining and Smelting Company, 332
Park City Mining District, *274*, 318, *319*, 320, 329–33, 337, 340
Park City Mountain Resort, 319
Park City Public Library, 328
Park City Ventures, 340
Park Galena Mining Company, 329, 337

Park Utah Mining Company, 331–38
Parker, I. N., 120
Parley's Canyon, 327
Parley's Park Mine, 320, 321, 329
Parliament, 278
Parowan, Utah, 126, 198, 199
Parsons, Ralph M., 113
Paunsagunt Plateau, 211
Paxton, John, 65
Payne, George, 107
PCS. *See* Potash Corporation of Saskatchewan
Pearl Harbor, 207
Pearson, Marshall, 367
Peerless mining camp, 129, 138
Pen Salt, 109
Pennsylvania, 128, 134, 137, 224, 265, 376
Pennsylvanian Period, *10*, *13*, 25, 28, 30, 34, 101, 121, 125
penny stocks, 161, 163, 164
Penrose, R. A. F., 228
Penrose, Spencer, 43, 227
Permian Period, *10*, 15, 25, 28
Peru, 150, 265
Peruvian Mines, 145
Peruvian Gulch, 282, 291
Petrified Forest, Arizona, 15
petroleum, 14, 20, 33, 34, 53, 247, 293, 383, 387, 443
Phelps, E. R., 374
Philadelphia, Pennsylvania, 43, 263, 274
Philippine Islands, 290
Phillips, Samuel G., 70
Phoenix mining camp, 225, 231, 357
phosphate, 15, 378, 390, 391, 408
Picacho Range, 70
Pick, Vernon, 154–57, 160, 161, *180*
Picnic Canyon, 29
Pine Valley, Utah, 32
Pineview, 124
Pinto Iron Works, 203
Pinto Mining District, 204, 211
Pinyon Ridge, 319
Pioche, Francois Louis Alfred, 256
Pioche, Nevada, 61, 71, 84, 215, 251, 254, 256–62, 266, 267, 271, 363, 367
Pioche and Bullionville Railroad, 204
Pioche Mine, 205–6,
Pioneer Ridge, 319, 322
Pioneer Smelting Works, 64
Pisco Mine, 152
pitchblende, 144, 148, 153, 154, 156
Pitchfork Creek, 362
Piute, Utah, 42, 53
plate tectonics, 4, 318, 415, 418
Pleasant Valley, Utah, 128, 131, 133, 138
Pleasant Valley Coal Company, 128, 133
Pleasant Valley Railroad Company, 128
Plutus Mine, 342
Poison Creek, 325
Poison Springs, 155
Poland, 146, 265
Polar Mesa District, 145, 147
polygamy, 70, 204, 285
Poole, H., 273
Pope deposit, 382
Porter, Maggie Tolman, 89
Portneuf River, 22
Portugal, 231
potash, 14, 37, 53, 101, 107, 115, 117–19, 121–23, 362, 409, 428, 429
Potash Corporation of Saskatchewan (PCS), 118, 123
potassium, 23, 101, 102, 104, 115, 117, 118, 124, 125, 167, 400, 409
Poulson, Blaine, 120
Poulson, Jewell, 120
Poulson, Wallace, 120
Poulson Brothers Salt Company, 120, 121
Powell, John Wesley, 142
Powell, Robert A., 87
Pratt, Addison, 198
Pratt, Parley P., 198, 320
Precambrian Era, 5, 6, 9, *10*, 20, 25, 27, 28, 33, 362, 409
Preston, Idaho, 22
Preston, William B., 349
Preuss Salt Zone, 101, 124, 429
Preuss Sandstone, 124
Price, Dan, 385
Price, Fred, 288
Price, George F., 62
Price, Tommy, 237
Price, Utah, 3, 32, 87, 129, 132
Price Canyon, 128, 134, 138
Price River, 127
Price tunnel, 28
Pride of the West Ledge, 258
Prince of Wales Mine, 92, 277
Progressive Metalworkers' Council, 337
Promontory Point, 39, 110, 111, 115
prospectors, 84–87, 151, 153, 154, 156, 158, *179*, 215, 250, 252, 257, *305*, 323, 373, 375
Prospector Mine, 148
Proterozoic Period, *10*, 11, 418
Provo, Utah, 129, 133, 205, 236, 285
Provo Canyon, 42
Provo level, 22, 24
Provo River, 23, 325
Prussia, 231
Public Health Service, 156–60
Pueblo, Colorado, 204, 206, 210, 343
Puerto Ricans, 242
Pulsipher, William, 252, 439
Purgatory Hill, 263
pyrite, 215, 294, 364, 409

Q

Quadmetals Mine, 372, 374
Quail Creek, 32, 258, 259
Quaker Crystal Salt Company, 111
Quandary Mine, 63, 67, 76, 77
quartz, 59–61, 66, 197, 215, 220, 222, 372, 373, 409
Quaternary, *10*, 20, 22
Queen Esther Mine, 329
Queen of the Hills and Flavilla Mine, 72, 73
Queen of the West Mine, 209
Quincy Mine, 290, 322, 324
Quinney, S. Joseph, 296

R

Raddatz, Emil, *312*
Raddon, Sam, 330
radiation, 150, 155–60, 164, *179*, 409
Radiation Exposure Compensation Act, 160
radium, 144, 145, 156
Radorock Incorporated, 163
Raft River, 11, 24, 60
railroad, 38, 39, 42, 43, 51, 52, 56, 65, 70, 72, 73, 75, 76, 80, 102, 109–11, 114, 117, 123, 126–29, 132, 135, 138, 141, 146, 152, 163, *179*, *189*, *190*, 203–10, 215, 218, 222, 223, 227–29, 231, 232, 236, 234, 240, 241, 243, 244, 275–77, 282–84, 290–93, 295, 296, *306*, *315*, 320, 325, 329, 332, 333, 337, 344, 346, 347, 356, 362, 363, 365–67, 370, 371, 373, 389, 390
Rainbow, Utah, 389
Rains, L. F., 205
Rains, Utah, 138
Rains Mine, 138
Ranchers Exploration and Development Corporation, 217
Rand and Rankin, 218
Rangley, Colorado, 53
Rankin, Robert B., 218
Rathbone, E. H., 354
Ratliff, J. H., 390
Rattler Mine, 366
Rattlesnake Mountain, 149
Raven Mining Company, 387
Raven Ridge, 386
Ray Consolidated Copper Company, 230
Ray, George, 218
Ray and Gila Valley Railroad, 241
Ray Mine, 244
Raymond, Rossiter W., 41, 69, 256, 277, 284

Raymond and Ely Mine, 254, 261, 439
RealSalt, 121
Red Creek quartzite, 11
Red Cross, 239
Red Hill, 213
Red Jacket Mine, 90
Red Narrows, 32
Red Rock Pass, 22
Redd family, 145, 148
Redmond, Utah, 18, 19, 101, 119–21
Redmond Clay and Salt Company, 121
Redmond Minerals Inc., 121
Redwood City, California, 66, 78
Regulator Mine, 284, 285
Reid, Robert K., 59, 272, 436
Reilly Chemicals at Wendover, 113, 114, 117–19
Relief Society, 355
Renco Inc., 114
Reno, Nevada, 343
Rescue Consolidated, 271
Rettich, Fritz, 285
Revolution Mine, 276
Rex Ore body, 208, 211
Reynolds, C. C., 268
Rhoades, Caleb, 87
Rhoades, Gale, 89
Rhoades, Thomas, 87
Rhoades Mine, 87–89, 379
Riblet Tramway Company, 292
Rice, George, 216
Rice, George Arthur, 346
Rice, George Graham, 290
Rice, John, 264
Rice, Sophie, 354
Rice, V. W., 325
Rich, Charles C., 198
Richards, Franklin, 108, 200
Ridgeley, William Barrett, 290
Rife, C. E., 354
Rifle, Colorado, 146
Right Hand Canyon, 212, 214
Rim Rocks oil deposit, 386
Rio Grande Western Railroad, 38, 122, 127, 128, 131, 132, 228, 325, 333, 344, 389
Rio Tinto PLC, 248
Rio Tinto Zinc (RTZ), 247, 248
Ritz Salt Company, 110
Roark, Hugh, 350
Rob Roy Mine, 372
Robb, William, 257
Roberts, D., 261
Roberts, David, 109
Roberts, Rhett, 121
Roberts Mountain, 25
Robinson, Earl, 92
Robinson, James E., 376
Robinson, L. L., 261
Rock Creek, 379
Rock Springs, Wyoming, 137, 391
Rockefeller, William, 42, 56, 225
Rocky District, 370, 377
Rocky Mountain Fur Company, 101
Rocky Mountains, 8, 13–15, 25, 29, 36, 69, 378, 419
Rodinia, 11, 13, 15
Rogers, Henry H., 42, 225
Roosevelt, Franklin D., 137, 145
Roosevelt, Utah, 384
Roosevelt oil fields, 383
roscoelite, 145
Rosie the Riveter, 243
Roundy, Emil, 218
Rowley, Utah, 55
Roxie Mine, 76, 79
Royal Crystal Salt Company, 105, 108, 110, 120
Royal Street Land Company, 340
RTZ. *See* Rio Tinto Zinc
Rural Free Delivery, 231
Rush Lake, 63, 75
Rush Valley, 62–70, 74–77, 79, 424
Rush Valley lode, 63
Ryan, Dennis, 363, 364
Ryan, James, 363
Rydalch, W. C., 72

S

S. F. Phosphate, 391
Safeway, 357
Salduro Salt Marsh, 101, 102, 117, 118
Sale, Thomas C. W., 440
Salina, Utah, 119, 120
salt, 13, 18, 19, 23, 37–39, 42, 44, 53, 63, 87, 88, 96, 101–25, 142, 164, *185*, 198–216, 220, 225, 226, 234, 238, 241, 247, 250, 258–60, 269, 280, 293, 298, *306*, 318, 335, 338–40, 343–45, 350, 351, 364, 371, 376, 395, 402
Salt Cave Hollow, 119
Salt Creek, 119
Salt Creek Canyon, 119
salt flats, 101, 102, 109, 115–18
Salt Lake and Alta Railroad, 290
Salt Lake and Los Angeles Railroad, 106
Salt Lake City, 11, 14, 27, 54, 59, 61, 62, 64–68, 71–80, 91, 98, 117, 118, 126, 128, 129, 132, 137, 147, 154, 156, 158, 159, 161, 162, 203, 205, 217, 223, 228, 232, 252–54, 259, 263, 266, 267, 272, 273, 277, 281, 285–88, 295, 296, 319, 322–29, 331, 341, 353, 359, 363, 375, 383
Salt Lake City Council, 322
Salt Lake Stock and Mining Exchange, 288–90
Salt Lake Stock Exchange, 162
Salt Lake Winter Sports Association, 296
Salt Lake, Sevier Valley and Pacific Railroad, 72
Salt Valley, 122, 145
Salt Wash, 144
Saltair, 105, 108, 112
Salzdetfurth, 115
Sampson Mine, 324
San Andreas Fault, 4, 20, 405
San Bernardino, California, 198
San Francisco, California, 66, 113, 208, 223, 260, 261, 277, 321, 343
San Francisco Chemical Company, 390
San Francisco Mines, Inc., 375
San Francisco Mining District, 44, *313*, 359–77
San Francisco Mountain Range, 359
San Jose, California, 261
San Juan County, Utah, 53, 121, 125, 145, 147, 164
San Juan Mountains, 289
San Juan River, 13
San Miguel, Colorado, 121, 149
San Pedro, Los Angeles and Salt Lake Railroad, 344
San Rafael Swell, 20, 25, 34, 35, 127, 154, 165
Sanderson, David, 254
sandstone, 7, 11, 13, 14, 15, *17*, 29, 32, 34, 35, 84, 85, 119, 124, 144, 147, 150, 151, 153, 201, 202, 250, 257, 259, 261, 262, *300*, 362, 385, 393, 409, 411, 415
Sanpete County, Utah, 38, 55, *116*
Sanpete Valley, 126, 127, 130
Sanpitch Mountains, 126
Santa Clara, Utah, 61, 252, 254
Santa Fe Railroad, 163
Santaquin, Utah, 85, 344
Sap Gulch, 237
Sardine Canyon, 85
Saul, Edward, 160
Savage Shaft, 263, *301*
Sawmill Canyon, 362
Saxony, 144
Scandinavia, 231
Scanlan, Lawrence, 266, 267, 348
Schenck, Robert C., 278–80
Scheuner, T. L., 203
Schilling, J. D., 229
Schoo, Edward, 374
Scofield, Utah, 44, 46, 128, 130, 131, 133, 136, 138, 139, *173*, 232
Scottish, 200, 284
Scotia Mine, 345
Scotland, 200, 265, 320
scrip, 138, 281, 296
Sears Utah Salt Company, 106, 107
SEC. *See* Securities and Exchange Commission
Second Cavalry, 60, 61, 63
Securities and Exchange Commission, 123, 125, 161, 162, 164

Sedalia, Missouri, 227
sediment, 14, 15, 23, 29, 33, 118, 124, 198, 392, 398, 411, 415
sedimentary, 6, 11, 12, 14–16, 20, 24, 27, 29–31, 167, 362, 393, 394, 397, 398, 400, 401, 404, 411, 412, 414, 415
Seegmiller, Pratt, 148, *183*, 197
Serbian, 132
Seven Cities of Cibola, 81
Seven Mile Canyon, 122, 165
Sevier Coal, *10*
Sevier County, Utah, 50, 52, 55, 121, 141
Sevier Desert, 101, *116*, 124
Sevier fault, 6, 8
Sevier Lake, 101, *116*, 123, 124
Sevier Mountains, 19, 20, 21, 25, 28, 36, 412
Sevier Valley Salt Company, 120
Sevier Valley, 72, 73, *116*
Shaky, Robert, 259
Shamrock Mine, 374
Sharp, Joseph C., 233, 234, *315*, 327, 388
Shauntesy, 264
Shauntie mining camp, 70
Shaunty, Utah, 257
Shay steam engine, *186*
Shea, P. H., 268, 346
Sheep Rock Mining and Milling Company, 374
Shell Oil Company, 385
Shenandoah Mine and mining camp, 70, 71, 73
Shepherd, 261
Sherman Silver Purchase Act, 349
Sherwood, Stephen, 61, 62, 252–54
Shinarump conglomerate, 144, 148, 151, 152, 155
Shirts, Peter, 203, 252
Shoaff, P. L., 71
Shoal Creek, 216, 217
Shoebridge mill, 119, 345
Short, C. A, 216
Short Line deposit, 208
Shoshoni, 58
Shot Able, 149, 150
Shot Harry, 159–60, *181*
Shumway, Arah, 145
Shumway, Harris, 145
Shumway, Lee, 145
Shumway, Seth, 145
Shutt, Samuel S., 254
Sibley County, Minnesota, 281
Sierra Nevada Mountains, 20, 25
Sigurd, Utah, 18, 19
Sihler, Charles A., 374
Silliman, Benjamin, 278
Silsbee, J. L., 118
Silurian Period, *10*, 28, 31
silver, *10*, 12, 15, 20, 22, 24, 27–29, 31, 32, 36–40, 42–44, 53, 58–69, 71–77, 79, 81, 83–86, 90, 92, 102, 107, 120, 156, 165, 197, 198, 203, 212, 213, 215–19, 222–25, 229, 241, 244, 246, 248, 250–57, 259–78, 282–85, 287–91, 294, 295, 297, *299–303*, *305*, *310–12*, *317*–19, 321–26, 329–38, 340–45, 349, 350, 353, 357–59, 362–64, 366, 367, 370–76, 379, 382, 393, 395, 398, 399, 401, 404, 405, 407, 409, 412
Silver Butte Hills, 260
Silver City Mine, 42, 343–45, 358
Silver Creek, 321, 325, 329
Silver Flat Mine, 260, 262
Silver Fork, 289, *305*
Silver Hill, 223
Silver King Coalition Mine, 329, 331–39, 341
Silver King Consolidated, 330
Silver King Mine (Park City), 29, 285, *310*, *311*, 324–26, 329–38, 341, 382
Silver King Mine (Rush Valley), 67, 72–74, 76, 77
Silver King Mine Company, 329, 382
Silver Mountain, 254
Silver Peak, 218
Silver Queen Mine, 63, 72, 74
Silver Reef, Nevada, 203
Silver Reef, Utah, 32, 36, 83–85, 165, 212, 213, 250–71, *300–302*
Silver Shield mining camp, 225
Silver Springs, Utah, 318
Silveropolis Mine, 67, 68, 71, 259
Simmons, Louis, 319
Simons and Company, 69
Sioux, 281
Sisters of the Holy Cross, 348
skiing, 272, 295, 296, *308*, 341, 445
Skliris, Leonidas G., 132, 135, 232, 233, 235, 437
Skougard Mine, 217
Slavic, 135, 238
Slick Rock formation, 151
Slovenian, 132
smelters, 38, 41, 42–44, 53, 55, 62, 64–66, 69, 70, 74–76, *185*, *190*, *194*, 206, 223, 225–27, 229, 233, 242–45, 247, 248,267, 270, 274, 275, 282, 283, 286–88, 292–94, 296, 326, 335, 337, 343, 345, 347, 348, 357, 358, 364, 366, 370–72, 374, 379, 382, 405, 406, 410
Smith, A.T., 110
Smith, Alton, 365
Smith, George A., 199, 200
Smith, H. H., 257
Smith, Jesse A., 59
Smith, John W., 92
Smith, Samuel P., 60, 63
Smith, W. H., 350
Smith Canning Company, 110
Smithfield, Utah, 85
Smoot, Abraham O., 133
Snake Creek, 320
Snake Range, 20
Snake River, 22
Snow and Bryant Company, 216
Snowbird, Utah, 272, 289, 294, 297, *306*
Snow, Erastus, 200, 202, 252, 253, 257, 258, 266
Snyder, George, 320
Snyder, Rhonda, 320
Snyder, Samuel C., 318
Snyderville Basin, 325
Socialists Club of Eureka, 355
Society Islands, 198
Society of the Blessed Virgin Mary, 355
Society of St. Patrick's Church, 355
Socony Vacuum Oil Company, 150
sodium sulfate, 101, 102, 115
SOHIO. *See* Standard Oil of Ohio
Solar Crystal Salt Company, 107
Solar Salt Company, 108–10, 112
Soldier Canyon, 70, 75, 76, 388
Soldier Summit, 128
Solitude Resort, 297
Solvay Process Company, 117, 118
Sonomia, 15, 25
Sons of Utah Pioneers, 244, 423, 433
South Africa, 374
South America, 13, 15, 29
South Camp mining camp, 70
South Columbus Mining Company, 295
South Hecla Mine, 290, 294
South Star District, 70, 92
South Utah Mines and Smelter Company, 364
Southern Pacific Railroad, 110
Southern Utah Power, 214
Southern Utah University, *182*, *183*, 211, 216
Soviet Union, 149, 163
Spanish Fork Canyon, 32
Spanish Trail, 198, 255
Sparta, Greece, 132
Spencer, Walter, 271
Spiro, Solon, 329–31, 341
Spiro Tunnel, *309*, 335
Split Mountain, 382
Spor Mountains, 53, 166, 167
Spring Bay, 110, 111
Spring Canyon mining camp, 74, 77, 129, 132, 136, 138, 347, 430
Spring Glen, Utah, 137
Springville, Utah, 128, 205, 344
Spry, William, 234, 436
St. George, Utah, 32, 61, 62, 83, 160, 202, 253, 254, 257, 265–70
St. George and Washington Irrigation Company, 203
St. George Tabernacle, 266
St. George Temple, 203, 258, 267
St. Joseph's School, 355

St. Louis, Missouri, 109, 111, 156, 218, 432
St. Mary's School, 266
Stagg Field, University of Chicago, 146
Standard Coal Company, 129
Standard Oil Company, 150, 225, 247, 383
Standard Oil of Ohio (SOHIO), 247, 248
Standardville Mine and mining camp, 129, 136, 138
Standish, Mr., 63
Stannard, E. T., 243
Stansbury Basin, 22,111–14
Stansbury Island, 25, 108, 109, 112
Stansbury Salt Company, 109
Stanton, Edwin M., 66, 424
Star District, 70–73, 371, 452
Star of Utah Mine, 329, 337
Starr, H., 69
State Constitution, 133
State Normal School, 227
State Tax, 209
Stateline Canyon, 215, 216, 435
Stateline Mining District, 215, 216
Stauffer Chemical Company, 390
Steele, Axel, 233, 450
Steen, Charlie, 150–54, 156, 157, 160, 161, 162, *179*, 294
Steen, Rector, 42, 321
Steen, Rose, 151–54
Steen's folly, 151, 152
Steptoe, E. J., 62
Stewart, William M., 255, 277–79
stock market crash, 375
Stockton, Utah, 29, 30, 63–67, 69–80, 96, 98, *312*
Stoddard, Arvin, 365
Stoddard Hotel, 371
Stormont Mines and Mining Company, 259, 263, 268–70, *301*
Stormy King Mine, 262
Strategic Metals Incorporated, 375
Strawberry Pinnacles, 390
Strawberry Valley, 323
strikebreakers, 91, 133–36, 232, 233, 335, 350, 411
Stump Creek, Idaho, 124
Sublette, Wyoming, 124
SUFCO Mine, 52
sulfate, 23, 101, 102, 104, 115, 124, 125, 402
sulfur, 22, 44, 55, 122, 202, 211, 225, 226, 248, 283, 362, 373, 392, 401, 405, 409, 414, 417
Sullivan, Dennis, 351
Sullivan, William, 339
Sultan, Louis, 261
Sultana smelter, 282, 285
Summit County, Utah, 34, 38, 44, 320, 334, 335, 338, 339
Sunbeam Mine, 342, 343
Sundance Sea, 36
Sundown Mine, 90
Sunnyside Mine and mining camp, 90, 128–34, 139, *171*, 232, 358, 386
Sunup Mine, 90
Superfund, 54, 358
Sutter's Mill, California, 258
Swansea Mine, 345
Swansea, Wales, 343
Swedes or Swedish, 91, 112, 161, 231, 285, 291, 436
Sweets mining camp, 138
Swindlehurst, Bernett, 372
Switzerland, 52
Swoboda, Alois Phil, 290
sylvinite, 122, 123
syncline, 9
Syracuse, Utah, 106, 107, 110
Syracuse Resort, 107
Syracuse Salt Company, 107
Szilard, Leo, 145

T

Tanner's Flat, 287, *303*
Tavaputs Plateau, 378, 453
taxes, 199, 209, 210, 216, 296, 339, 357, 385
Taylor, John, 73, 204
Taylor, Thomas, 203, 204
Te Quiero Mine, 152
Tecumseh Hill, 260
Tecumseh Mine, 250, 256, 258–60, 262, 263
Telegraph mining camp, 59, 67, 225, 235, 236, 372, 436
Telluride Power Company, 42, 236
Temperance Mine, 70
Tenas, John, 136
Tenderfoot Mine, 325
Terrace Heights mining camp, 225
Tertiary Period, 6, *10*, 20, 21, 25, 34, 35, 167, 197, 362, 409, 415, 418
Tevis, Lloyd, 321
Texas, 91, 149, 150, 286, 383
Texas Gulf Sulphur Company, 122
Texasgulf (TGS), 122, 123
TGS. *See* Texasgulf
Thatcher, Moses, 349
Thaynes Canyon, 322, 325, 328, 329, 335
Thaynes formations, *10*, 15, 28, 34
Third District Court, 71, 75, 76, 78
Thistle, Utah, 32
Thomas Mountains, 167
Thompson, Ezra, 286, 288
Thompson, Utah, 147
Thompson-McNally Mine, 263, 264
Three Mile Island, 164
Three Nephites, 87, 427
Three Peaks, 197
thrust fault, 6, 7, 8, 15, 18, 25, 28, 415
Tilley, Martha, 215
Timco Uranium, 162
TIMET. *See* Titanium Metals Corporation of America
Timpie Salt Plant, 111, 112
Tintic Drain Tunnel, 347
Tintic Historical Society, 357, 358
Tintic mining district, *10*, 12, 27, 31, 35, 42, 43, 85, 129, 285–88, *312*, 342–58
Tintic Mountains, 21, 27, 167, 342
Tintic quartzite, 7, 27, 28
Tintic Smelter, 358
Tintic Standard Mine, 32, *312*, 356, 357
Tintic Valley, 344
Titanium Metals Corporation of America (TIMET), 113
Toledo Mine, 276
Tolton, John Franklin, 370, 451
Tombstone, Arizona, 323
Tommy Knockers, 82, 83
Tomsich Butte, 35
Tonopah, Nevada, 271, 292
Tonopah Mining Company of Nevada, 292
Tooele, Utah, 62, 63, 72, 77, 259, 286, 298, 335
Tooele County, Utah, 42, 43, 54, 55, 79, 80, 119
Tooele Smelter, 286, 294
Topaz Mining Properties, 166
Topaz Mountain, 165, 166
Toquerville, Utah, 257, 268
tramway, 42, 213, 263, 288, 290–93, 295, 296, *306*, 325, 326, 331, 338
transcontinental railroad, 39, 68, 70, 80, 110, 223, 275, 276, 320
Traver, William M., 147, 431
Treasure Hill, 323, 331
Treasure Mountain Resort, 339
Treasury Department, 290
Trenor Park, 279
Triassic Period, *10*, 15, 25, 32, 34, 36, 144
Trixie Mine, 358
Truman, Harry S., 207, 218, 339
Tub Springs Creek, 362
Tucker and Webster mines, 214
Tucker, Floyd, 214
Tucker, Grant, 214
Tucker, Guy C., 213
Tucson, Arizona, 152
Tullidge, Edward, 77
tungsten, 294, 375, 376, 452
tunnels, 28, 29, 32, 60, 67, 68, 72, 76–80, 83, 136, 205, 213, 218, 222, 224, 244,279, 283, 287–91, 294, 297, *305*, *307*, *309*, 320, 323–25, 327, 331, 332, 335, 337, 339, 341, 347, 373–75, 392, 394, 398, 405, 416
Tushar Mountains, 22, 27
Tuttle, Daniel S., 65
Twin Creek, 34
Tyng, George, 286

U

U. S. Vanadium Corporation, 145, 163, 375
UCC. *See* Utah Construction Company
Uintah Arch, 20
Uinta Basin, 20, 33, 34, 53, 101, *116*, 125, *314*, *315*, *317*, 378–91
Uinta Valley, 61
Uintah County, Utah, 50, 87, 125, 378–80, 382–85, 388
Uintah Mining District, 320, 378–91
Uinta Mountains, 5, 8, 11, 15, 25, 87, 124, 318, 378, 379, 391
Uintah Railway, 389
Uintah-Little Cottonwood formation, 22, 125
UMWA. *See* United Mine Workers of America
Uncle Jesse. *See* Knight, Jesse
Union Carbide, 145
Union Concentrator, 325
Union Iron Works, 203
Union Mining District, 258
Union Oil, 383
Union Pacific Coal Company, 38, 128
Union Pacific Railroad, 65, 66, 111, 126, 205, 210, 332–37, 344
unions, 47, 133, 137, 207–10, 231–33, 235, 242, 243, 349, 351, 431
United Kingdom, 52
United Mine Workers of America (UMWA), 134–37, 140
United Park City Mines Company, 338
United States Annual Mining Review, 364
United States District Court, 44
United States Mining Company, 225
United States Reduction and Refining Company, 227
United States Smelter, 44, 287
United States Smelting, Refining and Mining Company, 225, 233
United States Steel Corporation (USX), 206, 210
United Steelworkers of America, 208
University of Michigan, 229
University of Utah, 4, 7, 19, 21, *26*, 40, 244, 246
uranium, 3, *10*, 15, 32, 34–36, 53–55, 142–66, *180*, 219, 294, 297, 379
uranium mill tailings, 164
Uranium Oil and Trading Company, 161
uranium reduction mill (URECO), 162, 163
Uranium Study Advisory Committee, 159
Uravan, Colorado, 146
Urban, Rachel, 327
URECO. *See* uranium reduction mill
USPC. *See* Utah Saludro Potash Company
USX. *See* United States Steel Corporation
Utah Apex Mine and Mill, 43, 91
Utah Bureau of Immigration, 45, 231, 420
Utah Central Coal Company, 128
Utah Central Railway, 68, 128, 325, 366
Utah Coal Producers and Operators Association, 137
Utah Consolidated Mining Company, 225
Utah Construction Company (UCC), *182*, 207–9
Utah Copper Company, 42, 43, 46, 132, *184*, 227, 228–36, 241, 264, 376
Utah Copper Mine, *193*, 228, 234
Utah County, Utah, 42, 44, 129, 343, 348, 357
Utah Division of Oil, Gas and Mining, 168
Utah Eastern Railroad, 325
Utah Federation of Labor, 135
Utah Fuel Company, 38, 135
Utah Industrial Commission, 45, 46, 212
Utah International, 209, 210
Utah Iron Manufacturing Company, 203
Utah Iron Ore Corporation, 206
Utah Lake, 23, 91, 205
Utah Mined Land Reclamation Act, 168
Utah Mining Association, 50
Utah National Guard, 136, 233
Utah Ore Concentrator, 43
Utah Power and Light, 41, 242
Utah Saludro Potash Company (USPC), 118
Utah Securities Commission, 161
Utah Southern Railroad, 203, 282, 344, 363, 365, 448
Utah State Capitol, 244
Utah State Senate, 322
Utah Steel Day, 205
Utah Territory, 58, 60, 69, 75, 80, 98, 220, 222, 250, 251, 254, 255, 278, 352, 359
Utah Valley, 127, 282
Ute, 87, 89 142, 342, 387, 388, 440, 453
UV Industries, 54

V

Valdez, Juan, 91
Van Rensselaer, Schuyler, 263
vanadium, 35, 53–54, 145–48, 156, 163
Vanadium Corporation of America (VCA), 53, 55, 145, 148, 163
Vanderbilt Mine, 264
Vandermark, Jacob N., 252, 253
VCA. *See* Vanadium Corporation of America
Vermillion, Utah, 205
Vernal, Utah, *317*, 379, 382, 385–90, 453–55
Vernon, Utah, 344
Verona Gulch mining camp, 225
Victoria Mine, 43, 342
Virgin Anticline, 32
Virgin City, Utah, 271
Virgin Dome, *271*
Virgin River, *17*, 203, 263
Virginia City, Nevada, 268, 322, 323
Virginia Miner's Union, 268
Vitro Chemical Company, 53, 54, 145, 163
volcano, 22, *26*, 27, 30, 414, 416

W

W. F. & Co's Bank, 80
Wabash Mine, 329
Wagner Labor Relations Act, 47
Wah Wah District, 376
Wah Wah Mountains, 22, 71, 165, 218
Wah Wah Trend, 21
Wakara (Ute Chief), 87, 88
Wales, 131, 200, 223, 265, 288, 289
Wales, Utah, 126
Walker, David, 259
Walker, Joseph R., 202, 259, 280, 281
Walker, M. M., 375
Walker, Matthew, 259
Walker, Rufus, 319
Walker, Samuel, 259
Walker brothers, 68, 259, 262, 277, 280, 281, 296
Walker Brothers Store, 281
Walker House, 71, 79
Walker Mine, 212
Walker War, 201
Walker-Webster Mine, 319, 324, 328
Walkerville, Montana, 79
Wall, Enos A., *184*, 224, 227, 230, 263, 286, 293
Wall Street, 136, 160, 162, 164, 277
"Wall Street of uranium," 162, 164
Wallace, F., 69
Wallace, Idaho, 217, 339
Walters, J., 161
Wanship, Utah, 320
Ward, Edith, 375
Ward, M. M., 375
War Production Administration (WPA), 337
Wasatch, Utah, 42
Wasatch and Jordan Valley mule line, 290

Wasatch County, 80, 320, 336, 339
Wasatch Drain Tunnel, 289, 294, 297
Wasatch fault, 6, 8, *10*, 11, 15, 20–22, 24, 27–29, 33, 36
Wasatch Front, 50, 290, 325
Wasatch Mine, 329
Wasatch Mountains, 6, 7, *10*, 44, 272–76, 278, 282, 284–86, 288–90, 292, 294, 297, 298, 318
Wasatch Plateau, 32, 35, 127, 128, 138
Wasatch Power Company, 287, *303*
Wasatch Store, 129
Washington, D.C., 66, 77, 88, 146, 149, 157, 158, 255, 257, 326
Washington County, Utah, 42, 44, 203, 211, 213, 250, 251, 253–55, 257, 264, 271
Washington Irrigation Company, 203
Watson, George H., 290, 295, 296
Wattis, Utah, 138, *177*
Wattis, W. H., 129
Wattis Mine, 129
Wayne, John, 298
Webb, Arthur, 136
Weber River, 23, 77, 126
Webster, Francis, 435
Webster, Lewis, 214
Webster Mine, 214
Weir, Thomas, 42, 224
Weir Salt, 106
Wellington, Utah, *314*, 388
Wells Fargo, 264, 299, 442
Wellsville, Utah, *316*
Welsh, 126, 130, 288, 347, 350, 415
Wendover, Utah, 112, 113, 117–19
West, Caleb B., 127
West Desert Pumping Project, 113, 114, 428
West Jordan, Utah, 59, 220
West Mountain Quartz Mining District, 59, 61, 220, 222
West Mountains, 372, 375, 452
West Side Auditorium, 92
West Star District, 70
Westenskow, John, *171*
Western Federation of Labor (WFL), 232, 233, 235
Western Federation of Miners, 330
Western Pacific Railroad, 109, 117, 232
Western Union Telegraph, 236
Westinghouse, 164
Westminster College, 322
WFL. *See* Western Federation of Labor
White, Charley, 102
White Canyon, 147, 148, 165
White Elephant Saloon, *186*
White Fawn Mine, 145
White Pine District (Nevada), 67, 257
White Reef, 258, 263, 264
Whiterocks oil deposit, 386
Whitney, Frank, 161
Wiest, Harry, 265
Wilberg Coal Mine, 55
Williams and Cusick Store, 345
Williams, Jesse, 212
Williams Mine, 214
Williams Smelter, 364
Wilson, Woodrow, 91
Winamuck mining camp, 225
Windlass Mine, 85, 86, 276
Wingate Cliffs, 34, 35
Wingate Formation, 15, 35, 152
Winter Olympics, 2002, 248
Winter Quarters Mine, 128, 130, 131, 133, 135, 138, *173*
Wisconsin, 216
wobblies. *See* Industrial Workers of the World
Wolf, Bernie, 157
Wolf, Isaac C., 216
Woman Lode, 60
Woodbury, John S., 216
Woodman, Emma, 276
Woodman, James, 274–77
Woodman's Shaft, 275, 276
Woodmen of the World, 232, 330, 328
Woodruff, Wilford, 350, 351
Woodside Mine, 15, 245
Woodside Gulch, 323, 325, 328
Woodward, Amos, 72
Woodward, Benjamin Charles, 85, 86
Woodward, Charles, 85
Wood, W. H., 353
Wooley, Lund, and Judd, 270
Woolley, E. G., 257
Wop Town mining camp, 139
Workers Compensation Act, 44, 235
Workman, Don, 373
Works Progress Association (WPA), 296
World War I, 39, 117, 118, 135, 136, 144, 145, 207 213, 232, 239, 240, 242, 289, 290, 330, 371
World War II, 34, 37, 48, 52, 56, 113, 140, 156, 157, *194*, 218, 242, 243, 271, 290, 293, 294, 334, 337, 357
Worthington, John S., 72
Wren, James H., 371
Wyoming, 33, 38, 51, 57, 124, 131, 133, 137, 209, 210, 232, 380, 383, 390, 391
Wyoming Craton, 5, 22, 25
Wyoming Mine, 345
Wyoming Thrust Belt, 101

Y

Yale University, 144, 278
Yampa Mining Company, 225
Yankee Blade Mine, 287
Yellow Cat Wash, 151
Yellow Circle claims, 145, 147
Yellow Jacket Mine, 90, 319
Yellowstone Lake Fork oil deposit, 386
Yellowstone Park, 22, 27, 31
Yorston, Walter K., 42
Young, Brigham, 58, 59, 61, 64–66, 70, 73, 80, 86–88, 102, 126, 198–202, 252, 253, 258, 259, 267, *311*, 319, 439
Young America Mine, 319

Z

ZCMI, 331
Ziegler Chemicals, 389
zinc, *10*, 12, 14, 20, 22, 24, 29–31, 37, 39, 40, 43, 44, 53, 90, 163, 217, 218, 286–89, 318, 329, 333–39, 356, 359, 362, 373, 375, 376, 382, 393, 407
Zion Canyon, *17*
Zion's Board of Trade, 203